Nutritional Neurosciences

Series Editor

Mohamed Essa, Sultan Qaboos University, Qaboos, Oman

This book series aims to publish volumes focusing on both basic and clinical research in the field of nutritional neuroscience with a focus on delineating the effect of nutrition on brain function and behavior. The books will examine the role of different nutrients, food agents and supplements (both macro and micro) on brain health, neurodevelopment, neurochemistry, and behaviour. The books will examine the influence of diet, including phytochemicals, antioxidants, dietary supplements, food additives, and other nutrients on the physiology and metabolism of neurons, neurotransmitters and their receptors, cognition, behavior, and hormonal regulations.

The books will also cover the influence of nutrients and dietary supplements on the management of neurological disorders. It details the mechanism of action of phytonutrients on signaling pathways linked with protein folding, aggregation, and neuroinflammation. The books published in the series will be useful for neuroscientists, nutritionists, neurologists, psychiatrists, and those interested in preventive medicine.

Arokiasamy Justin Thenmozhi •
Thamilarasan Manivasagam
Editors

Nutraceuticals for Alzheimer's Disease: A Promising Therapeutic Approach

Springer

Editors
Arokiasamy Justin Thenmozhi
Department of Biochemistry
School of Biological Sciences
Madurai Kamaraj University
Madurai, Tamil Nadu, India

Department of Biochemistry and
Biotechnology, Faculty of Science
Annamalai University
Chidambaram, Tamil Nadu, India

Thamilarasan Manivasagam
Department of Biochemistry and Biotechnology
Annamalai University
Chidambaram, Tamil Nadu, India

ISSN 2730-6712 ISSN 2730-6720 (electronic)
Nutritional Neurosciences
ISBN 978-981-99-0676-5 ISBN 978-981-99-0677-2 (eBook)
https://doi.org/10.1007/978-981-99-0677-2

This Springer imprint is published by the registered company Springer Nature Singapore Pte Ltd.
The registered company address is: 152 Beach Road, #21-01/04 Gateway East, Singapore 189721, Singapore

Foreword

The book *Nutraceuticals for Alzheimer's Disease: A Promising Therapeutic Approach* mainly focuses on the curative properties of natural ingredients and bioactive compounds termed as nutraceuticals for managing Alzheimer's disease. The estimated health care cost for the treatment of AD in 2020 is estimated about 305 million dollars. The basic concept arises from the quotation of the famous Greek Physician Hippocrates: "Let food be thy medicine and medicine be thy food." The book reveals the conclusions obtained from various scientific experiments involving the in vitro, in vivo, and clinical studies. In addition, the regulation of non-coding RNA by nutraceuticals and their clinical applications along with advantages and disadvantages of using these food gradients are also discussed.

Department of Biochemistry and
Biotechnology, Annamalai University,
Chidambaram, Tamil Nadu, India

B. Raja

Foreword

Preface

There is clear evidence that a diet rich in specific nutritional food groups (fruits, fish, vegetables) can reduce the incidence and prevalence of some of the primary clinical outcomes, such as neurodegenerative disorders, cardiovascular diseases, diabetes, and cancer. The particular group of plant and food components, the so-called nutraceuticals, has displayed the ability or strong potential to act as neuro-protectants and/or delay cognitive impairment over the years. However, there is no recommendation for nutraceuticals in dementia-related therapeutic guidelines. Nevertheless, the strong potential for their neuroprotective action warrants further studies in the field. Alzheimer's disease (AD) is regarded as a progressive and devastating neurodegenerative disorder. It is the most prevalent cause of dementia in aged individuals and is characterized by gradual loss of cognitive functions. It is essential to review the therapeutic effects and regulation of molecular targets of different natural products. The uniqueness of a specific volume will consist in collecting various aspects of the topic ranging from the description of the disease and nutraceuticals present in nature to the characteristics of the brain, equipped with a BBB representing a limit to the nutraceutical bioavailability, to the microbiota–gut–brain (MGB) axis, also considering epigenetic changes necessary to regulate gene expression.

Madurai, Tamil Nadu, India — Arokiasamy Justin Thenmozhi
Chidambaram, Tamil Nadu, India — Thamilarasan Manivasagam

Acknowledgement

First and foremost, we would like to praise and thank God, the Almighty, who has poured countless blessings, knowledge, and opportunity to us and all the contributors of the chapters for the successful completion of this book.

We are very much thankful to our family members for their unfailing support, understanding, and encouragement for allowing us to spend extra time on completing the book.

Our sincere thanks to Springer Nature staff for their valuable guidance and constant support during the different phases of this book publication.

Contents

Editors and Contributors

About the Editors

Arokiasamy Justin Thenmozhi is working as an Assistant Professor in the Department of Biochemistry, School of Biological Sciences, Madurai Kamaraj University, India. Earlier she served as an Assistant Professor in the Department of Biochemistry and Biotechnology, Annamalai University and Idhaya College for Women, Kumbakonam. She has an expertise in the field of Nutritional Neuroscience and Neuropharmacology, has published more than 60 research articles in peer-reviewed international journals, and has edited 2 books and 21 book chapters in Neuroscience. Her books entitled *Food and Huntington's disease* and *Food and Parkinson's disease* were awarded as best books in the world by GOURMAND Cookbook Awards. She is the Fellow of "Academy of Sciences," Chennai. She is holding membership in various international and national neuroscience societies. She has received the Best Researcher awards from the Annamalai University (2017–2018 and 2018–2019) and has served as an editor/reviewer for various peer-reviewed international journals.

Thamilarasan Manivasagam is an Assistant Professor in the Department of Biochemistry and Biotechnology, Annamalai University. His research interests are in neuroscience with specialization in cellular and animal models of Parkinson's disease, Alzheimer's disease, and Autism, and drug discovery from plant products. He has published more than 100 research articles in the peer-reviewed international journals, edited 3 books, and authored or co-authored 50 book chapters. He has more than 15 years of teaching experience in Biochemistry and has received Young Scientist Award from the Annamalai University and the Academy of Sciences, Chennai.

Contributors

A. Aarthy Department of Immunology, School of Biological Sciences, Madurai Kamaraj University, Madurai, Tamil Nadu, India

Hanan Fahad Alharbi College of Computer, Mathematics and Natural Sciences, University of Maryland, College Park, MD, USA
Department of Maternity and Child Health Nursing, College of Nursing, Princess Nourah Bint Abdurrahman University, Riyadh, Saudi Arabia

K. Amali Department of Biochemistry and Biotechnology, Annamalai University, Chidambaram, Tamil Nadu, India

Kanadan Anjali Department of Microbiology, Mahe Institute of Dental Sciences, Puducherry, India

P. Annapoorani Department of Biochemistry, V.V. Vanniaperumal College for Women, Virudhunagar, Tamil Nadu, India

E. Arockia Jeya Yasmi Prabha Department of Physics, N.M.S.S.V.N. College, Madurai, Tamil Nadu, India

P. Arokia Mary Department of Biochemistry and Biotechnology, Annamalai University, Chidambaram, Tamil Nadu, India

S. Aruna Devi Department of Biochemistry and Biotechnology, Annamalai University, Chidambaram, Tamil Nadu, India

Anandan Balakrishnan Departments of Dental Anatomy and Oral Histology, Mahe Institute of Dental Sciences, Puducherry, India

M. Dhivya Bharathi Department of Biochemistry and Biotechnology, Annamalai University, Chidambaram, Tamil Nadu, India

P. Brundha Department of Immunology, School of Biological Sciences, Madurai Kamaraj University, Madurai, Tamil Nadu, India

Venugopal Buvarahamurthy Dr. A.L.M. Post-Graduate Institute of Basic Medical Sciences, University of Madras, Chennai, Tamil Nadu, India

Samadder Chandrani Department of Molecular Microbiology, School of Biotechnology, Madurai Kamaraj University, Madurai, Tamil Nadu, India

Daniela S. Cosio Departments of Obstetrics, Gynecology and Reproductive Sciences, University of Miami Miller School of Medicine, Miami, FL, USA

Ponmari Guruvaiya Department of Animal Sciences, Manonmaniam Sundaranar University, Tirunelveli, Tamil Nadu, India

T. Hemalatha Department of Biochemistry and Biotechnology, Annamalai University, Chidambaram, Tamil Nadu, India

Hussain Hussain Department of Internal Medicine and Infectious Disease, Larkin Community Hospital, South Miami, FL, USA

Joven Isaac The University of South Florida, Tampa, FL, USA

S. Iswaryalakshmi Department of Immunology, School of Biological Sciences, Madurai Kamaraj University, Madurai, Tamil Nadu, India

U. Janakiraman Department of Surgery, Virginia Commonwealth University, Richmond, VA, USA

R. Jasmine Juliet Department of Biochemistry, V.V. Vanniaperumal College for Women, Virudhunagar, Tamil Nadu, India

Arumugam R. Jayakumar Department of Obstetrics, Gynecology and Reproductive Sciences, University of Miami Miller School of Medicine, Miami, FL, USA
Division of Neuropathology, Miami VA Medical Centre, Miami, FL, USA

M. Jayalakshmi Department of Immunology, School of Biological Sciences, Madurai Kamaraj University, Madurai, Tamil Nadu, India

J. Kiruba Department of Immunology, School of Biological Sciences, Madurai Kamaraj University, Madurai, Tamil Nadu, India

R. Kiruthika Department of Biochemistry and Biotechnology, Annamalai University, Chidambaram, Tamil Nadu, India

M. Aishwarya Lakshmi Department of Immunology, School of Biological Sciences, Madurai Kamaraj University, Madurai, Tamil Nadu, India

T. Manivasagam Department of Biochemistry and Biotechnology, Annamalai University, Chidambaram, Tamil Nadu, India

Miranda A. Melians School of Nursing and Health Studies, University of Miami, Coral Gables, Miami, FL, USA

Jonathan Newell NSU University School, Davie, FL, USA

Jayakumar Nithura Medical Academy for Science and Technology, Homestead, FL, USA

Michael J. Paidas Department of Obstetrics, Gynecology and Reproductive Sciences, University of Miami Miller School of Medicine, Miami, FL, USA

Fay F. Pon Departments of Obstetrics, Gynecology and Reproductive Sciences, University of Miami Miller School of Medicine, Miami, FL, USA

A. Prema Department of Chemistry, Government Arts and Science College, Cuddalore, Tamil Nadu, India

M. Walid Qoronfleh Q3CG Research Institute (QRI), Research & Policy Division, Ypsilanti, MI, USA

Ramamoorthy Rajalakshmi Departments of Obstetrics, Gynecology and Reproductive Sciences, University of Miami Miller School of Medicine, Miami, FL, USA

Yoga Rajamani Department of Maternity and Child Health Nursing, College of Nursing, Princess Nourah Bint Abdurrahman University, Riyadh, Saudi Arabia
College of Computer, Mathematics and Natural Sciences, University of Maryland, College Park, MD, USA

P. Ramalakshmi Department of Biochemistry, V.V. Vanniaperumal College for Women, Virudhunagar, Tamil Nadu, India

U. Ramesh Department of Molecular Biology, School of Biological Sciences, Madurai Kamaraj University, Madurai, Tamil Nadu, India

Suset Rodriguez Department of Obstetrics, Gynecology and Reproductive Sciences, University of Miami Miller School of Medicine, Miami, FL, USA

A. K. V. Sama Department of Biochemistry, V.V. Vanniaperumal College for Women, Virudhunagar, Tamil Nadu, India

Natarajan Sampath School of Chemical and Biotechnology, SASTRA Deemed University, Thanjavur, Tamil Nadu, India

Bastian T. Sebastian Departments of Oral Pathology, Mahe Institute of Dental Sciences, Puducherry, India

Nagarajarao Shamaladevi Molecular Analytics, Miami, FL, USA

Kalidass Subramaniam Department of Animal Sciences, Manonmaniam Sundaranar University, Tirunelveli, Tamil Nadu, India

R. Surya Department of Biochemistry and Biotechnology, Annamalai University, Chidambaram, Tamil Nadu, India

Arokiasamy Justin Thenmozhi Department of Biochemistry, School of Biological Sciences, Madurai Kamaraj University, Madurai, Tamil Nadu, India
Department of Biochemistry and Biotechnology, Faculty of Science, Annamalai University, Chidambaram, Tamil Nadu, India

Ramakrishnan Thiruchelvi Department of Bioengineering and Biotechnology, Vels Institute of Science Technology and Advanced Studies (VISTAS), Chennai, India

Ganesh Vaishnavi Department of Molecular Microbiology, School of Biotechnology, Madurai Kamaraj University, Madurai, Tamil Nadu, India

T. G. Vardhani Department of Immunology, School of Biological Sciences, Madurai Kamaraj University, Madurai, Tamil Nadu, India

Chapter 1
Introduction to Alzheimer's Disease

Thamilarasan Manivasagam, Ravi Surya, Mariakuttikan Jayalakshmi, Arokiasamy Justin Thenmozhi, Asokan Prema, M. Walid Qoronfleh, Hanan Fahad Alharbi, and Yoga Rajamani

Abstract Alzheimer's disease (AD) is a progressive neurodegenerative ailment associated with intracellular amyloid aggregates and extracellular neurofibrillary tangles. In the early stage, it is characterized by short term memory loss, and in advanced stages, it is manifested by confusion, aggression, mood changes, long term memory loss, and social withdrawal. The cholinergic system, particularly projections from the basal forebrain to hippocampus and cortex, is accountable for memory and learning. The etiological factors of AD are not clearly known, and it is probably the result of a multifactorial process including head trauma, oxidative stress, genetics, infectious agents, and environmental factors including

T. Manivasagam · R. Surya
Department of Biochemistry and Biotechnology, Annamalai University, Chidambaram, Tamil Nadu, India

M. Jayalakshmi
Department of Immunology, School of Biological Sciences, Madurai Kamaraj University, Madurai, Tamil Nadu, India

A. Justin Thenmozhi (✉)
Department of Biochemistry, School of Biological Sciences, Madurai Kamaraj University, Madurai, Tamil Nadu, India

Department of Biochemistry and Biotechnology, Faculty of Science, Annamalai University, Chidambaram, Tamil Nadu, India

A. Prema
Department of Chemistry, Government Arts and Science College, Tittagudi, Cuddalore, Tamil Nadu, India

M. W. Qoronfleh
Q3CG Research Institute (QRI), Research & Policy Division, Rachel Drive, Ypsilanti, MI, USA

H. F. Alharbi
College of Computer, Mathematics and Natural Sciences, University of Maryland, College Park, MD, USA

Y. Rajamani
Department of Maternity and Child Health Nursing, College of Nursing, Princess Nourah bint Abdurrahman University, Riyadh, Saudi Arabia

A. Justin Thenmozhi, T. Manivasagam (eds.), *Nutraceuticals for Alzheimer's Disease: A Promising Therapeutic Approach*, Nutritional Neurosciences,
https://doi.org/10.1007/978-981-99-0677-2_1

Al toxicity. Acetylcholinesterase inhibitors (donepezil, rivastigmine, etc.), antihyperammonemic drugs (carvedilol), anti-inflammatory drugs (rofecoxib), NMDA antagonists (memantine), and secretase inhibitors (memoquin) were shown to have therapeutic effects with significant side effects and could not be used successfully.

Keywords Types · Symptoms · Risk factors · Pathology · Therapy · Alzheimer's disease · Symptoms · Risk factors · Pathology · Drugs

1.1 Introduction

Neurodegenerative diseases (NDDs) are the most common and uprising etiology of cognitive dysfunction and affect millions of elderly population worldwide. These disorders are often associated with progressive neuronal degeneration in the particular regions of the brain (Bredesen et al. 2006). The main pathology of NDDs arises due to unusual accumulation of specific proteins that are retained as extracellular aggregate or intracellular inclusion specific for each disease. More studies indicated that mutations or environmental factors can enhance misfolding and accumulation of proteins in these diseases (Takalo et al. 2013). The general risk factors of NDDs are exposure to various toxins, genetic polymorphisms, increasing age, oxidative stress, inflammation, apoptosis, depression, and nutritional problems (Launer et al. 1999). Almost all NDDs share common symptomatic features at different stages of disease progression. Such disorders are untreatable, varied in their pathophysiology with few leading to memory impairment and others affecting a person's motor functions.

1.2 Alzheimer's Disease

Alzheimer's disease (AD) is the prevalent type of NDD, which is categorized by the presence of two pathological hallmarks including extracellular senile plaques (accumulation of amyloid precursor protein (APP)) and intracellular neurofibrillary tangles (consisting hyperphosphorylated tau protein) (Good et al. 1992). The prevalence is higher in women than in men. It leads to the progressive loss of memory, disruption in thinking and reasoning, and changes in personality and behavior, followed by malfunction of the entire body (Anand et al. 2014).

1.3 History

AD was found first by Alois Alzheimer, a German doctor, in his patient Auguste Deter in 1906. While observing the autopsy result of his patient, Dr. Alzheimer found fatty deposits in the blood vessels, shrunk cerebral cortex, and brain tissue injuries (Alzheimer et al. 1995; Alzheimer's Association 2010). He also noted protein deposits later known as SP and NFT inside the brain (Alzheimer 1907). The major component of NFTs is phosphorylated tau protein, and SP contains β-amyloid protein (Selkoe 1991). β-Amyloid was discovered in 1984, and after 2 years, NFT was discovered in AD patients. Both the proteins may be involved in initiating brain cell damage.

1.4 Types of AD

Almost all AD patients exhibit similar set of symptoms including cognitive dysfunction, confusion, and trouble with familiar tasks and making decisions. The are two main types of AD:

1.4.1 Familial AD (FAD)

This type affects 1% AD patients which is raised due to the defect in chromosome number 21. Early onset AD in most people develops before 65 years of age. The FAD arises mostly due to mutations in three major genes: APP, PSEN1, and PSEN2.

1.4.2 Sporadic AD (SAD)

Many AD patients are affected by sporadic AD, which arises due to unknown causes (Bertram et al. 2007) and usually develops after the age of 65, which is often referred to as late-onset AD. It is a multifaceted disease, as many genetic and environmental factors may contribute to determine the sporadic form of AD. The risk factors involved in AD are aging, neuroinflammation, head trauma, diabetes, and brain ischemia (Corder et al. 1993).

1.5 Signs and Symptoms

AD begins with mild symptoms, gets worse over time, and starts to intervene with daily life. The progression of AD involves three main stages: mild, moderate, and severe. The symptoms can be diagnosed at any stage of AD and vary from person to person. In AD patients, signs and symptoms of late onset seem in 60 s, whereas early onset appears between 30 s and 60 s. AD symptoms in mild stage are less clear and its average time frame is 2–4 years. Generally, complications in memory such as inability to take up and retrieve new data, repetitive queries or talk, omitting personal properties, forgetting recent happenings or activities, and loss of a familiar route are the first signs of cognitive impairment in AD. Deterioration in cognitive functions like spelling and writing; difficulty in speaking, word finding, and reading; and impairment in judgment may also indicate the early stages of AD. Some patients may be identified with mild cognitive dysfunction. During its progression, patients experience more memory loss and various cognitive impairments.

Moderate stage is the longest stage in AD and its average time frame is 2–10 years. Signs that appear in this stage include impaired ability to perform challenging mental arithmetic and decreasing awareness of current events. In this stage, AD patients face greater difficulty in performing complex tasks such as shopping, handling finances, and paying bills. They become prone to forgetfulness or unable to remember about one's own personal history and often remain moody and start to withdraw from society. In severe stage of AD, worsened memory with behavior changes were found, and a person needs extensive assistance for regular activities, and its average time frame is 1–3 years. At this stage, general symptoms are speech declines to only one word or limited to fewer words. Patients always need help with almost all their daily tasks such as bathing, eating, dressing properly, and toileting. They experience major changes in personality, behavior, and sleep patterns, and over time they will be no longer able to hold their heads up and unable to sit properly.

1.6 Clinical Symptoms

Since AD progresses, some key cognitive symptoms are acknowledged such as amnesia, apraxia, aphasia, agnosia, and administrative dysfunction, and these symptoms become more prominent over the course of time. There are also psychological and behavioral impairments like depression, hallucination, delusion, and apathy, which are obviously displayed during the AD progression. These signs affect the quality of life, increase woes to both caregivers and patients, and lead to early hospitalization of patients (Gonzalez-Salvador et al. 2000). Symptoms in AD are broadly classified as mild, moderate, and severe and are evaluated by employing scales. AD is often diagnosed in mild stage. Although a person looks healthy in this stage, he/she feels more inconvenient in making sense of the objects around him. In moderate stage, more care is needed that will become a burden to many families. People having severe AD symptoms are unable to communicate and totally reliant on others for their routine work. During the end stage, the patient is mostly bed ridden as the body functions nearly shut down.

1.7 Risk Factors

1.7.1 Age and Gender

AD is more prominent in elderly population, but it can arise in younger population as well, and its prevalence rises predominantly with progressing age, with greater increase reported among people aging from 65 to 85 years (Mayeux and Stern 2012). In the United States, it was estimated that annual occurrence of AD among people aged ~ 85 and above were 37 of 1000 persons and among aged 75–84 were 13 of 1000 persons. Enhanced prevalence and mortality rate in elderly population were due to AD validated by numerous epidemiologic experiments globally. More AD cases are found in women (Alzheimer's Association 2014), and increasing hazard of developing AD seems to be greater in women due to negative influence of the ApoEe4 gene (Farrer et al. 1997).

1.7.2 Genetics

The AD risk is intensely enhanced by genetic factors and increases with family history of the disease. The common and strong risk factor is the presence of apolipoprotein (APOE-ε4) gene particularly on chromosome 19 (Corder et al. 1993). APOE gene carries different alleles ε2, ε3, and ε4, and the occurrence of a copy of APO-ε4 allele enhances the AD risk for about 4-fold, and people carrying both copies of APOE-ε4 have 12-fold increased risk as compared to people having both copies of APOE-ε3. APOE-ε4 allele carriers get Aβ plaques earlier as compared to noncarriers, by influencing Aβ aggregation and clearance in the brain (Karch and Goate 2015). APOE-ε4 is reported to link with hippocampal volume loss, enhanced density of neuritic plaques, and cognitive impairment (Drzezga et al. 2009). AD is usually caused by mutations in presenilin 1 (PSEN1) gene, presenilin 2 (PSEN2), and amyloid precursor protein (APP). These genetic forms were found prior to sporadic AD, with an average age of 46 years (Ryman et al. 2014).

1.7.3 Environmental Factors

Several studies have revealed that harmful heavy metal or chemical exposure like aluminum (Al), copper, iron, and zinc causes neuronal degeneration and may increase the incidence of AD (Zatta et al. 2009). Exposure to Al induces neuro-chemical, neurobehavioral, and neuropathological changes similar to AD (Kawahara and Kato-Negishi 2011). Several risk factors linked with the AD are the presence of cerebrovascular disease, diabetes, hyperlipidemia, smoking, obesity, and traumatic brain injury.

1.7.4 Cholesterol Metabolism

Enhanced cholesterol levels are reported to influence the activity of enzymes involved in the production of Aβ. Statins (cholesterol-lowering drugs) are reported to lower the risk of AD development (Jick et al. 2000). Apolipoprotein E is the main transporter of cholesterol and APOE-ε4 allele is a common marker, which is found in AD (Corder et al. 1993). APOE-ε4 is involved in the formation and aggregation of tau and amyloid proteins (Reiman et al. 2009).

1.7.5 Smoking and Alcohol Consumption

The association between cognitive dysfunction and smoking remains ambiguous. Smoking induces free radical formation and elevates oxidant-antioxidant imbalance (Traber et al. 2000). Alcohol is known to have toxic effect and can cause "alcoholic dementia." Middle agers carrying the APOE-ε4 allele having a habit of heavy alcohol drinking had about threefold risk of dementia and AD as compared to APOE-ε4 allele carriers (Anttila et al. 2004). Light-to-moderate alcohol consumption has been associated with enhanced brain atrophy and reduced brain masses (Tongsong et al. 2014).

1.7.6 Traumatic Brain Injury

AD risk is enhanced in patients with history of traumatic brain injury (TBI) (Rasmusson et al. 1995). Previous experiments demonstrated the risk of AD was higher among TBI patients (Fleminger et al. 2003). The animal experiments have revealed the link between AD and TBI (Guo et al. 2000). Humans with brain injury showed the enhanced Aβ pathology (Hartman et al. 2002), tau pathology, and overproduction of APP in various brain regions (Franz et al. 2003).

1.7.7 Obesity

Individuals with more BMI are having increased AD and dementia risk (Gustafson 2006). Enhanced cognitive dysfunction was found in obese people aged ~ 50 years (Kivipelto et al. 2005). Reports suggested that low BMI and weight loss found in late-life are considered as the preliminary AD marker, especially assessed during few years before the diagnosis of AD.

1.7.8 *Diabetes*

Individuals with type 2 diabetes mellitus (DM) are having twofold enhanced AD risk (Luchsinger et al. 2001). DM directly increases the accumulation of Aβ and diminishes its clearance by competing with the activity of insulin-degrading enzyme (Farris et al. 2003). Excessive presence of adipose tissue may lead to type 2 DM by synthesizing cytokines and adipokines that play a key role in inducing inflammation. Adiponectin, resistin, leptin, tumor necrosis factor-α (TNF-α), and interleukin-6 (IL-6) are synthesized during insulin resistance and hyperinsulinemia and may cause AD directly or indirectly (Trujillo and Scherer 2005).

1.7.9 *Cerebrovascular Disease*

The cerebrovascular disease (CVD) produces similar symptoms resembling dementia and it mostly occurs along with AD (Schneider and Bennett 2010). Stroke leads to cognitive dysfunction by inducing damage of brain parenchyma (Jellinger 2002), damaging the regions that leads to amnestic syndromes and enhancing the Aβ accumulation (Blennow et al. 2006). In rodent model, an enhancement in Aβ expression was found, as the p25 and cdk5 overexpression leads to enhanced BACE1 expression, which in turn enhanced the amyloid precursor protein (APP) processing (Song et al. 2015). Hyper densities of the white matter are found in dementia patients, but the reasons behind the white matter changes leading to cognitive decline are not clear (Snowdon et al. 1997).

1.8 Pathophysiology of AD

The following hypotheses clarify the causes of AD:

1. Cholinergic hypothesis
2. Amyloid plaque hypothesis
3. Tau tangle hypothesis

1.8.1 *Cholinergic Hypothesis*

The cholinergic hypothesis indicates that the damage of the cholinergic neuronal pathway in the forebrain region leads to a loss of cholinergic neurons that synthesize and secrete acetylcholine. Cholinergic neurons projected into the various brain regions like the cortex and hippocampus that are reported to be involved in the memory and cognitive symptoms (Bartus 2000). Acetylcholine is synthesized from acetyl coenzyme A and choline by the enzyme acetylcholinesterase (AChE). Their levels were diminished in moderate and severe AD conditions in comparison with mild conditions, whereas the inhibition of enzyme activity increases the function of neurotransmitter and also reprieved the symptoms of AD (Terry and Buccafusco 2003).

Cholinergic system also includes the parasympathetic and sympathetic nervous system and the motor neurons of the spinal cord. Cholinergic fibers are ascertained within different regions of the hippocampus, and the cortex acquires the cholinergic signals from the forebrain. Impaired cholinergic neurons were found in the aged population, as aging is the effective risk factor for AD. Previously, it is reported that aged cholinergic neurons were more deficient to liberate the neurotransmitter even after stimulation with potassium (Gibson and Peterson 1981). The perception that the use of various methods improves acetylcholine output or inhibits the damage of cholinergic neurons in the aged brain is the main target in AD therapy (Gilad et al. 1987). Therefore, any constant abuse to cholinergic neurons could reduce the capability of these cells.

1.8.1.1 The Hippocampus and Its Function

The hippocampus is present in the temporal lobe, which forms a C shaped structure that has a characteristic laminar organization. In rodents, the hippocampus is configured as CA1, CA2, and CA3, while in humans there is an additional fourth area called CA4 region. The hippocampus is disparagingly a vital structure needed for learning and memory functions, enabling rapid programming of new information, alliance, and recovery of the memory function (Preston and Eichenbaum 2012). The neural aggregation found in the hippocampus regulates a circuit-like fashion in modifying the emotion and cognitive functions. The hippocampal segmentation is found as either dorsoventral or posterior-anterior throughout their axis in rodents and primates, respectively, which is accountable for its double functionality in controlling the cognition and emotion. In rodents, the dorsal part is responsible for the cognitive function, and the ventral is for anxiety and mood (Fanselow and Dong 2010).

Various experiments involving the psychological testing, imaging, and postmortem analysis indicated the strong relationship of the hippocampus with pathology of AD. Even during the mild AD, a smaller amount of pathological changes was found in the hippocampus. Amyloid plaques and neurofibrillary tangles formed by abnormal folding and impaired processing of protein in the AD brain were found initially in the cortex and then slowly spread to the hippocampus (Braak and Braak 1991). This phenomenon leads to the interruption of hippocampal mediated neural transmission (Supekar et al. 2008). The reduced hippocampal connectivity is also found in patients with mild cognitive impairment (MCI), who are having AD risk (Sorg et al. 2007).

1.8.1.2 Cerebral Cortex

The cerebral cortex is mainly involved in thinking, perception, and understanding the language. During the initial phase of AD, plaques are found in the superficial part of the cortex and spread deeply on its progression. Even in the moderate AD, plaques were spread to both superficial and deep layers of the cortex and also found in the primary visual cortex, a less affected area. Similar to plaques, neurofibrillary tangles are another pathological condition involving the accumulation of proteins predominantly found in the superficial layers of the visual cortex even during the initial stage of AD. In the other cortical areas like parietal, temporal, and prefrontal cortical regions that are more prone to AD, the tangles are present in superficial and deep layers, which is progressive during the advancement of disease. As the superficial layers of the cortex involved in connection of cortico-cortical neurons and deeper layers project out of the cortex, both the intra- and interneural information were affected in AD.

1.8.2 Amyloid Hypothesis

Amyloid precursor protein (APP) is an intermembrane glycoprotein belonging to the APP family that also comprises two amyloid-like proteins—APLP1 and APLP2. It is ubiquitously expressed in numerous tissues but particularly abundant in the brain. It functions like a trophic factor, which is needed for synapse remodeling, synaptogenesis, and neurite outgrowth (Tyan et al. 2012). Its expression is enhanced in physiological (maturation and differentiation of neurons) and pathological conditions like AD and head injury (Buoso et al. 2010). On enzymatic cleavage, APP produces various derivatives such as soluble extracellular domain of APP-α (sAPPα), sAPPβ, APP intracellular domain (AICD), and Aβ (Müller et al. 2017). APP contains a C-terminal region (intracellular), a Aβ peptide domain, and an N-terminal region (extracellular). In human beings, APP gene consists of 18 exons, which produces 3 predominant isoforms of APP—APP695, APP751, and APP770—by several alternative splicing processes. APP695 is predominantly expressed in the CNS (Van der Kant and Goldstein 2015).

In ectodomain shedding processes, the α-secretase, a metalloprotease, cleaves APP at L688 residue within the Aβ domain (16 and 17 residues of Aβ domain) and discharges the sAPPα (Lammich et al. 1999). Then the γ-secretase cleaves the remaining C-terminal region of APP (C83) to discharge non-toxic AICD and p3 peptide regions (Nathalie and Jean-Noël 2008). The smaller peptide p3 lacks the N-terminal part of Aβ and is unable to form amyloid fibrils and is generally cleared from the neurons. Therefore, APP processed by α- and γ-secretase is known as the non-amyloidogenesis (Fig. 1.1).

The β-secretase expression is found abundantly in lysosomes, endosomes, and the Golgi bodies of the pancreas and neurons of the brain (Haass et al. 1995) that is

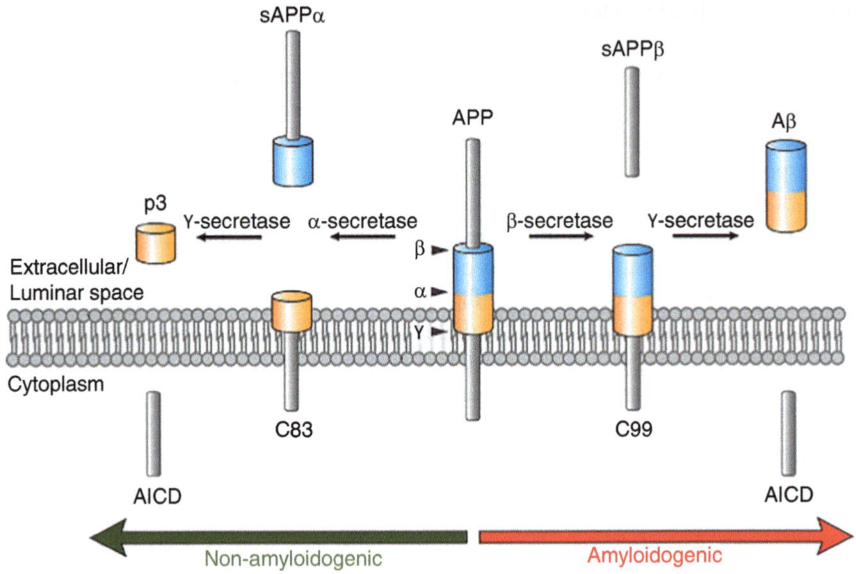

Fig. 1.1 Processing of APP by α-secretase, β-secretase, and γ-secretase: In amyloidogenic pathway, β-secretase and γ-secretase cut APP, whereas in non-amyloidogenic processing, the substrate is cleaved by α-secretase and subsequently by γ-secretase (Lichtenthaler et al. 2011)

effectively working at low pH (Knops et al. 1995). In the β-secretase mediated amyloidogenic pathway, the enzyme is cut at Aβ domain initially and liberated sAPPβ and C-terminal domain C99. γ-Secretase cuts the C-terminal domain of APP and releases Aβ fragments with varying length (39, 40, 42, and 43) and ACID fragments. In amyloidogenic pathway, β-secretase execution is considered as the rate limiting reaction. In the normal conditions, this enzyme cleaves only 10% of entire cellular APP. The leftover 90% of APP is processed by α-secretase leading to non-amyloidogenic pathway. Enzymatic cleavage of APP by two membrane-bound endoprotease (β- and γ-secretases) enzymes results in the formation of $A\beta_{1-40}$ and/or $A\beta_{1-42}$. $A\beta_{1-42}$ is longer and less soluble, and fibrinogenic protein predominantly accumulated in the brain (Lambert et al. 1998) (Fig. 1.1). Due to their fibrillogenic nature, $A\beta_{1-42}$ is susceptible to oligomer formation and following formation of fibril (Jarrett et al. 1993). $A\beta_{1-42}$ accumulates as a key part of SPs, whereas $A\beta_{1-40}$ deposits as a soluble and diffusible plaques. $A\beta_{1-42}$ oligomers have been shown to initiate direct lipid peroxidation and indirect protein oxidation events (Butterfield et al. 2002) by integrating into the membrane and/or binding to its surface and increasing bilayer permeability and conductance by altering its dielectric structure. Substitution of the Met-35 sulfur (S)-atom with a norleucine (Nle) CH_2 group in human $A\beta_{1-42}$ abrogates the oxidative and neurotoxic effects of the resulting peptide in cultured neurons while maintaining similar length, hydrophobicity, and tendency to aggregate as native human $A\beta_{1-42}$ (Boyd-Kimball et al. 2005). In *Caenorhabditis elegans*, substitution of Met codon with a cysteine codon in DNA encoding human $A\beta_{1-42}$ prevents in vo protein oxidation while not affecting $A\beta_{1-42}$ accumulation and plaque deposition (Yatin et al. 1999). While Aβ generation is pertained as amyloidogenesis, leading to instability between production and clearance of amyloid-β peptide and amyloid β protein is no longer regulated. Dal Pra et al. (Dal Prà et al. 2014) have indicated that $A\beta_{1-42}$ oligomers are synthesized in neurons and astrocytes. Aβ are produced in large quantities during the progression of AD, eventually accumulating as amyloid deposits within the brain. $A\beta_{1-42}$ oligomers destabilize microtubules by hyperphosphorylation of tau protein which assemble to produce neuronal filaments, without which the skeleton is degenerated. This will lead to the loss of neuronal connections between themselves and results in loss of neuronal synapses and mitochondria (Kurz and Perneczky 2011).

1.8.2.1 β-Secretase

The β-secretase is the membrane spanning protease that cuts APP and other substrates. BACE 1 and BACE 2 are the two homologous forms of β-secretase (Rachel et al. 2010), among them BACE1 is abundantly present in the brain and is responsible for Aβ generation (Marcinkiewicz and Seidah 2000). Transgenic mice (BACE1 gene knockout) showed a decrease in the neuronal levels of Aβ, CTFβ, and β-secretase (McConlogue et al. 2007). The activity and levels of β-secretase are enhanced in SAD and are linked to both plaque accumulation and disease progression (Yang et al. 2003).

1.8.2.2 γ-Secretase

The γ-secretase, a multienzyme complex, contains at least a copy of subunits like APH1, nicastrin, PEN2, and presenilin (PS1 or PS2) and cuts various membrane proteins including APP. Presenilin contains the active site (Li et al. 2000) and nicastrin helps in binding of substrate (Shah et al. 2005). The combination of all subunits is needed for the functioning and maturation of enzyme (Kimberly et al. 2003). This enzyme is comparatively unique in its mechanism, as it cuts the substrate within the lipid bilayer to remove a large ectodomain region subsequently after being cleaved by another protease (Selkoe 2003). However, modifications in this enzymatic activity result in the formation of several Aβ forms, which are the main etiology of AD (Scheuner et al. 1996).

1.8.2.3 Aβ Degradation

Aβ was degraded mainly by insulin-degrading enzyme (IDE) and neprilysin (NEP) (Marr et al. 2003). NEP is a membrane linked metalloprotease, which actively participates in the extracellular degradation of several peptides, whereas IDE acts both intra- and extracellularly (Qiu and Folstein 2006). IDE has more affinity for insulin almost 20-fold higher as that of Aβ, but lysing insulin at a low rate. So the insulin is considered as efficient inhibitor of IDE-mediated Aβ cleavage, which links the hyperinsulinemia, type 2 diabetes mellitus, and AD (Qiu and Folstein 2006). There is downregulation of NEP and IDE expression in various brain regions of AD patients (Sun et al. 2009). In the normal brain, a substantial amount of Aβ is not degraded, crossed the blood-brain barrier, and reached the bloodstream like other metabolites. Interference in this mechanism induces the further accumulation of Aβ (Van Uden et al. 2002). Soluble Aβ fibrils can cross the BBB either on the abluminal (brain) side by low-density lipoprotein receptor-related protein (LRP) (Shibata et al. 2000) or on the luminal (blood) side by receptor for advanced glycation end products (RAGE) (Deane et al. 2003).

1.8.3 Tau Hypothesis

The tau proteins of neurons are implicated in the stabilization of neuronal microtubule network and assembly and enhance axonal transport and synaptic and neuronal functions (Gong et al. 2006). Microtubule linked protein tau gene codes the tau protein, which contains 16 exons on chromosome 17q21 that produces 6 isoforms of tau by alternative splicing (Guo et al. 2017). The carboxy terminal of tau protein contains highly conserved three or four repeats of amino acids called microtubule binding domain, leading to polymerization and stabilization of microtubules (Goedert et al. 1989). The amino-terminal consists of numerous basic amino acids like threonine (T), proline (P), and serine (S), which help to bind with plasma membrane and premembranous structures (Brandt et al. 1995). Tau protein is altered posttranslationally by ubiquitination, glycosylation, and oxidation (Goedert et al. 2006). About ~7 Pi/mol and ~ 2 Pi/mol of phosphorylated tau were found in fetal and normal healthy adult brain, respectively, whereas ~8 Pi/mol of PHF-tau (paired helical filament) was found in AD patients (Kenessey and Yen 1993). Tau protein is irregularly hyperphosphorylated, at Ser 214 and Ser 262 positions; it loses its MT binding capability and may also set apart normal tau (Zhou et al. 2006). When tau is abnormally hyperphosphorylated, it loses its capability of MT binding, thereby leading to disturbance of the MT stability (Zhou et al. 2006). Various protein kinases like cell division cycle 2 kinase, mitogen-activated protein kinase (MAPK), cdc2, glycogen synthase kinase 3β (GSK-3β), etc. may be induced by pathological indices of AD pathology such as Aβ, oxidative stress, altered cell cycle, and inflammation (Arnaud et al. 2006). These enzymes cut the tau proteins in specific sites, generating precise fragments. Fragments of tau protein generated upon proteolysis can display an enhanced tendency for self-association, preceding the synthesis of aggregates and oligomers. These oligomers were not cleaned by the cell resulting in neuronal dysfunction. Hence synchronization among cleavage of tau and its clearance is key for maintaining the amount of functional tau (Fig. 1.2). In AD patients, enhanced levels of tau were found (Barton et al. 1990) that could be toxic to neurons since its

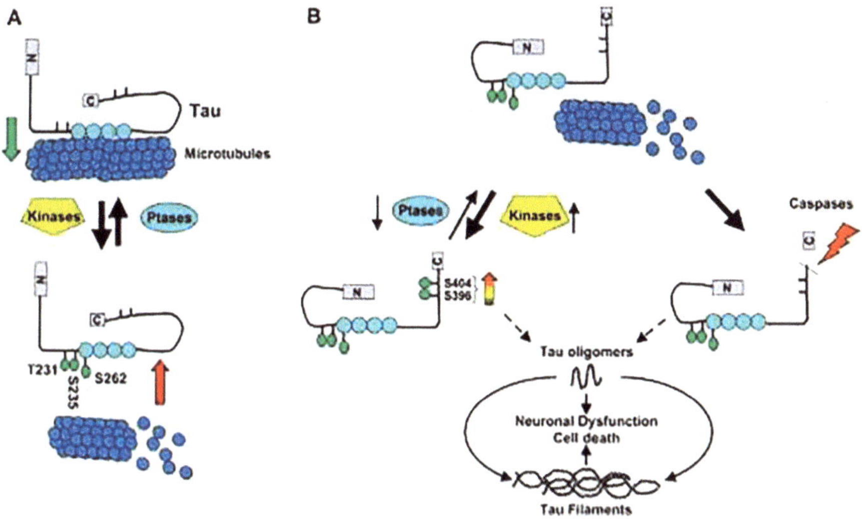

Fig. 1.2 Tau protein pathology in AD

decline has a shielding impact in experiment model of AD and its reduction may be therapeutically important (Gotz et al. 2013).

Two chief proteolytic systems involved in the degradation of proteins in cells are autophagy-lysosome and ubiquitin-proteasome systems. Autophagy means "self-eating" that occurs during starvation conditions to degrade the substrates in cells for generating energy. Autophagy is mainly involved in the clearance of proteins and aggregates and damaged organelles like peroxisomes and mitochondria (Johansen and Lamark 2011) and pathogenic bacteria (Von Muhlinen et al. 2012). Damaged proteins or organelles enclosed into a vesicle (autophagosome), trafficked to lysosome, and fused to form an autophagic vacuole. The lytic enzymes digest the damaged materials and the inner membrane of the autophagosome with the help of enzymes called cathepsins. Functional lysosomal compartment is critical for the autophagy process. In an in vitro assay, the activity of cathepsin Don tau proteins was studied by examining the role of enzyme on partially purified tau obtained from rat brain. The enzyme treatment reduced the length of tau protein and leads to formation of cleaved fragments of variable lengths (Bednarski and Lynch 1996).

The proteasome is a multiple rounded structure that is mainly involved in the clearance of cytosolic proteins. The 26S proteasome contains a regulatory cap on the edge of catalytic site (responsible for proteolytic activities) and degrades substrates linked with poly-ubiquitin chains. The regulatory cap along with chaperone proteins relaxes the substrate proteins and eliminates the ubiquitin tag by ATP-dependent process thereby degraded by proteasome. The catalytic core degrades the natural unfolded proteins directly by an ubiquitin and ATP-independent process (Hamano et al. 2009). Accumulation of proteins in the brain of AD patient may be due to diminished function of proteasome. It is suggested that activity of proteasome is diminished in AD-susceptible regions of the brain like the hippocampus and cortex as compared to normal brain (Keck et al. 2003). The PI3K/AKT/GSK-3β signaling pathway is playing a critical role in AD pathology because it induces tau hyperphosphorylation. The enzyme, glycogen synthase kinase-3β (GSK-3β) phosphorylates cAMP response element binding protein (CREB) thereby inducing the synthesis of the several neurotrophins including brain-derived neurotrophic factor (BDNF), which are needed for the maintenance of synaptic plasticity and long-term memory (Hu et al. 2013).

1.9 Mitochondrial Dysfunction in AD

Mitochondria sustain the regulation of various functions of the cells like the metabolism, protein folding, cell survival, and proliferation (Galluzzi et al. 2012). Oxidative metabolism is the key function of the mitochondria, and deficiency in ATP generation leads to deprivation of energy, loss of cellular function, and ultimately apoptotic death of cells in various NDDs (Beal 1996). It has been documented that degeneration of mitochondria by oxidative stress is an earliest sign of AD, appearing prior to neurofibrillary tangles, which may result in AD (Readnower et al. 2011). Literature evidences indicated that ROS mediated damage induces glial cells to pro-inflammatory cytokine synthesis and NO with subsequent generation of RNS (Jekabsone et al. 2006). Glial cell activation liberates H_2O_2, which is changed to hypochlorous acid, a highly toxic prooxidant (Halliwell 1992). Thus, a cruel cycle of Ca^{2+} overload, oxidative stress, and inflammation mediates the progression of NDDs (Fig. 1.3).

1.10 Oxidative Stress

Oxidative stress occurs due to the imbalance between the formation of free radicals and antioxidant defenses and the possibility of the body to counter their detrimental effects by action of antioxidants. A free radical is oxygen or its containing molecule having extra unpaired electrons, converting it into extremely reactive with DNA, protein, or lipid, and attains their electrons to obtain their stability. Hence, it weakens the cell components by triggering a large and continuous chain of reactions. The brain is mostly prone to oxidative imbalance because of enhanced levels of its lipid content and oxygen content. Oxidative stress is a key factor in the etiology and progression of several NNDs including AD (Nunomura et al. 2001). Accumulation of oxidative derivatives of lipids, proteins, and DNA was found in postmortem tissues of AD patients (Jenner 2003), indicating the pathological effect of oxidative stress. Therefore, antioxidant is considered as key therapeutic agents for AD, as it offers protective effects by nullifying ROS production and prevents ROS mediated neuronal damage (Samuel et al. 2005).

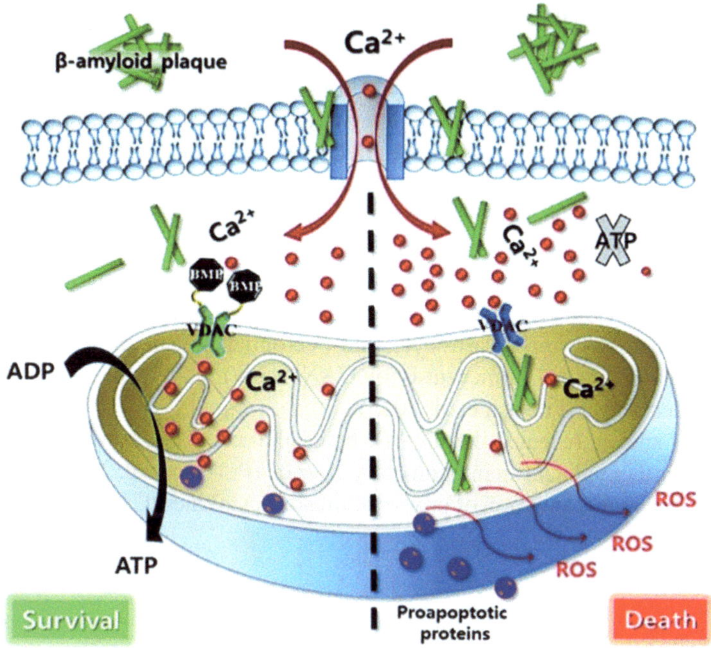

Fig. 1.3 Mitochondrial dysfunction in AD

1.11 Antioxidants

The antioxidants are the key molecules that play protective role against oxidative damage at primary level to intracellular contents (Halliwell and Whiteman 2004). There are many endogenous antioxidant defense systems and mechanisms the cell employs to scavenge ROS and RNS mediated cellular damage and protect cellular homeostasis. The levels of these antioxidant defense systems become impaired with age and neurodegenerative disease, since many antioxidant enzymes are redox sensitive and therefore get easily oxidized and become inefficient to combat with degeneration of free radicals (Mates et al. 1999).

Some of the major endogenous antioxidant enzymes are superoxide dismutase, glutathione peroxidase, catalase, and glutathione S-transferase to suppress the actual line of threat to the organs and tissues. Antioxidants with nonenzymatic action include α-tocopherol, uric acid, ascorbic acid, reduced glutathione, β-carotene, and bilirubin. However, many dietary antioxidants that have been explored for their neuroprotective effects include polyphenolic flavonoids (e.g., resveratrol and quercetin), coenzyme Q10, curcumin, ferulic acid and ferulic acid ethyl ester, and N-acetyl-L-cysteine (Halliwell and Gutteridge 1999).

1.12 Inflammation

Neuroinflammation is a key factor involved in the etiology and progression of AD and present approach for therapeutic interventions. The astrocytes, microglia, and neurons in the brain of AD patients are affected by inflammatory process (Akiyama et al. 2000). Cagninet al. (2001) indicated that the activation of microglia is found in the early stage of AD and leading to synaptic and memory dysfunction. Both astrocyte and microglia are getting activated during reactive gliosis that occurs due to the response of toxins that leads to neuroinflammation and neurodegeneration. Glial Fibrillary Acidic Protein and ionized calcium-binding adapter molecule 1 (allograft inflammatory factor-1), astrocyte, and microglial specific proteins needed for maintenance for normal homeostasis of CAN were reported to get enhanced in rodent model of AD (Prakash et al. 2013).

Previous studies revealed that in AD, brain glial cells and astrocytes can exaggerate and further validate an inflammatory cascade (Prakash et al. 2013; Prema et al. 2017) and enhance expression of pro-inflammatory cytokines (Campbell et al. 2004) (Table 1.1).

1.13 Apoptosis

Apoptosis is a well-characterized process with precise morphological alterations, during which a cell undergoes self-destruction (Kerr et al. 1972). Apoptosis is needed for the development and preservation of normal cellular homeostasis. But, the apoptotic dysregulation was associated with several diseased conditions like chronic inflammation, cancer, heart and lung diseases, and NDDs including AD (Fulda et al. 2010). Apoptosis could be initiated via two major pathways: the intrinsic or internal signaling pathway, where mitochondria contribute a central

Table 1.1 Function of cytokines

S. no.	Interleukins	Significance
1.	IL-1β	Key regulator of inflammation during host defense response
2.	IL-2	Several effect on hippocampal neurons, thereby enhancing cognitive performances
3.	IL-4	Enhanced during the sustained IL-1β neuroinflammation
4.	IL-6	Mediator of the inflammatory and immune responses
5.	TNF-α	Trigger other cytokines
6.	NF-κB	Involved in numerous physiological processes like neurogenesis, maintenance of synaptic plasticity, and enhancement in learning and memory processes
7.	COX-2	Trigger other cytokines

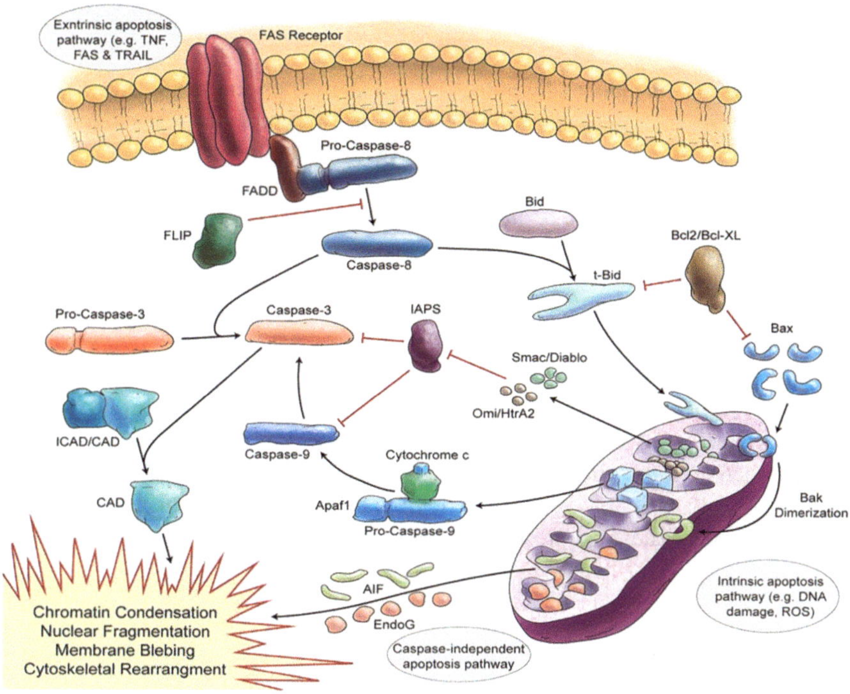

Fig. 1.4 Extrinsic and intrinsic pathway of apoptosis (Mohammad et al. 2015)

role, and the extrinsic or external signaling pathway, where plasma membrane receptors induce apoptotic process. Both intrinsic and extrinsic pathways leads to the activation of caspases that play a key role in the death of neurons (Fig. 1.4).

1.13.1 Extrinsic Pathway

TNF-Rs found in plasma membrane contains death domain (DD) and have various death receptors such as DR1, DR2, DR3, DR4, DR5, and DR6 (Mahmood and Shukla 2010). In DR2-mediated apoptosis, Fas ligand binds with its receptor (Fas) that triggers trimerization of receptors. It induces the caspase (8 or 10) cascade activation. Activated caspase-8 splits Bid belonging to Bcl-2 family into tBid which translocates into mitochondria and releases cytochrome-c (Kantari and Walczak 2011). The tBid also induces mitochondrial apoptosis by inducing the imbalance among the proapoptotic and antiapoptotic members of Bcl-2 family (Kantari and Walczak 2011).

1.13.2 Intrinsic Pathway

Intrinsic pathway is induced by mitochondrial alterations and finally carried out by activated caspase-3 (Adams and Cory 2002). Ca^{2+} accumulation leads to mitochondrial swelling and membrane potential loss that permits low weight molecules less than 1500 D from outer membrane into the mitochondrial matrix. This leads to opening of mitochondrial permeability transition pore (MPTP) (Kruman and Mattson 1999). Increased permeation of outer mitochondrial membrane induces the release of proapoptotic substances like mitochondrial activator of caspases, cyto c, direct inhibitor of apoptosis protein binding protein, and apoptosis inducing factor-1 (Apaf-1). In the cytosol, cyto c forms the "apoptosome" complex after binding with the Apaf-1, dATP, and procaspase-9 leading to the activation of caspase-9, which subsequently activates the procaspase-3 to caspase-3 (Adams and Cory 2002). Caspase-3 cleaves and converts the procaspase-6 and procaspase-7 to its active form, thereby activating DNase resulting in DNA degradation. Moreover, endonuclease G (endoG) and AIF translocate into nucleus and participate in caspase-independent apoptosis (Orrenius et al. 2015).

1.14 Diagnosis

Research on new strategies for earlier diagnosis, including ongoing efforts to identify and validate biomarkers for AD, is among the most active areas in Alzheimer's science. The diagnosis of AD could be confirmed only with autopsy. However, it could be diagnosed with 90% of the assurance by analyzing the patient's history and neurological examination. To identify the disease progression, assessment of person's memory, language, visuospatial attention, and problem-solving ability could be done by simple screening test (Mini-Mental State Examination—MMSE) (Folstein et al. 1975). Physicians use the Alzheimer's Disease Assessment Scale (Emilien et al. 2004) to determine the severity of disease by measuring the patient's orientation, memory, reasoning, and language (Table 1.2).

Blood tests may be used to avoid memory difficulties caused by other conditions such as vitamin deficiency or thyroid disorder. A brain scan (MRI or CT scan) could be done to diagnose memory problems caused by brain tumors and cerebrovascular damages due to accidents, trauma and infections, and strokes if any by accessing the brain structure and volume. These techniques are used to identify the tangles and plaques seen in AD brains. The MRI and PET scans facilitate the AD diagnosis by excluding other forms of dementia (Khachaturian 1985). Neuroimaging scans the brain shrinkage before the manifestations of symptoms and thereby enables the early diagnosis (Table 1.3).

Table 1.2 Psychological tests used to measure cognitive function in AD patients

Neuropsychological assessment method	Cognitive function measured
Mini-Mental State Examination	Orientation to place and time
Free-recall test, recognition span test, and Brown-Peterson distractor test	Memory
Boston naming test, Boston diagnostic aphasia examination, Western Aphasia test, Token test, Reporter's test	language skills
Picture-copying test, Wechsler Adult Intelligence Scale subtest	Praxis
Reaction-time task, continuous-performance test	Attention
Gollin incomplete-pictures test, Hooper test	Visual perception
Wisconsin Card Sorting Test, the poisoned food problem task of Arenberg	Problem-solving skills

Table 1.3 Laboratory techniques used in the diagnosis of AD

Laboratory assessment method	Observation in AD patients
Electroencephalography	Increased slow-wave activity
Evoked potentials	Increased latency of P300 potentials
Computerized tomography	Increased volume of the ventricular system, increased width of the third ventricle, narrowed gyri, and widened sulci
Regional cerebral blood flow	Decreased rCBF and cerebral metabolic rate
Positron emission tomography	Cerebral hypometabolism

1.15 Treatment

Five drugs are currently approved for the treatment of AD by the US Food and Drug Administration (FDA), which includes acetylcholine esterase inhibitors (rivastigmine, galantamine, tacrine, and donepezil) and NMDA receptor antagonist (memantine) that target symptoms at its best. The mechanism of these drugs is differing between each other but involves in the slowdown of Ach breakdown, a chemical needed for memory. Tacrine is hardly recommended to AD patients because of its adverse effects including liver damage. In general, rivastigmine, galantamine, and donepezil are most effective when treatment is begun in the early stages. In clinical studies, all these three cholinesterase inhibitors improve the memory and thinking of people taking the medications than those taking a placebo. The degree of improvement was small, and they may delay or slow worsening of symptoms, which varies from person to person. Memantine (Namenda) is the effective drug used in the later stages of AD. Although these drugs unassumingly slow down the progress of cognitive signs and diminish problematic behaviors in few patients, approximately 50% do not show response to them (Massoud and Gauthier 2010) (Table 1.4).

Table 1.4 Drugs used for the treatment of AD and its action

Drug name	Indication	Action	Side effects
Donepezil Brand name: Aricept	All stages	Prevents the breakdown of acetylcholine (ACh) by inhibiting the action of acetylcholinesterase Treats cognitive symptoms of AD	Nausea, vomiting, loss of appetite, and increased frequency of bowel movements
Galantamine Brand name: Razadyne	Mild to moderate AD	Prevents the breakdown of acetylcholine and stimulates receptors to release excess Ach Treats cognitive symptoms	Nausea, vomiting, loss of appetite, and increased frequency of bowel movements
Rivastigmine Brand name: Exelon	Mild to moderate AD Also used to treat dementia from Parkinson's disease	Prevents the breakdown of acetylcholine by inhibiting the enzymes that degrade ACh Treats cognitive symptoms of AD	Nausea, vomiting, loss of appetite, and increased frequency of bowel movements
Memantine Brand name: Namenda	Moderate to severe AD	Blocks glutamatergic (NMDA) receptors and regulates the action of glutamate Treats cognitive systems of AD	Headache, constipation, confusion, and dizziness

References

Adams JM, Cory S (2002) Apoptosomes: engines for caspase activation. Curr Opin Cell Biol 14: 715–720

Akiyama H, Arai T, Kondo H, Tanno E, Haga C, Ikeda K (2000) Cell mediators of inflammation in the Alzheimer disease brain. Alzheimer Dis Assoc Disord 14:47–53

Alzheimer A (1907) Uber eineeigenartige Erkrankung der Hirnrinde. Allgemeine Zeitschrift Fur Psychiatrie 64:146–148

Alzheimer A, Stelzmann RA, Schnitzlein HN, Murtagh FR (1995) An English translation of Alzheimer's 1907 paper, Uber eineeigenartigeerkankung der hirnrinde. Clin Anat 8:429–431

Alzheimer's Association (2010) Alzheimer's disease facts and figures. Alzheimers Dement 6:158–194

Alzheimer's Association (2014) Alzheimer's disease facts and figures. Alzheimers Dement 10:e47–e92

Anand R, Gill KD, Mahdi AA (2014) Therapeutics of Alzheimer's disease: past, present and future. Neuropharmacology 76:27–50

Anttila T, Helkala EL, Viitanen M, Kareholt I, Fratiglioni L, Winblad B, Soininen H, Tuomilehto J, Nissinen A, Kivipelto M (2004) Alcohol drinking in middle age and subsequent risk of mild cognitive impairment and dementia in old age: a prospective population based study. BMJ 329: 539

Arnaud L, Robakis NK, Figueiredo-Pereia ME (2006) It may take inflammation, phosphorylation and ubiquitination to 'tangle' in Alzheimer's disease. Neurodegener Dis 3:313–319

Barton AJ, Harrison PJ, Najlerahim A, Heffernan J, McDonald B, Robinson JR, Davies DC, Harrison WJ, Mitra P, Hardy JA (1990) Increased tau messenger RNA in Alzheimer's disease hippocampus. Am J Pathol 137:497–502

Bartus RT (2000) On neurodegenerative diseases, models and treatment strategies: lessons learned and lessons forgotten a generation following the cholinergic hypothesis. Exp Neurol 163:495–529

Beal MF (1996) Mitochondria, free radicals, and neurodegeneration. Curr Opin Neurobiol 6:661–666

Bednarski E, Lynch G (1996) Cytosolic proteolysis of tau by cathepsin D in hippocampus following suppression of cathepsins B and L. J Neurochem 67:1846–1855

Bertram L, McQueen MB, Mullin K, Blacker D, Tanzi RE (2007) Systematic meta-analyses of Alzheimer disease genetic association studies: the Alz gene database. Nat Genet 39:3917–3923

Blennow K, de Leon MJ, Zetterberg H (2006) Alzheimer's disease. Lancet 368:387–403

Boyd-Kimball D, Sultana R, Poon HF, Mohammad-Abdul H, Lynn BC, Klein JB, Butterfield DA (2005) Gamma-glutamylcysteine ethyl ester protection of proteins from Abeta(1-42)-mediated oxidative stress in neuronal cell culture: a proteomics approach. J Neurosci Res 79:707–713

Braak H, Braak E (1991) Neuropathological stageing [sic] of Alzheimer related changes. Acta Neuropathol 82:239–259

Brandt R, Leger J, Lee G (1995) Interaction of tau with the neural plasma membrane mediated by tau's amino-terminal projection domain. J Cell Biol 131:1327–1340

Bredesen DE, Rao RV, Mehlen P (2006) Cell death in the nervous system. Nature 443:796–802

Buoso E, Lanni C, Schettini G, Govoni S, Racchi M (2010) Beta amyloid precursor protein metabolism: focus on the functions and degradation of its intracellular domain. Pharmacol Res 62:308–317

Butterfield DA, Castegna A, Lauderback CM, Drake J (2002) Evidence that amyloid beta-peptide-induced lipid peroxidation and its sequelae in Alzheimer's disease brain contribute to neuronal death. Neurobiol Aging 23:655–664

Cagnin A, Brooks DJ, Kennedy AM, Gunn RN, Myers R, Turkheimer FE, Jones T, Banati RB (2001) In-vivo measurement of activated microglia in dementia. Lancet 358:461–467

Campbell A, Becaria A, Lahiri DK, Sharman K, Bondy SC (2004) Chronic exposure to aluminium in drinking water increases inflammatory parameters selectively in the brain. J Neurosci Res 75:565–572

Corder EH, Saunders AM, Strittmatter WJ, Schmechel DE, Gaskell PC, Small GW, Roses AD, Haines JL, Pericak-Vance MA (1993) Gene dose of apolipoprotein E type 4 allele and the risk of Alzheimer's disease in late on set families. Science 261:921–923

Dal Prà I, Chiarini A, Gui L, Chakravarthy B, Pacchiana R, Gardenal E, Whitfield JF, Armato U (2014) Do astrocytes collaborate with neurons in spreading the "infectious" Aβ and tau drivers of Alzheimer's disease? Neuroscientist 21:9–19

Deane R, Du Yan S, Submamaryan RK, LaRue B, Jovanovic S, Hogg E, Welch D, Manness L, Lin C, Yu J, Zhu H, Ghiso J, Frangione B, Stern A, Schmidt AM, Armstrong DL, Arnold B, Liliensiek B, Nawroth P, Hofman F, Kindy M, Stern D, Zlokovic B (2003) RAGE mediates amyloid-beta peptide transport across the blood-brain barrier and accumulation in brain. Nat Med 9:907–913

Drzezga A, Grimmer T, Henriksen G, Muhlau M, Perneczky R, Miederer I, PrausC SC, Wohlschlager A, Riemenschneider M, Wester HJ, Foerstl H, Schwaiger M, Kurz A (2009) Effect of APOE genotype on amyloid plaque load and gray matter volume in Alzheimer disease. Neurology 72:1487–1494

Emilien G, Durlach C, Minaker KL, Winblad B, Gauthier S, Maloteaux JM (2004) Alzheimer disease: neuropsychology and pharmacology. BirkhauserVerlag, Basel, p 288

Fanselow MS, Dong HW (2010) Are the dorsal and ventral hippocampus functionally distinct structures? Neuron 65:7–19

Farrer LA, Cupples LA, Haines JL, Hyman B, Kukull WA, Mayeux R, VanDuijn CM (1997) Effects of age, sex, and ethnicity on the association between apolipoprotein E genotype and Alzheimer disease. A meta analysis. APOE and Alzheimer disease meta analysis consortium. JAMA 278:1349–1356

Farris W, Mansourian S, Chang Y, Lindsley L, Eckman EA, Frosch MP, Eckman CB, Tanzi RE, Selkoe D, Guenette S (2003) Insulin-degrading enzyme regulates the levels of insulin, amyloid betaprotein, and the beta-amyloid precursor protein intracellular domain in vivo. Proc Natl Acad Sci U S A 100:4162–4167

Fleminger S, Oliver DL, Lovestone S, Rabe-Hesketh S, Giora A (2003) Head injury as a risk factor forAlzheimer's disease: the evidence 10 years on, a partial replication. J Neurol Neurosurg Psychiatry 74:857–862
Folstein MF, Folstein SE, McHugh PR (1975) "mini-mental state". A practical method for grading the cognitive state of patients for the clinician. J Psychiatr Res 12(3):189–198
Franz G, Beer R, Kampfl A, Engelhardt K, Schmutzhard E, Ulmer H, Deisenhammer F (2003) Amyloid β 1–42 and tau in cerebrospinal fluid after severe traumatic brain injury. Neurology 60: 1457–1461
Fulda S, Gorman AM, Osamu H, Afshin S (2010) Cellular stress responses: cell survival and cell death. Int J Cell Biol 10:1
Galluzzi L, Keppo O, Trojel-Hansen C, Kroemer G (2012) Mitochondrial control of cellular life, stress, and death. Circ Res 111:1198–1207
Gibson GE, Peterson C (1981) Aging decreases oxidative metabolism and the release and synthesis of acetylcholine. J Neurochem 37:978–984
Gilad GM, Rabey JM, Tizabi Y, Gilad VH (1987) Age-dependent loss and compensatory changes of septohippocampal cholinergic neurons in two rat strains differing in longevity and response to stress. Brain Res 436:311–322
Goedert M, Spillantini MG, Potier MC, Ulrich J, Crowther RA (1989) Cloning and sequencing of the cDNA encoding an isoform of microtubule-associated protein taucontaining four tandem repeats: differential expression of tau protein mRNAs in human brain. EMBOJ 8:393–399
Goedert M, Klug A, Crowther RA (2006) Tau protein, the paired helical filament and Alzheimer's disease. J Alzheimers Dis 9:195–207
Gong CX, Liu F, Grundke-Iqbal IK (2006) Dysregulation of protein phosphorylation/dephosphorylation in Alzheimer's disease: a therapeutic target. J Biomed Biotechnol 104:1–11
Gonzalez-Salvador T, Lyketsos CG, Baker A, Hovanec L, Roques C, Brandt J, Steele C (2000) Quality of life in dementia patients in long term case. Int J Geriatr Psychiatry 15:181–189
Good PF, Perl DP, Bierer LM, Schmeidler J (1992) Selective accumulation of aluminum and iron in the neurofibrillary tangles of Alzheimer's disease: a laser microprobe (LAMMA) study. Ann Neurol 31:286–292
Gotz J, Xia D, Leinenga G, Chew YL, Nicholas HR (2013) What renders TAU toxic. Front Neurol 4:72
Guo Z, Cupples LA, Kurz A, Auerbach SH, Volicer L, Chui H, Green RC, Sadovnick AD, Duara R, DeCarli C, Johnson K, Go RC, Growdon JH, Haines JL, Kukull WA, Farrer LA (2000) Head injury and the risk of AD in the MIRAGE study. Neurology 54:1316–1323
Guo T, Noble W, Hanger DP (2017) Roles of tau protein in health and disease. ActaNeuropathologica 133:665–704
Gustafson D (2006) Adiposity indices and dementia. Lancet Neurol 5:713–720
Haass C, Lemere CA, Capell A, Citron M, Seubert P, Schenk D, Lannfelt L, Selkoe DJ (1995) The Swedish mutation causes early-onset Alzheimer's disease by beta-secretase cleavage within the secretory pathway. Nat Med 1:1291–1296
Halliwell B (1992) Reactive oxygen species and the central nervous system. J Neurochem 59:1609–1623
Halliwell B, Gutteridge JMC (1999) Free radicals in biology & medicine. Oxford University Press, New York, pp 1–936
Halliwell B, Whiteman M (2004) Measuring reactive species and oxidative damage in vivo and in cell culture: how should you do it and what do the results mean? Br J Pharmacol 142(2):231–255
Hamano T, Gendron TF, Ko LW, Yen SH (2009) Concentration-dependent effects of proteasomal inhibition on tau processing in a cellular model of tauopathy. Int J Clin Exp Pathol 2:561–573
Hartman RE, Laurer H, Longhi L, Bales KR, Paul SM, McIntosh TK, Holtzman DM (2002) Apolipoprotein E4 influences amyloid deposition but not cell loss aftertraumatic brain injury in a mouse model of Alzheimer's disease. J Neurosci 22:10083–10087

Hu YS, Long N, Pigino G, Brady ST, Lazarov O (2013) Molecular mechanisms of environmental enrichment: impairments in AKT/GSK3β, neurotrophin-3 and CREB signaling. PLoS One 8: e64460

Jarrett JT, Berger EP, Lansbury PT Jr (1993) Thecarboxy terminus of the beta amyloid protein is critical for the seeding of amyloid formation: implications for the pathogenesis of Alzheimer's disease. Biochemistry 32:4693–4697

Jekabsone A, Mander PK, Tickler A, Sharpe M, Brown GC (2006) Fibrillar beta-amyloid peptide Abeta1-40 activates microglial proliferation via stimulating TNF-alpha release and H_2O_2 derived from NADPH oxidase: a cell culture study. J Neuroinflammation 3:24

Jellinger KA (2002) Vascular-ischemic dementia: an update. J Neural Transm Suppl 43:1–23

Jenner P (2003) Oxidative stress in Parkinson's disease. Ann Neurol 53:26–36

Jick H, Zornberg GL, Jick SS, Seshadri S, Drachman DA (2000) Statins and the risk of dementia. Lancet 356:1627–1631

Johansen T, Lamark T (2011) Selective autophagy mediated by autophagic adapter proteins. Autophagy 7:279–296

Kantari C, Walczak H (2011) Caspase-8 and bid: caught in the act between death receptors and mitochondria. BiochimicaetBiophysicaActa 1813:558–563

Karch CM, Goate AM (2015) Alzheimer's disease risk genes and mechanisms of disease pathogenesis. Biol Psychiatry 77:43–51

Kawahara M, Kato-Negishi M (2011) Link between aluminum and the pathogenesis of Alzheimer's disease: the integration of the aluminum and amyloid cascade hypotheses. Int J Alzheimers Dis 10:1–17

Keck S, Nitsch R, Grune T, Ullrich O (2003) Proteasome inhibition by paired helical filament-tau in brains of patients with Alzheimer's disease. J Neurochem 85:115–122

Kenessey A, Yen SHC (1993) The extent of phosphorylation of fetal tau is comparable to that of PHF-tau from Alzheimer paired helical filaments. Brain Res 629:40–46

Kerr JF, Wyllie AH, Currie AR (1972) Apoptosis: a basic biological phenomenon with wide ranging implications in tissue kinetics. Br J Cancer 26:239–257

Khachaturian ZS (1985) Diagnosis of Alzheimer's disease. Arch Neurol 42:1097–1105

Kimberly WT, LaVoie MJ, Ostaszewski BL, Ye W, Wolfe MS, Selkoe DJ (2003) Gamma-secretase is a membrane protein complex comprised of preselin, nicastrin, Aph-1, and Pen-2. Proc Natl Acad Sci U S A 100:6382–6387

Kivipelto M, Ngandu T, Fratiglioni L, Viitanen M, Kareholt I, Winblad B, Helkala EL, Tuomilehto J, Soininen H, Nissinen A (2005) Obesity and vascular risk factors at midlife and the risk of dementia and Alzheimer disease. Arch Neurol 62:1556–1560

Knops J, Suomensaari S, Lee M, McConlogue L, Seubert P, Sinha S (1995) Cell-type and amyloid precursor protein-type specific inhibition of a beta release by bafilomycin A1, a selective inhibitor of vacuolarATPases. J Biol Chem 270:2419–2422

Kruman II, Mattson MP (1999) Pivotal role of itochondrial calcium uptake in neural cell apoptosis and necrosis. J Neurochem 72:529–540

Kurz A, Perneczky R (2011) Novel insights for the treatment of Alzheimer's disease. Prog Neuro-Psychopharmacol Biol Psychiatry 35:373–379

Lambert MP, Barlow AK, Chromy BA, Edwards C, Freed R, Liosatos M, Morgan TE, Rozovsky I, Trommer B, Viola KL, Wals P, Zhang C, Finch CE, Krafft GA, Klein WL (1998) Diffusible, nonfibrillar ligands derived from Abeta1-42 are potent central nervous system neurotoxins. Proc Natl Acad Sci U S A 95:6448–6453

Lammich S, Kojro E, Postina R, Gilbert S, Pfeiffer R, Jasionowski M, Haass C, Fahrenholz F (1999) Constitutive and regulated alpha-secretase cleavage of Alzheimer's amyloid precursor protein by a disintegrin metalloprotease. Proc Natl Acad Sci U S A 96:3922–3927

Launer LJ, Terwindt GM, Ferrari MD (1999) The prevalence and characteristics of migraine in a population based cohort: the GEM study. Neurology 53:537–542

Li YM, Lai MT, Xu M, Huang Q, DiMuzio-Mower J, Sardana MK, Shi XP, Yin KC, Shafer JA, Gardell SJ (2000) Presenilin 1 is linked with γ-secretase activity in the detergent solubilized state. Proc Natl Acad Sci U S A 97:6138–6143

Lichtenthaler SF, Haass C, Steiner H (2011) Regulated intramembrane proteolysis-lessons from amyloid precursor protein processing. J Neurochem 117:779–796

Luchsinger JA, Tang MX, Stern Y, Shea S, Mayeux R (2001) Diabetes mellitus and risk of Alzheimer's disease and dementia with stroke in a multiethnic cohort. Am J Epidemiol 154: 635–641

Mahmood Z, Shukla Y (2010) Death receptors: targets for cancer therapy. Exp Cell Res 316:887–899

Marcinkiewicz M, Seidah NG (2000) Coordinated expression of beta-amyloid precursor protein and the putative beta-secretase BACE and alpha-secretase ADAM10 in mouse and human brain. J Neurochem 75:2133–2214

Marr RA, Rockenstein E, Mukherjee A, Kindy MS, Hersh LB, Gage FH, Verma IM, Masliah E (2003) Neprilysin gene transfer reduces human amyloid pathology in transgenic mice. J Neurosci 23:1992–1996

Massoud F, Gauthier S (2010) Update on the pharmacological treatment of Alzheimer's disease. Curr Neuropharmacol 8(1):69–80

Mates JM, Segura JM, Gomez CP, Rafael R, Olalla L (1999) Antioxidant enzymatic activities inhuman blood cells after an allergic reaction to pollen house dust mite. Blood Cells Mol Dis 25: 103–109

Mayeux R, Stern Y (2012) Epidemiology of Alzheimer disease. Cold Spring Harb Perspect Medicine 2:a006239. https://doi.org/10.1101/cshperspect.a006239

McConlogue L, Buttini M, Anderson JP, Brigham EF, Chen KS, Freedman SB, Games D, Johnson-wood K, Lee M, Zeller M, Liu W, Motter R, Sinha S (2007) Partial reduction of BACE1 has dramatic effects on Alzheimer's plaque and synaptic pathology in APP transgenic mice. J Biol Chem 282:26326–26334

Mohammad G, Alam K, Nawaz MI, Siddiquei MM, Mousa A, Abu El-Asrar AM (2015) Mutual enhancement between high-mobility group box-1 and NADPH oxidase-derived reactive oxygen species mediates diabetes-induced upregulation of retinal apoptotic markers. J Physiol Biochem 71(3):359–372

Müller UC, Deller T, Korte M (2017) Not just amyloid: physiological functions of the amyloid precursor protein family. Nat Rev Neurosci 18:281–298

Nathalie P, Jean-Noël O (2008) Processing of amyloid precursor protein and amyloid peptide neurotoxicity. Curr Alzheimer Res 5:92–99

Nunomura A, Perry G, Aliev G, Hirai K et al (2001) Oxidative damage is the earliest event in Alzheimer's disease. J Neuropathol Exp Neurol 60:759–767

Orrenius S, Gogvadz V, Zhivotovsky B (2015) Calcium and mitochondria in the regulation of cell death. Biochem Biophys Res Commun 460:72–81

Prakash D, Gopinath K, Sudhandiran G (2013) Fisetin enhances behavioral performances and attenuates reactive gliosis and inflammation during aluminium chloride-induced neurotoxicity. NeuroMolecular Med 15:192–208

Prema A, Justin Thenmozhi A, Manivasagam T, Mohamed Essa M, Guillemin GJ (2017) Fenugreek seed powder attenuated aluminum chloride-induced tau pathology, oxidative stress, and inflammation in a rat model of Alzheimer's disease. J Alzheimers Dis 60(s1):S209–S220

Preston AR, Eichenbaum H (2012) Interplay of hippocampus and prefrontal cortex in memory. Curr Biol 23:764–773

Qiu WQ, Folstein MF (2006) Insulin, insulin-degrading enzyme and amyloid-beta peptide in Alzheimer's disease: review and hypothesis. Neurobiol Aging 27.190–198

Rachel RA, Christopher JH, Robin LW, Li F, Tina LB, Paul MM (2010) BACE1 and BACE2 enzymatic activities in Alzheimer's disease. J Neurochem 112:1045–1053

Rasmusson DX, Brandt J, Martin DB, Folstein MF (1995) Head injury as a risk factor in Alzheimer's disease. Brain Inj 9:213–219

Readnower RD, Sauerbeck AD, Sullivan PG (2011) Mitochondria, amyloid β, and Alzheimer's disease. Int J Alzheimers Dis 10:1042–1045

Reiman EM, Chen K, Liu X, Bandy D, Yu M, Lee W, Ayutyanont N, Keppler J, Reeder SA, Langbaum JB, Alexander GE, Klunk WE, Mathis CA, Price JC, Aizenstein HJ, DeKosky ST, Caselli RJ (2009) Fibrillar amyloid-beta burden in cognitively normal people at 3 levels of genetic risk for Alzheimer's disease. Proc Natl Acad Sci U S A 106:6820–6825

Ryman DC, Acosta-Baena N, Aisen PS, Bird T, Danek A, Fox NC, GoateA FP, Ghetti B, Langbaum JB, Ringman JM, Salloway S, Schofield PR, Sperling R, Tariot PN, Xiong C, Morris JC, Bateman RJ (2014) Symptom onset in autosomal dominant Alzheimer disease: a systematic review and meta-analysis. Neurology 83:253–260

Samuel S, Kathirvel R, Jayavelu T, Chinnakkannu P (2005) Protein oxidative damage in arsenic induced rat brain: influence of DL-alpha-lipoic acid. Toxicol Letters 155:27–34

Scheuner D, Eckman C, Jensen M et al (1996) Secreted amyloid beta-protein similar to that in the senile plaques of Alzheimer's disease is increased in vivo by the presenilin 1 and 2 and APP mutations linked to familial Alzheimer's disease. Nat Med 2:864–870

Schneider JA, Bennett DA (2010) Where vascular meets neurodegenerative disease. Stroke 41: 144–146

Selkoe DJ (1991) The molecular pathology of Alzheimer's disease. Neuron 6:487–498

Selkoe DJ (2003) Folding proteins in fatal ways. Nature 426(6968):900–904

Shah S, Lee SF, Tabuchi K, Hao YH, Yu C, LaPlant Q, Ball H, Dann CE III, Sudhof T, Yu G (2005) Nicastrin functions as a γ-secretase-substrate receptor. Cell 122:435–447

Shibata M, Yamada S, Kumar SR, Calero M, Bading J, Frangione B, Holtzman DM, Miller CA, Strickland DK, Ghiso J, Zlokovic BV (2000) Clearance of Alzheimer's amyloid-ss(1-40) peptide from brain by LDL receptor-related protein-1 at the blood-brain barrier. J Clin Invest 106:1489–1499

Snowdon DA, Greiner LH, Mortimer JA, Riley KP, Greiner PA, Markesbery WR (1997) Brain infarction and the clinical expression of Alzheimer's disease. The Nun study. JAMA 277:813–817

Song WJ, Son MY, Lee HW, Seo H, Kim JH, Chung SH (2015) Enhancement of BACE1 activity by p25/cdk5-mediated phosphorylation in Alzheimer's disease. PLoS One 10:e0136950

Sorg C, Riedl V, MuhlauMCalhoun VD, Eichele T, Läer L, Drzezga A, Förstl H, Kurz A, Zimmer C, Wohlschläger AM (2007) Selective changes of resting-state networks in individuals at risk for Alzheimer's disease. Proc Natl Acad Sci U S A 104:18760–18765

Sun ZZ, Chen ZB, Jiang H, Li LL, Li EG, Xu Y (2009) Alteration of Abeta metabolism-related molecules in predementia induced by $AlCl_3$ and D-galactose. Age 31:277–284

Supekar K, Menon V, Rubin D, Musen M, Greicius MD (2008) Network analysis of intrinsic functional brain connectivity in Alzheimer's disease. PLoS Comput Biol 4(6):e1000100

Takalo M, Salminen A, Soininen H, Hiltunen M, Haapasalo A (2013) Protein aggregation and degradation mechanisms in neurodegenerative diseases. Am J Neurodegener Dis 2:1–14

Terry RD, Buccafusco JJ (2003) The cholinergic hypothesis of age and Alzheimer's disease-related cognitive deficits: recent challenges and their implications for novel drug development. J Pharmacol Exp Ther 306:821–827

Tongsong T, Puntachai P, Mekjarasnapha M, Traisrisilp K (2014) Severe fetal brain shrinkage following heavy maternal alcohol consumption. Ultrasound Obstetrics Gynecol 44:245–247

Traber MG, Vandervliet A, Reznick AZ, Cross CE (2000) Tobacco-related diseases. Is there a role for antioxidant micronutrient supplementation? Clin Chest Med 21:173–187

Trujillo ME, Scherer PE (2005) Adiponectin journey from an adipocyte secretory protein to biomarker of the metabolic syndrome. J Intern Med 257:167–175

Tyan SH, Shih AY, Walsh JJ, Maruyama H, Sarsoza F, Ku L, Eggert S, Hof PR (2012) Amyloid precursor protein (APP) regulates synaptic structure and function. Mol Cell Neurosci 51:43–52

Van der Kant R, Goldstein LS (2015) Cellular functions of the amyloid precursor protein from development to dementia. Dev Cell 32:502–515

Van Uden E, Mallory M, Veinbergs I, Alford M, Rokenstein E, Masliah E (2002) Increased extracellular amyloid deposition and neurodegeneration in human amyloid precursor protein transgenic mice deficient in receptor-associated protein. J Neurosci 22:9298–9304

Von Muhlinen N, Akutsu M, Ravenhill BJ, Foeglein A, Bloor S, Rutherford TJ, Freund SM, Komander D, Randow F (2012) LC3C, bound selectively by a noncanonical LIR motif in NDP52, is required for antibacterial autophagy. Mol Cell 48:329–342

Yang LB, Lindholm K, Yan R, Citron M, Xia W, Yang XL, Beach T, Sue L, Wong P, Price D, Li R, Shen Y (2003) Elevated beta-secretase expression and enzymatic activity decetd in sporadic Alzheimer's disease. Nat Med 9:3–4

Yatin SM, Aksenov M, Butterfield DA (1999) The antioxidant vitamin E modulates amyloid beta-peptide-induced creatine kinase activity inhibition and increased protein oxidation: implications for the free radical hypothesis of Alzheimer's disease. Neurochem Res 24:427–435

Zatta P, Drago D, Bolognin S, Sensi SL (2009) Alzheimer's disease, metal ions and metal homeostatic therapy. Trends Pharmacol Sci 30:346–355

Zhou ZX, Zeng ZY, Du JT, Zhao YF, Li YM (2006) The self-assembly ability of the first microtubule-binding repeat from tau and its modulation by phosphorylation. Biochem Biophys Res Commun 348:637–642

Chapter 2
Role of Vitamins in Alzheimer's Disease

Jayaraj Kiruba, Arokiasamy Justin Thenmozhi, Mariakuttikan Jayalakshmi, and Ephrem Arockia Jeya Yasmi Prabha

Abstract Alzheimer's disease (AD) is a severe neurodegenerative ailment of the brain, affecting millions of elderly people globally. It has also created a great health concern in the future. Several acquired disease conditions and some other factors enhance the chance of developing AD. No treatment can completely cure AD. Scientific literature has established that vitamins have a number of targets in the aetiology of AD through which they act to prevent the neuronal dysfunction in the disease. Thus, by understanding the role of vitamins, we are able to state that vitamins can be a good choice to overcome the detrimental effects of AD. Antioxidant vitamins play a major role because they act by reducing the degree of oxidative stress in the brain. The categories of vitamins having promising effects in declining the course of AD and its symptoms are reviewed. Low vitamin intake can increase the chances of acquiring AD. Fortunately, these significant vitamins that protect the brain can be acquired through various fruits and vegetables that are discussed here.

Keywords Alzheimer's disease · Vitamins · Antioxidants · Neurodegenerative diseases · Diet

J. Kiruba · M. Jayalakshmi
Department of Immunology, School of Biological Sciences, Madurai Kamaraj University, Madurai, Tamil Nadu, India

A. Justin Thenmozhi
Department of Biochemistry, School of Biological Sciences, Madurai Kamaraj University, Madurai, Tamil Nadu, India

Department of Biochemistry and Biotechnology, Faculty of Science, Annamalai University, Chidambaram, Tamil Nadu, India

E. Arockia Jeya Yasmi Prabha (✉)
Department of Physics, N.M.S.S.V.N. College, Madurai, Tamil Nadu, India

A. Justin Thenmozhi, T. Manivasagam (eds.), *Nutraceuticals for Alzheimer's Disease: A Promising Therapeutic Approach*, Nutritional Neurosciences,
https://doi.org/10.1007/978-981-99-0677-2_2

2.1 Introduction

A severe neurodegenerative ailment of the brain known as Alzheimer's disease (AD) is characterized by extracellular beta-amyloid plaques, intraneuronal tau fibrillary tangles, cell death of cholinergic neurons, vascular pathology, and inflammation (Terry et al. 1994). The chance of developing AD is increased by acquired ailments such cerebrovascular disease, diabetes, hypertension, obesity, and dyslipidemia. Globally, there are approximately 50 million AD patients, and this number is expected to double every 5 years to reach 152 million by 2050 (Vogel et al. 2009). Many factors, including nutritional factors, influence the disease, but its cause has not been established. It is reported that the deficiency of certain vitamins also leads to the progression of AD. Consuming these vitamin supplements plays a crucial role in the decline of AD.

Vitamins are collections of highly complex molecules. They are organic in nature and present in food in small amounts and are necessary for regular metabolism. When these nutrients are lacking, it results in various disorders, but when these nutrients are replenished, they cure those deficiency symptoms. Compared to lipids, carbohydrates, and proteins, vitamins are diverse in nature. The organic nature of vitamins sets them apart from other categories, and their classification is based on their chemical makeup and function. Vitamins are required in extremely minute quantities for development, health, and reproduction (Maqbool et al. 2018). AD is incurable today, although treatments exist to ease its symptoms (Livingston et al. 2020). The use of appropriate supplements and a properly balanced diet can contribute to improving the clinical condition of patients with AD.

2.2 AD and Antioxidant Vitamins

Antioxidant vitamins preserve the body from "free radicals," which harm cells and can result in cancer, heart disease, and Alzheimer's disease. Numerous antioxidants exist, including beta-carotene, vitamins C and E, and resveratrol. They can be found in meals that come from plants, such bell peppers, berries, and greens.

Grapes, red wine, peanuts, and certain dark chocolate all contain resveratrol. It may provide antiaging benefits and reduce the chance of developing certain diseases, according to scientific studies. For a while, researchers hypothesized that resveratrol could be able to shield the brain from the symptoms of AD. A recent study found that daily resveratrol dosages did decrease the disease's progression.

During the aging, free radicals frequently accumulate in the nerve cells. One of the elements causing the development and progression of neurodegenerative processes is increased oxidative stress. According to research on the brains of Alzheimer's patients, the body has been attempting to prevent free radical damage by showing indicators of oxidative stress (Swaminathan and Jicha 2014).Therefore, consuming more antioxidants would seem to be beneficial. However, there isn't currently a simple solution. Antioxidants may be better obtained through food than from supplements; however, researchers are unsure if particular antioxidants are superior to others.

A probable connection between dietary antioxidants and AD may be explained by a number of biological processes. Age-related oxidative stress of the brain tissue and lymphocytes, increased free radical buildup, and impaired antioxidant mechanisms are all reported. Antioxidants may, in the first instance, reduce the degree of oxidative stress in the brain. Thus, antioxidants may lessen the quantity of DNA deterioration, neuronal cell death, and amyloid accumulation in the brain (Swaminathan and Jicha 2014; Christen 2000). Second, a large intake of antioxidants may lessen the risk of dementia by lowering the risk of atherosclerosis since AD is linked to both cardiovascular risk factors and atherosclerosis (Hofman et al. 1997; Breteler 2000) and oxidative processes are implicated in atherogenesis (Witztum and Steinberg 1991). The risk of dementia may be decreased by halting the creation of these events, which are all significant neuropathological aspects of AD. Antioxidants' pleiotropic mechanisms, which include reducing oxidized lipid membranes, preventing nucleic acid damage, and preventing protein carbonylation, impede the discovery of a broken particular route that leads to the development of AD. There are a group of vitamins showing promising effects in AD prevention and management (Fig. 2.1).

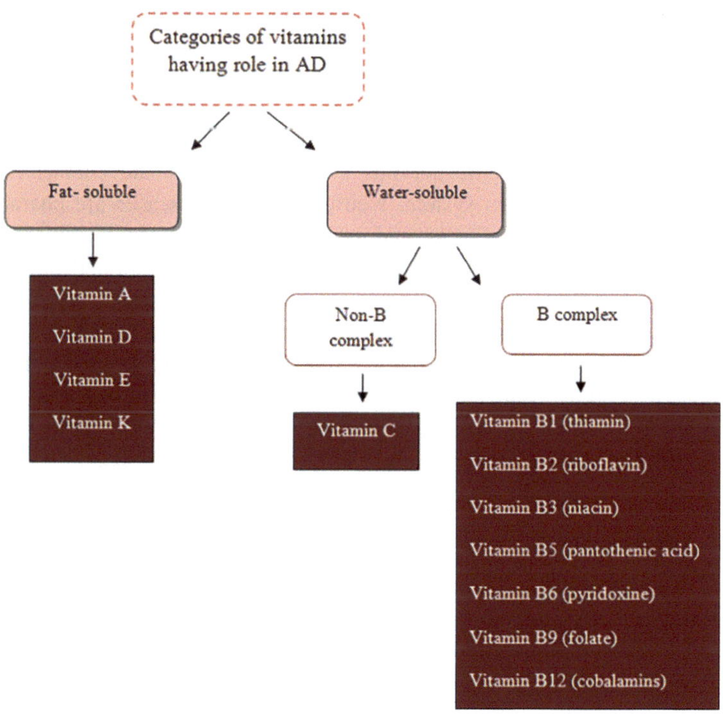

Fig. 2.1 Some of the categories of vitamins known to have promising effects in getting rid of AD and its symptoms

2.3 Fat-Soluble Vitamins

2.3.1 Vitamin A

Vitamin A, a fat-soluble micronutrient, plays a crucial role in a number of physiological processes, including the generation of red blood cells, immunology, and reproduction. The early development of brain form and function, neuronal proliferation and differentiation, neurite outgrowth, and synaptogenesis are the crucial functions that vitamin A plays in the central nervous system (CNS) (Ono and Yamada 2012). The amounts of vitamin A and beta-carotene in the CSF and blood of AD patients were found to be significantly lowered. It is observed that the average alpha-tocopherol and retinol serum concentration was lower in AD patients as compared to that of the control participants after screening by age, sex, and cardiovascular comorbidities (Bourdel-Marchasson et al. 2001).Vitamin A and beta-carotene have been demonstrated to impact the onset of neurodegenerative illnesses. In comparison with the control group, patients with dementia had significantly lower plasma levels of antioxidants that break chains, such as vitamins A, C, and E (Foy et al. 1999).

A key process that frequently results in neuronal toxicity in AD is the oligomerization of Aß fibrils. Another action of vitamin A and beta-carotene is the inhibition of the production and instability of Aβ fibrils (Ono and Yamada 2012). The aggregation and oligomerization of Aβ40 and Aβ42 fibrils have been demonstrated to be reduced by vitamin A administration, however. Beta-carotene and vitamin A have also been found to stop the decrease of cognitive function in AD. Additionally, these individuals' improved memory function and spatial learning have been linked to greater amounts of these vitamins (Takasaki et al. 2011; Rivière et al. 1998). Activation of microglia in the nervous system is one of the characteristics of inflammation of the nervous system. Neuropsychiatric diseases are characterized by inflammation both chronically and acutely. According to credible research, microglia activation is one of the factors that lead to AD (Sodhi and Singh 2014). The local concentrations of retinoic acid may also alter as a result of the microglia malfunctioning. Moreover, through interactions with retinoic acid and retinoid X receptors, vitamin A and its derivatives play a crucial role in the differentiation of nerve cells, as well as in the production of neurotransmitters in the brain and gene expression.

2.3.2 Vitamin D

The significance of vitamin D in preserving strong bones has long been established. Recent findings by scientists show a connection between vitamin D and a number of mental processes, including memory and learning. Neurons in the brain and spine that are involved in the production of neurotransmitters and the formation of new nerves are activated by vitamin D. Inflammation is a prevalent feature of Alzheimer's; thus, researchers also think that vitamin D protects brain neurons and lowers inflammation (Rogers et al. 1988). Especially for people over 65, vitamin D is one of the most crucial vitamins. The aged people over 65 are thought to be vitamin D deficient for a certain cause that is unique and is covered in the section below. Vitamin D insufficiency is associated with significant health complications.

Because it is increasingly difficult for human bodies to produce vitamin D in the aged people, vitamin D deficiency poses a special concern. In comparison with people in their 20s, individuals over the age of 70 make 70% less vitamin D. Ninety-five percent of seniors are thought to be vitamin D deficient. People having a higher risk of vitamin D insufficiency include those who live in regions with shorter days and those who spend less time outside. It is thought that vitamin D assists in regulating the immune system. Vitamin D supplementation is thought to improve one's ability to fight off illnesses and reduce the danger of contracting them. Numerous studies have shown an association between cognitive decline, AD, and dementia and vitamin D insufficiency. According to a research from the University of Cambridge, those who have severe vitamin D deficiency are twice as likely to have cognitive impairment (Granic et al. 2015). This is in contrast to people who have optimal vitamin D levels.

2.3.3 Vitamin E

The body utilizes vitamin E for a diverse range of purposes, making it a crucial nutrient. Because of its antioxidative qualities, vitamin E may be able to treat Alzheimer's, according to researchers. The central nervous system has a place to store vitamin E, which lowers lipid peroxidation and amyloid buildup. Antioxidant-based therapy should be started as soon as the condition may be diagnosed. Cholinesterase levels in the brain are decreased by vitamin E.

Dementia and blood levels of vitamins C and E are associated with AD patients. Because of its capacity to scavenge free radicals, lower amyloid activity, and involvement in the chelation of iron, zincs, and copper, ascorbic acid has a neuroprotective impact. It protects the central nervous system from free radical damage. A pro-oxidative diet has been demonstrated to raise the quantity of amyloid precursor protein in a rat research. In individuals with mild to severe AD, vitamin E alone demonstrated to reduce cognitive impairment. It's interesting to note that memantine treatment when combined with α-tocopherol had fewer favorable results (Liao et al. 2010). Humans require vitamin E as a micronutrient because it helps keep cell membranes healthy. Depending on age, many nations suggest a daily dietary consumption of between 3 and 15 mg of vitamin E. Tocopherols and tocotrienols are abundant in seeds and edible oils like those made from almond, peanut, olive, palm, canola, corn, and soybean while being sparse in plant foods like fruit and vegetables that have low lipid levels.

Vitamin E is the most significant lipophilic radical scavenger in vivo; its primary function is as an antioxidant. In order to neutralize free radicals, vitamin E primarily uses a hydrogen atom transfer process. This reaction yields a non-radical product and a vitamin E radical, which may then attack lipids or react with a reducing agent like vitamin C or ubiquinol to renew vitamin E. As the predominant form of vitamin E in tissues, α-tocopherol has been the subject of most research. In lipoproteins and cell membranes, α-tocopherol functions as a chain-breaking antioxidant, reducing lipid peroxidation and protecting membrane integrity. Tocotrienols, in contrast to α-tocopherol, may have distinct health-improving effects and a higher antioxidant capacity, according to certain research (Serbinova et al. 1991).

"Oxidative agents" are toxic substances that damage the brain and cause AD. Vitamin E typically guards against the brain's damage from these oxidative radicals. But studies have revealed that those who have Alzheimer's have particularly low levels of vitamin E (Persson et al. 2014). According to some scientists, the low levels of vitamin E allow oxidative agents to harm the brain. The brain is also harmed by cholinesterase, which is thought to be a factor in AD. Researchers determine vitamin E slows the growth of Alzheimer's.

2.3.4 Vitamin K

It is now well acknowledged that vitamin K has strong evidence supporting its crucial functions in the neurological system. One theory is that vitamin K may also contribute to the pathogenesis of AD, a crippling condition for which there is now no treatment. Sphingolipids are a significant group of lipids that are abundantly found in the membranes of brain cells. Alterations in sphingolipid metabolism have been associated with neurodegenerative diseases like AD and age-related cognitive impairment. Numerous studies have demonstrated the function of vitamin K in brain physiology by participating in sphingolipid metabolism and the biological activation of the vitamin K-dependent protein Gas6. Cell development, survival, and apoptosis are just a few of the many cellular activities that Gas6 is functionally engaged with. New evidence also points to the K vitamin menaquinone-4 (MK-4)'s distinct protective effects against oxidative stress and inflammation (Ferland 2012). Overall, the evidence suggests that vitamin K has an impact on cognition and psychomotor behavior.

According to a recent study, people with early-stage AD consumed less vitamin K than control subjects with normal cognitive function. Dietary vitamin K intakes were evaluated using food records from 5 nonconsecutive days gathered from 31 community-dwelling early-stage AD patients and 31 age- and sex-matched cognitively intact control participants. On a person-day basis, the mean vitamin K intake was 63 ± 90 μg/day in patients and 139 ± 233 μg/day in controls. The principal sources of vitamin K are green vegetables. Green vegetables thus contributed 33% and 49%, respectively, to total intakes in patients and control participants. Subjects with AD consumed fewer green vegetables overall, which contributed to their reduced vitamin K intakes (Presse et al. 2008). The main finding of the abovementioned study is that individuals with likely early-stage AD had considerably lower vitamin K intakes than healthy people who were age- and sex-matched. Nevertheless, they are consistent with the most current studies on vitamin K and AD, which contend that inadequate vitamin K intakes may increase the chance of developing AD or hasten its course.

Elderly people and apolipoprotein E4 carriers have an increased risk of developing AD. In older men and women, a relative vitamin K deficit that affects the vitamin's extrahepatic actions is prevalent. Comparatively to those with other APOE genotypes, APOE4 carriers have decreased levels of vitamin K in their blood circulation. In the brain, vitamin K has crucial functions. According to mounting evidence, vitamin K regulates sulfotransferase activity and the activation of a growth factor/tyrosine kinase receptor (Gas 6/Axl). It is currently believed that vitamin K insufficiency plays a role in the etiology of AD and that taking vitamin K supplements may help with either disease prevention or treatment. Additionally, cardiovascular disease-related neuronal damage may also be lessened by vitamin K (Allison 2001).

Recent research has emphasized the value of vitamin K2 (VK2) for maintaining human health. However, there have been no clinical studies looking into VK2's potential for treating or preventing AD. An increasing body of data indicating that VK2 has the ability to retard the advancement of AD and contribute to its prevention is found in reviewing basic science research and clinical trials that have linked VK2 to elements implicated in its pathogenesis. Numerous studies have noted VK2's antiapoptotic and antioxidant activities as well as its impacts on cardiovascular health, mitochondrial dysfunction, cognition, neuroinflammation, and comorbidities in AD (Fig. 2.2). Dysbiosis and VK2 are related in the context of the microbiome's role in the development of AD (Popescu and German 2021).

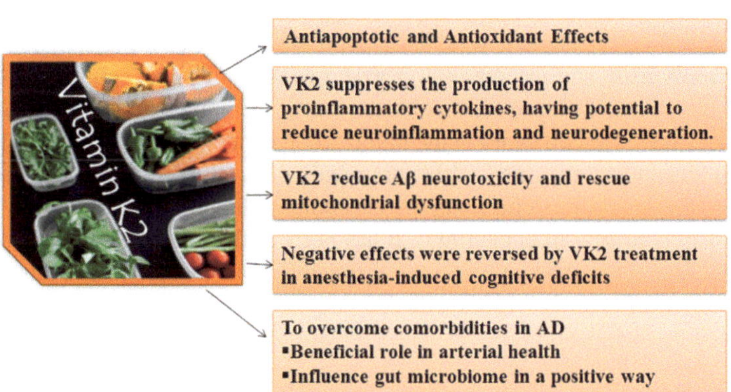

Fig. 2.2 Role of vitamin K2 (VK2) in AD

2.4 Water Soluble Vitamins

2.4.1 B-Complex

The B vitamins with the most conclusively known links to cognitive deterioration are vitamin B12, niacin, and thiamine. The nerve system and brain of the human play a critical function for B vitamins. Vitamin B6 is required to produce the chemicals that neurons use to communicate with one another, while vitamins B3 and B9 are necessary to generate and repair DNA. The development and repair of vital components of the human neurological system are just a few of the tasks that the B vitamins do, in addition to providing the brain with the energy it needs to function.

B vitamin supplement users had a lower risk of AD than nonusers, according to research from the University of Oxford and the Centers for Disease Control and Prevention (CDC). B vitamins have important functions in several regions of the neurological system and brain.

2.4.1.1 Vitamin B1

Thiamine deficiency in human leads to widespread beriberi, which has a substantial neurological impact. Many researches suggest that neurological issues such as cognitive impairments and encephalopathy are attributed to thiamine inadequacy. The etiology of AD is also related to vitamin B1 (Lu'o'ng and Nguyen 2011; Hazell et al. 2013). Three important enzymes, pyruvate dehydrogenase (PDH), alpha-ketoglutarate dehydrogenase, and alpha-transketolase, are the cofactors for thiamine pyrophosphate. Alpha-ketoglutarate dehydrogenase and alpha-transketolase, enzymes that depend on vitamin B for activity in the brain and peripheral tissues, have been found to have lower plasma levels of vitamin B1 in Alzheimer's patients (Fessel 2021). The three enzymes decarboxylate their substrates and transfer the resulting acyl groups to coenzymes A (CoA), which results in the regeneration of nicotinamide adenine dinucleotide hydride. A deficiency of these enzymes impairs brain metabolism and decreases synapse metabolism. The submedial thalamic nucleus constitutes the most susceptible portion of the brain with thiamine deficiency and causes selective cell death in the brain. AD patients' brains have lower levels of thiamine-dependent enzymes, which are required for glucose metabolism and their decrease, and thiamine deficit may be accountable for the deterioration in glucose metabolism (Calingasan et al. 1999). It has been well explored that oxidative stress and inflammation contribute in the preferential neurons that are lost in this zone and the activation of other cell types. For AD patients, increasing thiamine levels in the brain may be therapeutic.

2.4.1.2 Vitamin B2

Animals and vegetables both contain large amounts of vitamin B2, often known as riboflavin. Through a variety of biological oxidation processes, vitamin B6 metabolism, energy production, and cell development, it benefits human health. Moreover, it influences how iron is absorbed, stored, and mobilized by the body (Powers 2003). Additionally, B2 exhibit antioxidant properties. Riboflavin is also significant as an antioxidant in the glutathione redox cycle. Flavin adenine dinucleotide (FAD) and flavin mononucleotide (FMN) are precursors of vitamin B2. Through the reduction of FAD by decreased nicotinamide adenine dinucleotide phosphate (NADPH), glutathione reductase is activated. For the glutathione redox cycle to function continuously, it is necessary to complete this important phase (Saedisomeolia and Ashoori 2018). So it is common to monitor glutathione reductase activity in order to determine the condition of cellular riboflavin. Therefore, riboflavin deficiency may contribute to the development and stabilization of neurodegenerative disorders where oxidative stress is enhanced. There is mounting evidence that oxidative stress plays a significant role in the pathogenesis of neurodegenerative disorders (Mariani et al. 2005). The brain endures excessive oxidative stress as a result of oxidants being released during the advancement of AD (Mangialasche et al. 2009). Reactive oxygen species are remnants of aerobic metabolism, and intracellular enzymes and antioxidants regulate their level (Schaffer and Bronnikova 2012; Apel and Hirt 2004). Reactive oxygen species are transient signaling molecules that are at low levels during normal physiological circumstances. The peroxidation process, which significantly damages proteins, phospholipids, and DNA, is led on by an excessive generation of reactive oxygen species. The oxidase activity of mitochondrial cytochrome C can be reduced by these damages, which can lead to metabolic disruption and cell death (Joodi et al. 2011). In humans, vitamin B2 supplementation may activate conserved signaling pathways that prevent the emergence and progression of neurodegenerative disorders like AD.

2.4.1.3 Vitamin B3

The development of nerve cells, the production of useful energy for the brain, and DNA synthesis and repair are all facilitated by vitamin B3. B3 also acts as a potent antioxidant and aids in brain cell regeneration.

Niacin, commonly known as nicotinic acid, is a kind of vitamin B3 that has subsequently drawn the attention of some scientists. Over the past 20 years, a few epidemiological studies that suggested a link between niacin intake and a lower risk of general cognitive deterioration have raised interest in the vitamin. Recently, scientists have started investigating if the substance benefits people with neurodegenerative illnesses. In mouse models for Parkinson's, glioblastoma, and multiple sclerosis, recent research findings have revealed that niacin can influence the activity of microglia, crucial immune cells in the brain (Dursun and Gezen-Ak 2019). An AD mouse model has recently demonstrated the first signs of this protective effect.

Researchers are investigating the vitamin B3 niacinamide, also known as nicotinamide, for its potential to reduce tau, a protein that builds up in tangles in the brains of those with AD. A potential approach for the treatment of AD and other dementias is the supplementation of nicotinamide adenine dinucleotide (NAD) to restore the damaged mitochondria. A safe NAD precursor with a high oral bioavailability, nicotinamide ribose slows down the aging process (Bachurin et al. 2018). According to CDC researchers, those who consume fewer vitamin B3 are more likely to acquire AD. Additionally, the study found that vitamin B3 levels above normal reduced the growth of Alzheimer's.

2.4.1.4 Vitamin B5

Coenzyme A (CoA) and its subsequent derivatives, such as acetyl-CoA, succinyl-CoA, malonyl-CoA, and 3-hydroxy-3-methylglutaryl (HMG)-CoA, are all precursors to pantothenic acid, also known as vitamin B5. In many biological processes, including the metabolism of nucleic acids, carbohydrates, proteins, and lipids, CoA is an essential cofactor. It is involved in the process that creates the vital anti-inflammatory hormone cortisol. It is crucial for myelinization and synthesis of the neurotransmitter acetylcholine. Additionally, vitamin B5 may lessen oxidative stress by encouraging the formation of glutathione. Numerous foods made from plants and animals contain pantothenic acid. Furthermore, this vitamin can be obtained via fermented foods. It has been calculated that human gut bacteria could contribute 0.078% of the recommended daily intake of this vitamin and that 51% of the human gut microbiota genomes include the genes for the pathway that synthesize vitamin B5.

Humans with experimentally induced pantothenic acid shortage had sleeplessness and other health problems. It's interesting to note that significant vitamin B5 deficiency was found in the brain tissues of AD patients in a recent postmortem investigation. These alterations were particularly noticeable in the hippocampus, entorhinal cortex, and middle temporal gyrus, three brain regions linked to the pathogenesis of AD. The scientists concluded that a vitamin B5 deficiency may be a factor in the development of dementia and neurodegeneration in AD since CoA is involved in the formation of acetylcholine and myelin in the brain. Additionally, a postmortem investigation of patients with Huntington's disease found that the vitamin B5 levels in their brain tissues had significantly decreased globally. Animal research proved cerebral pantothenate was primarily localized to myelin-containing regions, supporting the aforementioned findings (Rudzki et al. 2021).

2.4.1.5 Vitamin B6

The three chemically different molecules pyridoxal, pyridoxamine, and pyridoxine that make up vitamin B6 are important in the control of mood and cognitive function. A lack of vitamin B6 is linked to an increase in blood homocysteine levels because it is a crucial cofactor in homocysteine re-methylation. Homocysteine may directly damage central nervous system neurons in addition to being a risk factor for cerebrovascular disease. A vitamin B6 deficiency has been associated with neuropsychiatric illnesses such as seizures, migraines, chronic pain, and depression. Older persons frequently have low vitamin B6 levels, according to epidemiological research. A reason or mechanism for the onset of AD and other kinds of dementia has been proposed: hyperhomocysteinaemia. It has been shown that taking supplements of B vitamins, especially vitamin B6, lowers blood homocysteine levels (Malouf and Grimley Evans 2003). In the generation of neurotransmitters, vitamin B6 is utilized. Nerves interact with one another through neurotransmitters, which are chemicals. The digestion of foods to produce energy for the human brain requires B6.

2.4.1.6 Vitamin B9

The water-soluble, naturally occurring form of vitamin B9 is called folate, and it may be found in a variety of foods. Amino acids and DNA are two substances required for the correct operation of practically every system in the human body, and vitamin B9 is crucial for their production. In addition to its role in DNA and RNA synthesis, folate also helps to the metabolism of proteins. Homocysteine, an amino acid that may be harmful to the body in high doses, must be broken down in order for it to function. In addition to being necessary for the production of healthy red blood cells, folate is also crucial during periods of rapid growth, such as during pregnancy and the development of the fetus.

An increasing number of researches suggest that the crucial nutrient folate is essential for the progression of AD (Robinson et al. 2018). Folate concentrations in the plasma and serum should not exceed 45.3 nmol/L. When plasma/serum folate levels are below 6.8 nmol/L and above 13.5 nmol/L, respectively, folate deficiency and probable deficiency are considered to exist (WHO 2015). Previous researches suggested that low folate levels are correlated to all kinds of dementia, including vascular dementia and AD, in addition to particular areas of cognitive performance such as episodic recall and recognition (Wahlin et al. 1996). Additionally, supplementing with folate and folic acid helps to improve cognitive abilities in elderly people and in situations of moderate cognitive impairment (Ma et al. 2019). In addition, a lack of folate causes an increase in tau phosphorylation, which is a key component of neurofibrillary tangles (Chan and Shea 2006). Compared to healthy controls, folate levels were decreased in AD patients. A suspected or actual folate deficit may raise the risk of AD. The chance of developing AD may be lowered by getting enough folate each day. Studies have already revealed that people with moderate cognitive impairment who take folic acid supplements experience a slower rate of brain shrinkage and cognitive loss (Ma et al. 2019). According to these results, getting enough folate can help prevent AD. Randomized controlled studies are required to establish the link between folic acid supplementation or adequate folate consumption and the prevention of AD.

2.4.1.7 Vitamin B12

It is widely known that vitamin B12 functions as a cofactor in the one carbon cycle process where the neurotoxic homocysteine is converted to methionine by the enzyme methionine synthase (Chan and Shea 2006). New aspects of vitamin B12's role as a gene regulator and an epigenetics modifier impacting brain intellectual abilities have been revealed in the last 10 years (Vogel et al. 2009). The essential function of B12 in the brain is implied by the presence of the cubam receptor, a particular transporter receptor that is identical to that found in the gut (Douaud et al. 2013). Moreover, there is broad agreement that proven AD sufferers have a B12 inadequacy (Dayon et al. 2017), emphasizing the potentially beneficial function of B12 consumption in the treatment of recognized AD. Accordingly, it was shown that older individuals with moderate dementia who received vitamin B supplements saw a sevenfold reduction in brain gray matter shrinkage in AD susceptible areas (Morris 2003). Although research investigating the efficacy of B12 and its mechanistic function in verified AD is few, they seldom relate B12 impact to cholinergic pathways and barely ever link B12 effect to homocysteine, despite seeming clinically beneficial.

Making, maintaining, and repairing the myelin sheath are one of vitamin B12's most crucial jobs. The layer that shields and surrounds nerve cells is called the myelin sheath. For nerves to communicate with one another, it is essential. Nerve cells are unable to interact when the myelin sheath is destroyed, as it is in Alzheimer's and dementia (Liao et al. 2010). As a result, a variety of mental issues arise, including the inability to create new memories and communication issues. Vitamin B12 deficiency has a well-known neurologic condition that is defined by cognitive and mental problems, as well as by subacute combined degeneration of the spinal cord and peripheral neuropathy (Goetz and Pappert 1999; Savage and Lindenbaum 1995). Additionally, cognitive problems caused by the neurologic condition can be improved with high-dose vitamin B12 treatment (Allen et al. 1990; Kuzminski et al. 1998). Also, vitamin B12 is required to develop and sustain nerve cells.

2.4.2 *Vitamin C*

The proper function of the brain is significantly influenced by vitamin C, according to earlier in vivo and in vitro research. The assumption that vitamin C has preventive benefits throughout the spectrum of neurodegenerative illnesses was further supported by decreased plasma levels in patients despite appropriate consumption (Montilla-López et al. 2002). In accordance with a study by Polidori et al. (2015), a certain serum levels of vitamin C may be essential for the prevention of AD and other clinical signs of vascular and cognitive aging. Therefore, it may be established that antioxidant vitamins offer defense against damage caused by oxidative stress in AD. Vitamin C can prevent the progression of AD because of the effects it has on the disease's pathophysiology in several ways (Fig. 2.3). Numerous investigations, both in vivo and in vitro, came to the conclusion that vitamin C prevents the oligomerization of the Aß peptide, hence assisting in reducing oxidative stress.

Oxidative stress in the tissue is brought on by brain damage, which lowers the amounts of antioxidants like SOD and vitamin C. Supplementing with vitamin C appears to increase SOD levels, which subsequently reduce oxidative stress and stop additional brain damage from occurring (Harrison 2012). Even a typical diet's consumption of vitamin C may have a neuroprotective impact in AD patients, according to a theory. In addition, it has been shown that Alzheimer's patients (Fig. 2.3) who consume enough vitamin C experience much less cognitive deterioration (Zandi et al. 2004).

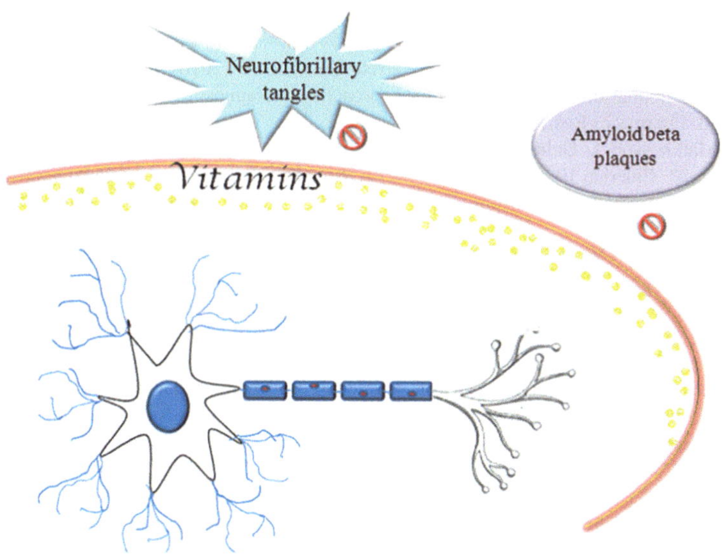

Fig. 2.3 Therapeutic potential of vitamins in AD. Vitamins prevent Aβ plaque formation by inhibiting the aggregation of beta-amyloid plaques. Vitamins also prevented the tau protein aggregation and its oligomerization into neurofibrillary tangles

2.5 Conclusion

A serious neurological disorder associated with aging is AD. Since there are more people suffering from AD, it is necessary to find a medication that can stop or slow the disease's growth. Although there is no known cure for Alzheimer's, there are therapies that may slow down the disease's course as well as medication and non-medication approaches that might assist to manage symptoms. In order to manage symptoms and enhance quality of life, people with the condition and those who care for them might benefit from being aware of their alternatives. Understanding available alternatives can aid those suffering with the condition and those who are caring for them in managing symptoms and enhancing quality of life. The risk of AD may be reduced by consuming a diet high in vitamins (Fig. 2.3). Consuming enough fruits and vegetables will make up for vitamin deficiencies, which are typically seen in dementia patients.

References

Allen RH, Stabler SP, Savage DG, Lindenbaum J (1990) Diagnosis of cobalamin deficiency I: usefulness of serum methylmalonic acid and total homocysteine concentrations. Am J Hematol 34(2):90–98

Allison AC (2001) The possible role of vitamin K deficiency in the pathogenesis of Alzheimer's disease and in augmenting brain damage associated with cardiovascular disease. Med Hypotheses 57(2):151–155

Apel K, Hirt H (2004) Reactive oxygen species: metabolism, oxidative stress, and signal transduction. Annu Rev Plant Biol 55:373–399

Bachurin SO, Gavrilova SI, Samsonova A, Barreto GE, Aliev G (2018) Mild cognitive impairment due to Alzheimer disease: contemporary approaches to diagnostics and pharmacological intervention. Pharmacol Res 129:216–226

Bourdel-Marchasson I, Delmas-Beauvieux MC, Peuchant E, Richard-Harston S, Decamps A, Reignier B, Emeriau JP, Rainfray M (2001) Antioxidant defences and oxidative stress markers in erythrocytes and plasma from normally nourished elderly Alzheimer patients. Age Ageing 30(3):235–241

Breteler MM (2000) Vascular risk factors for Alzheimer's disease: an epidemiologic perspective. Neurobiol Aging 21(2):153–160

Calingasan NY, Uchida K, Gibson GE (1999) Protein-bound acrolein: a novel marker of oxidative stress in Alzheimer's disease. J Neurochem 72(2):751–756

Chan A, Shea TB (2006) Dietary and genetically-induced oxidative stress alter tau phosphorylation: influence of folate and apolipoprotein E deficiency. J Alzheimer's Dis 9(4):399–405

Christen Y (2000) Oxidative stress and Alzheimer disease. Am J Clin Nutr 71(2):621S–629S

Dayon L, Guiraud SP, Corthésy J, Da Silva L, Migliavacca E, Tautvydaitė D, Oikonomidi A, Moullet B, Henry H, Métairon S, Marquis J, Descombes P, Collino S, Martin FJ, Montoliu I, Kussmann M, Wojcik J, Bowman GL, Popp J (2017) One-carbon metabolism, cognitive impairment and CSF measures of Alzheimer pathology: homocysteine and beyond. Alzheimers Res Ther 9(1):43

Douaud G, Refsum H, de Jager CA, Jacoby R, Nichols TE, Smith SM, Smith AD (2013) Preventing Alzheimer's disease-related gray matter atrophy by B-vitamin treatment. Proc Natl Acad Sci U S A 110(23):9523–9528

Dursun E, Gezen-Ak D (2019) Vitamin D basis of Alzheimer's disease: from genetics to biomarkers. Hormones (Athens) 18(1):7–15
Ferland G (2012) Vitamin K, an emerging nutrient in brain function. Biofactors 38(2):151–157
Fessel J (2021) Supplemental thiamine as a practical, potential way to prevent Alzheimer's disease from commencing. Alzheimer's dementia (N. Y.) 7(1):e12199
Foy CJ, Passmore AP, Vahidassr MD, Young IS, Lawson JT (1999) Plasma chain-breaking antioxidants in Alzheimer's disease, vascular dementia and Parkinson's disease. QJM 92(1):39–45
Goetz CG, Pappert EJ (1999) Textbook of clinical neurology. W.B. Saunders, Philladelphia
Granic A, Hill TR, Kirkwood TB, Davies K, Collerton J, Martin-Ruiz C, von Zglinicki T, Saxby BK, Wesnes KA, Collerton D, Mathers JC, Jagger C (2015) Serum 25-hydroxyvitamin D and cognitive decline in the very old: the Newcastle 85+ study. Eur J Neurol 22(1):106–1e7
Harrison FE (2012) A critical review of vitamin C for the prevention of age-related cognitive decline and Alzheimer's disease. J Alzheimers Dis 29(4):711–726
Hazell AS, Faim S, Wertheimer G, Silva VR, Marques CS (2013) The impact of oxidative stress in thiamine deficiency: a multifactorial targeting issue. Neurochem Int 62(5):796–802
Hofman A, Ott A, Breteler MM, Bots ML, Slooter AJ, van Harskamp F, van Duijn CN, Van Broeckhoven C, Grobbee DE (1997) Atherosclerosis, apolipoprotein E, and prevalence of dementia and Alzheimer's disease in the Rotterdam study. Lancet 349(9046):151–154
Joodi G, Ansari N, Khodagholi F (2011) Chitooligosaccharide-mediated neuroprotection is associated with modulation of Hsps expression and reduction of MAPK phosphorylation. Int J Biol Macromol 48(5):726–735
Kuzminski AM, Del Giacco EJ, Allen RH, Stabler SP, Lindenbaum J (1998) Effective treatment of cobalamin deficiency with oral cobalamin. Blood 92(4):1191–1198
Liao WC, Chen JR, Wang YJ, Tseng GF (2010) Methylcobalamin, but not methylprednisolone or pleiotrophin, accelerates the recovery of rat biceps after ulnar to musculocutaneous nerve transfer. Neuroscience 171(3):934–949
Livingston G, Huntley J, Sommerlad A, Ames D, Ballard C, Banerjee S, Mukadam N (2020) Dementia prevention, intervention, and care: 2020 report of the lancet commission. Lancet 396(10248):413–446
Lu'o'ng KV, Nguyen LT (2011) Role of thiamine in Alzheimer's disease. Am J Alzheimers Dis Other Dement 26(8):588–598
Ma F, Zhou X, Li Q, Zhao J, Song A, An P, Du Y, Xu W, Huang G (2019) Effects of folic acid and vitamin B12, alone and in combination on cognitive function and inflammatory factors in the elderly with mild cognitive impairment: a single-blind experimental design. Curr Alzheimer Res 16(7):622–632
Malouf R, Grimley Evans J (2003) The effect of vitamin B6 on cognition. Cochrane Database Syst Rev 4:CD004393
Mangialasche F, Polidori MC, Monastero R, Ercolani S, Camarda C, Cecchetti R, Mecocci P (2009) Biomarkers of oxidative and nitrosative damage in Alzheimer's disease and mild cognitive impairment. Ageing Res Rev 8(4):285–305
Maqbool MA, Aslam M, Akbar W, Iqbal Z (2018) Biological importance of vitamins for human health: a review. J Agric Basic Sci 2(3):50–58
Mariani E, Polidori MC, Cherubini A, Mecocci P (2005) Oxidative stress in brain aging, neurodegenerative and vascular diseases: an overview. J Chromatogr B Analyt Technol Biomed Life Sci 827(1):65–75
Montilla-López P, Muñoz-Agueda MC, Feijóo López M, Muñoz-Castañeda JR, Bujalance-Arenas I, Túnez-Fiñana I (2002) Comparison of melatonin versus vitamin C on oxidative stress and antioxidant enzyme activity in Alzheimer's disease induced by okadaic acid in neuroblastoma cells. Eur J Pharmacol 451(3):237–243
Morris MS (2003) Homocysteine and Alzheimer's disease. Lancet Neurol 2(7):425–428
Ono K, Yamada M (2012) Vitamin a and Alzheimer's disease. Geriatr Gerontol Int 12(2):180–188

Persson T, Popescu BO, Cedazo-Minguez A (2014) Oxidative stress in Alzheimer's disease: why did antioxidant therapy fail? Oxidative Med Cell Longev 2014:1

Polidori MC, Ruggiero C, Croce MF, Raichi T, Mangialasche F, Cecchetti R, Pelini L, Paolacci L, Ercolani S, Mecocci P (2015) Association of increased carotid intima-media thickness and lower plasma levels of vitamin C and vitamin E in old age subjects: implications for Alzheimer's disease. J Neural transm 122(4):523–530

Popescu A, German M (2021) Vitamin K2 holds promise for Alzheimer's prevention and treatment. Nutrients 13(7):2206

Powers HJ (2003) Riboflavin (vitamin B-2) and health. Am J Clin Nutr 77(6):1352–1360

Presse N, Shatenstein B, Kergoat MJ, Ferland G (2008) Low vitamin K intakes in community-dwelling elders at an early stage of Alzheimer's disease. J Am Diet Assoc 108(12):2095–2099

Rivière S, Birlouez-Aragon I, Nourhashémi F, Vellas B (1998) Low plasma vitamin C in Alzheimer patients despite an adequate diet. Int J Geriatr Psychiatry 13(11):749–754

Robinson N, Grabowski P, Rehman I (2018) Alzheimer's disease pathogenesis: is there a role for folate? Mech Ageing Dev 174:86–94

Rogers J, Luber-Narod J, Styren SD, Civin WH (1988) Expression of immune system-associated antigens by cells of the human central nervous system: relationship to the pathology of Alzheimer's disease. Neurobiol Aging 9(4):339–349

Rudzki L, Stone TW, Maes M, Misiak B, Samochowiec J, Szulc A (2021) Gut microbiota-derived vitamins–underrated powers of a multipotent ally in psychiatric health and disease. Prog Neuro-Psychopharmacol Biol Psychiatry 107:110240

Saedisomeolia A, Ashoori M (2018) Riboflavin in human health: a review of current evidences. Adv Food Nutr Res 83:57–81

Savage DG, Lindenbaum J (1995) 11 neurological complications of acquired cobalamin deficiency: clinical aspects. Baillieres Clin Haematol 8(3):657–678

Schaffer WM, Bronnikova TV (2012) Peroxidase-ROS interactions. Nonlin Dyn 68(3):413–430

Serbinova E, Kagan V, Han D, Packer L (1991) Free radical recycling and intramembrane mobility in the antioxidant properties of alpha-tocopherol and alpha-tocotrienol. Free Radic Biol Med 10(5):263–275

Sodhi RK, Singh N (2014) Retinoids as potential targets for Alzheimer's disease. Pharmacol Biochem Behav 120:117–123

Swaminathan A, Jicha GA (2014) Nutrition and prevention of Alzheimer's dementia. Front Aging Neurosci 6:282

Takasaki J, Ono K, Yoshiike Y, Hirohata M, Ikeda T, Morinaga A, Takashima A, Yamada M (2011) Vitamin a has anti-oligomerization effects on amyloid-β in vitro. Alzheimer's Dis 27(2):271–280

Terry RD, Masliah E, Hansen LA (1994) Structural basis of the cognitive alterations in Alzheimer disease, Alzheimer Dis 179–196

Vogel T, Dali-Youcef N, Kaltenbach G, Andres E (2009) Homocysteine, vitamin B12, folate and cognitive functions: a systematic and critical review of the literature. Int J Clin Pract 63(7):1061–1067

Wahlin A, Hill RD, Winblad B, Bäckman L (1996) Effects of serum vitamin B12 and folate status on episodic memory performance in very old age: a population-based study. Psychol Aging 11(3):487–496

WHO (2015 Serum and red blood cell folate concentrations for assessing folate status in populations. https://apps.who.int/iris/handle/10665/75584

Witztum JL, Steinberg D (1991) Role of oxidized low density lipoprotein in atherogenesis. J Clin Invest 88(6):1785–1792

Zandi PP, Anthony JC, Khachaturian AS, Stone SV, Gustafson D, Tschanz JT, Norton MC, Welsh-Bohmer KA, Breitner JC, Cache County Study Group (2004) Reduced risk of Alzheimer disease in users of antioxidant vitamin supplements: the Cache County study. Arch Neurol 61(1):82–88

Chapter 3
Therapeutic Potential of Extra-virgin Olive Oil Against Alzheimer's Disease Progression

Ponmari Guruvaiya and Kalidass Subramaniam

Abstract Olive oil is obtained from olive tree fruits, so it is called natural oil, and extra-virgin olive oil (EVOO) is the most nutritious among the types of olive oil, and depending on the method of production, olive oil could be graded which is further identified by the percentage of saturated fatty acids in the final product. Besides, the saturated fatty acid of EVOO contains modest level of vitamin K and E. Olive oil is considered as one of the most ideal antioxidants as part of the Mediterranean diet, and the antioxidants reduce the risk of several chronic diseases including Alzheimer's disease. Alzheimers Disease (AD) is characteristic steady progression of cognitive and behavioral abnormalities which has co-existing earlier with dementia later other symptoms could be developed and the rate of AD prevalence is pars with age. This chapter would like to emphasize extra-virgin olive oil and further explored the relationship between EVOO intake and AD progression and also validates the theory behind this relationship as therapeutic effect of EVOO on AD.

Keywords Extra-virgin olive oil · Constituents · Anti-amyloid activity · Tau proteins · Anti-inflammatory role

3.1 Introduction

Extra-virgin olive oil (EVOO) is a food component, widely used in the Mediterranean diet (MD) and believed to be a source of healthy fat. EVOO contains various health-beneficial components, including high proportion of monounsaturated fatty acids and biologically active phenolic compounds, which act either individually or synergistically. However, these compounds exert neuroprotective potential that could be endorsed by its well-known anti-inflammatory and antioxidant activities.

Alzheimer's disease has the characteristic of gradual progression of cognitive and behavioral abnormalities. In the year 2020, around the globe, 50 million people have

P. Guruvaiya · K. Subramaniam (✉)
Department of Animal Science, Manonmaniam Sundaranar University, Tirunelveli, Tamil Nadu, India

A. Justin Thenmozhi, T. Manivasagam (eds.), *Nutraceuticals for Alzheimer's Disease: A Promising Therapeutic Approach*, Nutritional Neurosciences,
https://doi.org/10.1007/978-981-99-0677-2_3

a most common clinical manifestation of AD and dementia; this figure doubled in every 5 years, and by the year 2050, this will reach about 152 million cases (Yiannopoulou and Papageorgiou 2020). Whereas, to concern with the pathology of AD, increase in amyloid β peptide deposition and uncontrolled accumulation of hyper phosphorylated tau proteins in brain tissue being the mechanism demonstrated in various studies (DeTure and Dickson 2019). To date, there is no effective disease prevention or therapeutic measures to treat AD. Nevertheless, studies demonstrated that lifestyle modifications, including healthy diet intake, can strongly persuade the disease progression, thereby reducing the burden of people living with dementia and AD (Nichols et al. 2019). In this review, we present comprehensive information on extra-virgin olive oil (EVOO) and their major constituents pertaining to AD progression.

3.2 Olive Oil

Olive oil (OO) is a juice extracted from *Olea europaea* L. fruits; the product is obtained by successive predefined physical processes such as washing, decantation, centrifugation, and filtration, under specific thermal condition (Fig. 3.1) which have not undergone any further treatment could be known as virgin olive oil (VOO) (Oliveras-López et al. 2014; Foscolou et al. 2018). Further, the European Commission (EC) Regulation No. 1234/2007 defines on its Article 118 six main types of olive oils and its features (Council of the European Union Council Regulation (EC) No 1234/2007 of 22 October 2007 2007) that were further delineated by the International Olive Council (IOC) trade standard notifications (International Olive Council 2019).

3.2.1 Extra-virgin Olive Oil

Among the six types of olive oils, the EC categorized VOOs into three subtypes: lampante olive oil, VOO, and extra-virgin olive oil (EVOO). Furthermore, IOC has its own subgroup called ordinary virgin oil. In order to certify OO as EVOO, both organizations have some consensus code in terms of chemical composition: The acidity level in the product is ≤0.8% expressed as equivalent to free oleic acid, according to their acidity level which provides the information on free fatty acids in the product. Minimal acidity level of the product authenticates the product obtained from healthy olive fruits processed under ideal conditions showing it could be high standard oil. The level of peroxide value ≤20 m Eq relative to O_2/kg, minimal oxidation values (K270 ≤0.22; K232 ≤2.50; ΔK ≤ .01) obtained from UV spectrometry analysis of the product, and organoleptic or sensory assessment were expressed as median of 0 denotes defects and fruity flavor denotes >0, percentage of fatty acid ethyl esters (FAEEs) could be ≤30 mg/kg in EVOO. Hence, IOC

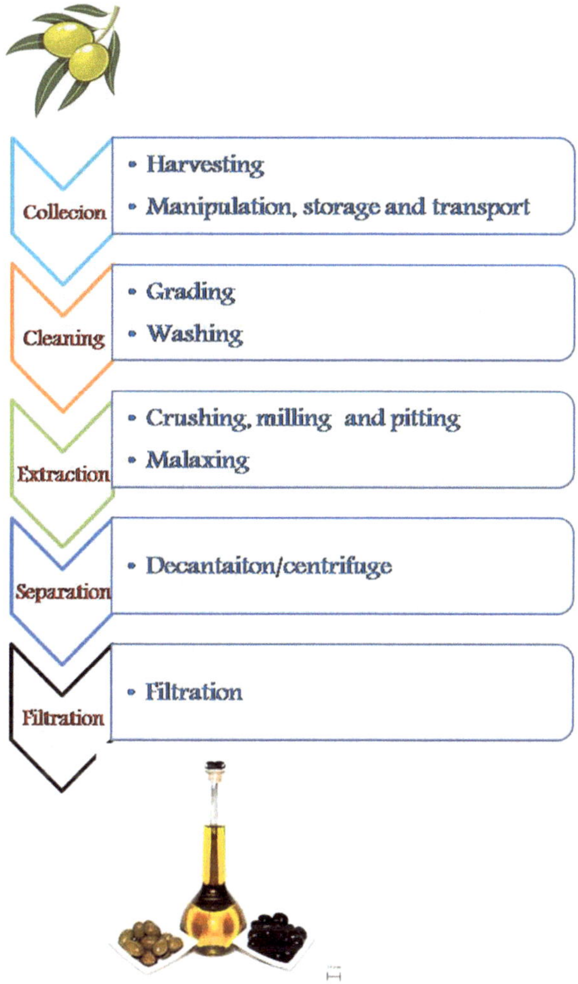

Fig. 3.1 Summary of extra-virgin olive oil (EVOO) extraction process

specifies high FAEE threshold values indicates the product being EVOO (≤35 mg/kg of EVOO). Further, IOC has some additional qualifying measures, such as moisture and volatile matter ≤0.20% and ≤0.10% respectively, and least level of trace metals such as iron ≤3.0 and ≤0.1 for copper, insoluble impurities in light petroleum. Hence, OO meets the terms of all quality criteria declared by both the EU and IOC to be defined as EVOO. Considering the standards that EVOO is required to satisfy, the quality of the products must follow both the IOC and EC defined standards in terms of purity that represents the oil composition and quality that represents the sensory profile, acidity, and fat content, among others. However, the sensory analysis is aimed to confirm there is no flaw in complying with standard operating protocol during the oil production process (Seçmeler and Galanakis 2019).

Additionally, another outstanding aspect is traceability. Traceability is guaranteed during the entire process of EVOO production, including harvesting, milling, storage, and packaging. This procedure allows access to the production chain of the final product and also supports the certification process. Finally, these procedures provide the knowledge to customers about the rigorous process and systems employed during the production of EVOO. The chemical composition of final product may change depending on the extraction process followed during the oil production from fruits. In general, extraction of oil starts from the process of crushing and separating the oil from olive fruit pulp under certain pressure. The olive oil extracted following this process has a characteristic of high in color intensity, minimal aroma, and high in content of free fatty acids. After chemical extraction process, refining process can be performed to purify the oil, and this process removes the impurities and residual matters present in the oil that can be ready for consumption. Refined oil is minimal in vitamins, polyphenols, sterols, and other natural ingredients that have low molecular weight (Kamm et al. 2001).

The process followed in the production of EVOO facilitates the conservation of more number of compounds that would further strengthen the place of EVOO among the plant oils available in the market. The presence of fatty acids such as PUFA and MUFA with many bioactive molecules including phenols in hydrophilic nature, phytosterols, tocopherols, and carotenes endorses the disease prevention and therapeutic potential of EVOO which also provides longer shelf life due to its strong oxidative stability. This uniqueness that promotes the usage of EVOO is considered top among the oils used for good health (Rotondi et al. 2004; Servili et al. 2009). Oils with more saturated fat composition such as plant derived palm oil become more stable during the process of cooking or frying compared to oils with more unsaturated fat. Hence, they do not have a health-beneficial effect as it's the unsaturated one. In contrary to the above, sunflower oil contains high amount of linoleic and oleic acids as unsaturated fats, which enhance the health-beneficial effect of the product, but the thermal stability of the compounds is limited (Echarte et al. 2013).

An ideal ratio of PUFA:MUFA in EVOO could provide more stability against oxidation mediated thermal degradation, in particular volatile aldehyde formation, than other counterparts. So, EVOO is ideal and highly recommended to use in frying (Molina-Garcia et al. 2017). In addition, the appropriate ratio of PUFA and MUFA of unsaturated fats makes EVOO as one of the best plant oils to suggest for consumption when ingested raw since it may lower LDL cholesterol levels in humans (Sun et al. 2015). EVOO is more expensive than other types of olive counterparts due to its low yield and more intense standards in production, but it carries highest amount of polyphenols than others (Kalogeropoulos and Tsimidou 2014).

During the EVOO extraction process, the free fatty acids can be removed resulting in less color intensity, flavor, and aroma (Lynch and Rozema 2013; Šarolić et al. 2015). Interestingly, the process of filtration affects the properties of virgin olive oil; upon filtration, the high polar phenolic compounds are vanished while removing the water. Due to the involvement of multiple technological processes with respect to the type of olive oil, the content of polyphenols may differ.

3.3 Constituents of Extra-virgin Olive Oil

3.3.1 Primary Metabolites

3.3.1.1 Lipids in EVOO

In the edible oil, triacylglycerols constitute a large part of saponifiable fraction falls in the category of MUFA (Gavahian et al. 2019). The major part (~50%) of triacylglycerol present in the EVOO is made up of oleic-oleic-oleic (OOO), while palmitic-O-O (POO), O-O-linoleic (OOL), stearic-O-O (SOO), and P-O-L are the other triacylglycerol found in VOO (Boskou et al. 2006; Aranda et al. 2004). In VOO, 1–2.8% of diacylglycerols and 0.25% of monoacylglycerols have been identified (Alu et al. 2017).

3.3.1.2 Sterols in EVOO

Sterols are also found in olive oil, which are used to identify the authenticity of the product because they are connected to the quality of the oil. The major classes of sterols in EVOO are found as both free and esterified forms like 4-desmethylsterols, 4α-methylsterols, and triterpene alcohols (4, 4-dimethylsterols) (Boskou et al. 2006). It is found that campesterol, β-sitosterol, and Δ5-avenasterol are the principal components of the sterol fraction (Boskou and Morton 1975). The European Commission specifies that the minimal range of sterol in EVOO is 1000 mg/kg, but it varies up to 2000 mg/kg (Boskou et al. 2006). In the total sterol content, β-sitosterol is the major compound found with about 75–90%, while Δ5-avenasterol was found in the range of 5–20% (Itoh et al. 1981).

3.3.1.3 Tocopherols

Three tocopherol isoforms such as α-, β-, and γ-tocopherol were found in EVOO. α-Tocopherol is the principal type among three and found in free form. α-Tocopherol occurs more than 90% in the identified tocopherol section, and the content varies from 206.5 to 270.9 mg/kg of oil to 191.5–292.7 mg/kg of oil (Itoh et al. 1981).

3.3.1.4 Carbohydrates

There are two principal hydrocarbon molecules present in OO including squalene and β-carotene. In addition, the remaining portion of carbohydrates in EVOO is composed of triterpene and diterpene, isoprenoid polyolefins, hydrocarbons, and n-paraffins (Boskou et al. 2006; Lanzón et al. 1994).

3.3.2 Secondary Metabolites

3.3.2.1 Phenolic Compounds

A number of studies substantiated that more than 30 phenolic compounds were found in EVOO and the hydrophilic phenolic compounds are high in number which is highly linked to determining the oil quality related to sensory parameters like taste, pungency, and stability (Servili et al. 2009; Sánchez de Medina et al. 2013) Further, it can determine the organoleptic character including aroma and flavor of EVOO (Alu et al. 2017). The olive variety and quality being the factor determines the oxidative potential of the final product, besides time of harvest and area of cultivation; the levels of antioxidants from tocopherols, carotenes, and hydrophilic phenolic compounds and degree of unsaturation also influence the quality of EVOO (Gimeno et al. 2002).

Further, the polyphenols found in EVOO can be classified as phenolic acids, flavonoids, lignans, secoiridoids, phenolic alcohols, and hydroxy-isocromans (Fig. 3.2). Moreover, phenolic acids were present in larger quantity than other group of polyphenolic portion and identified as hydroxybenzoic, *p*-ferulic, coumaric, gallic, vanillic, syringic, caffeic, *o*-coumaric, and sinapic acids (Alu et al. 2017). In addition, derivative of luteolin, apigenin, is the principal flavonoid found in EVOO, while (+)-pinoresinol and (+)-1-acetoxypinoresinol are the major lignans present in EVOO, and usually the content of the lignans (Bendini et al. 2007) in EVOO is 1 and 100 mg/kg (Murkovic et al. 2004; Brenes et al. 2000).

Interestingly, secoiridoids are rare phenolic compounds found in large quantity in Oleaceae plant species; hence, they are found in abundance exclusively in *O. europaea* leaves and fruits. Nevertheless, only a small amount of secoiridoids are present in the final EVOO product after the mechanical oil extraction system due to its insoluble nature of secoiridoids in oil. In addition, secoiridoids are the key micronutrients present in EVOO implicated for their sensorial and health-beneficial properties (Servili et al. 2009; Cicerale et al. 2009). Secoiridoids including demethyloleuropein, oleuropein, and ligstroside and aglycones are the most common, and approximately 90% of phenolic constituents were identified as ligstroside and their aglycones in EVOO (De La Torre-Carbot et al. 2005). In the EVOO extraction process, crushing and malaxation led to enzymatic hydrolysis of secoiridoids to form their aglycones by endogenous β-glucosidases (Rovellini and Cortesi 2002). In addition, secoiridoids specifically oleuropein aglycone are considered as the source associated to the bitterness of EVOO (Mateos et al. 2004).

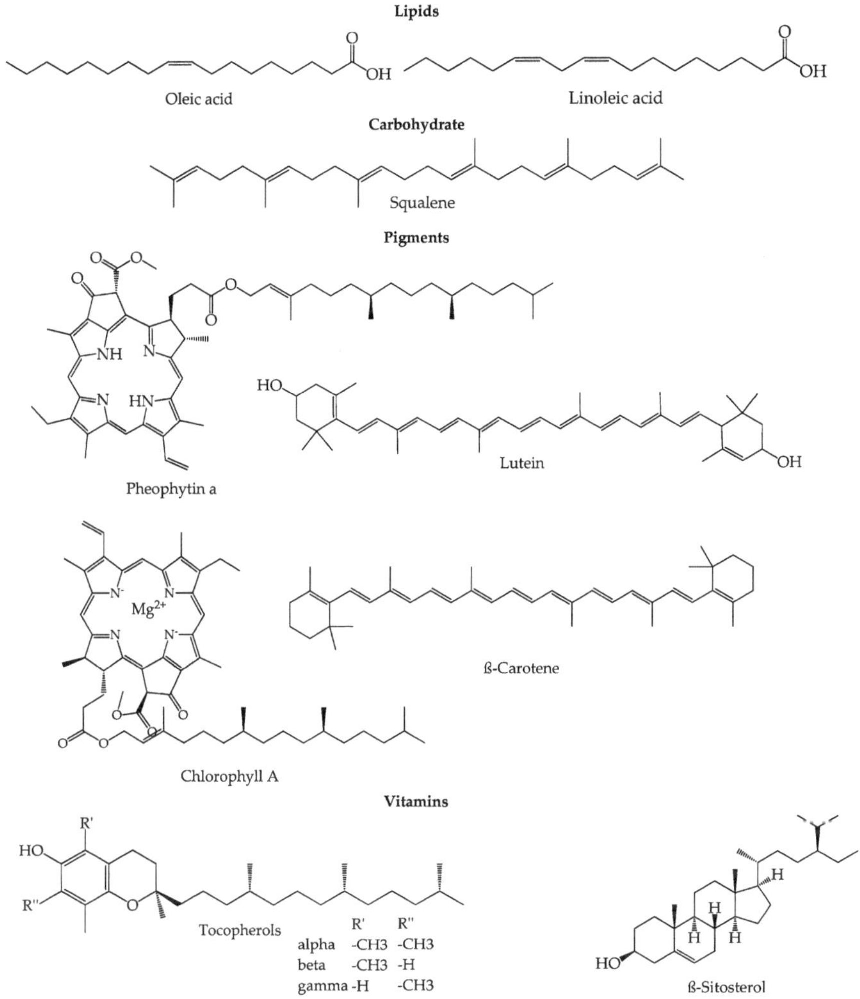

Fig. 3.2 Representative chemical structures of some relevant compounds present in EVOO (Image courtesy: Jimenez-Lopez et al. 2020)

In EVOO, isochromans are present in low concentration, and the two mainly found isochromans in EVOO are 1-Phenyl-6,7-dihydroxy-isochroman and 1-(3′ Methoxy-4′-hydroxy) phenyl-6,7-dihydroxy isochromans (Bianco et al. 2002). Moreover, two major phenolic alcohols identified in EVOO are tyrosol and hydroxytyrosol (Gómez-Alonso et al. 2002).

3.3.2.2 Pigments

The color of olive oil is highly influenced by chlorophyll and lipophilic carotenoid pigments preset in the raw olive fruits used for production (Montealegre et al. 2010).

The coloration of EVOO is greener in the presence of green olives with higher chlorophyll content, whereas using mature olives with higher carotenoid content results in more yellowish oil, so the color of final product mostly depends on the ratio of these pigments (Lazzerini et al. 2016).

3.4 Effect of EVOO on AD

Numerous studies showed the preventive or therapeutic effect of the phenolic compounds of EVOO against the neurodegenerative diseases such as AD (Roman et al. 2019; Dinu et al. 2018). Studies in humans also substantiated that Mediterranean diet with EVOO can considerably reduce the risk of AD (Walters et al. 2018; Berti et al. 2018). Further, studies on rodent model also agreed with these findings, and oral administration of EVOO and their specific phenolic compounds can decrease the β-amyloid (Aβ) accumulation and tau protein deposition in neuropathology examination, followed by improved memory and cognition (Qosa et al. 2015a, b; Lauretti et al. 2017, 2020; Al Rihani et al. 2019).

In randomized trial with 522 participants by PREDIMED-NAVARRA, consumption of either EVOO rich Mediterranean diet or nuts (30 g/day) as a supplement results in an improvement in cognition when compared to participants who consume low-fat diet supplement over 6.5 years of diet intervention. In addition, volunteers having Mediterranean diet supplemented with EVOO or nuts had observed significant improvement in Mini-Mental State Examination and Clock Drawing Test score compared to the low-fat control diet volunteers (Martinez-Lapiscina et al. 2013). In addition, in another sub-study by PREDIMED in which some additional neuropsychological tests were performed to evaluate the cognitive health of volunteers, after 4.1 years of record, researchers found that volunteers who take Mediterranean diet with either EVOO or nuts supplement had observed excellent cognitive function than those who consumed a low-fat diet (Valls-Pedret et al. 2015).

In addition, in a French multicenter cohort study called Three-City Study, researchers examined 6947 volunteers and correlated the olive oil intake and cognitive health of participants by Mini-Mental State Examination, Isaacs Set Tests, and Benton Visual Retention Test. The results of this study indicate that moderate or high olive oil intake reduced the likelihood of cognitive deficit for verbal fluency and visual memory loss compared to those who never used olive oil (Berr et al. 2009). In a number of animal model studies, oleocanthal secoiridoid found in EVOO halted the AD progression and sustained the disease prevention strategy (Qosa et al. 2015a, b; Lauretti et al. 2017, 2020; Al Rihani et al. 2019). Further, researchers aimed to assess tau metabolism and synaptic mechanisms after chronic EVOO supplementation on transgenic mice with human tau protein overexpression, which showed an increase in complexin 1, involved in presynaptic mechanism and a considerable decrease in oligomers of tau resulting in improved short-term plasticity and memory (Lauretti et al. 2020).

In a previous study by the same authors, a triple transgenic mice model which contains three mutated genes linked to AD (*PS1M146V*, *tauP301L*, and *APPSwe*)

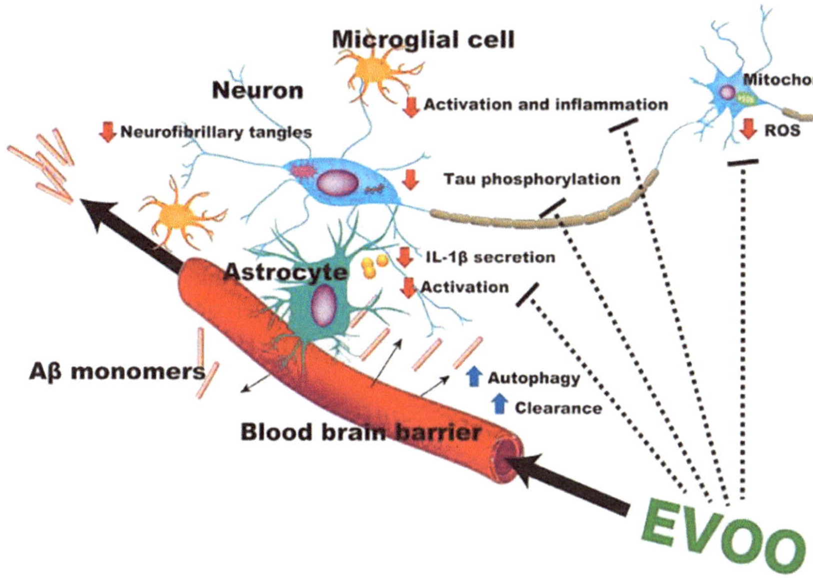

Fig. 3.3 Possible therapeutic role of extra-virgin olive oil (EVOO) on Alzheimer's disease pathology. The ideal antioxidant potential of EVOO attenuates the brain antioxidant enzyme level, thereby counteracting the reactive oxygen species (ROS). EVOO may prevent or halt the progression of AD that could hinder the deposition of amyloid-β (Aβ) and also mitigate the neuropathology of tau proteins by promoting autophagy and sustaining the mechanisms across the blood-brain barrier. In addition, EVOO reduced the astrocyte and microglial activation in brain cells, thereby reducing the production of inflammatory cytokines. (Image courtesy; Millman et al. 2021)

was fed with EVOO diet for 6 months. This study shows extensive reduction in Aβ peptide formation as a result of stimulation of autophagy mechanism linked to improve amyloid plaques, and neurofibrillary tangle formation also reduces the tau neuropathology (Lauretti et al. 2017). In an in vivo study, the effect of intraperitoneal injection (4-week) of oleocanthal secoiridoid was studied in TgSwDI mice having human amyloid-β precursor protein under the mouse promoter of Thy-1 with Swedish (K670N/M671L), Dutch (E693Q), and Iowa (D694N) mutations that accelerate the Aβ peptide deposition (Van Vickle et al. 2008). Further, the result of this study showed the reduction of Aβ in microvessels and hippocampal parenchyma, and overexpression of Aβ clearance proteins linked to the blood-brain barrier in human model also decreased the astrocytes and IL-1β expression in the brain (Qosa et al. 2015b). In an another study, consumption of EVOO for 6 months by TgSwDI mice shows depletion in Aβ level, due to crucial changes in processing of Aβ precursor protein (Qosa et al. 2015a).

Furthermore, supplementation of oleocanthal-rich EVOO to TgSwDI mice with advanced stage of AD resulted in considerable decrease in NLRP3 inflammasome activation, whereas further significant Aβ clearance through activating autophagy and AMPK– Unc-51-like kinase 1 (ULK1) pathway (Fig. 3.3) (Al Rihani et al. 2019).

3.5 Conclusion

Extra-virgin olive oil is most beneficial to individual well-being and top ranked among the kinds of diet element used in Mediterranean basin, and most of the studies reported that individuals practicing intake of olive oil diet in their early of life have apparently reduced the risk of Alzheimer's associated complications at adulthood. However, the preventive effect of olive oil is highly conceived than the therapeutic potential. This beneficial effect is mostly instigated by the antioxidant capability of the product, thereby providing good health for the individual and making them less prone to Alzheimer's pathological condition. This chapter suggested that extra-virgin olive oil is the foremost element of interest for its ideal nature and be included as part of the diet for all age groups.

References

Al Rihani SB, Darakjian LI, Kaddoumi A (2019) Oleocanthal-rich extra-virgin olive oil restores the blood-brain barrier function through NLRP3 inflammasome inhibition simultaneously with autophagy induction in TgSwDI mice. ACS Chem Neurosci 10:3543–3554

Alu MH, Rababah T, Alhamad MN (2017) Application of olive oil as nutraceutical and pharmaceutical food: composition and biofunctional constituents and their roles in functionality, therapeutic, and nutraceutical properties. Elsevier, Amsterdam

Aranda F, Gómez-Alonso S, Rivera Del Álamo RM, Salvador MD, Fregapane G (2004) Triglyceride, total and 2-position fatty acid composition of Cornicabra virgin olive oil: comparison with other Spanish cultivars. Food Chem 86:485–492

Bendini A, Cerretani L, Carrasco-Pancorbo A, Gómez-Caravaca AM, Segura-Carretero A, Fernández-Gutiérrez A et al (2007) Phenolic molecules in virgin olive oils: a survey of their sensory properties, health effects, antioxidant activity and analytical methods. An overview of the last decade. Molecules 12:1679

Berr C, Portet F, Carriere I, Akbaraly TN, Feart C, Gourlet V et al (2009) Olive oil and cognition: results from the Three-City study. Dement Geriatr Cogn Disord 28:357–364

Berti V, Walters M, Sterling J, Quinn CG, Logue M, Andrews R et al (2018) Mediterranean diet and 3-year Alzheimer brain biomarker changes in middle-aged adults. Neurology 90:e1789–e1798

Bianco A, Coccioli F, Guiso M, Marra C (2002) The occurrence in olive oil of a new class of phenolic compounds: Hydroxy-isochromans. Food Chem 77:405–411

Boskou D, Morton ID (1975) Changes in the sterol composition of olive oil on heating. J Sci Food Agric 26:1149–1153

Boskou D, Blekas G, Tsimidou M (2006) Olive oil composition, 2nd edn. AOCS Press, Urbana

Brenes M, García A, García P, Garrido A (2000) Rapid and complete extraction of phenols from olive oil and determination by means of a coulometric electrode array system. J Agric Food Chem 48:5178–5183

Cicerale S, Conlan XA, Sinclair AJ, Keast RSJ (2009) Chemistry and health of olive oil phenolics. Crit Rev Food Sci Nutr 49:218–236

Council of the European Union Council Regulation (EC) No 1234/2007 of 22 October 2007 (2007) Establishing a common organisation of agricultural markets and on specific provisions for certain agricultural products (single CMO regulation) off. J Eur Union 299:1–149

De La Torre-Carbot K, Jauregui O, Gimeno E, Castellote AI, Lamuela-Raventós RM, López-Sabater MC (2005) Characterization and quantification of phenolic compounds in olive oils by solid-phase extraction, HPLC-DAD, and HPLC-MS/MS. J Agric Food Chem 53:4331–4340

DeTure MA, Dickson DW (2019) The neuropathological diagnosis of Alzheimer's disease. Mol Neurodegener 14:32

Dinu M, Pagliai G, Casini A, Sofi F (2018) Mediterranean diet and multiple health outcomes: an umbrella review of meta-analyses of observational studies and randomised trials. Eur J Clin Nutr 72:30–43

Echarte MM, Puntel LA, Aguirrezabal LAN (2013) Assessment of the critical period for the effect of intercepted solar radiation on sunflower oil fatty acid composition. F Crop Res 149:213–222

Foscolou A, Critselis E, Panagiotakos D (2018) Olive oil consumption and human health: a narrative review. Maturitas 118:60–66

Gavahian M, Mousavi Khaneghah A, Lorenzo JM, Munekata PES, Garcia-Mantrana I, Collado MC et al (2019) Health benefits of olive oil and its components: impacts on gut microbiota antioxidant activities, and prevention of non communicable diseases. Trends Food Sci Technol 88:220–227

Gimeno E, Castellote AI, Lamuela-Raventós RM, La Torre MCD, López-Sabater MC (2002) The effects of harvest and extraction methods on the antioxidant content (phenolics, α-tocopherol, and β-carotene) in virgin olive oil. Food Chem 78:207–211

Gómez-Alonso S, Salvador MD, Fregapane G (2002) Phenolic compounds profile of Cornicabra virgin olive oil. J Agric Food Chem 50:6812–6817

International Olive Council (2019) Trade standard applying to olive oils and olive pomace oils. International Olive Council, Madrid, Spain. COI/T.15/NC No 3/Rev. 13. English version, Original: French version

Itoh T, Yoshida K, Yatsu T, Tamura T, Matsumoto T, Spencer GF (1981) Triterpene alcohols and sterols of Spanish olive oil. J Am Oil Chem Soc 58:545–550

Jimenez-Lopez C, Carpena M, Lourenço-Lopes C, Gallardo-Gomez M, Lorenzo JM, Barba FJ (2020) Bioactive compounds and quality of extra virgin olive oil. Foods 9:1014

Kalogeropoulos N, Tsimidou MZ (2014) Antioxidants in Greek virgin olive oils. Antioxidants 3: 387–413

Kamm W, Dionisi F, Hischenhuber C, Engel KH (2001) Authenticity assessment of fats and oils. Food Rev Int 17:249–290

Lanzón A, Albi T, Cert A, Gracián J (1994) The hydrocarbon fraction of virgin olive oil and changes resulting from refining. J Am Oil Chem Soc 71:285–291

Lauretti E, Iuliano L, Pratico D (2017) Extra-virgin olive oil ameliorates cognition and neuropathology of the 3xTg mice: role of autophagy. Ann Clin Transl Neurol 4:564–574

Lauretti E, Nenov M, Dincer O, Iuliano L, Patrico D (2020) Extra virgin olive oil improves synaptic activity, short-term plasticity, memory, and neuropathology in a tauopathy model. Aging Cell 19:E13076

Lazzerini C, Cifelli M, Domenici V (2016) Pigments in extra-virgin olive oil: authenticity and quality. In: Boskou D, Clodoveo ML (eds) Products from Olive Tree. IntechOpen, McFarland, WI

Lynch B, Rozema A (2013) Olive oil: conditions of competition between us and major foreign supplier industries. United States International Trade Commission, Washington, DC

Martinez-Lapiscina EH, Clavero P, Toledo E, Estruch R, Salas-Salvado J, San Julian B et al (2013) Mediterranean diet improves cognition: the PREDIMED-NAVARRA randomised trial. J Neurol Neurosurg Psychiatry 84:1318–1325

Mateos R, Cert A, Carmen Pérez-Camino M, García JM (2004) Evaluation of virgin olive oil bitterness by quantification of secoiridoid derivatives. J Am Oil Chem Soc 81:71–75

Millman JF, Okamoto S, Teruya T, Uema T, Ikematsu S, Shimabukuro M (2021) Extra-virgin olive oil and the gut-brain axis: influence on gut microbiota, mucosal immunity, and cardiometabolic and cognitive health. Nutr Rev 79:1362–1374

Molina-Garcia L, Santos CSP, Cunha SC, Casal S, Fernandes JO (2017) Comparative fingerprint changes of toxic volatiles in low PUFA vegetable oils under deep-frying. J Am Oil Chem Soc 94:271–284

Montealegre C, Alegre MLM, García-Ruiz C (2010) Traceability markers to the botanical origin in olive oils. J Agric Food Chem 58:28–38

Murkovic M, Lechner S, Pietzka A, Bratacos M, Katzogiannos E (2004) Analysis of minor components in olive oil. J Biochem Biophys Methods 61:155–160

Nichols E, Szoeke CE, Vollset SE, Abbasi N, Abd-Allah F, Abdela J (2019) Global, regional, and national burden of Alzheimer's disease and other dementias, 1990–2016: a systematic analysis for the global burden of disease study 2016. Lancet Neurol 18:88–106

Oliveras-López MJ, Berná G, Jurado-Ruiz E, López-García de la Serrana H, Martín F (2014) Consumption of extra-virgin olive oil rich in phenolic compounds has beneficial antioxidant effects in healthy human adults. J Funct Foods 10:475–484

Qosa H, Batarseh YS, Mohyeldin MM, El Sayed KA, Keller JN, Kaddoumi A (2015a) Oleocanthal enhances amyloid-beta clearance from the brains of TgSwDI mice and in vitro across a human blood-brain barrier model. ACS Chem Neurosci 6:1849–1859

Qosa H, Mohamed LA, Batarseh YS, Alqahtani S, Ibrahim B, LeVine H 3rd et al (2015b) Extra-virgin olive oil attenuates amyloid-beta and tau pathologies in the brains of TgSwDI mice. J Nutr Biochem 26:1479–1490

Roman GC, Jackson RE, Reis J, Roman AN, Toledo JB, Toledo E (2019) Extra-virgin olive oil for potential prevention of Alzheimer disease. Rev Neurol (Paris) 175:705–723

Rotondi A, Bendini A, Cerretani L, Mari M, Lercker G, Toschi TG (2004) Effect of olive ripening degree on the oxidative stability and organoleptic properties of cv. Nostrana di Brisighella extra virgin olive oil. J Agric Food Chem 52:3649–3654

Rovellini P, Cortesi N (2002) Liquid chromatography-mass spectrometry in the study of oleuropein and ligstroside aglycons in virgin olive oil: Aldehydic, dialdehydic forms and their oxidized products. Riv Ital Sostanze Grasse 79:1–14

Sánchez de Medina V, Riachy ME, Priego-Capote F, Luque de Castro MD (2013) Mass spectrometry to evaluate the effect of the ripening process on phenols of virgin olive oils. Eur J Lipid Sci Technol 115:1053–1061

Šarolić M, Gugić M, Friganović E, Tuberoso CIG, Jerković I (2015) Phytochemicals and other characteristics of Croatian monovarietal extra virgin olive oils from Oblica, Lastovka and Levantinka varieties. Molecules 20:4395–4409

Seçmeler Ö, Galanakis CM (2019) Chapter 8-olive fruit and olive oil; innovations in traditional foods. Woodhead Publishing, Sawston, pp 193–220

Servili M, Esposto S, Fabiani R, Urbani S, Taticchi A, Mariucci F et al (2009) Phenolic compounds in olive oil: antioxidant, health and organoleptic activities according to their chemical structure. Inflammopharmacology 17:76–84

Sun Y, Neelakantan N, Wu Y, Lote-oke R, Pan A, Dam RM (2015) Van palm oil consumption increases LDL cholesterol compared with vegetable oils low in saturated fat in a meta-analysis of clinical. J Nutr Nutr Epidemiol 145:1549–1558

Valls-Pedret C, Sala-Vila A, Serra-Mir M, Corella D, Torre R, Angel Martinez-Gonzalez M et al (2015) Mediterranean diet and age-related cognitive decline: a randomized clinical trial. JAMA Intern Med 175:1094–1103

Van Vickle GD, Esh CL, Daugs ID, Kokjohn TA, Kalback WM, Patton RL et al (2008) Tg-SwDI transgenic mice exhibit novel alterations in AβPP processing, Aβ degradation, and resilient amyloid angiopathy. Am J Pathol 173:483–493

Walters MJ, Sterling J, Quinn C, Ganzer C, Osorio RS, Andrews RD et al (2018) Associations of lifestyle and vascular risk factors with Alzheimer's brain biomarker changes during middle age: a 3-year longitudinal study in the broader New York City area. BMJ Open 8:E023664

Yiannopoulou KG, Papageorgiou SG (2020) Current and future treatments in Alzheimer's disease: an update. J Cent Nerv Syst Dis 12:1–12

Chapter 4
The Role of Omega-3 Fatty Acids in Alzheimer's Disease Prevention and Therapy

Ponmari Guruvaiya and Kalidass Subramaniam

Abstract Omega-3 fatty acids represent a group of fatty acids characterized by the double bond located at the third carbon atom from the terminal methyl end of the fatty acid chain. Omega-3 fatty acids are most essential to some important bodily functions of humans and must be obtained through diet which means the body can't produce them on its own. Omega-3 fatty acids are largely present in some natural foods such as seafood, nuts, seeds, and plant oils. The three major types of omega-3 fatty acids such as alpha-linolenic acid (ALA), eicosapentaenoic acid (EPA), and docosahexaenoic acid (DHA) are present in higher ratio in food components than other counterparts. Several studies substantiate omega-3's role on health and demonstrated that appropriate intake of omega-3-rich foods such as fish and other seafood reduces the risk of several chronic disease progression. Hence, the intake of foods rich in omega-3 fatty acids mediated beneficial effect on AD progression is largely uncertain. In this first part of the chapter, details about omega-3 fatty acids were discussed, and then the final part deals with available studies on omega-3 fatty acids on AD therapy.

Keywords Alzheimer's disease · Omega-3 fatty acids · Sources · Types · Clinical studies

4.1 Introduction

Fatty acids with double bond linking the third and fourth carbon from the methyl end of the fatty acid chain is called as ω-3 or n-3 fatty acids that represent heterogeneous group of fatty acids (n-3 FAs). In general, polyunsaturated fatty acids (PUFAs) and monounsaturated fatty acids (MUFAs) are distinguished by the number of double bonds present in the carbon chain of the molecule. With respect to nomenclature, the MUFs have one double bond and PUFs have more than one bond in its carbon chain.

P. Guruvaiya · K. Subramaniam (✉)
Department of Animal Science, Manonmaniam Sundaranar University, Tirunelveli, Tamil Nadu, India

A. Justin Thenmozhi, T. Manivasagam (eds.), *Nutraceuticals for Alzheimer's Disease: A Promising Therapeutic Approach*, Nutritional Neurosciences,
https://doi.org/10.1007/978-981-99-0677-2_4

Hence, conjugated fatty acids (CFAs) are one of its kinds of PUFs with at least one pair of double bonds in a conjugate form which means the double bonds are not separated, but single bond is separated by methyl bridges (Nagao and Yanagita 2005).

Here, we also discussed some modified form of omega-3 fatty acids including hydroxy (HFAs), oxo or keto fatty acids, and hydroperoxy (-OOH) fatty acids. Among the HFAs, saturated fatty acids have carboxyl group with long unbranched carbon chain, while unsaturated fatty acids consist of one or more hydroxy groups. Oxo or keto fatty acids contain both carboxy group and aldehydic or ketonic group in its molecular structure. However, one hydroperoxy group must be present in the hydroperoxy fatty acid structure. In some articles, researchers cited the term omega-3 fatty acids and long-chain polyunsaturated fatty acids (LC-PUFs) as identical and but omega-3 fatty acids is more broader (Scorletti and Byrne 2013) than LC-PUFs.

Omega-3 fatty acids *cis-trans* isomerism (geometric) extended to *Z* configuration (Chatgilialoglu et al. 2014). In order to confirm the type of geometrical isomerism, constituents present in the right and left part of the double bond are vital. The term "cis" and "trans" describe the positions of the two hydrogen atoms located next to the carbon atoms where the double bond exists. In the case of *cis*-isomer these two groups are located on the same side of the reference plane; in the *trans*-isomer they are occur in converse arrangement (IUPAC 1997a, b).

According to earlier studies, the *cis* (Z) or *trans* (E) isomerism of fatty acids play a crucial role in shaping the chemical structure and biological activities of the lipids (Nagao and Yanagita 2005). Fatty acids found in nature usually have up to 28 carbon atoms; nevertheless, fatty acids found in the brain, retina, and spermatozoa have longer carbon chain (Poulos et al. 1986; Aveldaño and Sprecher 1987; IUPAC. Fatty Acids 1997a). According to the length of the carbon chain, fatty acids can be classified into four basic groups:

1. Short-chain fatty acids (SCFAs), one to six carbon atoms (C1–6) and found in the digestive tract of mammals which are formed as a result of carbohydrate fermentation by the gut microbiota (Rombeau et al. 1990).
2. Medium-chain fatty acids (MCFAs), 7–12 carbon atoms (C7–12) (Schönfeld and Wojtczak 2016); according to some studies, 8–14 carbon atoms (Beermann et al. 2003) too fall in this category.
3. Long-chain fatty acids (LCFAs) (C14–18) found in various diets (Braverman and Eichler 2009).
4. Very long-chain fatty acids (VLCFAs) have more than 20 carbon atoms (C >20) (Sassa and Kihara 2014) and more than 22 carbon atoms (C >22) (Poulos et al. 1987; Hardy et al. 1994a, b).

Fatty acids having carbon atoms up to nine or less are in a state of liquid at room temperature (Schönfeld and Wojtczak 2016). Essential fatty acids (EFAs) are vital, but small group of fatty acids cannot be synthesized by the organism, otherwise, and cannot be synthesized sufficiently to maintain the homeostasis and must be supplied with food (Singh 2005; Arts et al. 2001). Some authors believed that linoleic acid

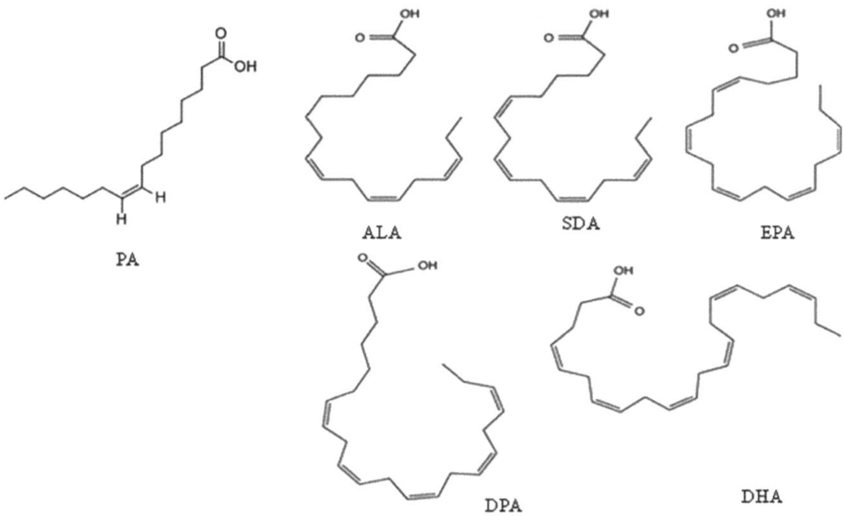

Fig. 4.1 Basic chemical structures of some omega-3 fatty acids. *PA* palmitoleic acid, *ALA* α-linolenic acid, *DHA* docosahexaenoic acid, *DPA* docosapentaenoic acid, *EPA* eicosapentaenoic acid, *SDA* stearidonic acid

(LA) and *alpha*-linolenic acid (ALA) are the most vital among all essential PUFs, LA and ALA being called as parent of all essential fatty acids (Semba 2007; van Goor et al. 2011).

In situ with studies mammalian literature which substantiates 23 acids as essential, in contrary the aquatic literature signify only two EFAs—EPA (5Z,8Z,11Z,14Z,17Z)-Icosa-5,8,11,14,17-pentaenoic acid) and DHA (4Z,7Z,10Z,13Z,16Z,19Z)-Docosa-4,7,10,13,16,19-hexaenoic acid (cervonic acid)). However, consider the crucial role of ARA; it is concluded that ARA, DHA, and EPA (Fig. 4.1) are the most essential long-chain PUFAs in mammals and fish. Some animals can synthesize LC-PUFs by using LA and ALA as precursors when it is available in adequate quantity (Parrish 2009).

4.2 Sources of Omega-3 Fatty Acids

Average intake of people in the United States in the years 2003–2008 is 0.17 g/day of long-chain omega-3 fatty acids from diet; among omega-3 FA total consumption, the primary sources of omega-3 fatty acids in to the people are cereal products and fish. The quantity of omega-3 fatty acids in food is highly dependent on fish supply, which is the major source of EPA and DHA for human (Mori 2017). Algae rich in EPA and DHA are the food source for many fish and some organisms like marine invertebrates (Fialkow 2016; Monroig et al. 2013). Microalgae found in seawater are the main source to play an essential role in primary production of PUFAs. Due to their ability to synthesize some of them de novo, marine invertebrates are also another important and essential source of production of PUFAs; by consuming microalgae, the marine invertebrate oyster *Crassostrea gigas* can produce EPA and DHA (Monroig et al. 2013). However, seafood is the major source of omega-3 fatty acids, but frequent consumption of seafood often exposed to methyl mercury led neurotoxic effect, specifically harmful to fetus during development of the central nervous system (Puri et al. 2016).

In total lipid content EPA constitutes 18.8% in pollock roe (Shirai et al. 2006), 13.6% of total muscle fatty acids is made up of EPA in wild sardine (Bandarra et al. 2018), and 15% of total lipids is made by EPA in herring (Shirai et al. 2006). In plants, the EPA is rare, and to the extent, *Undaria pinnatifida* is made of 13% of EPA in total essential oil content (Kang et al. 2016). The total lipid content in flying fish, pollock, herring, and salmon roe is constituted by DHA of 27.9%,22.2%, 22.6%, and 17.4% respectively (Shirai et al. 2006). The muscle tissue fatty acids of *Cirrhinus mrigala* and *Catla catla* contain 18.07 and 17.98 g of DHA, respectively (Memon et al. 2011). The content of this DHA in the jackalberry is 4.65 g FA per 100 g oil and is found in *Diospyros mespiliformis* (Ezeagua et al. 1998). ALA is using enzymes such as desaturases and elongases and synthesizes the EPA and DHA (Singh 2005; Richter et al. 2018); this process is inadequate in humans, and the conversion rate of this fatty acids is 10% in men and 14% in women, respectively (Fialkow 2016; Yang et al. 2017).

ALA is a more common form of fatty acids and largely found in algae and higher plants (Das 2011). The sources of ALA in plants include *Linum usitatissimum* seeds, and the content of ALS is varying from 1.1 to 65.2% against total fatty acids which is depending on the genotype (Bjelková et al. 2012). The seed oils of chia *Salvia hispanica* contain 64.04% fatty acids (Jin et al. 2012; Souza et al. 2017), and *Trichosanthes kirilowii* seed contains 33.77–38.66% (Yang et al. 2012). The amount of ALA present in fish (*Sardina pilchardus*) is minimal i.e., about 1.1% of the muscle total FAs (Bandarra et al. 2018).

Turkish sage species were also found to contain (all *trans*)-9,12,15-octadecatrienoic acid (linolenelaidic acid), and the content of seed fatty acids of some varieties includes *Salvia virgata* (0.4), *Salvia potentillifolia* (0.7), *Salvia recognita* (1.1), and *Salvia tomentosa* (1.4) percentile equivalent to total seed fatty acids (Goren et al. 2006). Interestingly, linolenelaidic acid is found to be very minimal in *Nicotiana tabacum (tobacco)*, which is about 0.03% of seed oil FAs (Giannelos et al. 2002). Seed oil of *Salvia nilotica* contains 5.4% of (9Z, 12Z, 15Z)-2-Hydroxyoctadeca-9, 12, 15-trienoic acid which is a hydroxy acid (Bohannon and Kleiman 1975).

Animal sources such as hen egg yolk contain (11Z,14Z,17Z)-Icosa-11,14,17-trienoic acid of only 0.15–0.16% in total lipid content; this acid is sometimes named as homo-*alpha*-linolenic acid (Pintea et al. 2012), and more amount of this acid was found in *Torreya grandis* kernel oil of 6.78–8.73% in total FA content (He et al. 2016a, b). However, the highest of homo-*alpha*-linolenic acid is found in

Pittosporum undulatum seed oil which contains 31.44% of the total FA content (Kobelnik et al. 2017).

Modifications in omega-3 fatty acids could play a vital role as hormones or produced in unusual condition in plants. In this review, only few of them were mentioned that may occur in dietary sources. In plants, the genus *Lesquerella* has been found to have 2-hydroxylinolenic acid and two hydroxy acids including densipolic and auricolic acid. The densipolic acid was found in *Linum usitatissimum*, and (9Z)-12-hydroxy-9-dodecenoic acid (HDA) is present in plants, which is the principal end product of lipoxygenase pathway. Hydroperoxide lyase is one of the fatty acids that catalyze the reaction in plants and produce 12-oxo-*cis*-9-dodecenoic acid, which is present in plants in higher concentration, with mature soy beans among others (Cholewski et al. 2018).

4.3 Omega-3 Fatty Acids on Alzheimer's Disease Prevention and Therapy

Several studies related to disease occurrence pattern revealed that higher intake of ω-3 FAs is linked to minimal or less prone to cognitive decline or Alzheimer's symptoms (Cole et al. 2009). MacLean et al. (2004) explored numerous of pharmacological and preclinical studies validating the significance of omega-3 fatty acids on Alzheimer's disease. DHA is the vital component of phospholipids present in the brain tissue, including the cerebral cortex, mitochondria, and synaptic vesicles of brain cells (Connor 2000).

In the normal brain activity, the mechanistic role of omega-3 fatty acids can be identified as (1) maintaining the membrane fluidity, (2) meditation of membrane-bound enzymes, (3) receptor affinity, (4) appropriate function of ion channels, (5) neurotransmitter production and activity, and (6) regulation of signal transduction, thereby controlling the neuronal growth factors and neurotransmitters (Yehuda et al. 2005).

It has been found that omega-3 fatty acids could downregulate the lymphocyte proliferation, natural killer cell activation, and inflammatory mediators such as tumor necrosis factor-alpha (TNF-α) and interleukin-1 and interleukin-2 (IL-1 and IL-2) production in humans (Boudrault et al. 2009; Singer and Richter-Heinrich 1991). Studies on brain tissue about fatty acid turnover could suggest that deficiency in essential fatty acid ratio increases the risk of attention deficit hyperactivity disorder (ADHD) (Farooqui and Horrocks 2001). In addition, studies on child with ADHD also exposed that inappropriate level of some essential fatty acids specifically ω-3 PUFAs leads to neurological diseases (Ross et al. 2003). Michael et al. (2004) reported that deficiency of omega-3 fatty acids in early neonatal life could lead to abnormal appetite signaling and increased the chance of schizophrenia and Huntington's disease during adulthood. Further, Dijck-Brouwer et al. (2005) studied the relationship between neurological conditions and fatty acid composition of umbilical artery and vein in 317 infants indicating the lesser amount of fetal DHA, AA, and essential fatty acids responsible for negative regulation. Number of reviews also reported the therapeutic effect of ω-3 s on Alzheimer's disease (Cole et al. 2009; Cunnane et al. 2009).

Yehuda et al. (2011) demonstrated that administration of omega-3 fatty acids to children with ADHD and iron deficiency could significantly improve the sleep, life quality, concentration, and hemoglobin level. In situ, literature explored that intake of omega-3 fatty acids or foods rich in ω-3 fatty acids has improved the cognitive health in older adults, thereby supporting the good cognitive health related to AD. In contrary to the preventive or therapeutic results, consumption of omega-3 fatty acids seems to be not beneficial when considering patients with AD.

However, analysis available during 2015–2016 related to older adults with memory related abnormalities including mild cognitive impairment and Alzheimer's disease progression reported that ω-3 fatty acid supplementation has attributed to the significant improvement from disease condition (Cederholm 2017). Studies on the effects of ω-3 s in Alzheimer's disease/dementia are provided in Table 4.1 as summary.

4.4 Conclusion

Omega-3 fatty acids are essential fatty acids mainly supplied through diet. Studies shown that people received adequate level of omega-3s are associated with lower rate of incidence and development of AD, dementia, and other complications. Available studies are not sufficient to make a conclusion on the potential of omega-3 on AD management. More studies are needed to confirm the effects of omega-3 on brain function.

Table 4.1 Studies on the effects of omega-3 fatty acids in Alzheimer's disease/dementia as summary (Shahidi and Ambigaipalan 2018)

Details of the study	No. of persons	Outcome of the study	References
Intake of phosphatidylserine enriched with DHA (100 mg/day) could improve or maintain cognitive status in elderly subjects with memory complaints	122 elderly individuals	ND	Vakhapova et al. (2014)
Low levels of red blood cell DHA were associated with smaller brain volumes and a vascular pattern of cognitive impairment even in persons free of clinical dementia	1575 participants (854 women) aged 67 ± 9 years	ND	Tan et al. (2012)
Increased DHA intake from marine sources reduced the risk of dementia	266 participants	42 dementia and 30 AD	Lopez et al. (2011)
Supplementation with algal DHA (2 g/day) did not slow down the rate of cognitive and functional decline in patients with mild to moderate AD	295 individuals with mild to moderate AD	ND	Quinn et al. (2010)
The cognitive function did not decline over 2 years of study in healthy adults with administration of 200 mg EPA plus 500 mg DHA	748 cognitively healthy adults (55% men), aged 70–79 years	ND	Dangour et al. (2010)
Intake of ω-3 PUFAs was not associated with dementia or AD in the Canadian Study of Health and Aging	663 non-dementia subjects aged more than 65 years	149 were incident cases of dementia, including 105 with AD	Kröger et al. (2009)
Supplementation with DHA (800 mg/day) and lutein (12 mg/day) significantly improved verbal fluency scores, memory scores, and rate of learning in elderly women	49 women (aged 60–80 years)	ND	Johnson et al. (2008)
High consumption of fish (unprocessed lean fish and fatty fish) and fish products (>10 g/day) was associated with better cognitive performance in a dose-dependent manner in elderly people	2031 subjects (55% women) aged 70–74 years	80 poor cognitive performance who had low fish consumption (<10 g/day)	Nurk et al. (2007)
Intake of fatty fish and marine ω-3 PUFAs reduced the risk of impaired cognitive function in this middle-aged population, whereas intake of cholesterol	1613 subjects ranging from 45 to 70 years	ND	Kalmijn et al. (2004)

(continued)

Table 4.1 (continued)

Details of the study	No. of persons	Outcome of the study	References
and saturated fat showed an increased risk			
Intake of dietary ω-3 (DHA) PUFAs and fish reduced the risk of incident AD, but EPA did not show any significant effect	815 residents (65–94 years), who were initially unaffected by AD	131 participants developed AD	Morris et al. (2003)
Consumption of fish (weekly) reduced the risk of AD	8085 non-dementia participants aged 65	281 incident cases of dementia, including 183 AD	Barberger-Gateau et al. (2002)

AD Alzheimer's disease, *DHA* docosahexaenoic acid, *EPA* eicosapentaenoic acid, *ND* no data, *PUFAs* polyunsaturated fatty acids

References

Arts MT, Ackman RG, Holub BJ (2001) "Essential fatty acids" in aquatic ecosystems: a crucial link between diet and human health and evolution. Can J Fish Aquat Sci 58:122–137

Aveldaño MI, Sprecher H (1987) Very long chain (C24 to C36) polyenoic fatty acids of the n-3 and n-6 series in dipolyunsaturated phosphatidylcholines from bovine retina. J Biol Chem 262: 1180–1186

Bandarra NM, Marçal A, Cordeiro AR, Pousão-Ferreira P (2018) Sardine (*Sardina pilchardus*) lipid composition: does it change after one year in captivity? Food Chem 244:408–413

Barberger-Gateau P, Letenneur L, Deschamps V, Pérès K, Dartigues JF, Renaud S (2002) Fish, meat, and risk of dementia: cohort study. BMJ 325:932–933

Beermann C, Jelinek J, Reinecker T, Hauenschild A, Boehm G, Klör HU (2003) Short term effects of dietary medium-chain fatty acids and n-3 long-chain polyunsaturated fatty acids on the fat metabolism of healthy volunteers. Lipids Health Dis 2:10

Bjelková M, Nôžková J, Fatrcová-Šramková K, Tejklová E (2012) Comparison of linseed (*Linum usitatissimum* L.) genotypes with respect to the content of polyunsaturated fatty acids. Chem Pap 66:972–976

Bohannon MB, Kleiman R (1975) Unsaturated C18 α-hydroxy acids in *Salvia nilotica*. Lipids 10: 703–706

Boudrault C, Bazinet RP, Ma DW (2009) Experimental models and mechanisms underlying the protective effects of n-3 polyunsaturated fatty acids in Alzheimer's disease. J Nutr Biochem 20: 1–10

Braverman N, Eichler F (2009) Peroxisomal disorders and neurological disease. In: Squire LR (ed) Encyclopedia of neuroscience, vol 7. Academic, Oxford, pp 579–588

Cederholm T (2017) Fish consumption and omega-3 fatty acid supplementation for prevention or treatment of cognitive decline, dementia or Alzheimer's disease in older adults: any news? Curr Opin Clin Nutr Metab Care 20:104–109

Chatgilialoglu C, Ferreri C, Melchiorre M, Sansone A, Torreggiani A (2014) Lipid geometrical isomerism: from chemistry to biology and diagnostics. Chem Rev 114:255–284

Cholewski M, Tomczykowa M, Tomczyk MA (2018) Comprehensive review of chemistry, sources and bioavailability of Omega-3 fatty acids. Nutrients 10:1662

Cole GM, Ma QL, Frautschy SA (2009) Omega-3 fatty acids and dementia. Prostaglandins Leukot Essent Fat Acids 81:213–221

Connor WE (2000) Importance of n-3 fatty acids in health and disease. Am J Clin Nutr 71:171S–175S

Cunnane SC, Plourde M, Pifferi F, Bégin M, Féart C, Barberger-Gateau P (2009) Fish, docosahexaenoic acid and Alzheimer's disease. Prog Lipid Res 48:239–256

Dangour AD, Allen E, Elbourne D, Fasey N, Fletcher AE, Hardy P et al (2010) Effect of 2-y n-3 long-chain polyunsaturated fatty acid supplementation on cognitive function in older people: a randomized, double-blind, controlled trial. Am J Clin Nutr 91:1725–1732

Das U (2011) Essential fatty acids—biochemistry, physiology and clinical significance. In: Das U (ed) Molecular basis of health and disease, 1st edn. Springer, Dordrecht, pp 101–151

Dijck-Brouwer DJ, Hadders-Algra M, Bouwstra H, Decsi T, Boehm G, Martini AI et al (2005) Lower fetal status of docosahexaenoic acid, arachidonic acid and essential fatty acids is associated with less favorable neonatal neurological condition. Prostaglandins Leukot Essent Fat Acids 72:21–28

Ezeagua IE, Petzkeb KJ, Langeb E, Metges CC (1998) Fat content and fatty acid composition of oils extracted from selected wild-gathered tropical plant seeds from Nigeria. J Am Oil Chem Soc 75:1031–1035

Farooqui AA, Horrocks LA (2001) Plasmalogens, phospholipase A_2, and docosahexaenoic acid turnover in brain tissue. J Mol Neurosci 16:263–272

Fialkow J (2016) Omega-3 fatty acid formulations in cardiovascular disease: dietary supplements are not substitutes for prescription products. Am J Cardiovasc Drugs 16:229–239

Giannelos PN, Zannikos F, Stournas S, Lois E, Anastopoulos G (2002) Tobacco seed oil as an alternative diesel fuel: physical and chemical properties. Ind Crop Prod 16:1–9

Goren AC, Kilic T, Dirmenci T, Bilsel G (2006) Chemotaxonomic evaluation of Turkish species of *salvia*: fatty acid compositions of seed oils. Biochem Syst Ecol 34:160–164

Hardy SJ, Ferrante A, Poulos A, Robinson BS, Johnson DW, Murray AW (1994a) Effect of exogenous fatty acids with greater than 22 carbon atoms (very long chain fatty acids) on superoxide production by human neutrophils. J Immunol 153:1754–1761

Hardy SJ, Ferrante A, Robinson BS, Johnson DW, Poulos A, Clark KJ (1994b) In vitro activation of rat brain protein kinase C by polyenoic very-long-chain fatty acids. J Neurochem 62:1546–1551

He XX, Wu XL, Chen RP, Chen C, Liu XG, Wu BJ, Huang ZM (2016a) Effectiveness of omega-3 polyunsaturated fatty acids in non-alcoholic fatty liver disease: a meta-analysis of randomized controlled trials. PLoS One 11(10):e0162368

He Z, Zhu H, Li W, Zeng M, Shengfang W, Chen S (2016b) Chemical components of cold pressed kernel oils from different *Torreya grandis* cultivars. Food Chem 209:196–202

IUPAC (1997a) Fatty Acids. In: Compendium of Chemical Terminology—The "Gold Book", 2nd edn. Blackwell, Oxford. http://goldbook.iupac.org/html/F/F02330.html. Accessed 10 Dec 2017

IUPAC (1997b) Cis-trans isomers. In: Compendium of Chemical Terminology—The "Gold Book", 2nd edn. Blackwell, Oxford. http://goldbook.iupac.org/html/C/C01093.html. Accessed 10 December 2017

Jin F, Nieman DC, Sha W, Xie G, Qiu Y, Jia W (2012) Supplementation of milled chia seeds increases plasma ALA and EPA in postmenopausal women. Plant Foods Hum Nutr 67:105–110

Johnson EJ, McDonald K, Caldarella SM, Chung HY, Troen AM, Snodderly DM (2008) Cognitive findings of an exploratory trial of docosahexaenoic acid and lutein supplementation in older women. Nutr Neurosci 11:75–83

Kalmijn SV, Van Boxtel MPJ, Ocke M, Verschuren WMM, Kromhout D, Launer LJ (2004) Dietary intake of fatty acids and fish in relation to cognitive performance at middle age. Neurology 62: 275–280

Kang JY, Chun BS, Lee MC, Choi JS, Choi IS, Hong YK (2016) Anti-inflammatory activity and chemical composition of essential oil extracted with supercritical CO_2 from the brown seaweed *Undaria pinnatifida*. J Essent Oil Bear Plants 19:46–51

Kobelnik M, Fontanari GG, Ribeiro CA, Crespi MS (2017) Evaluation of thermal behavior and chromatographic characterization of oil extracted from seed of *Pittosporum undulatum*. J Therm Anal Calorim 131:371–378

Kröger E, Verreault R, Carmichael PH, Lindsay J, Julien P, Dewailly E et al (2009) Omega-3 fatty acids and risk of dementia: the Canadian study of health and aging. Am J Clin Nutr 90:184–192

Lopez LB, Kritz-Silverstein D, Barrett-Connor E (2011) High dietary and plasma levels of the omega-3 fatty acid docosahexaenoic acid are associated with decreased dementia risk: the rancho Bernardo study. J Nutr Health Aging 15:25–31

MacLean CH, Mojica WA, Morton SC, Pencharz J, Garland RH, Tu W et al (2004) Effects of omega-3 fatty acids on lipids and glycemic control in type II diabetes and the metabolic syndrome and on inflammatory bowel disease, rheumatoid arthritis, renal disease, systemic lupus erythematosus, and osteoporosis. Evid Rep Technol Assess (Summ) 89:1–4

Memon NN, Talpur FN, Bhanger MI, Balouch A (2011) Changes in fatty acid composition in muscle of three farmed carp fish species (*Labeo rohita*, *Cirrhinus mrigala*, *Catla catla*) raised under the same conditions. Food Chem 126:405–410

Michael L, Mathai MS, Chen N, Jayasooriya AP, Andrew J et al (2004) Does perinatal ω-3 polyunsaturated fatty acid deficiency increase appetite signaling? Obesity 12:1886–1894

Monroig Ó, Tocher DR, Navarro JC (2013) Biosynthesis of polyunsaturated fatty acids in marine invertebrates: recent advances in molecular mechanisms. Mar Drugs 11:3998–4018

Mori TA (2017) Marine OMEGA-3 fatty acids in the prevention of cardiovascular disease. Fitoterapia 123:51–58

Morris MC, Evans DA, Bienias JL, Tangney CC, Bennett DA, Wilson RS et al (2003) Consumption of fish and n-3 fatty acids and risk of incident Alzheimer disease. Arch Neurol 60:940–946

Nagao K, Yanagita T (2005) Conjugated fatty acids in food and their health benefits. J Biosci Bioeng 100:152–157

Nurk E, Drevon CA, Refsum H, Solvoll K, Vollset SE, Nygård O et al (2007) Cognitive performance among the elderly and dietary fish intake: the Hordaland health study. Am J Clin Nutr 86:1470–1478

Parrish CC (2009) Essential fatty acids in aquatic food webs. In: Lipids in aquatic ecosystems. Springer, New York, pp 309–326

Pintea A, Dulf FV, Bunea A, Matea C, Andrei S (2012) Comparative analysis of lipophilic compounds in eggs of organically raised ISA Brown and Araucana hens. Chem Pap 66:955–963

Poulos A, Sharp P, Johnson D, White I, Fellenberg A (1986) The occurrence of polyenoic fatty acids with greater than 22 carbon atoms in mammalian spermatozoa. Biochem J 240:891–895

Poulos A, Johnson DW, Beckman K, White IG, Easton C (1987) Occurrence of unusual molecular species of sphingomyelin containing 28-34-carbon polyenoic fatty acids in ram spermatozoa. Biochem J 248:961–964

Puri R, Mahajan M, Sahajpal NS, Singh H, Singh H, Jain SK (2016) Self-nanoemulsifying drug delivery system of docosahexanoic acid: development, in vitro, in vivo characterization. Drug Dev Ind Pharm 42:1032–1041

Quinn JF, Raman R, Thomas RG, Yurko-Mauro K, Nelson EB, Van Dyck C et al (2010) Docosahexaenoic acid supplementation and cognitive decline in Alzheimer disease: a randomized trial. J Am Med Assoc 304:1903–1911

Richter CK, Bowen KJ, Mozaffarian D, Kris-Etherton PM, Skulas-Ray AC (2018) Total long-chain n-3 fatty acid intake and food sources in the United States compared to recommended intakes. NHANES 2003–2008. Lipids 52:917–927

Rombeau JL, Kripke SA, Settle RG (1990) Short-chain fatty acids production, absorption, metabolism, and intestinal effects. In: Kritchevsky D, Bonfield C, Anderson JW (eds) Dietary fiber. Chemistry, physiology, and health effects, 1st edn. Springer, New York, pp 317–337

Ross BM, McKenzie I, Glen I, Bennett CPW (2003) Increased levels of ethane, a non-invasive marker of n-3 fatty acid oxidation, in breath of children with attention deficit hyperactivity disorder. Nutr Neurosci 6:277–281

Sassa T, Kihara A (2014) Metabolism of very long-chain fatty acids: genes and pathophysiology. Biomol Ther 22:83–92

Schönfeld P, Wojtczak L (2016) Short- and medium-chain fatty acids in energy metabolism: the cellular perspective. J Lipid Res 57:943–954

Scorletti E, Byrne CD (2013) Omega-3 fatty acids, hepatic lipid metabolism, and nonalcoholic fatty liver disease. Annu Rev Nutr 33:231–248

Semba RD (2007) Essential fatty acids and visual development in infants. In: Semba RD (ed) Nutrition & Health: handbook of nutrition and Ophtalmology, 1st edn. Humana Press, Totowa, pp 415–441

Shahidi F, Ambigaipalan P (2018) Omega-3 polyunsaturated fatty acids and their health benefits. Annu Rev Food Sci Technol 9:345–381

Shirai N, Higuchi T, Suzuki H (2006) Analysis of lipid classes and the fatty acid composition of the salted fish roe food products, Ikura, Tarako, Tobiko and Kazunoko. Food Chem 94:61–67

Singer P, Richter-Heinrich E (1991) Stress and fatty liver: possible indications for dietary long-chain n-3 fatty acids. Med Hypotheses 36:90–94

Singh M (2005) Essential fatty acids, DHA and human brain. Indian J Pediatr 72:239–242

Souza AL, Martinez FP, Ferreira SB, Kaiser CR (2017) A complete evaluation of thermal and oxidative stability of chia oil. The richest natural source of a-linolenic acid. J Therm Anal Calorim 130:1307–1315

Tan ZS, Harris WS, Beiser AS, Au R, Himali JJ, Debette S et al (2012) Red blood cell omega-3 fatty acid levels and markers of accelerated brain aging. Neurology 78:658–664

Vakhapova V, Cohen T, Richter Y, Herzog Y, Kam Y, Korczyn AD (2014) Phosphatidylserine containing omega-3 fatty acids may improve memory abilities in non demented elderly individuals with memory complaints: results from an open-label extension study. Dement Geriatr Cogn Dis 38:39–45

Van Goor SA, Dijck-Brouwer DAJ, Muskiet FAJ (2011) Mother–child long chain polyunsaturated fatty acid relationships: implications for diet and behavior. In: Preedy VR (ed) Handbook of behavior, food and nutrition, vol 2, 1st edn. Springer, New York, pp 1139–1156

Yang J, Zhou C, Yuan G, Li D (2012) Effects of geographical origin on the conjugated linolenic acid of *Trichosanthes kirilowii* maxim seed oil. J Am Oil Chem Soc 89:401–407

Yang N, Sampathkumar K, Loo SCJ (2017) Recent advances in complementary and replacement therapy with nutraceuticals in combating gastrointestinal illnesses. Clin Nutr 36:968–979

Yehuda S, Rabinovitz S, Mostofsky DI (2005) Essential fatty acids and the brain: from infancy to aging. Neurobiol Aging 26:98–102

Yehuda S, Rabinovitz-Shenkar S, Carasso RL (2011) Effects of essential fatty acids in iron deficient and sleep-disturbed attention deficit hyperactivity disorder (ADHD) children. Eur J Clin Nutr 65:1167–1169

Chapter 5
Fat and Alzheimer's Disease

**Suset Rodriguez, Ramakrishnan Thiruchelvi,
Arokiasamy Justin Thenmozhi, Daniela S. Cosio, Nagarajarao Shamaladevi,
Hussain Hussain, Natarajan Sampath, Michael J. Paidas,
and Arumugam R. Jayakumar**

Abstract Alzheimer's disease is one of the devastating causes of increased mortality and morbidities worldwide. There are various causes and hypothetical factors embedded in the development of the pathological process in terms of aging, idiopathy, genetics, and others. However, incremental changes in the lipid profile are considered an important factor involved in the pathological process. Given the characteristic feature of the neurons, the brain microvessel endothelial cells (BMVECs) are not fully understood with regard to their involvement in the pathogenesis, yet particular investigations occurred in vivo demonstrating the potential changes in the function of these cells. The plasma membrane cholesterol could affect the physiological properties of various structures including the receptors, enzymes, and ion gates. Until now, there has been no clear evidence showing why

---

S. Rodriguez · D. S. Cosio · M. J. Paidas · A. R. Jayakumar (✉)
Department of Obstetrics, Gynecology and Reproductive Sciences, University of Miami, Miller School of Medicine, Miami, FL, USA
e-mail: ajayakumar@med.miami.edu

R. Thiruchelvi
Department of Bioengineering and Biotechnology, Vels Institute of Science Technology and Advanced Studies (VISTAS), Chennai, Tamil Nadu, India

A. Justin Thenmozhi
Department of Biochemistry, School of Biological Sciences, Madurai Kamaraj University, Madurai, Tamil Nadu, India

Department of Biochemistry and Biotechnology, Faculty of Science, Annamalai University, Chidambaram, Tamil Nadu, India

N. Shamaladevi
Molecular Analytics, Miami, FL, USA

H. Hussain
Department of Internal Medicine and Infectious Disease, Larkin Community Hospital, Miami, FL, USA

N. Sampath
School of Chemical and Biotechnology, SASTRA Deemed University, Thanjavur, Tamil Nadu, India

A. Justin Thenmozhi, T. Manivasagam (eds.), *Nutraceuticals for Alzheimer's Disease: A Promising Therapeutic Approach*, Nutritional Neurosciences,
https://doi.org/10.1007/978-981-99-0677-2_5

hypercholesterolemia contributes to AD. Given the association of hypercholesterolemia with various diseases including sodium resistance in diabetes mellitus II, investigations were conducted to examine whether hypercholesterolemia had negative impacts on BMVECs in increased age and in mice that received a high-fat-content diet. Decreased cholesterol levels aid in the delay of this impact. Additionally, hyperphosphorylation of neuronal pTDP-43, a transactive response DNA-binding protein, attributed to AD as well as abnormal Tau protein accumulation. Furthermore, various studies demonstrated the negative impact of hypercholesterolemia on NMDAR, Aβ40, Aβ42, synaptophysin, and others. Thus, proper management of hypercholesterolemia established a reduction in the accumulation of beta-amyloid.

Keywords Alzheimer's disease · Cholesterol · Dietary fats · Saturated fatty acids · Trans-fatty acids

Abbreviations

2-HβC	2-hydroxypropyl-β cyclodextrin
ACAT	Acyl-CoA cholesterol acyltransferase
AD	Alzheimer's disease
ApoE	Apolipoprotein E
ApoE4	Apolipoprotein E4
APP	Amyloid-β protein precursor
Aβ	Amyloid beta
BACE1	β-secretase
BBB	Blood-brain barrier
BMVECs	Brain microvascular endothelial cells
CNS	Central nervous system
CSF	Cerebrospinal Fluid
CVD	Cardiovascular disease
DDS	Drug delivery system
FAs	Fatty acids
FAD	Familial Alzheimer's disease
HFD	High-fat diet
HSPs	Heat shock proteins
LDL	Low-density lipoprotein
NCD	Normal chow diet
NMDA	N-methyl-D-aspartate
NPS	Nanoparticles
TGF-β	Transcription growth factor beta

5.1 Introduction

Alzheimer's disease (AD) is a type of dementia that causes problems with memory, thinking, and behavior. It accounts for approximately 60–80% of dementia cases and doubles in prevalence every 5 years after age 65 (Prince et al. 2014). It is the sixth leading cause of death in the United States and the only top 10 cause that is still rising (Alzheimer's Association 2018). Projections show that AD will continue to increase its impact as the domestic and world populations age (Hurd et al. 2013). Symptoms typically begin with mild memory difficulties and progress to cognitive impairment, dysfunctions in complex activities of daily living, and impairment in other areas of daily living (Kukull and Bowen 2002). There is a progressive decline in more than one cognitive domain, including language, memory, executive and visuospatial function, personality, and behavior, which results in a loss of capacity to perform activities of daily living. AD's impact on patients, their families, and the economy is estimated to result in approximately US$1 trillion annually. Clinically, AD is classified into two phenotypes, familial AD (fAD) and late-onset AD (LOAD). Early-onset familial AD is seen in 5%–10% of AD cases (Bekris et al. 2010). It is a rare form of AD that occurs in the fourth or fifth decade of life and is due to the autosomal dominant inheritance of PSEN1, APP, or PSEN2 genes (Zekanowski et al. 2003). LOAD is the more common phenotype and is not linked to any known gene mutations. Diagnosis involves a comprehensive evaluation which includes pathologic AD biomarkers in the cerebrospinal fluid (CSF) and PET scan; however, in typical cases, an individualized approach can be used including history, exam, select lab tests, and neuroimaging to provide a high-confidence diagnosis.

5.2 AD Mechanisms

AD is characterized by extracellular amyloid plaques, intracellular neurofibrillary tangles, and nerve cell death (Selkoe and Hardy 2016; Iqbal et al. 2005; Mufson et al. 2008). Amyloid-β protein precursor (APP) is serially cleaved by β-secretase (BACE1) and then γ-secretase to create amyloid-β (Aβ) (Yan and Vassar 2014). Aβ monomers aggregate to form oligomers, fibrils, and insoluble amyloid plaques (Selkoe and Hardy 2016). On the other hand, tau protein when hyperphosphorylated results in neurofibrillary tangles. Tau normally promotes the stabilization of microtubules; however, when hyperphosphorylated, it accumulates into paired helical fragments which become tangles (Iqbal et al. 2005). In addition, Aβ and tau deposits begin to occur at synaptic sites as a result of defects in axonal transport, mitochondrial damage, oxidative stress, and other processes which ultimately result in loss of dendritic spines, presynaptic terminals, and axonal dystrophy (Overk and Masliah 2014).

Several hypotheses are postulated for the pathogenesis of AD. The first hypothesis is the amyloid cascade which proposes the main pathogenic event as the aggregation of Aβ as neuritic plaques, diffuse plaques, or oligomeric forms (Iqbal et al. 2005). Next, the tau hypothesis posits tau hyperphosphorylation as the main event (Šimić et al. 2016). Furthermore, the cholinergic hypothesis assumes reduced choline acetyltransferase activity and acetylcholine levels in several areas of the brain (Contestabile 2011). In addition, the impairment of brain mitochondria as the instigating event constitutes the mitochondrial cascade hypothesis (Selkoe and Hardy 2016). Finally, the vascular hypothesis focuses on decreased cerebral blood flow as the main factor (Di Marco et al. 2015).

Furthermore, insulin is also implicated in AD. Insulin is a peptide hormone made in the pancreatic β-cells and serves to regulate glucose metabolism in peripheral tissues; however, it also serves major roles within the central nervous system (CNS). Glucose transport from the periphery across the endothelial cells of the blood-brain barrier (BBB) is insulin dependent, as well as the transport into the neuron across the cell membrane (Benarroch 2014; Simpson et al. 2007; Joost and Thorens 2001). Insulin crosses into the BBB via a receptor-mediated active transport system found in endothelial cells (Bosco et al. 2011). Insulin regulates the metabolism of glucose, neuronal integrity, and cognition via receptor-mediated mechanisms, including calcium influx, neurotransmitter accumulation, synaptic connections, apoptosis, and neurogenesis (Chiu et al. 2008).

5.3 AD Treatments

Overall, the molecular mechanisms underlying the increased phosphorylation of tau and Aβ are poorly understood; therefore, there is no cure for AD. Currently available drugs are treatments targeting symptoms of AD. The two main categories are cholinesterase inhibitors and memantine. Cholinesterase inhibitors – donepezil, rivastigmine, and galantamine – are used for mild, moderate, or severe AD (Howard et al. 2012). Memantine is a noncompetitive N-methyl-D-aspartate (NMDA) receptor antagonist as well as a dopamine agonist, which is used for patients with moderate to severe AD (Grossberg et al. 2013). These medications can increase the quality of life for the patient if not given too late, but they do not alter the disease progression or rate of decline (Mossello and Ballini 2012). Most therapeutic agents under development over the past 15 years have been unsuccessful. Immune therapy with antibodies targeting Aβ has resulted in clinical trials that have not reached clinically significant improvement in memory performance or have led to adverse effects with a subsequent dropout of subjects (Karran and De Strooper 2016). Thus, current treatment options aim to decrease symptom progression and disability. There is no cure for AD as of yet that has been identified.

5.4 Role of Fats in AD

Increased accumulation of brain levels of cholesterol is an evolving concept in the pathogenies of AD and its associated dementia in the aged population. An abnormal level of lipids, including cholesterol, has been identified as a risk factor for AD and has been used as an index for the prediction of AD (Liu et al. 2020; Wu et al. 2019). Cholesterol in the brain is found in myelin, astrocytes, and nerve cell membranes. It helps support brain function, neuron development, and synaptic plasticity. Brain cholesterol is mostly synthesized in situ as the BBB prevents peripheral cholesterol from entering the brain. Oligodendrocytes and astrocytes produce the greatest amount of cholesterol in the brain. Glial cells synthesize lipoproteins containing apolipoprotein E (ApoE), which then allows for the transfer of cholesterol into the nerve cells.

Meanwhile, developing neurons can regulate their cholesterol needs by synthesizing more cholesterol; however, mature neurons mostly depend on exogenous cholesterol as they can synthesize cholesterol only in small quantities. Apolipoprotein E4 (ApoE4) transports cholesterol made from astrocytes which is then transferred into the extracellular membrane through ATP-binding cassette transporters A1 and F1 (Jeong et al. 2019). The low-density lipoprotein (LDL) receptor on the nerve cell membrane then binds the cholesterol which is used for dendritic growth and synaptic formation. Unused cholesterol is collected in the endochylema after esterification. Elevated cholesterol also increases the formation of free radicals, which creates a disruption of the BBB and ultimately increases levels of cholesterol in the brain. This was evidenced in a study in which rabbits fed with high-cholesterol diet as compared to a normal diet had increased BBB permeability (Jiang et al. 2012). Reduced expression of tight junction proteins was also noted in the high-cholesterol-diet-fed mice which subsequently contributes to the decreased integrity of the BBB and ultimately affects the progression of AD (Jiang et al. 2012).

Cholesterol in the brain affects amyloidogenesis by creating more Aβ plaques, ultimately resulting in neurodegeneration. Increased levels of intracellular cholesterol result in increased levels of cell membrane cholesterol which then aggregate into membrane microdomains. These microdomains promote the binding of the transmembrane protein APP. APP is synthesized via two pathways – a non-amyloid metabolic pathway and the amyloid metabolic pathway. The amyloid metabolic pathway involves cleavage of APP by BACE1; subsequently, the remaining product is hydrolyzed by γ-secretase in two different areas to create Aβ40 and Aβ4 (Beel et al. 2010). Aβ42 is the main peptide in senile plaques. Aβ can exist as monomers, amyloid fibers, and oligomers, of which the latter is the main cause of impaired cognitive function in patients with AD and therefore the most toxic. Consequently, disequilibrium in the production and clearance of Aβ results in increased levels of Aβ in the brain and ultimately results in the onset of AD. The amount of Aβ produced is regulated by the levels of intracellular cholesterol, whereby higher levels of cholesterol result in higher activity of BACE1 and γ-secretase (Grimm et al. 2008). In the case of decreased BACE1 and γ-secretase,

APPs are hydrolyzed via the non-amyloid metabolic pathway, and therefore, there is no production of Aβ. In addition, the production of Aβ is also influenced by the number of cholesterol esters. The amyloid metabolic pathway is stimulated by the conversion of a mass of cholesterol via acyl-CoA cholesterol acyltransferase (ACAT) into cholesterol esters. Therefore, suppression of Aβ can be achieved by inhibiting ACAT (Huttunen et al. 2010). Evidence of cholesterol's effect on the plasma membrane was provided by Hardy et al. (2006) who found that increased levels of cholesterol in the plasma membrane affected the function of synapses and consequently cognitive degeneration in AD.

LDL is also implicated in AD. Elevated LDL has vascular and neurotoxic effects (Lesser 2012; Wu et al. 2019; Xu et al. 2015). LDL is a 20–25-nm-sized particle and contains the highest cholesterol content. Oxidized LDL binds to Aβ which increases the neurotoxicity of neuronal cell cultures (Keller et al. 1999). Furthermore, disruptions in the BBB allow for increased amounts of LDL to enter the brain parenchyma, which is then internalized by neurons and promotes APP internalization ultimately resulting in Aβ accumulation (Hui et al. 2012). Another proposed mechanism of the neurotoxicity of LDL is that increased amounts of LDL added to cultured neurons resulted in greater activity of BACE1 which led to Aβ accumulation in the endolysosome. A meta-analysis of nine studies analyzing LDL trends in patients with AD compared to healthy controls demonstrated significantly higher levels in AD patients (Liu et al. 2020).

Other types of fats have also been associated with AD. For instance, fatty acids (FAs) are fats found in membranes and differ by position and type of saturation and unsaturation, chain length, and lipid polar head group. Trans FAs are a type of unsaturated fat that naturally occurs in small amounts in meat and milk fat, however, became more commonly used in the food industry via hydrogenation which allowed for the stabilization of vegetable oils. Not only have they been associated with an increased risk for cardiovascular disease (CVD) but also have been found to increase the risk for AD (Morris et al. 2003). In addition, saturated fats also play a role in AD. Saturated FAs are synthesized by the body and the main dietary source is animal products, such as full-fat dairy, red meat, and poultry. Elevated levels of saturated FAs in the diet have been associated with an increased risk of AD (Morris et al. 2003) and with increased amounts of cholesterol in the blood, which independently is associated with greater AD risk.

5.5 Diseases Associated with Fat

Lipids are a class of molecules made of different subtypes: triglycerides, free FAs, sterols, phospholipids, and other groups. FAs in the diet are one of the main factors influencing cholesterol levels. The degree of saturation and configuration of the saturation determines FA composition. Having relatively more unsaturated than saturated FAs favors a desirable cholesterol profile. Dietary FAs are required for normal physiologic function as they cannot be synthesized in the body and must be consumed to maintain good health. FAs are needed for the absorption of lipid-soluble vitamins A, D, E, and K and carotenoids. Saturated FAs are mostly found in meat and dairy products, existing as solids at room temperature (Linscheer and Vergroesen 1994). Trans FAs are created through a hydrogenation process that incorporates hydrogen into mono- and polyunsaturated FAs to create a solid at room temperature. It can also occur naturally in food products from ruminant animals. Trans fat is popularized due to its ability to lengthen shelf life and improve food texture and flavor; however, trans fat increases LDL cholesterol and decreases HDL cholesterol (Mozaffarian et al. 2013).

Increased dietary fats are implicated in the development of CVD. Saturated FAs and trans FAs increase pro-inflammatory and oxidative stress (Munoz and Costa 2013) which contributes to the development of CVD (Ruparelia et al. 2017). For instance, the prospective study Lipids Research Clinics Prevalence Follow-up Study found that dietary consumption of trans fat and saturated fat increased the risk of CVD (Guasch-Ferre et al. 2015). Further studies have also shown that increased amounts of dietary saturated FAs are associated with a greater risk of ischemic heart disease (Mann et al. 1997; Boniface and Tefft 2002). Similarly, studies have demonstrated a positive association between LDL cholesterol and risk for CVD, also increasing with higher blood levels and time of exposure (Duncan et al. 2019; Tsao et al. 2012). Trans FAs increase the levels of LDL cholesterol and pose a significant risk for CVD, whereby a 2% absolute increase in energy intake from trans fat is associated with a 23% increase in CVD risk (Islam et al. 2019).

In addition, cholesterol has been found to play a role in cancer. A positive correlation between cancer risk and elevated levels of serum cholesterol has been shown in melanoma, prostate cancer, non-Hodgkin's lymphoma, and endometrial and breast cancer (Kuzu et al. 2016). Interestingly, the widely used class of cholesterol-lowering medications, statins, which work by inhibiting HMG-CoA reductase, has been shown to have anticancer activity in a variety of liquid and solid tumors (Clendening and Penn 2012). Cholesterol synthesis is modified by oncogenic and tumor suppressor factors through their ability to activate or inhibit, respectively, sterol regulatory element-binding proteins (Aylon and Oren 2016). A prospective cohort study of over 500,000 participants with 16-year follow-up demonstrated an association between saturated fat intake and cancer mortality (Zhuang et al. 2019). Kim et al. (2021) performed a systematic review and meta-analysis of prospective cohort studies assessing for risk of CVD and cancer with dietary fats and noted that diets with high saturated fat were correlated with greater mortality from CVD and cancer and also diets with high trans fat had higher mortality from all-causes and CVD.

Cholesterol and dietary FA ingestion have further been associated with diabetes risk. Many patients with type 2 diabetes are found to have increased levels of free FAs and decreased levels of LDL and HDL (Dunn 2010). Hu et al. (2001) conducted a large prospective cohort study of middle-aged women with a BMI of 25 or lower and demonstrated a positive correlation between trans-fat consumption and the risk of diabetes. Additionally, over 70,000 women without CVD and diabetes were followed for 20 years. It was noted that after adjusting for other risk factors, the incidence of diabetes was correlated with trans-fat dietary consumption (Oh et al. 2005). Interestingly, the effect of saturated FAs on diabetes has been met with equivocal findings. A literature review of randomized controlled trials and observational studies evaluating the risk of diabetes with saturated FAs demonstrated that saturated FAs inconsistently affect insulin resistance, and no association has been demonstrated (Micha and Mozaffarian 2010).

Overall, lipids are the basic component of neuronal cell membranes and therefore make up majority of dry weight in the brain. They also function as energy storage and molecular signaling (Linscheer and Vergroesen 1994). The characteristic neuritic plaques involved in AD contain cholesterol at their core. Furthermore, it has been proposed that one of the main functions of APP is to clear excess cholesterol from the brain (Puglielli et al. 2003). One of the most studied genes involved in fAD is APOE4, which creates a protein that aids in cholesterol transport between the gut, liver, and peripheral tissues and is the primary cholesterol transport protein in the brain (Martins et al. 2006).

The term Western diet refers to an energy-dense diet high in saturated fat and sugar. Prior human and animal studies have demonstrated impaired hippocampal-dependent memories as well as hippocampal pathologies in patients consuming a Western diet (Davidson et al. 2013). The hippocampus is noted to have Aβ and neurofibrillary tangles, possibly associated with cognitive impairment. Moreover, epidemiological studies have elucidated a relationship between dietary patterns and AD, focusing on the increased risk of AD from high consumption of saturated fats (Morris et al. 2003).

5.5.1 Human Studies Regarding Fat and AD

Cholesterol has been implicated in AD with a growing body of evidence. A study of over 400 subjects found elevated cholesterol in midlife to be associated with 3× the risk of developing AD (Solomon et al. 2009). Studies have demonstrated an association between increased cholesterol levels in mid to late life and the development of dementia (Liu et al. 2020; Xu et al. 2015). A recent study demonstrated an elevation of cholesterol levels by 10% in patients with AD compared to control subjects (Popp et al. 2013). Furthermore, a study by Helzner et al. (2009) found that patients with elevated levels of total cholesterol had a greater rate of cognitive decline as compared to patients with normal levels of cholesterol.

With the growing obesity epidemic, studies have demonstrated that obesity in midlife can be an early predictor of AD (Dahl et al. 2010; Chuang et al. 2016; Gottesman et al. 2017). A meta-analysis of four prospective cohort studies analyzing the effect of dietary fat intake on AD and dementia risk noted a significant association between higher saturated fat intake and risk for AD and dementia; however, no significant risk associated with total, monounsaturated, and polyunsaturated fat was observed (Ruan et al. 2018). Furthermore, Melo et al. (2020) found increased levels of palmitate, a saturated FA found in high-fat foods, to be increased in the CSF of overweight and obese patients with amnestic mild cognitive impairment. Other studies have noted that diets high in saturated fats and simple carbohydrates affected cognition levels of CSF biomarkers in normal older adults and adults with mild cognitive impairment (Bayer-Carter et al. 2011; Hanson et al. 2013; Hill et al. 2019).

A separate study noted increased Aβ deposition on ^{11}C Pittsburgh compound B amyloid positron emission tomography over 3 years in patients adhering to a Western, high-fat diet as compared to a Mediterranean diet (Berti et al. 2018). Another human study investigating the effect of a Western diet high in levels of saturated and trans FAs in dementia-free subjects in their sixth decade of life found an elevated odds ratio for having total tau pathology and preclinical AD as compared to other types of diets (Samuelsson et al. 2021).

A population-based study assessed fat intake in individuals via survey and subsequently followed up in 21 years. Their study demonstrated that saturated fat intake was associated with an increased risk of dementia; however, these associations were seen only among patients that were ApoE4 carriers (Laitinen et al. 2006). Fieldhouse et al. (2020) also sought to demonstrate an association between dietary fat and AD dementia. They assessed adherence to dietary guidelines and cognitive performance in individuals with AD dementia, mild cognitive impairment, and controls. Their results demonstrated that a high-trans-fat intake was associated with poorer executive functioning in patients with AD dementia. A systematic review by Barnard et al. (2014) noted a positive association between saturated fat intake and AD risk in three out of four studies and also positive associations between trans-fat intake and dementia in three studies.

Morris (2009) performed clinical assessments on patients aged greater than 75 years old without AD and followed patients over the course of 3.9 years. Results demonstrated a positive association between trans-fat intakes, with the top 80% in trans-fat consumption having a 4× risk of developing AD. In a separate study, a meta-analysis of prospective randomized controlled trials assessing the correlation between nutrition and cerebral function in AD patients noted that saturated FAs and trans FAs occur more frequently in patients with AD than in those without (Albrahim 2020).

Cell culture studies have also investigated the effect of trans FAs on APP. It was noted that compared to cis FAs, trans FAs increase amyloidogenic and decrease nonamyloidogenic processing of APP, therefore increasing the production of Aβ. In the study, it was also evident that trans FAs were associated with oligomerization and aggregation of Aβ (Grimm et al. 2012).

5.5.2 Animal Studies Regarding Fat and AD

Given the extensive evidence associating cholesterol with AD, our lab has been undergoing extensive work in the area. To test whether HFD feeding alters TDP-43 and tau phosphorylation, C57BL/6 mice aged 7 weeks were fed a normal chow diet (NCD) or custom-made HFD (60% lipid from our laboratory) for 5 weeks. Levels of

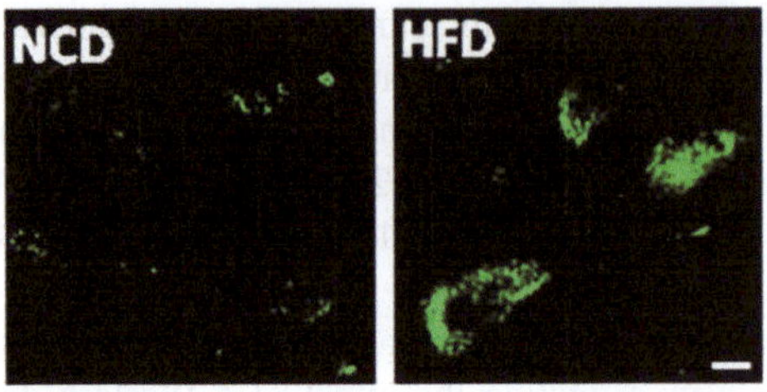

Fig. 5.1 Phosphorylated TDP-43 (p-TDP-43) level in the cerebral cortex of mice after high-fat diet (HFD). Normal chow diet (NCD)-fed mice show a basal level of pTDP-43. However, high-fat diet (HFD)-fed mice displayed hyperphosphorylation of cytoplasmic pTDP-43. Scale bar = 25 μm

Fig. 5.2 Phosphorylated tau (p-tau) level in the cerebral cortex of mice after HFD. NCD-fed mice show the absence of p-tau. However, HFD-fed mice displayed an increase in *p*-tau-positive cells. Scale bar = 25 μm

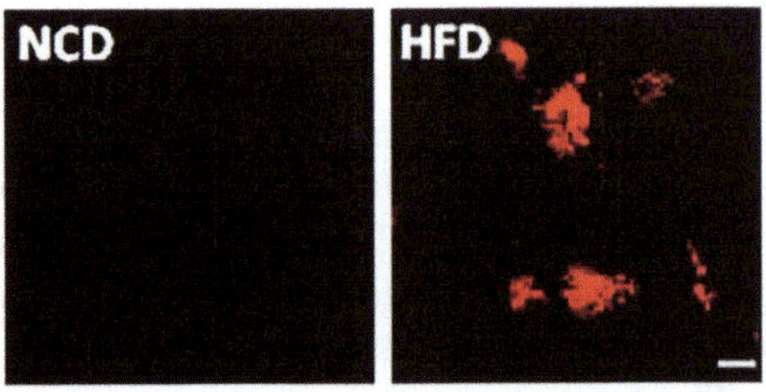

pTDP-43 and p-tau were determined in cortical sections from HFD and aged mice by immunofluorescence (IF), as previously described by us (Jayakumar et al. 2017). To identify changes in pTDP-43 and p-tau in neurons, astrocytes, and microglia, sections were co-immunostained with β-tubulin (neuronal marker), GFAP (astrocyte marker), and Iba-1 (activated microglia marker), and the intensity of IF was examined with a confocal laser-scanning microscope (Jayakumar et al. 2014). An increase in pTDP-43 fluorescence (3.7-fold) and an increase in p-tau-positive neurons were observed from HFD mice, as compared to NCD mice (Figs. 5.1 and 5.2), while no changes in levels of pTDP-43 and p-tau were detected in astrocytes or microglia (Figure not shown). Similar results were obtained when aged mice were stained with pTDP-43 and p-tau (data not shown). These findings suggest that TDP-43 and tau proteinopathy occurs in neurons after HFD or aging that may ultimately contribute to the defective neuronal integrity and neurobehavioral deficit in HFD or aging (see below for the role of cholesterol on p-TDP-43 and p-tau on neuronal proteins).

To test whether HFD feeding affected BMVEC's glucose transport, isolated cortical BMVECs from HFD and aged mice were subjected to immunoblotting as described previously by us (Jayakumar et al. 2006, 2011). We found decreased Glut-1 in isolated BMVECs from HFD mice (59.4% as compared to NCD mice) (Fig. 5.3). We also found a reduction in Glut-1 in isolated BMVECs from aged mice (37.1% as compared to young mice) (Fig. 5.4) indicative of reduced glucose uptake by the brain.

We also noted increased levels of Aβ, as well as soluble amyloid peptides (data not shown) and behavioral and cognitive abnormalities (Figs. 5.3 and 5.4) in mice post-HFD and aged mice, and that blocking cholesterol accumulation significantly

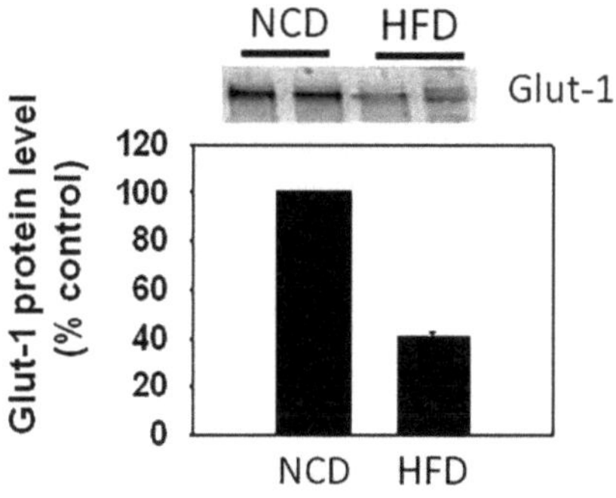

Fig. 5.3 *A reduction in glucose transporter-1 (Glut-1) in mice BMVECs post-HFD. Isolated mice BMVECs that fed with HFD for 5 weeks exhibited a reduction in Glut-1. NCD, normal chow diet; HFD, high-fat diet* ($n = 5$)

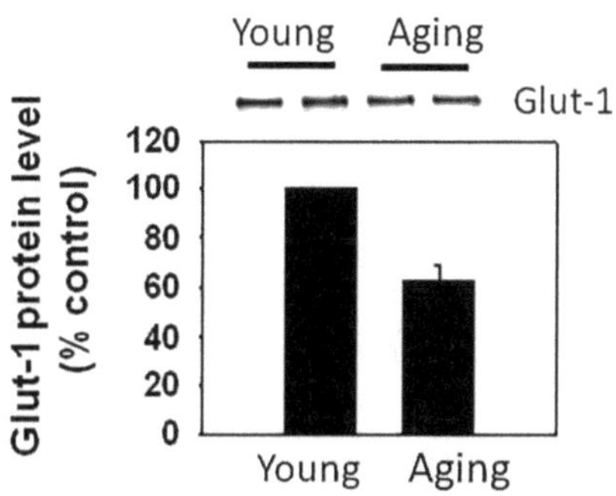

Fig. 5.4 *Differences in glucose transporter-1 (Glut-1) in young and aged mice BMVECs. Isolated mice BMVECs from aged mice exhibit a reduction in Glut-1* ($n = 5$)

reversed all these changes both in HFD and in aged mice. These important and promising preliminary findings strongly suggest that BMVEC's cholesterol level plays a major role in diabetic and/or age-related Alzheimer's dementia. Because tau and Aβ are involved in AD pathogenesis, this preliminary observation strongly supports that BMVECs play a crucial role in diabetes and aging-associated increased risk of onset and development of AD.

We conducted additional studies to assess cellular and plasma membrane cholesterol levels of BMVECs post-HFD and after aging. Cellular and plasma membrane fractions were obtained from isolated BMVECs from HFD and aged mice as described previously (Boland and Tweto 1980). The total cholesterol level in these fractions was determined using the Amplex Red cholesterol assay as described previously (Cummings et al. 2014). This colorimetric assay is based on the reaction of cholesterol with cholesterol oxidase to yield H_2O_2 which can be detected using the Amplex Red reagent. In preliminary studies, we found that the plasma membrane cholesterol concentration for NCD is 6.8 ± 0.3 pmol/μg protein. In the case of HFD, the membrane cholesterol concentration increased to 28.9 ± 1.4 pmol/μg protein. Similarly, the BMVECs plasma membrane cholesterol level increased in aged mice (16.8 ± 0.9 pmol/μg protein), as compared to young mice (5.9 ± 0.6 pmol/μg protein). Cellular cholesterol level was also increased in both HFD and after aging (data not shown).

Reduced levels of ABCG1 in BMVECs plasma membrane post-HFD and after aging: Cholesterol plays a critical role in maintaining proper membrane structure and

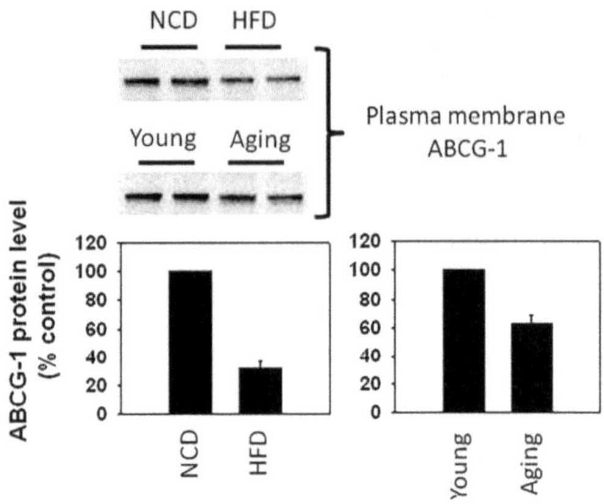

Fig. 5.5 Representative immunoblot shows a reduction in plasma membrane ABCG-1 in BMVECs post-HFD and after aging. ABCG-1 level decreased in both HFD-fed mice cortical BMVECs and in aged mice brain BMVECs. Bands are normalized with plasma membrane marker Na + K+ − ATPase (bans not shown). *NCD* Normal chow diet, *HFD* High-fat diet ($n = 5$)

function through its interactions with other membrane lipids and proteins. Many plasma membrane proteins in late endocytic compartments have been shown to play critical roles in intracellular cholesterol trafficking, in particular, the ABCA1 (31). Little, however, is currently known about the mechanism(s) by which these proteins alter the disposition of cholesterol in donor membranes to make it available for removal. The ABCA1 transporter is one of the better-characterized proteins that alter the distribution of membrane cholesterol. ABCG1 resides on the cell surface, as well as in late endosomes that shuttle back to the cell surface, and that ABCG1 mobilizes a pool of cholesterol on the cell surface. ABCG1 promotes the efflux of cellular lipids to mature HDL, as well as LDL, cyclodextrin, and liposomes. In pilot studies, we found a reduction in ABCG1 levels in plasma membrane fractions of BMVECs post-HFD and after aging (58 and 37%, respectively, as compared to NCD and young mice) (Fig. 5.5).

Since ABCG1-mediated enhancement of cellular cholesterol efflux requires delivery of ABCG1 from its site of synthesis (i.e., ER) to the plasma membrane and late endocytic compartments and that ABCG1 rapidly cycles between endosomes and the cell surface, we examined whether its level was altered in cellular fractions. We found a reduction in ABCG1 levels in cellular fractions of BMVECs post-HFD and after aging (46 and 41%, respectively, as compared to NCD and young mice (data not shown), suggesting that in addition to the reduction in the plasma membrane, changes in cellular ABCG1 may also play a role in the defective cholesterol efflux and subsequent cholesterol accumulation in BMVECs.

Inhibition of ABCG1 reduction diminished the accumulation of cholesterol in BMVECs PM post-HFD and after aging: Since anti-inflammatory cytokines such as IL-10 and TGF-β1 have been shown to increase the expression of ABCA1 in other conditions, we examined whether an infusion of human recombinant TGF-β1 (rTGF-β1) had any effect in diminishing or preventing the accumulation of cholesterol in BMVECs PM post-HFD and after aging. Accordingly, HFD and aged C57BL/6 mice received injections of either vehicle or rTGF-β1 (catalog# 240-B-010; R&D Systems, Minneapolis, MN) as noted above in Specific Aim 1. Mice

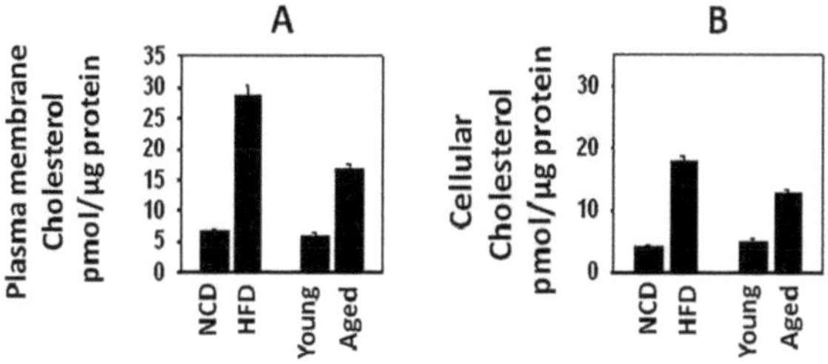

Fig. 5.6 Cholesterol concentration in the plasma membrane and cellular fractions of BMVECs. Increased levels of plasma membrane (**a**) and cellular (**b**) cholesterol were observed in BMVECs from HFD and aged mice, as compared to NCD and young mice. *NCD* Normal chow diet, *HFD* High-fat diet. Young, 6 weeks; aged, 24 weeks ($n = 5$)

received (every alternate day) intraperitoneal injections of 15 μg/kg for 5 weeks and were euthanized.

Cellular and plasma membrane fractions were obtained from isolated BMVECs from HFD and aged mice post-rTGF-β1 treatment, and total cholesterol levels in these fractions were determined as noted above in Specific Aim 1. In the pilot studies, we found that the increased plasma membrane cholesterol concentration in HFD (Fig. 5.6) was diminished by the administration of rTGF-β1 (from 28.9 ± 1.4 pmol/μg protein to 11.8 ± 0.9 pmol/μg protein; NCD is 6.8 ± 0.3 pmol/μg protein). Similarly, the BMVECs' plasma membrane cholesterol level increased in aged mice (16.8 ± 0.9 pmol/μg protein) and that was diminished by rTGF-β1 treatment (8.6 ± 0.6 pmol/μg protein), as compared to young mice (5.9 ± 0.6 pmol/μg protein). Cellular cholesterol level was also reduced in both HFD and after aging (data not shown).

Furthermore, the effect of HFD and aging on spatial memory deficits were assessed in mice using the water-maze task. Mice tended to spatial memory deficits as determined by a reduced number of entries into the target quadrant during the probe test (8.2 ± 0.6 in mice with HFD vs. 18.4 ± 0.5 in NCD controls) (Fig. 5.7). Treatment of mice with cholesterol uptake inhibitor 2-hydroxypropyl-β cyclodextrin (2-HβC were injected subcutaneously with 300 mg/kg), a modified cyclodextrin, which is well known to change the physicochemical properties of lipophilic compounds, and recombinant transcription growth factor beta (TGF-β, 10 μg/kg b.wt) diminished the HFD-induced cholesterol accumulation in BMVECs as well as diminished the spatial memory deficit. Similar findings were observed in aged mice (data not shown). These agents have been used to treat behavioral abnormalities in other conditions (Davidson et al. 2013).

Motor coordination was also measured by the endurance time using the rotarod apparatus. Endurance time was markedly lower in mice with HFD (29.8 ± 7 s) than that in the NCD controls (60 s) (Fig. 5.8). Similarly, endurance time was markedly lower in aged mice (Fig. 5.9). Additionally, the cerebral cortex from HFD and aged mice was also investigated. Brain amyloid peptides Aβ40 and Aβ42 were measured

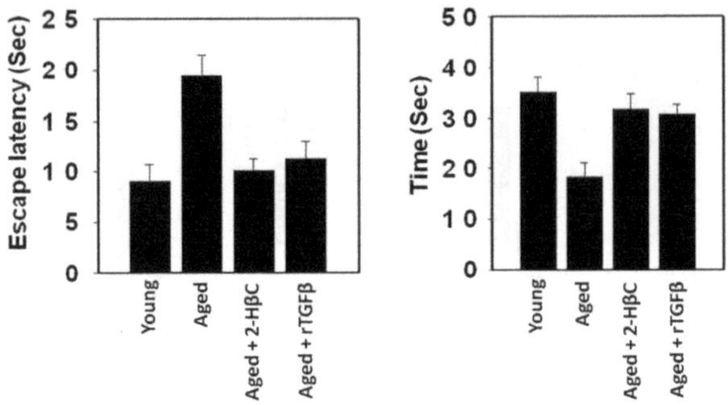

Fig. 5.7 Spatial memory impairments in aged mice (**a**) Water-maze task. Escape latency tests were performed for 4 days (i.e., 31–35 days). (**b**) Time spent in the target quadrant during the probe trial is indicative of a deficit in spatial memory. *2-HβC* 2-hydroxypropyl-β cyclodextrin, *rTGFβ-1* Human recombinant TGFβ-1 ($n = 5$)

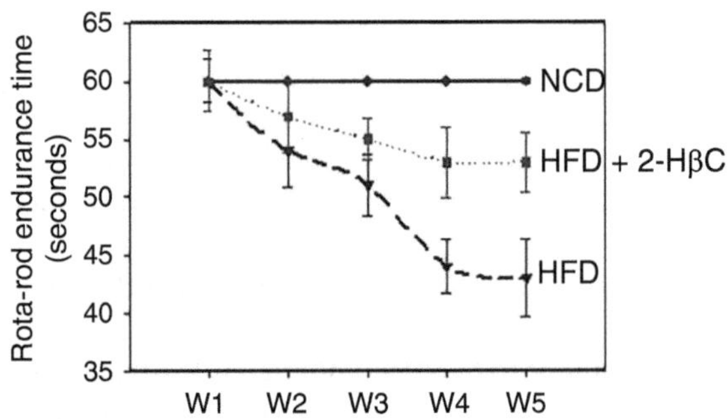

Fig. 5.8 Motor function in HFD-fed mice. Rota-rod endurance time in HFD-fed mice showed a marked reduction in endurance time as compared to NCD mice. HFD mice that are treated with, 2-hydroxypropyl-β cyclodextrin (2-HβC) attenuated the reduction in endurance time. *NCD* Normal chow diet, *HFD* High-fat diet ($n = 5$)

Fig. 5.9 Motor function in aged mice. A reduction in endurance time was observed when animals were in a gradual aging process as compared to young animals. Such changes were ameliorated by mice that are treated with 2-hydroxypropyl-β cyclodextrin (2-HβC) ($n = 5$)

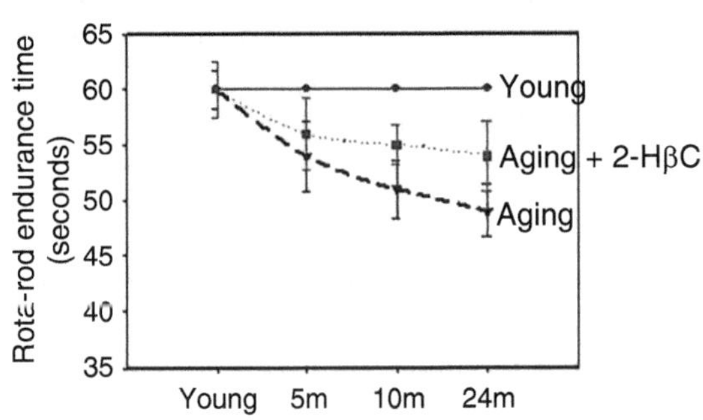

by specified ELISA on protein fractions differentially extracted from the brain of HFD and aged mice to represent soluble and plaque-associated amyloid peptides. Compared with NCD mice, the levels of soluble Aβ40 and Aβ42 were increased in

HFD mice. Similarly, the levels of soluble Aβ40 and Aβ42 were increased in older mice as compared to young mice (data not shown).

Defective neuronal integrity (i.e., loss of neuronal proteins, in particular, synaptophysin, synaptotagmin, PSD95, and NMDAr) is the characteristic feature of AD dementia (Guasch-Ferre et al. 2015), and such changes have been strongly implicated in the neurobehavioral and cognitive abnormalities in AD dementia (Hanson et al. 2013). Accordingly, we examined whether loss or reduction in key neuronal proteins that are implicated in AD dementia also occur post-HFD and aging and whether those changes are mediated by cholesterol accumulation as well as by changes in insulin resistance and glucose transporter or cholesterol metabolism impairments. Accordingly, changes in levels of these key neuronal proteins are evaluated post-HFD and after aging. In preliminary studies, we identified a reduction in the presynaptic protein synaptophysin, postsynaptic density protein (PSD95), NMDA-nr-1, and synaptotagmin by 56.9, 51.5, 39.6, and 34.8%, respectively ($n = 5$). Similar findings were observed in aged mice (49.2, 31.9, 30.7, and 29.4. %, respectively).

Similar studies have also been conducted with similar results. One animal study used amyloid transgenic mice to demonstrate the association of a high-fat diet on BBB integrity. Takechi et al. (2013) fed amyloid transgenic mice either a high-saturated-FA or high-cholesterol diet for 12 weeks. After the 12 weeks, the mice fed with high-saturated-FA diet experienced 30-fold increased BBB dysfunction compared to sevenfold dysfunction in high-cholesterol-diet-fed mice. It was also noted that administering an antihyperlipidemic drug, probucol, prevented the loss of BBB integrity or cerebrovascular inflammation.

Chan et al. (2012) observed significantly decreased levels of sphingomyelin and cholesterol esters in AD patients and transgenic familial AD mice. Barbero-Camps et al. (2013) studied the relationship between alterations in cholesterol homeostasis and AD and, using a mice model, demonstrated a cholesterol-mediated pathway via depletion of mitochondrial glutathione as a key event in accelerating the onset of neuropathological landmarks in AD. A separate study by Marwarha et al. (2019) analyzed the underlying cellular and molecular mechanisms of saturated FA palmitate in developing cognitive dysfunction and AD pathology. Triple transgenic AD mice and matched controls were fed a palmitate-enriched diet for 3 months. Compared to their matched controls, palmitate-fed mice had significant oxidative damage to lipids, proteins, and nucleic acids as well as increased activation of BACE1 activity and, subsequently, Aβ production.

Another study demonstrated a link between hypercholesterolemia and cognitive dysfunction via neuroinflammation and APP processing. A group of LDL receptor-deficient mice and normal controls were fed with a high-fat/cholesterol diet and subsequently tested for memory exercises, and the LDL receptor-deficient mice demonstrated impaired working memory as compared to healthy controls (Winocur and Greenwood 2005). Furthermore, immunohistochemical analysis revealed greater expression of cytokines and mediators, such as tumor necrosis factor-alpha, IL-1beta and -6, nitric oxide synthase 2, and cyclooxygenase 2, reflecting a neuroinflammatory response (Thirumangalakudi et al. 2008). Similar results were observed with rats in a study by Ullrich et al. (2010) where rats were fed with a cholesterol-enriched diet for 5 months. Impaired learning and detriments in long-term memory were observed in the cholesterol diet rats compared to a regular diet. Also, increased levels of APP, Aβ, and tau were seen in the cortex of cholesterol-diet rats compared to controls. The effect of saturated FAs on AD was also investigated using a mouse model. In the study, transgenic APPswe/PS1dE9 mice were fed diets with varying amounts of saturated and polyunsaturated FAs and cholesterol. It was noted that Aβ was increased in mice with greater levels of saturated FAs and cholesterol (Oksman et al. 2006). The effect of a chronic high-saturated-fat diet was further investigated using a rat model. Transgenic rats fed with a high saturated fat diet for 6 months were severely impaired in a range of learning and memory tasks (Martino Adami et al. 2017). Additionally, guinea pigs contain Aβ with a sequence identical to the human peptide. Sharman et al. (2013) accordingly used a guinea pig model to demonstrate the upregulation of BACE1 in guinea pigs that were fed with a high cholesterol diet, resulting in increased levels of Aβ, and also noted elevated levels of PSV2 (an isoform of human PSEN2) with a high-cholesterol diet.

Studies of rabbits fed with high-cholesterol diets demonstrated enlarged endolysosomes containing an accumulation of cholesterol, synaptophysin, Aβ, and phosphorylated tau in olfactory bulbs suggesting the major role of cholesterol in the pathogenesis of AD (Chen et al. 2010). A separate rabbit study compared rabbits fed with high-cholesterol diet for 4, 6, and 8 weeks to those fed with a control diet and found increasingly elevated levels of intracellular immunolabeled Aβ, whereas control diet mice had no accumulation (Sparks et al. 1994). Additionally, Ghribi et al. (2006) studied white rabbits fed with high-cholesterol diet and noted that dietary cholesterol supplementation resulted in cerebral amyloidosis in wild-type rabbits.

A canine model was also used to demonstrate the effect of saturated FA on cognition. Aged canines fed with a diet high in saturated FAs and low in monounsaturated FAs had greater errors in learning and worsened cognitive performance (Snigdha et al. 2012).

5.6 Consequences of Fat Deposition in the Brain in Experimental Models of AD

Experimental models have demonstrated the effect of HFD progression on AD. Theriault et al. (2016) fed Western Diet (WD) to young and old APPswe/PS1 mice for 4 months. Subsequently, cognitive assessments were performed using a water maze and a new object recognition paradigm for novelty. Significant impairments were noted in both young and old mice in the water-maze behavioral analysis, but only significant impairment was noted in young mice with WD for the new object recognition paradigm; however, overall demonstrated an acceleration in age-induced cognitive decline with WD. Similar results were demonstrated by

Weiss et al. (2022) whereby aging rabbits were fed with a diet enriched with cholesterol or fructose and had statistically significant impairments in tests of object location memory.

The effect of HFD on neurobehavior was also demonstrated using APP/PS1 transgenic mice that were fed with HFD and had poorer trends in memory performance, impaired social interactions, and a synergistic impairment of sensory-motor function (Bracko et al. 2020).

5.7 Current Therapeutic Recommendations Based on Experimental Studies

As previously noted, FDA-approved treatments for AD are solely for symptom management, with no ability to cure or halt disease progression. Some trials involving small molecules or immunotherapy did not demonstrate a significant difference between the drug and placebo and the possible risk of toxicity (Cummings et al. 2014). There is a dire need for additional treatment options. One obstacle to successful treatment is the BBB, which acts as a physiological and biochemical barrier. A proposed noninvasive mechanism to penetrate the BBB is via a drug delivery system (DDS). Invasive attempts at drug delivery via direct administration into the brain with high osmolar solutions, intracerebral injection, or via catheter pose risks of infection and damage to brain tissue. A noninvasive approach previously attempted is via intranasal route; however, this requires skilled administration, may damage the respiratory system, and also requires precise tailoring of pH, solubility, and salinity of the drug. For a noninvasive route using an intravenous approach, focused ultrasound is used to create sonication at the BBB for temporary permeability, allowing for intermittent drug delivery; however, more data is needed (Lipsman et al. 2018). DDSs have been created to overcome problems of adjusting the physical and chemical properties of drugs to pass through the BBB and also the problem of ligand-conjugated drugs and their dissociation rate.

Nano DDSs consist of biodegradable materials such as natural or synthetic polymers, lipids, and inorganic materials. The desired drug can be attached to the surface or contained inside the DDS. Existing DDSs include polymeric nanoparticles (NPs), liposomes, metallic NPs, and cyclodextrins. For instance, cyclodextrins are cyclic oligosaccharides that have a cage-like supramolecular structure and can carry lipophilic molecules. Preliminary studies conducted in our lab using cyclodextrin with cholesterol uptake inhibitor 2-HβC and recombinant TGF-β diminished the HFD-induced cholesterol accumulation in BMVECs as well as diminished the spatial memory deficit. Further work is still needed for DDS given the contradictory results of in vitro studies (Markoutsa et al. 2014).

Furthermore, other investigations have assessed the role of chaperone proteins in AD. Chaperones are proteins that are involved in protein folding and improve the protein quality control system within cells. Heat shock proteins (HSPs) are a type of molecular chaperone that can block the accumulation of misfolded proteins, Aβ and tau, as well as increase breakdown (Campanella et al. 2018; Wilhelmus et al. 2007).

5.8 Conclusion

There are various causes of Alzheimer's disease including aging, genetics, idiopathy, and recently hypercholesterolemia linked with the pathogenesis. Various studies have been conducted in the last decades concerning the findings of cholesterol involvement in particular cells, neurons, and neurotransmitter pathophysiology. Lipid-lowering agents as well as changing the lifestyle in terms of a low-fat-content diet promise a significant result in vivo and delay the accumulation of beta-amyloid in the neurons.

Acknowledgments This work was supported by a Merit Review from the Department of Veterans Affairs and by a Stanley J. Glaser grant (ARJ).

Conflict of Interest The authors declare no competing financial interests.

References

Albrahim T (2020) The potential role of nutritional components in improving brain function among patients with Alzheimer's disease: a meta-analysis of RCT studies. Neurosciences (Riyadh) 25(1):4–17

Alzheimer's Association (2018) 2018 Alzheimer's disease facts, and figures. Alzheimers Dement 14(3):367–429

Aylon Y, Oren M (2016) The hippo pathway, p53 and cholesterol. Cell Cycle 15:2248–2255

Barbero-Camps E, Fernández A, Martínez L, Fernández-Checa JC, Colell A (2013) APP/PS1 mice overexpressing SREBP-2 exhibit combined Aβ accumulation and tau pathology underlying Alzheimer's disease. Hum Mol Genet 22(17):3460–3476

Barnard ND, Bunner AE, Agarwal U (2014) Saturated and trans fats and dementia: a systematic review. Neurobiol Aging 35(Suppl 2):S65–S73

Bayer-Carter J, Green P, Montine T, VanFossen B, Baker L, Watson S (2011) Diet intervention and cerebrospinal fluid biomarkers in amnestic mild cognitive impairment. Arch Neurol 68:743–752

Beel AJ, Sakakura M, Barrett PJ, Sanders CR (2010) Direct binding of cholesterol to the amyloid precursor protein: an important interaction in lipid-Alzheimer's disease relationships? Biochim Biophys Acta 1801:975–982

Bekris LM, Yu CE, Bird TD, Tsuang DW (2010) Genetics of Alzheimer disease. J Geriatr Psychiatry Neurol 23(4):213–227

Benarroch EE (2014) Brain glucose transporters: implications for neurologic disease. Neurology 82:1374–1379

Berti V, Walters M, Sterling J, Quinn CG, Logue M, Andrews R (2018) Mediterranean diet and 3-year Alzheimer brain biomarker changes in middle-aged adults. Neurology 90:E1789–E1798

Boland JD, Tweto J (1980) Cellular fractionation and isolation of the plasma membrane of Burkitt's lymphoma cells. Biochim Biophys Acta 600(3):713–729

Boniface DR, Tefft ME (2002) Dietary fats and 16-year coronary heart disease mortality in a cohort of men and women in Great Britain. Eur J Clin Nutr 56:786–792

Bosco D, Fava A, Plastino M, Montalcini T, Pujia A (2011) Possible implications of insulin resistance and glucose metabolism in Alzheimer's disease pathogenesis. J Cell Mol Med 15: 1807–1821

Bracko O, Vinarcsik LK, Cruz Hernández JC, Ruiz-Uribe NE, Haft-Javaherian M, Falkenhain K, Ramanauskaite EM, Ali M, Mohapatra A, Swallow MA, Njiru BN, Muse V, Michelucci PE, Nishimura N, Schaffer CB (2020) High fat diet worsens Alzheimer's disease-related behavioral abnormalities and neuropathology in APP/PS1 mice, but not by synergistically decreasing cerebral blood flow. Sci Rep 10(1):9884

Campanella C, Pace A, Caruso Bavisotto C, Marzullo P, Marino Gammazza A, Buscemi S, Palumbo Piccionello A (2018) Heat shock proteins in Alzheimer's disease: role and targeting. Int J Mol Sci 19:2603

Chan RB, Oliveria TG, Cortes EP et al (2012) Comparative lipidomic analysis of mouse and human brain with Alzheimer disease. J Biol Chem 287:2678–2688

Chen X, Wagener JF, Morgan DH, Hui L, Ghribi O, Geiger JD (2010) Endolysosome mechanisms associated with Alzheimer's disease-like pathology in rabbits ingesting cholesterol-enriched diet. J Alzheimers Dis 22(4):1289–1303

Chiu SL, Chen CM, Cline HT (2008) Insulin receptor signaling regulates synapse number, dendritic plasticity, and circuit function in vivo. Neuron 58:708–719

Chuang YF, An Y, Bilgel M, Wong DF, Troncoso JC, O'brien RJ et al (2016) Midlife adiposity predicts earlier onset of Alzheimer's dementia, neuropathology and presymptomatic cerebral amyloid accumulation. Mol Psychiatry 21:910–915

Clendening JW, Penn LZ (2012) Targeting tumor cell metabolism with statins. Oncogene 31:4967–4978

Contestabile A (2011) The history of the cholinergic hypothesis. Behav Brain Res 221(2):334–340

Cummings JL, Morstorf T, Zhong K (2014) Alzheimer's disease drug-development pipeline: few candidates, frequent failures. Alzheimers Res Ther 6:37

Dahl A, Hassing LB, Fransson E, Berg S, Gatz M, Reynolds CA, Pedersen NL (2010) Being overweight in midlife is associated with lower cognitive ability and steeper cognitive decline in late life. J Gerontol A Biol Sci Med Sci 65(1):57–62

Davidson TL et al (2013) Inter-relationships among diet, obesity and hippocampal-dependent cognitive function. Neuroscience 253:110–122

Di Marco LY, Venneri A, Farkas E, Evans PC, Marzo A, Frangi AF (2015) Vascular dysfunction in the pathogenesis of Alzheimer's disease—a review of endothelium-mediated mechanisms and ensuing vicious circles. Neurobiol Dis 82:593–606

Duncan MS, Vasan RS, Xanthakis V (2019) Trajectories of blood lipid concentrations over the adult life course and risk of cardiovascular disease and all-cause mortality: observations from the Framingham study over 35 years. J Am Heart Assoc 8:01143

Dunn FL (2010) Management of dyslipidemia in people with type 2 diabetes mellitus. Rev Endocr Metab Disord 11:41–51

Fieldhouse JLP, Doorduijn AS, de Leeuw FA, Verhaar BJH, Koene T, Wesselman LMP, Schueren MV, Visser M, Rest OV, Scheltens P, Kester MI, Flier WMV (2020) A suboptimal diet is associated with poorer cognition: the NUDAD project. Nutrients 12(3):703

Ghribi O, Golovko MY, Larsen B, Schrag M, Murphy EJ (2006) Deposition of iron and beta-amyloid plaques is associated with cortical cellular damage in rabbits fed with long-term cholesterol-enriched diets. J Neurochem 99:438–449

Gottesman RF, Albert M, Alonso A, Coker LH, Coresh J, Davis SM et al (2017) Associations between midlife vascular risk factors and 25-year incident dementia in the atherosclerosis risk in communities (ARIC) cohort. JAMA Neurol 74:1246–1254

Grimm MO, Grimm HS, Tomic I, Beyreuther K, Hartmann T, Bergmann C (2008) Independent inhibition of Alzheimer disease beta- and gamma-secretase cleavage by lowered cholesterol levels. J Biol Chem 283:11302–11311

Grimm MO, Rothhaar TL, Grösgen S, Burg VK, Hundsdörfer B, Haupenthal VJ, Friess P, Kins S, Grimm HS, Hartmann T (2012) Trans fatty acids enhance amyloidogenic processing of the Alzheimer amyloid precursor protein (APP). J Nutr Biochem 23(10):1214–1223

Grossberg GT, Manes F, Allegri RF et al (2013) The safety, tolerability, and efficacy of once-daily memantine (28 mg): a multinational, randomized, double-blind, placebo-controlled trial in patients with moderate-to-severe Alzheimer's disease taking cholinesterase inhibitors. CNS Drugs 27(6):469–478

Guasch-Ferre M, Babio N, Martinez-Gonzalez MA, Corella D, Ros E, Martin-Pelaez S, Estruch R, Aros F, Gomez-Gracia E, Fiol M et al (2015) Dietary fat intake and risk of cardiovascular disease and all-cause mortality in a population at high risk of cardiovascular disease. Am J Clin Nutr 102:1563–1573

Hanson A, Bayer-Carter J, Green P, Montine T, Wilkinson C, Baker L (2013) Effect of apolipoprotein E genotype and diet on apolipoprotein E lip- idation and amyloid peptides: randomized clinical trial. JAMA Neurol 70:972–980

Hardy J (2006) Alzheimer’s disease: the amyloid cascade hypothesis: An update and reappraisal. J Alzheimers Dis 9:151–153

Helzner EP, Luchsinger JA, Scarmeas N et al (2009) Contribution of vascular risk factors to the progression in Alzheimer disease. Arch Neurol 66:343–348

Hill E, Goodwill A, Gorelik A, Szoeke C (2019) Diet and biomarkers of Alzheimer’s disease: a systematic review and meta-analysis. Neurobiol Aging 76:45

Howard R, McShane R, Lindesay J et al (2012) Donepezil and memantine for moderate-to-severe Alzheimer's disease. N Engl J Med 366(10):893–903

Hu FB, Manson JE, Stampfer MJ, Colditz G, Liu S, Solomon CG (2001) Diet, lifestyle, and the risk of type 2 diabetes mellitus in women. N Engl J Med 345(11):790–797

Hui L, Chen X, Geiger JD (2012) Endolysosome involvement in LDL cholesterol-induced Alzheimer's disease-like pathology in primary cultured neurons. Life Sci 91(23–24):1159–1168

Hurd MD, Martorell P, Delavande A, Mullen KJ, Langa KM (2013) Monetary costs of dementia in the United States. N Engl J Med 368:1326–1334

Huttunen HJ, Havas D, Peach C et al (2010) The acyl-coenzyme A: cholesterol acyltransferase inhibitor CI-1011 reverses diffuse brain amyloid pathology in aged amyloid precursor protein transgenic mice. J Neuropathol Exp Neurol 69:777–788

Iqbal K, Alonso Adel C, Chen S, Chohan MO, El-Akkad E, Gong CX, Khatoon S, Li B, Liu F, Rahman A, Tanimukai H, Grundke-Iqbal I (2005) Tau pathology in Alzheimer disease and other tauopathies. Biochim Biophys Acta 1739:198–210

Islam MA, Amin MN, Siddiqui SA, Hossain MP, Sultana F, Kabir MR (2019) Trans fatty acids and lipid profile: a serious risk factor to cardiovascular disease, cancer and diabetes. Diabetes Metab Syndr 13(2):1643–1647

Jayakumar AR, Bethea JR, Tong XY, Gomez J, Norenberg MD (2011) NF-κB in the mechanism of brain edema in acute liver failure: studies in transgenic mice. Neurobiol Dis 41(2):498–507

Jayakumar AR, Panickar KS, Murthy CR, Norenberg MD (2006) Oxidative stress and mitogen-activated protein kinase phosphorylation mediate ammonia-induced cell swelling and glutamate uptake inhibition in cultured astrocytes. J Neurosci 26(18):4774–4784

Jayakumar AR, Tong XY, Curtis KM, Ruiz-Cordero R, Abreu MT, Norenberg MD (2014) Increased toll-like receptor 4 in cerebral endothelial cells contributes to the astrocyte swelling and brain edema in acute hepatic encephalopathy. J Neurochem 128(6):890–903

Jayakumar AR, Tong XY, Shamaladevi N, Barcelona S, Gaidosh G, Agarwal A, Norenberg MD (2017) Defective synthesis and release of astrocytic thrombospondin-1 mediates the neuronal TDP-43 proteinopathy, resulting in defects in neuronal integrity associated with chronic traumatic encephalopathy: in vitro studies. J Neurochem 140(4):645–661

Jeong W, Lee H, Cho S, Seo J (2019) ApoE4-induced cholesterol dysregulation and its brain cell type-specific implications in the pathogenesis of Alzheimer's disease. Mol Cell 42:739–746

Jiang X, Guo M, Su J et al (2012) Simvastatin blocks blood–brain barrier disruptions induced by elevated cholesterol both in vivo and in vitro. Int J Alzheimers Dis 2012:109324

Joost HG, Thorens B (2001) The extended GLUT-family of sugar/polyol transport facilitators: nomenclature, sequence characteristics, and potential function of its novel members (review). Mol Membr Biol 18:247–256

Karran E, De Strooper B (2016) The amyloid cascade hypothesis: are we poised for success or failure? J Neurochem 139(Suppl 2):237–252

Keller JN, Hanni KB, Markesbery WR (1999) Oxidized low-density lipoprotein induces neuronal death: implications for calcium, reactive oxygen species, and caspases. J Neurochem 72(6): 2601–2609

Kim Y, Je Y, Giovannucci EL (2021) Association between dietary fat intake and mortality from all-causes, cardiovascular disease, and cancer: a systematic review and meta-analysis of prospective cohort studies. Clin Nutr 40(3):1060–1070

Kukull WA, Bowen JD (2002) Dementia epidemiology. Med Clin North Am 86:573–590

Kuzu OF, Noory MA, Robertson GP (2016) The role of cholesterol in cancer. Cancer Res 76:2063–2070

Laitinen MH, Ngandu T, Rovio S, Helkala EL, Uusitalo U, Viitanen M, Nissinen A, Tuomilehto J, Soininen H, Kivipelto M (2006) Fat intake at midlife and risk of dementia and Alzheimer's disease: a population-based study. Dement Geriatr Cogn Disord 22(1):99–107

Lesser GT (2012) Association of Alzheimer Disease Pathology with abnormal lipid metabolism: the Hisayama study. Neurology 78:1280

Linscheer WG, Vergroesen AJ (1994) Lipids. In: Shils M, Olson JA, Shike M (eds) Modern nutrition in health and disease. Lea & Febiger, Philadelphia, pp 47–88

Lipsman N, Meng Y, Bethune AJ, Huang Y, Lam B, Masellis M, Herrmann N, Heyn C, Aubert I, Boutet A et al (2018) Blood-brain barrier opeining in Alzheimer's disease using MR-guided focused ultrasound. Nat Commun 9:2336

Liu Y, Zhong X, Shen J, Jiao L, Tong J, Zhao W, Du K, Gong S, Liu M, Wei M (2020) Elevated serum TC and LDL-C levels in Alzheimer's disease and mild cognitive impairment: a meta-analysis study. Brain Res 1727:146554

Mann JI, Appleby PN, Key TJ, Thorogood M (1997) Dietary determinants of ischemic heart disease in health conscious individuals. Heart 78:450–455

Markoutsa E, Papadia K, Giannou AD (2014) Mono and dually decorated nanoliposomes for brain targeting, in vitro and in vivo studies. Pharm Res 31:1275–1289

Martino Adami PV, Galeano P, Wallinger ML, Quijano C, Rabossi A, Pagano ES, Olivar N, Reyes Toso C, Cardinali D, Brusco LI, Do Carmo S, Radi R, Gevorkian G, Castaño EM, Cuello AC, Morelli L (2017) Worsening of memory deficit induced by energy-dense diet in a rat model of early-Alzheimer's disease is associated to neurotoxic Aβ species and independent of neuroinflammation. Biochim Biophys Acta Mol basis Dis 1863(3):731–743

Martins IJ, Hone E, Foster JK et al (2006) Apolipoprotein E, cholesterol metabolism, diabetes, and the convergence of risk factors for Alzheimer's disease and cardiovascular disease. Mol Psychiatry 11:721–736

Marwarha G, Claycombe-Larson K, Lund J, Schommer J, Ghribi O (2019) A diet enriched in palmitate and deficient in linoleate exacerbates oxidative stress and amyloid-β burden in the hippocampus of 3xTg-AD mouse model of Alzheimer's disease. J Alzheimers Dis 68(1): 219–237

Melo HM, Seixas da Silva GDS, Sant'Ana MR, Teixeira CVL, Clarke JR, Miya Coreixas VS, de Melo BC, Fortuna JTS, Forny-Germano L, Ledo JH, Oliveira MS, Figueiredo CP, Pardossi-Piquard R, Checler F, Delgado-García JM, Gruart A, Velloso LA, Balthazar MLF, Cintra DE, Ferreira ST, De Felice FG (2020) Palmitate is increased in the cerebrospinal fluid of humans with obesity and induces memory impairment in mice via pro-inflammatory TNF-α. Cell Rep 30(7):2180–2194.e8

Micha R, Mozaffarian D (2010) Saturated fat and cardiometabolic risk factors, coronary heart disease, stroke, and diabetes: a fresh look at the evidence. Lipids 45(10):893–905

Morris MC (2009) The role of nutrition in Alzheimer's disease: epidemio- logical evidence. Eur J Neurol 16(Suppl 1):1–7

Morris MC, Evans DA, Bienias JL, Tangney CC, Bennett DA, Aggarwal N (2003) Dietary fats and the risk of incident Alzheimer disease. Arch Neurol 60:194–200
Mossello E, Ballini E (2012) Management of patients with Alzheimer's disease: pharmacological treatment and quality of life. Ther Adv Chronic Dis 3(4):183–193
Mozaffarian D, de Oliveira Otto MC, Lemaitre RN, Fretts AM, Hotamisligil G, Tsai MY, Siscovick DS, Nettleton JA (2013) Trans-Palmitoleic acid, other dairy fat biomarkers, and incident diabetes: the multi-ethnic study of atherosclerosis (MESA). Am J Clin Nutr 97(4):854–861
Mufson EJ, Counts SE, Perez SE, Ginsberg SD (2008) Cholinergic system during the progression of Alzheimer's disease: therapeutic implications. Expert Rev Neurother 8:1703–1718
Munoz A, Costa M (2013) Nutritionally mediated oxidative stress and inflammation. Oxidative Med Cell Longev 2013:1
Oh K, Hu FB, Manson JE, Stampfer MJ, Willet WC (2005) Dietary fat intake and risk of coronary heart disease in women: 20 years of follow-up of the Nurses' health study. Am J Epidemiol 161: 672–679
Oksman M, Iivonen H, Hogyes E, Amtul Z, Penke B, Leenders I, Broersen L, Lütjohann D, Hartmann T, Tanila H (2006) Impact of different saturated fatty acid, polyunsaturated fatty acid and cholesterol containing diets on beta-amyloid accumulation in APP/PS1 transgenic mice. Neurobiol Dis 23(3):563–572
Overk CR, Masliah E (2014) Pathogenesis of synaptic degeneration in Alzheimer's disease and Lewy body disease. Biochem Pharmacol 88:508–516
Popp J, Meichsner S, Kolsch H et al (2013) Cerebral and extracerebral cholesterol metabolism and CSF markers of Alzheimer's disease. Biochem Pharmacol 86:37–42
Prince M, Albanese E, Guerchet M, et al (2014) World alzheimer report: dementia and risk reduction an analysis of protective and modifiable factors
Puglielli L, Tanzi RE, Kovacs DM (2003) Alzheimer's disease: the cholesterol connection. Nat Neurosci 6:345–351
Ruan Y, Tang J, Guo X, Li K, Li D (2018) Dietary fat intake and risk of Alzheimer's disease and dementia: a meta-analysis of cohort studies. Curr Alzheimer Res 15(9):869–876
Ruparelia N, Chai JT, Fisher EA, Choudhury RP (2017) Inflammatory processes in cardiovascular disease: a route to targeted therapies. Nat Rev Cardiol 14:133–144
Samuelsson J, Kern S, Zetterberg H, Blennow K, Rothenberg E, Wallengren O, Skoog I, Zettergren A (2021) A Western-style dietary pattern is associated with cerebrospinal fluid biomarker levels for preclinical Alzheimer's disease-a population-based cross-sectional study among 70-year-olds. Alzheimers Dement (N Y) 7(1):e12183
Selkoe DJ, Hardy J (2016) The amyloid hypothesis of Alzheimer's disease at 25 years. EMBO Mol Med 8:595–608
Sharman MJ, Moussavi Nik SH, Chen MM, Ong D, Wijaya L, Laws SM, Taddei K, Newman M, Lardelli M, Martins RN, Verdile G (2013) The Guinea pig as a model for sporadic Alzheimer's disease (AD): the impact of cholesterol intake on expression of AD-related genes. PLoS One 8(6):e66235
Šimić G, Leko MB, Wray S et al (2016) Tau protein hyperphosphorylation and aggregation in alzheimer's disease and other tauopathies, and possible neuroprotective strategies. Biomol Ther 6(6):1–28
Simpson IA, Carruthers A, Vannucci SJ (2007) Supply and demand in cerebral energy metabolism: the role of nutrient transporters. J Cereb Blood Flow Metab 27:1766–1791
Snigdha S, Astarita G, Piomelli D, Cotman CW (2012) Effects of diet and behavioral enrichment on free fatty acids in the aged canine brain. Neuroscience 202:326–333
Solomon A, Kivipelto M, Wolozin B et al (2009) Midlife serum cholesterol and increased risk of Alzheimer's and vascular dementia three decades later. Dement Geriatr Cogn Disord 28:75–80
Sparks DL, Scheff SW, Hunsaker JC 3rd, Liu H, Landers T, Gross DR (1994) Induction of Alzheimer-like beta-amyloid immunoreactivity in the brains of rabbits with dietary cholesterol. Exp Neurol 126(1):88–94

Takechi R, Galloway S, Pallebage-Gamarallage MM, Lam V, Dhaliwal SS, Mamo JC (2013) Probucol prevents blood-brain barrier dysfunction in wild-type mice induced by saturated fat or cholesterol feeding. Clin Exp Pharmacol Physiol 40:45–52

Thériault P, ElAli A, Rivest S (2016) High fat diet exacerbates Alzheimer's disease-related pathology in APPswe/PS1 mice. Oncotarget 7(42):67808–67827

Thirumangalakudi L, Prakasam A, Zhang R, Bimonte-Nelson H, Sambamurti K, Kindy MS, Bhat NR (2008) High cholesterol-induced neuroinflammation and amyloid precursor protein processing correlate with loss of working memory in mice. J Neurochem 106(1):475–485

Tsao CW, Preis SR, Peloso GM et al (2012) Relations of long-term and contemporary lipid levels and lipid genetic risk scores with coronary artery calcium in the Framingham heart study. J Am Coll Cardiol 60:2364–2371

Ullrich C, Pirchl M, Humpel C (2010) Hypercholesterolemia in rats impairs the cholinergic system and leads to memory deficits. Mol Cell Neurosci 45(4):408–417

Weiss C, Bertolino N, Procissi D et al (2022) Diet-induced Alzheimer's-like syndrome in the rabbit. Alzheimers Dement (NY) 8(1):e12241

Wilhelmus MM, de Waal RM, Verbeek MM (2007) Heat shock proteins and amateur chaperones in amyloid-Beta accumulation and clearance in Alzheimer's disease. Mol Neurobiol 35:203–216

Winocur G, Greenwood CE (2005) Studies of the effects of high fat diets on cognitive function in a rat model. Neurobiol Aging 26(Suppl 1):46–49

Wu Y, Wang Z, Jia X, Zhang H, Zhang H, Li J, Zhang K (2019) Prediction of Alzheimer's disease with serum lipid levels in Asian individuals: a meta-analysis. Biomarkers 24:341–351

Xu W, Tan L, Wang HF, Jiang T, Tan MS, Tan L, Zhao QF, Li JQ, Wang J, Yu JT (2015) Meta-analysis of modifiable risk factors for Alzheimer's disease. J Neurol Neurosurg Psychiatry 86: 1299–1306

Yan R, Vassar R (2014) Targeting the beta secretase BACE1 for Alzheimer's disease therapy. Lancet Neurol 13:319–329

Zekanowski C, Styczyńska M, Pepłońska B et al (2003) Mutations in presenilin 1, presenilin 2 and amyloid precursor protein genes in patients with early-onset Alzheimer's disease in Poland. Exp Neurol 184(2):991–996

Zhuang P, Zhang Y, He W, Chen X, Chen J, He L, Mao L, Wu F, Jiao J (2019) Dietary fats in relation to Total and cause-specific mortality in a prospective cohort of 521 120 individuals with 16 years of follow-up. Circ Res 124:757–768

Chapter 6
Neuroprotective Effects of Garlic Against Alzheimer's Disease

Richard Jasmine Juliet, Ponnambalam Annapoorani, Ramarajan Kiruthika, Paul Joseph Arokia Mary, and Thamilarasan Manivasagam

Abstract Medicinal plants and their products are achieving more popularity in preventing several diseases recently. In previous days, garlic is used in our daily foodstuffs for flavor enhancers and also medicinal purposes. Various experiments indicated the several pharmacological properties of garlic such as antioxidant, renoprotective, anti-atherosclerotic, anticarcinogenic, antidiabetic, antimicrobial, and antihypertensive activities. It is reported to be a big source of sulfur-containing compounds like ajoenes, alliin, vinyldithiins, allicin, and flavonoids including quercetin. Several garlic extracts and isolated compounds, such as allyl mercaptan, S-allylcysteine, diallyl disulfide, diallyl trisulfide, S-allylmercaptocysteine, etc., of *garlic* have been evaluated for various biological activities including neuroprotective actions. In this chapter, the anti-alzheimeric role of several garlic extracts and components were discussed elaborately.

Keywords Alzheimer's disease · Garlic · Components · Formulations · Therapy

6.1 Introduction

Nowadays, about 80% of the people globally depend on alternative, complementary, and traditional medicines as an important resource for maintaining their primary health (Batiha et al. 2020). Garlic (*Allium sativum* L.) is a commonly used spice throughout the world. It is utilized for both cooking purposes as condiments and are enhancer of the flavor of foods. It has obtained a status of alarming prophylactic and

---

R. Jasmine Juliet · P. Annapoorani (✉)
Department of Biochemistry, V.V. Vanniaperumal College for Women, Virudhunagar, Tamil Nadu, India
e-mail: annapoorani@vvvcollege.org

R. Kiruthika · P. J. Arokia Mary · T. Manivasagam
Department of Biochemistry and Biotechnology, Annamalai University, Chidambaram, Tamil Nadu, India

A. Justin Thenmozhi, T. Manivasagam (eds.), *Nutraceuticals for Alzheimer's Disease: A Promising Therapeutic Approach*, Nutritional Neurosciences,
https://doi.org/10.1007/978-981-99-0677-2_6

beneficial medicinal substance in the tradition for many countries over the centuries. The therapeutic effect of garlic is documented in the Bible and also in the Egyptian Cordex Ebers belonging to 1550 BC (Moyers 1996). It's is reported to have origin from Central Asia more than 6000 years ago. Garlic-based therapy has been used in India and China since 5000 and 3000 years ago, respectively (Rivlin 2001).

Scientific classification of garlic

Kingdom	:	Plantae—Plants
Subkingdom	:	Viridiplantae—Green plants
Division	:	Tracheophyta—Vascular plants
Subdivision	:	Spermatophytina—Seed plants
Superorder	:	Lilianae—Monocots
Order	:	Asparagales
Family	:	Amaryllidaceae
Genus	:	*Allium*
Species	:	*Allium sativum* (cultivated garlic)

6.2 Composition of Garlic

The garlic consists of about 200 chemical substances (Table 6.1) with various pharmacological properties. The main and important components of the garlic are sulfur-containing compounds which are responsible for both its characteristic taste and flavor and taste and are also accountable for its pharmacological properties (Matsuura 1997).

Table 6.1 Composition of garlic

S. No	Constituents	Percentage
1.	Water	65%
2.	Carbohydrate mainly starch, glucose, sucrose, & fructose	28%
3.	Organosulfur compounds	2.3%
4.	Proteins mainly alliinase	2%
5.	Free amino acid particularly arginine, glutamic acid, aspartic acid, & leucine	1.2%
6.	Fiber	1.5%
7.	Phenolic compounds such as gallic acid, ferulic acid, isorhamnetin, quercetin, naringenin, rutin, apigenin, p-coumaric acid, and luteolin	0.05%
8.	Fatty acids like palmitic acid, oleic acid, linoleic acid, & linolenic acid	
9.	Fat-soluble vitamins (vitamins A, K, & E)	Trace amounts
10.	Water-soluble vitamins (vitamin C, B1, B2, B3, B6, & B8)	Trace amounts
11.	Minerals (Fe, Zn, P, Ca, K, Mg, & Na)	Trace amounts

6.2.1 Organosulfur Compounds

Garlic consists of mainly the volatile and nonvolatile substances. The important precursor for the nonvolatile sulfur compounds are γ-glutamyl-S-allyl-L-cysteine and its sulfoxide (alliin). Undamaged or intact garlic contains more amount of alliin, that is metabolized by the enzyme alliinase to other compound allicin, during the crashing or macerating of gloves. Allicin and other thiosulfinates found in garlic are unstable and converted to several compounds like dithiins, diallyl sulfide, diallyl disulfide, diallyl trisulfide, and ajoene. During garlic processing, three subgroups of volatile organosulfur compounds (Fig. 6.1) are formed: (a) thiosulfinates, synthesized from sulfoxides; (b) allicin (diallyl thiosulfinate), a key compound, which is not present in undisturbed or undamaged garlic but formed by enzyme and immediately destroyed to form methyl allyl disulfide, DAS, DADS, DATS, and/methyl allyl trisulfide; and (c) water-soluble organosulfur compounds, formed through the aqueous or alcoholic extract of garlic by γ-glutamyl-S-allyl-L-cysteine decomposition to S-allyl-L-cysteine and S-methyl-L-cysteine (Rodríguez et al. 2022).

6.3 Garlic Products

Garlic are also used in several forms including juices, liquid and dried extracts, volatile oils, and macerates in the pharmaceutical industry.

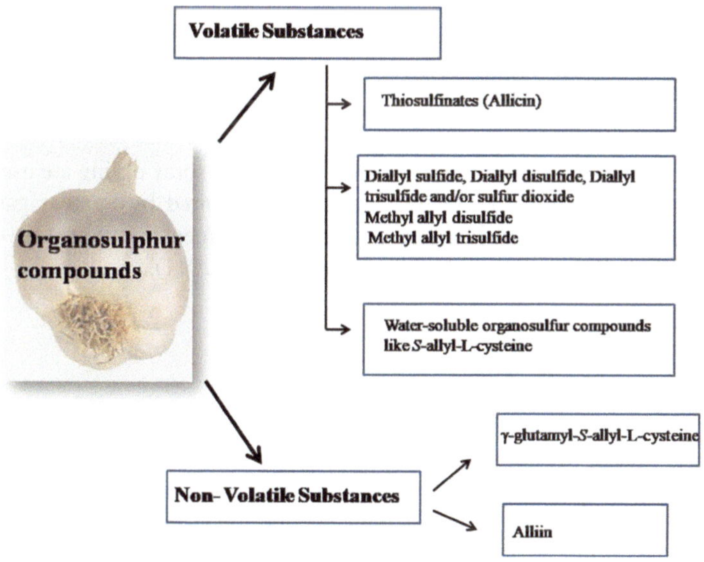

Fig. 6.1 Organosulfur compounds present in garlic

6.3.1 Dehydrated Garlic

The yellow or white varieties of garlic contain the most pungent smell and contain more amounts of solid particles. The garlic is chopped and kept in hot air tunnels. More decontamination processes are needed for the elimination of microbial load. The dehydrated garlic cloves are finally grounded, powdered, and stored.

6.3.2 Garlic Oil

Garlic oil is attained by using steam distillation of the crushed garlic materials. It contains about 0.1%–0.2% of essential oil on volume/weight basis. The flavor of about 900 mg of fresh garlic is almost equal to 200 g of dehydrated garlic powder. As it has a pungent flavor, oil cannot be directly utilized, but it was used in encapsulated form or diluted in vegetable oil.

6.3.3 Garlic Juice

Garlic is crushed into a medium-sized pulp. Then, the pulp is stored in a buffer tank for 15–60 min for completing the enzymatic actions. The juice is extracted by decantation or pressing processes. Finally, it is adjusted to obtain the desired pH and allowed to pass into the preheater for a few seconds.

6.3.4 Garlic Extracts

Various processes like evaporation, lyophilization, and spray drying are used for the extraction of garlic extracts. Garlic extracts are acquired by treating garlic with aqueous alcohol extraction. Methanol and ethanol treatment inactivated the enzymes like peroxidase isoenzymes and alliinase (Mane et al. 2011).

6.3.5 Aged Garlic Extract

Raw garlic intake can cause several adverse reactions such as growth retardation, changes in serum protein, anemia, and loss of gut microflora. Raw garlic consumption may induce indigestion and its odor may be found in the breath. Aged garlic extract (AGE) is rich in antioxidants and has less odor with less side effects. Garlic powder is extracted in a prolong period of time (~15 months) with a nontoxic solvent to form AGE. This process alters the allicin (unstable compound) to form organosulfur compounds like S-allylcysteine and di-allyl-disulfide (stable substance). It also contains trace amounts of ajoene, flavonoids, polyphenols, allixin, and thiosulfinates.

6.4 Pharmacological Properties of Garlic

Garlic and its components are reported to have several pharmacological properties such as antidiabetic, antifungal, anti-atherosclerotic, antibacterial, renoprotective (Davis 2005), antioxidant and anticarcinogenic (Rahman and Lowe 2006), and antihypertensive activities (Badal et al. 2019). Moreover, it is also used to improve digestion, treat the infections of respiratory and urinary tract, cure cardiac diseases and also have antipyretic, aphrodisiac, carminative, sedative, and diuretic effects (Souza et al. 2011).

6.4.1 Garlic Oil and AD

Yoshioka et al. (2021) studied the prohibitory role of garlic essential oil on AD-related enzymes and also estimated the distribution of oil components in the brain. Administration of various sulfur-containing compounds in oil and the oil itself significantly inhibited the enzymes like β- and γ-secretase in mouse model of AD. Six hours after the post administration of oil, sulfur elements were found in the circulation and the brain. The brain levels of allyl methyl sulfide and allyl mercaptan were increased significantly even than that of oil, and other components such as allyl methyl trisulfide, diallyl trisulfide, and dimethyl trisulfide were lowered in the brain than those of oil. This results indicated that the active sulfur compounds were transformed into allyl methyl sulfide or allyl mercaptan, which are responsible for the potent neuroprotective effect of garlic oil.

6.4.2 Other Components of Garlic and AD

Zilbeyaz et al. (2021) designed and synthesized the small library of garlic-associated asymmetrical thiosulfonates. The acetylcholinesterase and butyrylcholinesterase inhibitory role of thiosulfonates in in vitro conditions was analyzed. The inhibitory activity was shown in nanomolar concentrations (IC_{50}), and K_i values were obtained for both enzymes. Moreover, in silico experiments were conducted to find out probable binding relations between the cholinesterases and thiosulfonates. S-propargyl-cysteine, also called as ZYZ-802, is an S-allylcysteine analog, containing propargyl group instead of allyl group. Another analog of these compounds is S-propyl cysteine, containing propyl group instead of cysteine structure. Controlled release of S-propyl cysteine offered potent neuroprotective effect than S-allylcysteine on animal models of AD. The propagyl group has a strong bond with the cysteine than the other analogs. It also diminished the inflammatory cytokines, reduced the accumulation Ca^{2+}, enhanced the levels and activities of antioxidants, repressed STAT3, and upregulated the expression of p53 and Bax (Wen and Zhu 2015).

Neuroinflammation is reported to implicate in the process of amyloidogenesis. The thiacremonone, a sulfur-containing substance, showed the potent anti-inflammatory effects. Lin et al. (2012) studied the anti-inflammatory and anti-amyloidogenic properties of thiacremonone against lipopolysaccharide induced in in vitro and in vivo models of AD. Administration of thiacremonone diminished the LPS-mediated memory deficits, microglial activation, and the expression of pro-inflammatory and amyloidogenesis indices.

Manral et al. (2016) constructed analogues of DADS and studied the role of DADS analogues against Aβ-mediated toxicity in in vitro model of AD and scopolamine-induced rat model of AD. The two analogues of DADS (7 k and 7 l) significantly inhibited Aβ1–42-induced ROS generation and apoptosis (restoration of Bax/Bcl-2 ratio). Moreover, oral administration of DADS analogues attenuated scopolamine-induced neurochemical changes, alterations in acetylcholinesterase activity, memory impairment, and oxidant-antioxidant indices. The neuroprotective effect of DADS analogues (7 k and 7 l) in animal models of AD are confirmed by the histological analysis of the cerebral cortex and hippocampus.

6.4.3 Garlic Extract and AD

The anti-amyloidogenic properties of aqueous garlic extract (fresh and boiled) on Aβ aggregation and defibrillation in in vitro conditions were examined. It was performed by using thioflavin-T-based fluorescence assay, SDS-polyacrylamide gel electrophoresis, and transmission electron microscopy. The aqueous fresh garlic extract inhibited the formation of Aβ fibril and the defibrillated preformed ones in a dose- and time-dependent manner. The boiled aqueous garlic extract engaged partial anti-amyloidogenic activity, i.e., inhibition of Aβ fibril formation and but not the defibrillation of the preformed fibrils. As the proteases found in boiled garlic are denatured during boiling, it did not degrade Aβ in lab conditions. The activity of NLRP3 inflammasome is a key component of the innate immune response against tissue injury, and deregulation of NLRP3 activity is implicated in several neurological diseases including AD. Liu et al. (2021) synthesized the vesicle-like nanoparticles from *Allium* vegetables including garlic and analyzed their efficacy on NLRP3 inflammasome in primary macrophages. The garlic chive-derived nanoparticles exhibited the anti-NLRP3 inflammasome activity in in vitro conditions by inhibiting pathways enhancing NLRP3 inflammasome activation like cytokine release, activation of caspase-1, and pyroptosis. Chauhan (2003) analyzed the role of dietary garlic on the diminution of amyloid load in a transgenic mouse (K670N/M671L)-(Tg2576) model of AD. Dietary garlic administration to transgenic AD mice attenuated an Aβ burden by enhancing soluble APP-α levels and diminishing $A\beta_{1-40}$ and $A\beta_{1-42}$.

Promyo et al. (2017) studied the synergetic effect of *Artemisia scoparia* extract and garlic extract and *Artemisia scoparia* extract alone on rat model of AD. Oral administration of *Artemisia scoparia* extract reduced the levels of Aβ and phosphorylated tau proteins and expression of β-secretase and phosphorylated glycogen synthase kinase 3β. Although the combined extract showed a reduction in the Aβ accumulation and tau hyperphosphorylation, no synergistic effect was found as compared to *Artemisia scoparia* extract-alone-treated animals.

6.4.4 Aged Garlic Extract and AD

Chauhan (2006) compared the anti-amyloidogenic, anti-tangle, and anti-inflammatory role of dietary intervention of 2% aged garlic extract with its major components: S-allylcysteine and di-allyl-disulfide in Swedish double-mutant (Tg2576) mouse model of AD. The protective efficacy of dietary administration was found in the order of extract > S-allylcysteine > di-allyl-disulfide. The potent neuroprotective effect of the extract is may be due to both the cholesterol-dependent and/or independent mechanisms. Moreover, the author suggested that the greater therapeutic benefit of extract is due to its least adverse effects and pleiotropic properties instead of a single-component. The neuroprotective role of aged garlic extract against neurotoxicity induced by $A\beta_{25-35}$ in neuronal (PC12 cells) model of AD was analyzed. Exposure of extract to nonproliferating (neuronal) PC12 cells (pretreated with nerve growth factor before the experiment) enhanced the cell viability (MTS-based assay) and effectivity against Aβ toxicity in a dose-dependent manner (Griffin et al. 2000).

Joeng et al. (2013) analyzed the neuroprotective activities of aged garlic-ethyl acetate extract in both in vitro (PC-12 cells) and in vivo (mouse brain homogenate) models of AD. Exposure to ethyl acetate fractions of garlic to Aβ-intoxicated PC-12 cells showed improvement in cell viability (MTT and LDH assay). Administration of extract offered potent antioxidant function (enhanced inhibition of MDA formation and ABTS scavenging activity) in Aβ-intoxicated mouse brain. The learning and memory deficits (Y-maze and passive avoidance test) induced by Aβ infusion in ICR mice were attenuated by aged garlic extract. Li and Kim (2019) demonstrated the protective effect of ethyl acetate extract of aged black garlic on scopolamine-intervened cognitive deficits in mice models of AD. Oral administration of the extract enhanced the number of platform crossings (reduced by scopolamine), reduced the latency time (enhanced by scopolamine) in Morris water-maze test, and increased the latency time (decreased by neurotoxin) in passive avoidance test. The aged black garlic extract administration significantly attenuated the oxidative stress induced by scopolamine. It also ameliorated scopolamine-induced acetylcholine depletion by inhibiting activity of acetylcholinesterase (enzyme involved in degradation of acetylcholine) and enhancing activity of choline acetyltransferase (enzyme involved in the synthesis of acetylcholine).

In the neurodegenerative disease like AD, apoptosis leads to degeneration of neurons. The activation of caspase-3 results in apoptosis, and their enhanced levels were found during early phase of apoptosis. Further increase was found in the cells undergoing apoptosis and quickly detected in the final stage. Moreover, it induces the synthesis of Aβ, one of the main hallmarks of AD. Exposure to aged garlic extract in in vitro conditions significantly reduced the caspase-3 activity but not the caspase-8 activity. Jackson et al. (2002) indicated that the extract offered neuroprotection in attenuating apoptotic neuronal death possibly by diminishing caspase-3 activity. The neuroprotective role of aged garlic extract and S-allylcysteine on Aβ (25–35)-induced oxidative stress and apoptosis in an in vitro model of AD (rat pheochromocytoma cells) were studied. Both the extract- and S-allylcysteine-attenuated Aβ (25–35) induced ROS formation, DNA fragmentation, activation of caspase-3, breakdown of PARP and apoptosis (Peng et al. 2002). Throjak et al. (2017) investigated the neuroprotective effects of aged garlic extract on Aβ-mediated glutamatergic, cholinergic, and GABAergic systems along with cognitive dysfunction in rats. Oral administration of the extract significantly enhanced the working memory and reference memory (radial arm maze test) in Aβ-infused rats. It also attenuated the cholinergic neuronal degeneration and enhanced the vesicular glutamate transporter 1 and glutamate decarboxylase levels in the hippocampus of AD rats.

6.4.5 Allicin and AD

Allicin is a lipophilic compound that can efficiently transverse the cellular membranes and act like a reactive sulfur species within the cells. It is a very active substance with potent antioxidant property. Nadeem et al. (2021) reviewed the antioxidant, anti-inflammatory, and neuroprotective role of allicin against various neurodegenerative and psychological diseases. Allicin is capable of reducing the levels of reactive oxygen species by diminishing the activity of NADPH-oxidizing enzymes. It is also decreased the levels of various forms of reactive oxygen species synthesized by various types of peroxidases. The neuroprotective role is mostly due to the regulation of redox-dependent pathways. It exhibits the partial neuroprotective function due to the inhibition of neuroinflammation by reducingP38, TLR4/MyD88/NF-κB, and JNK signaling pathways. The potent inhibitory effect against the acetylcholinesterase and butyrylcholinesterase activity also enhances its usage for the management of Alzheimer's disease. Kumar et al. (2015) explored the anticholinesterase inhibitory effect of allicin on in vitro conditions. Allicin exposure inhibited the activity of AChE and BuChE enzymes in a dose-dependent manner. The neuroprotective effect of allicin is mostly by inhibiting cholinesterase enzymes and thereby enhances acetylcholine levels.

Allicin, an important garlic component, was shown to attenuate the learning and memory deficits in various animal models of AD. Zhang et al. (2018) studied the mechanism involved in its neuroprotection by using double transgenic APP/PS1 mice model of AD. Allicin administration attenuated the cognitive impairment in APP/PS1 transgenic mice by diminishing the oxidant-antioxidant imbalance and Aβ levels and enhancing mitochondrial function.

Endoplasmic reticulum (ER) stress particularly the dysregulation of DS RNA-dependent protein kinase)-like ER-resident kinase (PERK), are reported to play a key role in the pathology of AD. Zhu et al. (2015) investigated the protective role of allicin on ER stress-intervened cognitive impairments in rats. Endoplasmic reticulum stress is induced in rats by lateral ventricular infusion of ER stress stimulator, tunicamycin (TM). TM infusion enhanced the hyperphosphorylation of tau, deposition of $A\beta_{1-42}$, and oxidant-antioxidant imbalance in the rat hippocampus. TM infusion enhanced the expression of hippocampal PERK and nuclear factor erythroid-derived 2-like 2 (downstream substrate of PERK). Pretreatment of allicin attenuated hallmarks of AD (tau hyperphosphorylation and $A\beta_{1-42}$ deposition) and ER stress in the hippocampus.

In AD, Aβ deposition leads to neuroinflammation, a pathological process induced by host defense response. The microglial activation may induced the neurodegeneration by the liberation of several proinflammatory cytokines like interleukin-1β and tumor necrosis factor-α that may further induce neuronal damage. Nillert et al. (2017) studied the role of aged garlic extract on $A\beta_{1-42}$-mediated cognitive impairment and neuroinflammation. Oral administration of the extract significantly enhanced short-term recognition memory (novel object recognition test) in $A\beta_{1-42}$-infused AD rats. Moreover, it significantly reduced the inflammatory processes by diminishing the microglial activation (IL-1β and TNFα) in the cortex and hippocampus of AD rats.

6.4.6 AD and SAC

Keri et al. (2016) synthesized a library of natural-based hybrid therapeutic agents, by conjugating a tacrine moiety to S-allylcysteine or its analog S-propargyl-cysteine, and evaluated their neuroprotective ability by analyzing enzymes of the cholinergic system. In silico experiments were carried out for the selection of linkers that can bind with the active site of acetylcholinesterase enzyme. Then, the selected compounds were analyzed for inhibitory effect against acetylcholinesterase activity, ROS levels, and aggregation of Aβ fibrils in in vitro conditions. The compound 9 d showed potent acetylcholinesterase inhibitory activity, while the compound 9 l offered antioxidant activity by diminishing production of superoxide and neuroprotective effect by reducing Aβ-mediated toxicity. Ray et al. (2011) tested the protective role of aged garlic extract and S-allyl-L-cysteine by analyzing several biochemical and molecular indices in cellular and rodent models of AD. Pretreatment with the extract and SAC enhanced the viability of neuronal cells exposed to H_2O_2 in a dose-dependent manner. Both the extract and SAC exposure attenuated the expression of synaptosomal-associated protein of 25 kDa and synaptophysin in APP-transgenic mice.

Gupta and Rao (2007) investigated the neuroprotective role of S-allylcysteine on Aβ aggregation in in vitro conditions. The Aβ aggregation was measured by using thioflavin T, TEM, sodium dodecyl sulfate-polyacrylamide gel electrophoresis and size exclusion-high-performance liquid chromatography. SAC dose dependently repressed Aβ fibrillation and weakened preformed Aβ fibrils. The circular dichroism and fluorescence reduction experiment investigated the binding capacity of SAC to Aβ. Javed et al. (2011) analyzed the neuroprotective role of S-allylcysteine, a sulfur-containing amino acid of garlic, on the cognitive impairments and hippocampal oxidative stress in intracerebroventricular streptozotocin-intoxicated mice. An enhanced path length and latency were examined in streptozotocin lesioned mice, whereas SAC pretreatment attenuated STZ-induced cognitive impairment. Pretreatment of SAC significantly attenuated STZ-induced oxidative stress by reducing the level of thiobarbituric acid reactive substances and enhancing the levels of reduced glutathione (GSH) and activities of glutathione peroxidase and glutathione reductase. Moreover, the SAC exhibited the antiapoptotic properties by regulating DNA fragmentation and Bcl2 and p53 expression against STZ-induced neurotoxicity.

References

Badal DS, Dwivedi AK, Kumar V, Singh S, Prakash A, Verma S et al (2019) Effect of organic manures and inorganic fertilizers on growth, yield and its attributing traits in garlic (*Allium sativum* L.). J Pharmacogn Phytochemistry 8:587–590

Batiha GS, Alkazmi LM, Nadwa EH, Rashwan EK, Beshbishy AM (2020) Physostigmine: a plant alkaloid isolated from *Physostigma venenosum*: a review on pharmacokinetics, pharmacological and toxicological activities. J Drug Deliv Ther 10:187–190

Chauhan NB (2003) Anti-amyloidogenic effect of Allium sativum in Alzheimer's transgenic model Tg2576. J Herb Pharmacother 3(1):95–107

Chauhan NB (2006) Effect of aged garlic extract on APP processing and tau phosphorylation in Alzheimer's transgenic model Tg2576. J Ethnopharmacol 108(3):385–394

Davis SR (2005) An overview of the antifungal properties of allicin and its breakdown products-the possibility of a safe and effective antifungal prophylactic. Mycoses 48:95–100

Griffin B, Selassie M, Gwebu ET (2000) Effect of aged garlic extract on the cytotoxicity of Alzheimer β-amyloid peptide in neuronal PC12 cells. Nutr Neurosci 3(2):139–142

Gupta VB, Rao KSJ (2007) Anti-amyloidogenic activity of S-allyl-L-cysteine and its activity to destabilize Alzheimer's beta-amyloid fibrils in vitro. Neurosci Lett 429(2–3):75–80

Jackson R, McNeil B, Taylor C, Holl G, Ruff D, Gwebu ET (2002) Effect of aged garlic extract on caspase-3 activity, in vitro. Nutr Neurosci 5(4):287–290

Javed H, Khan MM, Khan A, Vaibhav K, Ahmad A, Khuwaja G et al (2011) S-allyl cysteine attenuates oxidative stress associated cognitive impairment and neurodegeneration in mouse model of streptozotocin-induced experimental dementia of Alzheimer's type. Brain Res 1389: 133–142

Jeong JH, Jeong HR, Jo YN, Kim HJ, Shin JH, Heo HJ (2013) Ameliorating effects of aged garlic extracts against Aβ-induced neurotoxicity and cognitive impairment. BMC Complement Altern Med 13:268

Keri RS, Quintanova C, Chaves S, Silva DF, Cardoso SM, Santos MA (2016) New tacrine hybrids with natural-based cysteine derivatives as multitargeted drugs for potential treatment of Alzheimer's disease. Chem Biol Drug Des 87(1):101–111

Kumar S (2015) Dual inhibition of acetylcholinesterase and butyrylcholinesterase enzymes by allicin. Indian J Pharm 47(4):444–446

Li F, Kim MR (2019) Effect of aged garlic ethyl acetate extract on oxidative stress and cholinergic function of scopolamine-induced cognitive impairment in mice. Prev Nutr Food Sci 24(2): 165–170

Lin GH, Lee YJ, Choi DY, Han SB, Jung JK, Hwang BY et al (2012) Anti-amyloidogenic effect of thiacremonone through anti-inflamation in vitro and in vivo models. J Alzheimers Dis 29(3): 659–676

Liu B, Li X, Yu H, Shi X, Zhou Y, Alvarez S et al (2021) Therapeutic potential of garlic chive-derived vesicle-like nanoparticles in NLRP3 inflammasome-mediated inflammatory diseases. Theranostics 11(19):9311–9330

Mane P, Mayee R, Atre K (2011) Medicinal properties of allium Sativum (garlic): a review. Int J Pharm Res Dev 3(2):17–27

Manral A, Meena P, Saini V, Siraj F, Shalini S, Tiwari M (2016) DADS analogues ameliorated the cognitive impairments of alzheimer-like rat model induced by scopolamine. Neurotox Res (30, 3):407–426

Matsuura H (1997) Phytochemistry of garlic horticultural and processing procedures. In: Lachance PA (ed) Neutraceuticals: designer foods III. Garlic, soy and licorice. Food and Nutrition Press, Trumbull, CT, pp 55–59

Moyers S (1996) Garlic in health, history and world cuisine. Suncoast Press, St Petersburg, FL, pp 1–36

Nadeem MS, Kazmi I, Ullah I, Muhammad K, Anwar F (2021) Allicin, an antioxidant and neuroprotective agent, ameliorates cognitive impairment. Antioxidants (basel) 11(1):87

Nillert N, Pannangrong W, Welbat JU, Chaijaroonkhanarak W, Sripanidkulchai K, Sripanidkulchai B (2017) Neuroprotective effects of aged garlic extract on cognitive dysfunction and neuroinflammation induced by β-amyloid in rats. Nutrients 9(1):24

Peng Q, Buz'Zard AR, Lau BHS (2002) Neuroprotective effect of garlic compounds in amyloid-beta peptide-induced apoptosis in vitro. Med Sci Monit 8(8):BR328–BR337

Promyo K, Cho JY, Park KH, Jaiswal L, Park SY, Ham KS (2017) Artemisia scoparia attenuates amyloid β accumulation and tau hyperphosphorylation in spontaneously hypertensive rats. Food Sci Biotechnol 26(3):775–782

Rahman K, Lowe GM (2006) Garlic and cardiovascular disease: a critical review. J Nutr 136:736s–740s

Ray B, Chauhan NB, Lahiri DK (2011) Oxidative insults to neurons and synapse are prevented by aged garlic extract and S-allyl-L-cysteine treatment in the neuronal culture and APP-Tg mouse model. J Neurochem 117(3):388–402

Rivlin RS (2001) Historical perspective on the use of garlic. J Nutr 131:951S–954S

Rodríguez L, García-Recio E, Ruiz C, Luna-Bertos ED, Illescas-Montes R, Costela-Ruiz VJ (2022) Biological properties and therapeutic applications of garlic and its components. Food Funct 13: 241–52426

Souza GA, Ebaid GX, Seiva FR, Rocha KH, Galhardi CM, Mani F et al (2011) N-acetylcysteine an *allium* plant compound improves high-sucrose diet-induced obesity and related effects. Evid Based Complement Alternat Med 2011:643269

Thorajak P, Pannangrong W, Welbat JU, Chaijaroonkhanarak W, Sripanidkulchai K, Sripanidkulchai B (2017) Effects of aged garlic extract on cholinergic, glutamatergic and gabaergic systems with regard to cognitive impairment in Aβ -induced rats. Nutrients 9(7):686

Wen YD, Zhu YZ (2015) The pharmacological effects of s-propargyl-cysteine, a novel endogenous H2S-producing compound. Handb Exp Pharmacol 230:325–336

Yoshioka Y, Matsumura S, Morimoto M, Takemoto Y, Kishi C, Moriyama T et al (2021) Inhibitory activities of sulfur compounds in garlic essential oil against Alzheimer's disease-related enzymes and their distribution in the mouse brain. J Agric Food Chem 69(35):10163–10173

Zhang H, Wang P, Xue Y, Liu L, Li Z, Liu Y (2018) Allicin ameliorates cognitive impairment in APP/PS1 mice via suppressing oxidative stress by blocking JNK signaling pathways. Tissue Cell 50:89–95

Zhu YF, Li XH, Yuan ZP, Li CY, Tian RB, Jia W (2015) Allicin improves endoplasmic reticulum stress-related cognitive deficits via PERK/Nrf2 antioxidative signaling pathway. Eur J Pharmacol 762:239–246

Zilbeyaz K, Oztekin A, Kutluana EG (2021) Design and synthesis of garlic-related unsymmetrical thiosulfonates as potential Alzheimer's disease therapeutics: in vitro and in silico study. Bioorg Med Chem 540:116194

Chapter 7
Role of Citrus Fruits in Alzheimer's Disease: A Current Perspective

Udaiyappan Janakiraman, Ravi Surya, Arokiasamy Justin Thenmozhi, Thamilarasan Manivasagam, Kulandaisamy Amali, Hanan Fahad Alharbi, and Yoga Rajamani

Abstract *Citrus* fruits are commonly consumed fruits that are known for its nutrient content, energy source, and health supplements. Citrus fruits are rich sources of several secondary metabolites like limonoids, alkaloids, coumarins, flavonoids, carotenoids, essential oils, and phenolic acids that are responsible for its beneficiary effect against various diseases. These include anti-inflammatory, cardiovascular-protective, anticancer, antioxidative, and neuroprotective effects. Present-day pharmacological therapies for the AD are deficient. Unfortunately, other strategies such as neural transplantation and stem cell transplantation remain in the experimental stage. Numerous cellular, animal, and human studies have delineated the neuroprotective effects of citrus fruits owning to their active components. Increased consumption of citrus fruits is associated with higher antioxidant status and phytochemical constituents, thus helpful against oxidative stress, mitochondrial dysfunction, inflammation, and apoptosis that play an important role in the cause and progression of neurological disorders. The multifactorial etiology of

U. Janakiraman
Department of Surgery, Virginia Commonwealth University, Richmond, VA, USA

R. Surya · T. Manivasagam (✉) · K. Amali
Department of Biochemistry and Biotechnology, Annamalai University, Chidambaram, Tamil Nadu, India

A. Justin Thenmozhi
Department of Biochemistry, School of Biological Sciences, Madurai Kamaraj University, Madurai, Tamil Nadu, India

Department of Biochemistry and Biotechnology, Faculty of Science, Annamalai University, Chidambaram, Tamil Nadu, India

H. F. Alharbi
Department of Maternity and Child Health Nursing, College of Nursing, Princess Nourah Bint Abdurrahman University, Riyadh, Saudi Arabia

Y. Rajamani
College of Computer, Mathematics and Natural Sciences, University of Maryland, College Park, MD, USA

A. Justin Thenmozhi, T. Manivasagam (eds.), *Nutraceuticals for Alzheimer's Disease: A Promising Therapeutic Approach*, Nutritional Neurosciences,
https://doi.org/10.1007/978-981-99-0677-2_7

neurodegenerative diseases suggests that drugs with multiple targets could have therapeutical potential for these pathologies.

Keywords Citrus fruits · Bioflavonoids · Amyloid · Tau · Signaling molecules

7.1 Introduction

7.1.1 Citrus Plants

The Citrus genus belongs to the Rutaceae family, consisting of approximately 140 genera and 1300 species. *Citrus sinensis* (orange), *Citrus aurantium* (sour orange), *Citrus medica* (Citron), *Citrus limon* (lemon), *Citrus paradisi* (grapefruit), *Citrus grandis* (shaddock), *Citrus reticulata* (tangerine), and *Citrus aurantifolia* (lime) are few main fruits belonging to the genus citrus (Singh et al. 1983). Citruses are produced globally in tropical and subtropical countries estimated at about 105 million metric tons/year. Major citrus producers are Brazil (19.2 million metric tons/year), followed by the United States of America, Japan, China, Mexico, Pakistan (1.76 million metric tons/year), and the Mediterranean countries (Mahmood 2005).

These fruits are having high economic and pharmacological value as they are used in the food industry, cosmetics, and folk medicine (Silalahi 2002; Saidani et al. 2004). Both the fresh fruit and its processed juice are used in large-scale consumption, while the waste product of citrus fruits excluding juice including seeds, peels, and pulps is used to treat agro-industrial waste, act as a source for essential oils and various secondary metabolites (Andrea et al. 2003).

7.1.2 Chemical Constituents

Citrus plants are considered a vital fruit crop globally because of the nutritional value and special flavor of citrus fruits. Most widely consumed citrus fruit juices are from lemons, limes, grapefruits, and sweet oranges. There is a reciprocal relationship between citrus fruit consumption and several diseases (Joshipura et al. 2001). The pharmacological properties of citrus fruit are mainly due to the occurrence of various bioactive compounds, like vitamin C, phenolics (e.g., glycosides, flavanone, and hydroxycinnamic acids), and carotenoids (Guimaraes et al. 2010).

7.1.3 Importance of Citrus Fruits in AD

The modern strategy in the treatment of AD is to enhance the brain's acetylcholine levels by using cholinesterase inhibitors. Due to the increased activities of cholinesterase, both the neurotransmitters, acetylcholine, and butyrylcholine get lysed, thereby leading to AD symptoms. As the uses of artificial cholinesterase inhibitors are associated with various side effects, more interest is given for the selection of natural cholinesterase inhibitors. The essential oil obtained from peels and juice of lemon, peels of lime and leaf extract, and peel of diamante citron are reported to have the source of cholinesterase inhibitors. In a clinical study, consumption of citrus fruit gave more aid as compared to intake of other vegetables or fruits to moderate, severe, and hospitalized AD patients.

7.1.4 Vitamin C

In contrast to other fruits, citrus fruits are usually having strong antioxidant functions. Apart from ascorbic acid (vitamin C), they are the source of various potent antioxidants like carotenoids, polyphenols (flavonoids), glutathione, and numerous enzymatic antioxidants. Due to the antioxidant function of vitamin C, citrus fruits are reported to have a preventive action against cataract, cancer, asthma, heart disease, and cognitive dysfunction. The maximum levels of vitamin C are present mainly in the brain cells, and it is a key antioxidant substance in the brain. The neurons contain about 10 mM of vitamin C, while the glia cells contain about 1 mM. Other functions of vitamin C are as follows: (1) serve as a cofactor for various enzymatic reactions (collagen, norepinephrine, and carnitine), (2) reduce the inflammatory reactions by inhibiting the TNF-α-mediated NF-κB activation through p38 MAPK activation, and (3) regenerate the intracellular antioxidant vitamin E. Rinaldi et al. (2003) indicated the diminished levels of several vitamins including vitamin C in the AD patients. Another study demonstrated that the diminished circulatory levels of vitamin C are related to the extent of cognitive dysfunction in AD (Harrison et al. 2009).

7.1.5 Citrus Bioflavonoids

Flavonoids are ubiquitous and abundant plant-based compounds having several pharmacological benefits and therapeutic activities. Based on their structure (Mulvihill and Huff 2010), they are subclassified into flavanones, flavonols, flavanols, flavones, isoflavones, and anthocyanidins (Table 7.1). The flavanones (di- and tri-O-glycosides), flavonols, flavone (di- and tri-O-glycosides and C-glycosides), and polymethoxyflavones are the four major subgroups of flavonoids found in citrus fruits.

The cause and progression of numerous neurodegenerative diseases like AD, PD, HD, and ALS occur due to the cumulative effect of oxidative stress and its associated mutations and dysfunction of the mitochondria. The abundant citrus flavonoids like hesperidin, neohesperidin, hesperetin, naringin, and naringenin showed moderate

Table 7.1 Types of flavonoids with examples and their sources

Flavonoids	Subclasses	Examples	Source
Flavanones-O-glycosides	Rutinosides	Narirutin, isorhoifolin, hesperidin, diosmin, and eriocitrin	Tasteless and found in oranges, lemons, and tangerines
	Neohesperidosides	Neoeriocitrin, neodiosmin, naringin, and neohesperidin	Bitter and present in grapefruit and pummelo
Flavone	Aglycons	Diosmetin	General citrus plants
		Luteolin	Lemons
Polymethoxyflavones		Tangeretin, nobiletin, natsudaidain, and heptamethoxyflavone	Rich in oranges, tangerines, and lemons and are less abundant in some grapefruits
		Hydroxylated polymethoxychalcones	Sweet oranges
Flavonols		Flavonols (l-o)	Lemons

antioxidant properties. Usually, aglycons showed more potent antioxidant activities than glycosides, but hesperidin has more inhibitory action on the formation of free radicals and lipid peroxidation products than hesperetin.

Various experimental studies demonstrated that the activation of microglia and its linked inflammation is associated with the etiology and progression of several neurodegenerative diseases. Accumulation or continuous exposure to cytokines, chemokines, endotoxins, and damaged or misfolded proteins leads to their activation. During this pathological condition, translocation of redox-perceptive cytosolic nuclear factor-kappa-B (NF-kB) to the nucleus of the cell occurs and leads to DNA adduct formation. Activation of NF-kB leads to synthesis and secretion of several pro-inflammatory markers like interleukin-1 beta (IL-1β), IL-6, cyclooxygenase-2 (COX-2), tumor necrosis factor-alpha (TNF-α), and inducible nitric oxide synthase (iNOS). Citrus flavonoids like hesperidin, nobiletin, naringenin, naringin, and 6-demethoxynobiletin offered neuroprotection by regulating immune functions in the brain. They offered an anti-inflammatory effect by attenuating microglial activation and its related secretion of interleukins, TNF-α, COX-2, and iNOS, which is mediated by several intracellular signaling pathways such as activation of NF-kB and mitogen-activated protein kinase pathway.

Mitochondria plays a key role in Ca^{2+} regulation, cellular aerobic respiration, ROS formation, and cellular apoptosis. Mitochondrial dysfunctions lead to disruption in physiological processes, which will result in the commencement of several neurodegenerative diseases like AD and PD. Mitochondrial dysfunction can be induced by inhibition of electron transfer from complex I, which is key for ATP production. Moreover, mitochondrial dysfunction is also induced by the deposition and aggregation of amyloid-beta peptides. Babylon et al. (2021) indicated that the soluble amyloid that is synthesized by α- and γ-secretases was able to pierce the neuronal mitochondrial membrane and induce apoptosis. Nano-hesperetin crystals are reported to recover mitochondrial function in an in vitro SH-SY 5Y model of AD (Babylon et al. 2021). In an Aβ_{1-42}-exposed rat pheochromocytoma cells, mitochondrial dysfunction was attenuated by limonene, a monoterpene of citrus fruits (Piccialli et al. 2021). In the rat experimental model, mitochondrial dysfunction induced by $AlCl_3$ was reverted by oral treatment of hesperidin (Thenmozhi et al. 2017).

Apart from improving cognitive impairment and antioxidant function, naringin was reported to attenuate the activities of the mitochondrial complexes I and III in a murine model. Flavonoids are reported to have an affinity to various regulatory kinases, receptors, and proteins linked with cellular physiology. They are linked with protein and lipid kinase signaling cascades like tyrosine kinase, phosphatidylinositol 3-kinase, Akt/PKB, protein kinase C, and mitogen-activated protein kinases that are connected with cellular survival and death. They combined with calcium membrane and mitochondria ATPase, protein kinases A and C, and topoisomerase (ATP-binding proteins), and GABA-A or adenosine receptors and regulate their activities. Ca^{2+}operate as second messengers and induce extracellular signal-regulated kinase Akt and c-JunN-terminal kinase signaling pathways. Both the naringenin and hesperetin hinder ATP-binding sites, inactivate the kinase activities, and account for their neuroprotective properties by the virtue of antiapoptotic function. Moreover, the administration of naringin enhances the activity of CaMKII and develops cognitive function in the transgenic rodent model of AD (Wang et al. 2013). In amyloid-β peptide-exposed PC12 cells, nobiletin obtained from citrus fruit diminishes c-Jun N-terminal kinase and p38 expression without disturbing the extracellular signal-regulated kinase 1/2 pathway (Youn et al. 2019).

7.1.6 Diosmin and Diosmetin

Diosmetin is the most abundant aglycone of Citrus species and derivative of glycoside diosmin (Patel et al. 2013). Diosmetin offered a neuroprotective effect through its anti-inflammatory effect (diminishing the expression of TNF-α) and antioxidant action (enhancing the expression of superoxide dismutase (Yang et al. 2017). Diosmetin enhanced the inhibitory GSK-3β phosphorylation while selectively reducing γ-secretase activity, Aβ generation, tau hyperphosphorylation, and pro-inflammatory activation of microglia in vitro, without altering Notch processing (Sawmiller et al. 2016).

Diosmin treatment diminished Aβ linked with toxicity in transgenic (Tg2576) mice (Rezai-Zadeh and Douglas 2009). Oral diosmin administration could reduce cerebral soluble and oligomer Aβ levels and AD-like tau pathology and ameliorate cognitive impairment in 3xTg-AD mouse models via modulation of GSK-3 activity and transient receptor potential canonical 6 (TRPC6)-related mechanisms (Sawmiller et al. 2016).

7.1.7 Hesperidin and Neohesperidin

Hesperidin, a flavonoid glycoside, is present abundantly in various citrus fruits like sweet orange, lemon, and grapefruits. It is also found in other species like *C. unshiu*, Ponderosa lemon, unripe oranges, and *C. mitis* (Ikan 2013). It is also reported to have anticarcinogenic, antioxidant, and anti-inflammatory properties (Roohbakhsh et al. 2015). It offered the neuroprotective function against several neurodegenerative diseases including stroke and Huntington's, Parkinson's, and Alzheimer's disease (Thenmozhi et al. 2015; Cho et al. 2006; Tamilselvam et al. 2013; Kumar et al. 2013).

Neohesperidin is a pungent bitter isomer of hesperidin that is present in bitter orange (Qin et al. 2012). It ameliorated the $A\beta_{25-35}$-mediated ROS formation and mitochondrial impairment in an in vitro model of PD. Its administration offered antiapoptotic properties against amyloid-induced toxicity and protected neuronal morphology. Chakraborty et al. (2021) indicated that the administration of neohesperidin in primary cultured hippocampal neurons inhibited the formation of fibrillar amyloid from soluble Aβ molecules against $A\beta_{25-35}$-mediated toxicity. Neohesperidin treatment attenuated $A\beta_{25-35}$-induced apoptosis by preventing neurotoxicity associated with lethal UPR and ER stress via blocking S-nitrosylation of protein-disulfide isomerase in $A\beta_{25-35}$-treated primary cultures of hippocampal neurons (Wang et al. 2018). Neohesperidin reduced the toxicity and ROS induced by both monomeric and oligomeric Aβ species in primary hippocampal neuronal cells.

Hesperidin administration enhances cognitive function due to its potent antioxidant properties. In the transgenic mouse model of AD, oral treatment of hesperidin (16 weeks) attenuated learning and memory deficits by improving mitochondrial dysfunctions and diminishing the levels of malondialdehyde and hydrogen peroxide and also enhancing levels of reduced glutathione and total antioxidants. It also offered mitochondrial protection by enhancing the activities of mitochondrial complex I–IV enzymes (Wang et al. 2014). GSK-3β, a protein kinase, is the key regulator of hyperphosphorylation and mitochondrial entry of tau protein and oxidative stress (Badalzadeh et al. 2015). Hesperidin improved the cognitive dysfunction, reduced the $A\beta_{1-40}$ level, and offered the mitochondrial protective role by diminishing the activity and expression of GSK-3β (Wang et al. 2014; DaRocha-Souto et al. 2012). In the aluminum chloride-induced rat model of AD, hesperidin administration rescued learning and memory deficits by its acetylcholine esterase inhibition activity. It also reduced the expression of amyloid precursor protein through NF-κB-dependent mechanism and reduced the expression of $A\beta_{1-40}$ and β- and γ-secretases (amyloidogenic enzymes) in the cortex and hippocampus of rats (Thenmozhi et al. 2015). Due to its antiapoptotic effect by enhancing B-cell lymphoma 2 (antiapoptotic protein) and reducing Bcl2-associated X proteins (pro-apoptotic protein) hesperidin reversed the cognitive dysfunction in aluminum chloride-induced AD rats (Thenmozhi et al. 2017).

Apart from cognitive dysfunction, AD is linked with various noncognitive and behavioral deficits. Tau aggregation, amyloid deposition, and neuro-inflammation processes are responsible for the noncognitive behavioral symptoms. In a transgenic mouse model of AD, this phytochemical restored both the cognitive and behavioral symptoms linked with AD. It inhibited the inflammation, diminished APP formation, and Aβ peptide accumulation in the cortex and hippocampus of the transgenic AD mouse and thus improved the social behavior and nesting ability (Li et al. 2015). Its anti-inflammatory role is associated with the downregulation of transforming growth factor β1 and NF-κB (linked with AD progression) in the cortex of transgenic mice (Ghorbani et al. 2012).

Oral administration of hesperidin diminished the oxidant levels and inflammation in the cortex of mouse model of AD by enhancing an increase of the heme oxygenase-1 expression and its levels that can suppress oxidative stress and inflammation. It also diminished the levels and activities of other inflammatory markers such as C-reactive protein, tumor necrosis factor-α, NF-κβ, and monocyte chemoattractant protein 1. It is reported to activate Akt/NF-E2-related factor 2 pathway and inhibit the advanced glycation end product-mediated NF-κB pathway (Hong et al. 2018), thereby favoring neuronal and behavioral function. In the future, hesperidin and neohesperidin may emerge as nontoxic and multifactorial scaffold for the treatment of AD.

7.1.8 Naringin

Naringin is a derivative of naringenin and the key active substance of *C. aurantium* and *C. medica L.* (Zhang et al. 2014; Yin et al. 2015). It is found in trace amounts in all citrus fruits (Wong et al. 2013) and gives a bitter taste to juices of citrus fruits (Chtourou et al. 2015). Wang et al. (2013) indicated that naringin is reported to have anti-ulcer, anti-inflammatory, antioxidant, antiapoptotic, anticarcinogenic, and anti-osteoporotic properties. Naringin act as a neuroprotective agent against various CNS disorders such as epilepsy, Parkinson's disease, and Alzheimer's disease (Saaby and Jager 2011). It enhanced learning and memory deficits induced by hydrocortisone in mice. It offered an anti-alzheimeric role by regulating Aβ metabolism, tau hyperphosphorylation, oxidative stress, excitotoxicity, and acetylcholinergic and neuronal apoptosis. The neuroprotective effect of naringin was induced by the regulation of phosphorylated-P38/P38 expression by involving the MAPK/P38 pathway. Moreover, naringin exerted neuroprotective effects by enhancing cognitive function in mice. Chronic intake of naringin reduced learning and memory problems, enhanced movement and activities, diminished Aβ deposition, and attenuated energy metabolism in the brain. It also significantly enhanced the phosphorylation of GSK-3β. Administration of naringin to transgenic AD mice decreased the formation of plaques and enhanced glucose uptake by GSK-3β inhibition (Wang et al. 2012).

Naringin ameliorated Aβ-mediated toxicity by enhancing mitochondrial integrity and functions in all the regions of the brain. It attenuated the Aβ toxicity by enhancing levels of cytoplasmic and mitochondrial calcium in the brain. Naringin reverted Aβ-mediated apoptosis, mitochondrial calcium uniporter, and hemeoxygenase-1 levels in an in vivo model of AD (Varshney and Garababu 2021). Naringin-pegylated nanoparticles showed an improvement in the neuroprotective effect against Aβ-mediated oxidant-antioxidant imbalance and neurodegeneration in rat primary hippocampal cultures. Aβ plaque in various regions of the hippocampus was significantly diminished after exposure to naringin. Naringin administration decreased the amyloid plaque formation owing to its potent iron chelation property.

The neuroprotective effect of naringin particularly the improvement of cognitive function is partially owing to its antioxidant activities thereby decreasing the levels of malondialdehyde and nitrite and enhancing the activities of glutathione S-transferase, catalase, and superoxide dismutase and the levels of reduced glutathione (Kumar et al. 2010).

7.1.9 Nobiletin

Nobiletin, a hexamethoxyflavone, is found in the peels of *C. reticulata, C. depressa, C. sinensis,* and *C. limon.* It contains six methoxyl groups (5, 6, 7, 8-positions, ring A; and 3′, 4′-positions, ring B), on a basic flavone pattern, and offered tremendous lipophilic properties (Chen et al. 1997; Nogata et al. 2006). This lipophilic property is related to several biological functions, particularly the capability to cross the blood-brain barrier and favor their entry into the CNS. Moreover, it is reported to have various pharmacological functions including anti-inflammatory (Cui et al. 2010), anticarcinogenic (Lam et al. 2011; Lee et al. 2011), antidiabetic (Lee et al. 2010), anti-atherogenic, and anti-obesity activities (Kanda 2012).

Oral administration of nobiletin recovered short-term memory impairment (Y-maze test) and associative memory (step-through passive avoidance test) and attenuated acetylcholinesterase (AChE) immunoreactivity in the hippocampus of olfactory-bulbectomized AD mice. In $A\beta_{1-40}$ fused rat model of AD, nobiletin treatment enhanced the reference and working memory dysfunction (eight-arm radial maze test) (Matsuzaki et al. 2006). Moreover, nobiletin administration diminished the levels of insoluble $A\beta_{1-40}$ and $A\beta_{1-42}$ and deposition of Aβ plaques in the hippocampus of transgenic AD mice. Nobiletin administration significantly enhanced the short-term (Y-maze test) and recognition (novel object recognition test) memory in 3XTg-AD mice (Nakajima and Ohizumi 2019).

Nobiletin attenuated the $A\beta_{25-35}$-induced apoptosis by regulating oxidant-antioxidant imbalance, nuclear morphology, cell cycle, and caspase activity (apoptosis inducer). Nobiletin repressed the synthesis of various inflammatory markers such as inducible NO synthase, tumor necrosis factor-α, interleukin-1β, cyclooxygenase-2, prostaglandin E2, and nitric oxide (NO) during Aβ induction attributed due to the obstruction of NF-κB and phosphorylation of its inhibitor, IκB-α. It diminishes the c-Jun N-terminal kinase and p38 expression without altering extracellular signal-regulated kinase 1/2 activation in in vitro model of AD (Youn et al. 2019).

7.1.10 Tangeretin

Tangeretin is an abundant polymethoxylated flavonoid that is present in *C. tangerina, C. sinensis,* and *C. aurantium* fruit peels (Braidy et al. 2017). Previous studies indicated that the transcription of the cAMP response element (CRE) is altered in various neurodegenerative diseases (Chalovich et al. 2006). The citrus flavonoids such as tangeretin, 5-demethylnobiletin, nobiletin, 6-demethoxytangeretin, sinensetin, and 6-demethoxynobiletin stimulated the CRE-mediated transcription and induced neurite outgrowth in PC12 cells and hippocampal neurons. The cAMP/protein kinase A/ERK/CREB signaling pathway is key for learning and memory processes (Kawahata et al. 2013).

7.2 Conclusion

In conclusion, citrus fruits are rich sources of phytochemicals, and they have a wide range of physiological and pharmacological effects. There is now unquestionable evidence from both in vitro studies and experimental models to demonstrate that these compounds possess anticholesterolemic, antioxidant, anti-inflammatory, and antimicrobial properties, as well as neuroprotective activities. Various research studies were carried out to find the neuroprotective role of the phytochemical constituents of citrus fruits against NDDs. Reported results suggested the clinical role of citrus fruits in treating various neurological disorders. Before entering the clinical trials, more studies are needed to know about its mechanisms, dosages and duration, absorption, distribution and excretion kinetics, long-term effects, and side effects. Additionally, phase II clinical trials in patients with neurodegenerative disorders are necessary to confirm whether the neuroprotective effects of citrus can be successfully translated to the clinic.

References

Andrea V, Nadia N, Teresa RM, Andrea A (2003) Analysis of some Italian lemon liquors (Limoncello). J Agric Food Chem 51(17):4978–4983

Babylon L, Grewal R, Stahr PL, Eckert RW, Keck CM, Eckert GP (2021) Hesperetin nanocrystals improve mitochondrial function in a cell model of early Alzheimer disease. Antioxidants (Basel) 10(7):1003

Badalzadeh R, Mohammadi M, Yousefi B, Farajnia S, Najafi M, Mohammadi S (2015) Involvement of glycogen synthase kinase-3β and oxidation status in the loss of cardioprotection by postconditioning in chronic diabetic male rats. Adv Pharm Bull 5:321

Braidy N, Behzad S, Habtemariam S, Ahmed T, Daglia M, Nabavi SM et al (2017) Neuroprotective effects of citrus fruit-derived flavonoids, nobiletin and tangeretin in Alzheimer's and Parkinson's disease. CNS Neurol Disord Drug Targets 16(4):387–397

Chakraborty S, Rakshit J, Bandyopadhyay J, Basu S (2021) Multi-target inhibition ability of neohesperidin dictates its neuroprotective activity: implication in Alzheimer's disease therapeutics. Int J Biol Macromol 15(176):315–324

Chalovich EM, Zhu JH, Caltagarone J, Bowser R, Chu CT (2006) Functional repression of cAMP response element in 6-hydroxydopamine-treated neuronal cells. J Biol Chem 281(26): 17870–17881

Chen J, Montanari AM, Widmer WW (1997) Two new polymethoxylated flavones, a class of compounds with potential anticancer activity, isolated from cold pressed Dancy tangerine peel oil solids. J Agric Food Chem 45:364–368

Cho J (2006) Antioxidant and neuroprotective effects of hesperidin and its aglycone hesperetin. Arch Pharm Res 29:699–706

Chtourou Y, Gargouri B, Kebieche M, Fetoui H (2015) Naringin abrogates cisplatin-induced cognitive deficits and cholinergic dysfunction through the down-regulation of AChE expression and iNOS signaling pathways in hippocampus of aged rats. J Mol Neurosci 56:349–362

Cui Y, Wu J, Jung SC, Park DB, Maeng YH, Hong JY et al (2010) Anti-neuroinflammatory activity of nobiletin on suppression of microglial activation. Biol Pharm Bull 33:1814–1821

DaRocha-Souto B, Coma M, Perez-Nievas B, Scotton T, Siao M, Pet S-F et al (2012) Activation of glycogen synthase kinase-3 beta mediates β-amyloid induced neuritic damage in Alzheimer's disease. Neurobiol Dis 45:425–437

Ghorbani A, Nazari M, Jeddi-Tehrani M, Zand H (2012) The citrus flavonoid hesperidin induces p53 and inhibits NF-κB activation in order to trigger apoptosis in NALM-6 cells: involvement of PPARγ-dependent mechanism. Eur J Nutr 51:39–46

Guimaraes R, Barros L, Barreira JCM, Sousa MJ, Carvalho AM, Ferreira ICFR (2010) Targeting excessive free radicals with peels and juices of citrus fruits: grapefruit, lemon, lime and orange. Food Chem Toxicol 48:99–106

Harrison FE, Hosseini AH, McDonald MP, May JM (2009) Vitamin C reduces spatial learning deficits in middle-aged and very old APP/PSEN1 transgenic and wild-type mice. Pharmacol Biochem Behav 93(4):443–450

Hong Y, An Z (2018) Hesperidin attenuates learning and memory deficits in APP/PS1 mice through activation of Akt/Nrf2 signaling and inhibition of RAGE/NF-kappaB signaling. Arch Pharm Res 41(6):655–663

Ikan R (2013) Natural products: a laboratory guide. Elsevier, Amsterdam

Joshipura KJ, Hu FB, Manson JE, Stampfer MJ, Rimm EB, Speizer FE (2001) The effect of fruit and vegetable intake on risk for coronary heart disease. Ann Intern Med 134:1106–1114

Kanda K, Nishi K, Kadota A, Nishimoto S, Liu MC, Sugahara T (2012) Nobiletin suppresses adipocyte differentiation of 3T3-L1 cells by an insulin and IBMX mixture induction. Biochim Biophys Acta 182:461–468

Kawahata I, Yoshida M, Sun W, Nakajima A, Lai Y, Osaka N (2013) Potent activity of nobiletin-rich Citrus reticulata peel extract to facilitate cAMP/PKA/ERK/CREB signaling associated with learning and memory in cultured hippocampal neurons: identification of the substances respon sible for the pharmacological action. J Neural Transm (Vienna) 120(10):1397–1409

Kumar A, Chaudhary T, Mishra J (2013) Minocycline modulates neuroprotective effect of hesperidin against quinolinic acid induced Huntington's disease like symptoms in rats: behavioral, biochemical, cellular and histological evidences. Eur J Pharmacol 720:16–28

Kumar A, Dogra S, Prakash A (2010) Protective effect of naringin, a citrus flavonoid, against colchicine-induced cognitive dysfunction and oxidative damage in rats. J Med Food 13(4): 976–984

Lam KH, Deepa A, Lam IK, Tsui SKW, Yang ZF, Lee MYS (2011) Nobiletin, a polymethoxylated flavonoid from citrus, shows anti-angiogenic activity in a zebrafish in vivo model and HUVEC in vitro model. J Cell Biochem 112:3313–3321

Lee YC, Cheng TH, Lee JS, Chen JH, Liao YC, Fong Y et al (2011) Nobiletin, a citrus flavonoid, suppresses invasion and migration involving FAK/PI3K/Akt and small GTPase signals in human gastric adenocarcinoma AGS cells. Mol Cell Biochem 347:103–115

Lee YS, Cha BY, Saito K, Yamakawa H, Choi SS, Yamaguchi K et al (2010) Nobiletin improves hyperglycemia and insulin resistance in obese diabetic Ob/Ob mice. Biochem Pharmacol 79: 1674–1683

Li C, Zug C, Qu H, Schluesener H, Zhang Z (2015) Hesperidin ameliorates behavioral impairments and neuropathology of transgenic APP/PS1 mice. Behav Brain Res 281(Suppl C):32–42

Mahmood MA (2005) Hurdles in way of citrus export Dawn. http://www.dawn.com/2005/10/31/ebr4.htm. Accessed 31 Oct 2005

Matsuzaki K, Yamakuni T, Hashimoto M, Haque AM, Shido O, Mimaki Y et al (2006) Nobiletin restoring beta-amyloid-impaired CREB phosphorylation rescues memory deterioration in Alzheimer's disease model rats. Neurosci Lett 400:230–234

Mulvihill EE, Huff MW (2010) Antiatherogenic properties of flavonoids: implications for cardiovascular health. Can J Cardiol 26(Suppl A):17A–21A

Nakajima A, Ohizumi Y (2019) Potential benefits of Nobiletin, a citrus flavonoid, against Alzheimer's disease and Parkinson's disease. Int J Mol Sci 20(14):3380

Nogata Y, Sakamoto K, Shiratsuchi H, Ishii T, Yano M, Ohta H (2006) Flavonoid composition of fruit tissues of citrus species. Biosci Biotechnol Biochem 70:178–192

Patel K, Gadewar M, Tahilyani V, Patel DK (2013) A review on pharmacological and analytical aspects of diosmetin: a concise report. Chin J Integr Med 19(10):792–800

Piccialli I, Tedeschi V, Caputo L, Amato G, De Martino L, De Feo V et al (2021) The antioxidant activity of limonene counteracts neurotoxicity triggered byAβ1-42 oligomers in primary cortical neurons. Antioxidants (Basel) 10(6):937

Qin XY, Cheng Y, Yu LC (2012) Potential protection of green tea polyphenols against intracellular amyloid beta-induced toxicity on primary cultured prefrontal cortical neurons of rats. Neurosci Lett 513:170–173

Rezai-Zadeh K, Douglas Shytle R, Bai Y, Tian J, Hou H, Tet M et al (2009) Flavonoid-mediated presenilin-1 phosphorylation reduces Alzheimer's disease beta-amyloid production. J Cell Mol Med 13(3):574–588

Rinaldi P, Polidori MC, Metastasio A, Mariani E, Mattioli P, Cherubini A et al (2003) Plasma antioxidants are similarly depleted in mild cognitive impairment and in Alzheimer's disease. Neurobiol Aging 24(7):915–919

Roohbakhsh A, Parhiz H, Soltani F, Rezaee R, Iranshahi M (2015) Molecular mechanisms behind the biological effects of hesperidin and hesperetin for the prevention of cancer and cardiovascular diseases. Life Sci 124:64–74

Saaby L, Jager AK (2011) Flavonoids and the CNS. Molecules 16:1471–1485

Saidani M, Dhifi W, Marzouk B (2004) Lipid evaluation of some Tunisian citrus seeds. J Food Lipids 11:242–250

Sawmiller D, Habib A, Li S, Darlington D, Hou H, Tian J et al (2016) Diosmin reduces cerebral Aβ levels, tau hyperphosphorylation, neuroinflammation, and cognitive impairment in the 3xTg-AD mice. J Neuroimmunol 299:98–106

Silalahi J (2002) Anticancer and health protective properties of citrus fruit components. Asia Pac J Clin Nutr 11:79–84

Singh U, Wadhwani AM, Johri BM (1983) Dictionary of economic plants in India, 2nd edn. New Delhi, Indian Council of Agricultural Research (ICAR), pp 51–53

Tamilselvam K, Braidy N, Manivasagam T, Essa MM, Prasad NR, Karthikeyan S et al (2013) Neuroprotective effects of hesperidin, a plant flavanone, on rotenone-induced oxidative stress and apoptosis in a cellular model for Parkinson's disease. Oxidative Med Cell Longev 2013:1

Thenmozhi AJ, Raja TRW, Janakiraman U, Manivasagam T (2015) Neuroprotective effect of hesperidin on aluminium chloride induced Alzheimer's disease in Wistar rats. Neurochem Res 40:767–776

Thenmozhi AJ, Raja TRW, Manivasagam T, Janakiraman U, Essa MM (2017) Hesperidin ameliorates cognitive dysfunction, oxidative stress, and apoptosis against aluminium chloride-induced rat model of Alzheimer's disease. Nutr Neurosci 20(6):360–368

Varshney V, Garabadu D (2021) Naringin exhibits mas receptor-mediated neuroprotection against amyloid Beta-induced cognitive deficits and mitochondrial toxicity in rat brain. Neurotox Res 39(4):1023–1043

Wang D, Gao K, Li X, Shen X, Zhang X, Ma C et al (2012) Long-term naringin consumption reverses a glucose uptake defect and improves cognitive deficits in a mouse model of Alzheimer's disease. Pharmacol Biochem Behav 102(1):13–20

Wang D, Liu L, Zhu X, Wu W, Wang Y (2014) Hesperidin alleviates cognitive impairment, mitochondrial dysfunction and oxidative stress in a mouse model of Alzheimer's disease. Cell Mol Neurobiol 34:1209–1221

Wang DM, Yang YJ, Zhang L, Zhang X, Guan FF, Zhang LF (2013) Naringin enhances CaMKII activity and improves long-term memory in a mouse model of Alzheimer's disease. Int J Mol Sci 14:5576–5586

Wang J, Yuan Y, Zhang P, Zhang H, Liu X, Zhang Y (2018) Neohesperidin prevents $A\beta_{25-35}$-induced apoptosis in primary cultured hippocampal neurons by blocking the S-Nitrosylation of protein-Disulphide isomerase. Neurochem Res 43(9):1736–1744

Wong KC, Pang WY, Wang XL, Mok SK, Lai WP, Chow HK (2013) Drynariafortunei-derived total flavonoid fraction and isolated compounds exert oestrogen-like protective effects in bone. Br J Nutr 110:475–485

Yang Y, Gong XB, Huang LG, Wang ZX, Wan RZ, Zhang P et al (2017) Diosmetin exerts anti-oxidative, anti-inflammatory and anti-apoptotic effects to protect against endotoxin-induced acute hepatic failure in mice. Oncotarget 8(19):30723–30733

Yin L, Cheng W, Qin Z, Yu H, Yu Z, Zhong M et al (2015) Effects of naringin on proliferation and osteogenic differentiation of human periodontal ligament stem cells in vitro and in vivo. Stem Cells Int 2015:758706

Youn K, Lee S, Jun M (2019) Discovery of Nobiletin from citrus Peel as a potent inhibitor of β-amyloid peptide toxicity. Nutrients 11(11):2648

Zhang J, Gao W, Liu Z, Zhang Z, Liu C (2014) Systematic analysis of main constituents in rat biological samples after oral administration of the methanol extract of Fructus aurantii by HPLC-ESI-MS/MS. Iran J Pharm Res 13:493–503

Chapter 8
Ameliorative Effect of Pomegranate on Alzheimer's Disease

Anna Kuppusamy iyer Venkatachari Sama, Ponnambalam Annapoorani, Thamilarasan Manivasagam, Arokiasamy Justin Thenmozhi, and Mathiyazahan Dhivya Bharathi

Abstract Pomegranate, a well-known edible fruit, having high nutritional values and taken as a part of Mediterranean diet. It is reported to possess enormous amount of polyphenols and vitamins that are responsible for its beneficial effect against several chronic diseases. It is also shown to have neuroprotective properties against various neurological diseases. Here, in this chapter, the possible mechanisms by which pomegranate may offer neuroprotective effect for the sake of AD patients was discussed.

8.1 Introduction

8.1.1 Why Pomegranate?

Pomegranate (*Punica granatum* L.) is a commonly consumed fruit globally from very ancient times. Its usage is also reported from the Biblical times, and evidence for its therapeutic properties were found in several cultured peoples (Longtin 2003). According to Babylonians, pomegranate seeds are considered as the "agent of resurrection." The Ancient Chinese people considered pomegranate seeds as the

A. K. i. V. Sama · P. Annapoorani (✉)
Department of Biochemistry, V.V. Vanniaperumal College for Women, Virudhunagar, Tamil Nadu, India
e-mail: annapoorani@vvvcollege.org

T. Manivasagam · M. Dhivya Bharathi
Department of Biochemistry and Biotechnology, Faculty of Science, Annamalai University, Chidambaram, Tamil Nadu, India

A. Justin Thenmozhi
Department of Biochemistry, School of Biological Sciences, Madurai Kamaraj University, Madurai, Tamil Nadu, India

Department of Biochemistry and Biotechnology, Faculty of Science, Annamalai University, Chidambaram, Tamil Nadu, India

A. Justin Thenmozhi, T. Manivasagam (eds.), *Nutraceuticals for Alzheimer's Disease: A Promising Therapeutic Approach*, Nutritional Neurosciences,
https://doi.org/10.1007/978-981-99-0677-2_8

symbol for immortality and longevity, while the Persian people thought that the seeds bring invincibility in the war (Aviram et al. 2000). The pomegranate tree itself is believed to have grown in the garden of Eden and reported to be utilized in complementary and alternative medicine. According to the oldest Greek myth, this fruit was considered as the "fruit of the dead" and also called as vestments of the high priest in the Hebrew tradition. Vidal et al. (2003) stated the pomegranate as "healing food" with various therapeutic functions against various diseases. The palatable portion of the fruit includes the highly juicy and red coloured arils, fully consisting of phytochemicals (ellagic acid, punigluconin, ellagitannins, punicafolin, gallic acid, punicacortein (A, B, C, & D), granatin (A & B), and punicalagin) and antioxidants (vitamins B6 & C) that are responsible for its therapeutic effects. The pharmacological properties of pomegranate fruit includes (1) antioxidant (Cam et al. 2009), (2) antimicrobial (Duman et al. 2009), (3) antihepatotoxic (Celik et al. 2009), (4) antitumor (Hamad and Al-Momene 2009), (5) anti-inflammatory (Lee et al. 2010), (6) cardiovascular protective (Davidson et al. 2009), (7) antiviral (Haidari et al. 2009), (8) oral and skin promotive (Di Silvestro et al. 2009; Aslam et al. 2006), (9) antidiabetic (Xu et al. 2009), (10) anti-alzheimeric (Singh et al. 2008), and male fertility promotive (Forest et al. 2007) functions.

8.2 Scientific Classification of Pomegranate

Kingdom: Plantae
Clade: Tracheophytes, Angiosperms, Eudicots, Rosids
Order: Myrtales
Family: Lythraceae
Genus: Punica
Species: P granatum
Binomial name: *Punica granatum L*
Synonyms
Punica grandiflora hort. ex. steud
Punica spinosa Lam.
Punica nana L.
Punica florida Salisb.

8.2.1 Chemical Composition

It mainly contains the following (Rahimia et al. 2012):

1. Ellagic acid (EA) and their derivatives such as 3,3′-di-O-methyl EA, Eschweilenol C, diellagic acid rhamnosyl(1–4) glucoside, 3,3′, 4′-tri-O-methyl EA, and 3′-O-methyl-3, 4-methylene EA

2. Ellagitannins and gallotannins such as casuariin; Cyclic 2,4:3,6-bis (4,4′,5,5′,6,6′-hexahydroxy [1,1′-biphenyl]- 2,2′-dicarboxylate) 1-(3,4,5-trihydroxybenzoate) b-D-Glucose; 2,3-(S)-HHDP-D-glucose; 2-O-galloyl-4,6 (S,S) gallagoyl-D-glucose; punicalin; tellimagrandin I; corilagin; casuarinin; punicalagin; castalagin; granatin A and B; punicafolin; strictinin; punicacortein A, B, C, and D; punigluconin; tercatain; terminalin/ gallayldilacton; pedunculagin; and 5-O-galloyl-punicacortein D
3. Catechin and procyanidins like (-)-catechin, gallocatechin, catechin-(4,8)-gallocatechin, gallocatechin-(4,8)-gallocatechin, gallocatechin-(4,8)-catechin, and Procyanidin B1 & B2
4. Flavonols such as kaempferol, myricetin, eriodictyol, apigenin, naringenin, quercetin, luteolin, and quercimeritrin
5. Gallyol derivatives like brevifolin and their derivatives (carboxylic acid/its 10-monosulphate), 1,2,3- / 1,2,4- / 1,2,6- / / 1, 3,4–/ 1,4,6- Tri- /, 1,2, 4, 6-Tetra- / 1,2,3,4, 6-Pent-O-galloyl-β-D-glucose
6. Organic acids like succinic acid, gallic acid, Chlorogenic acid, o- and p-Coumaric acid, ferulic acid, caffeic acid, L-malic acid, quinic acid, cinnamic acid, oxalic acid, citric acid, and protocatechuic acid
7. Fatty acids and triglycerides including linoleic acid, stearic acid, punicic acid, oleic acid, eicosenoic acid, linolenic acid, palmitic acid, and various derivatives of punicylglycerol
8. Alkaloids like pelletierine, sedridine, hygrine, N-methyl pelletierine, pseudopelletierine, norhygrine, and nor-pseudopelletierine
9. Anthocyanins and anthocyanidins like cyanidin and their glucoside/diglucoside/ rutinoside derivatives, delphinidin and their glucoside/diglucoside derivatives), and pelargonidin and its derivative (glucoside)
10. Sterols and terpenoids such as betulinic acid, daucosterol, estradiol, asiatic acid, estrone, cholesterol, β-sitosterol, testosterone, estriol, stigmasterol, Friedooleanan-3-one, and ursolic acid
11. Other compounds such as phenylethylrutinoside, icariside D1, and mannitol

The fruit is divided into various parts: peel, seeds, and arils (Table 8.1).

8.2.2 In Silico Experiments on Pomegranate and AD

Using in silico method, the inhibitory effect of various constituents of pomegranate juice and their metabolic products such as ellagic acid, punicalagin, delphinidin-3-glucoside, pelargonidin, ellagitannin, dimethylellagic acid-glucuronide, gallotannin, gallic acid, urolithins A and B, urolithin A-glucuronide, on antioxidant enzymes like glutathione peroxidase, superoxide dismutase, glutathione reductase, catalase, and glutathione-S-transferase were demonstrated. The results of the experiment stated that the juice components are potent to inhibit enzymatic antioxidant defence system (Mazumder et al. 2019). In silico experiments involved the protective role of

Table 8.1 Structure and composition of pomegranate fruit

Edible/non-edible part of fruit	Structure	Composition	References
Non edible part (~50% weight of total fruit)	Peel	Luteolin, punicalagin, kaempferol, quercetin, gallagic, & ellagic acid glycosides, pedunculagin, etc.	Viuda-Martos et al. (2010)
Edible part (~50% weight of total fruit)	Seeds (~10% weight of total fruit)	~12–20%—seed oil containing punicic, stearic, linoleic, oleic, palmitic, and linolenic acids, polyphenols, protein, phytoestrogen coumestrol, vitamins, isoflavones (including genistein), crude fibres, minerals, pectin, estrone, sugars, etc.	Viuda-Martos et al. (2010)
	Arils (~40% weight of fruit)	Water (~85%); sugars mainly glucose and fructose (approx. 10%), pectin (~1.5%); citric, ascorbic and malic acids; trace amount of phenolics and flavonoids (anthocyanins)	Viuda-Martos et al. (2010)

pomegranate urolithins (A and B), their methyl ether metabolites, and their analogues against the enzymes involved in the pathogenesis of AD. These compounds are very potent inhibitors of several enzymes including acetylcholinesterase, monoamine oxidase B, butyrylcholinesterase and cyclooxygenase 1 and 2, and antioxidant enzymes (Noshadi et al. 2020).

8.2.3 Role of Pomegranate Extract/Compounds in Neurotoxin-Induced Model of AD

The nonenzymatic glycation enable the synthesis of several advanced glycation end products (AGEs), which have an inhibitory effect in the cause and progression of several diseases such as AD and type-2 diabetes. Pomegranate fruit has been used in AD management due to its potent AGEs' inhibition effect. Liu et al. (2014) demonstrated the anti-glycation role of pomegranate and its constituents such as gallic acid, ellagic acid, punicalagin, and urolithin A and B in an in vitro condition. Administration of extract, positive control (aminoguanidine), and some of its constituents (gallic acid and ellagic acid) showed more significant activities than the remaining compounds. Punicalagin administration enhanced the protein expression and activity of methionine sulfoxide reductase A, an antioxidant enzyme, that attenuated oxidative stress and is regulated by oxidizable methionine (35th position) in amyloid protein against human IMR-32 cells' neuroblastoma treated with Aβ peptides.

The protective role of freeze-dried pomegranate seed extracts on IL-1β-induced in vitro (SK-N-SH cells) model of AD was analysed. The extracts reduced the levels of Aβ and BACE-1 induced by IL-1β exposure. Moreover, the expression of inflammatory markers like phospho-IκB, COX-2, and phospho-IKK proteins were reduced in IL-1β, and the extract co-exposed the cells. The results also indicated that the extract suppressed NF-κB activation (Velagapudi et al. 2016). Pomegranate urolithin and its methyl derivative attenuated the fibrillation of Aβ in in vitro conditions. Pre-exposure of these compounds protected the *Caenorhabditis elegans* injected with $A\beta_{1-42}$ by attenuating paralysis and neurotoxicity more significantly than pomegranate extract or its predominant ellagitannins (Yuan et al. 2016).

Morezelle et al. (2016) studied the role of pomegranate peel extract on spatial memory, indices of oxidative stress, neuroplasticity, and inflammation in a mouse (C57Bl/6) infused with Aβ1–42 using mini-osmotic pumps. Barnes maze test was used to analyse spatial memory, as mouse fed with peel extract easily found escape box, which is not found in Aβ group. The peel extract intake diminished the density of amyloid plaque, enhanced the expression of BDNF (neurotrophin), and lowered the acetylcholinesterase activity. Along with the reduced lipid peroxidation process and inflammation (lowered levels of pro-inflammatory cytokine—TNF-α), no hepatic lesions were found in mouse administered with pomegranate peel extract.

Kim et al. (2017) demonstrated the preventive role of punicalagin on memory deficits induced by lipopolysaccharide-induced neurotoxicity in cultured microglial BV-2, astrocytes, and mice. Punicalagin ameliorated lipopolysaccharide-induced memory dysfunction and inflammation. It inhibited the lipopolysaccharide-induced inflammation by diminishing Cox-2, iNOS, TNF-α, and IL-1β expression. Moreover, it lowered the NF-κB activation through IκB and translocation of p65 and p50 into the nucleus in both the in vitro and in vivo models of AD. Moreover, it also significantly lowered the $A\beta_{1-42}$ generation by down-regulating Aβ precursor protein and β-secretase expression. The neuroprotective role of urolithins and its methylated products ameliorated the neuroinflammation in both the in vitro (lipopolysaccharide exposed lBV-2 microglia and co-cultured with SH-SY5Y neuroblastoma) and in vivo (transgenic-R1.40 mice) models of AD. Exposure of urolithins diminished the levels of IL-6, nitric oxide, TNF-α, and prostaglandin E2 in the LPS-induced BV-2 microglia. Moreover, its exposure attenuated the cell viability by inhibiting the apoptosis and caspase activation in the co-cultured SH-SY5Y model. It also offered its neuroprotective effect by inhibiting the over-expression of inflammatory biomarkers such as COX-2, IL-1 and -6, and TNF-α in the hippocampus of transgenic mice (DaSilva et al. 2019). Almuhayawi et al. (2020) compared the neuroprotective role of pomegranate extract and its loaded nanoparticles in an $AlCl_3$-mediated rat model of AD. Both the extract and its nano-formulation offered similar pattern of neuroprotection against aluminium toxicity by enhancing the brain/body weight ratio and antioxidant activities and also by maintaining normal architecture of the brain. These findings were supported by the behavioural studies showing enhanced discrimination index in the novel object recognition test.

Ooi et al. (2020) determined the role of several fruit juice mixtures (pomegranate, Roselle, and white guava) on behavioural, biochemical, and anatomical alterations in intracerebroventricular *Aβ*-infused rats. Histological studies and neuronal counting in CA1 region of the hippocampus indicated that the administration of formulation prevented the Aβ-mediated neuronal damage and shrinkage. The behavioural studies involving open field and novel object recognition tests showed no significant changes between formulation- and water-administered Aβ-infused groups. Although the significant reduction in the hippocampal MDA level, plasma CRH level, and iNOS expression was found in formulation-fed Aβ-infused group, no significant alterations were found in SOD activity.

8.2.4 Effect of Pomegranate Extract/Compounds in Transgenic Mouse Model of AD

Hartman et al. (2006) demonstrated the role of dietary administration of pomegranate juice on the learning deficits and AD pathology in a transgenic (APP (sw)/Tg2576) mouse model. In the Morris water maze test, juice-administrated mice learned easily to escape in the hidden platform and swam more quickly compared to control AD mice. Moreover, mouse fed with juice showed significantly less hippocampal soluble Aβ-42 and amyloid deposition than that in the AD control. Rojanathammanee et al. (2013) determined the neuroprotective effect of a dietary intervention of pomegranate extract in drinking water (juice) on microgliosis against the transgenic (APP/PS1) mice model of AD. Three months of daily pomegranate administration to transgenic mice diminished the path length to escape in the Barnes maze test compared with the control transgenic mice. It also reduced the levels of inflammatory markers like nuclear factor of activated T-cell and tumour necrosis factor-α compared to AD controls. Immunocytochemistry indicated that pomegranate-fed mice had ameliorated the deposition of Aβ plaque and microgliosis. A month of pomegranate administration after the experimental period enhanced the spontaneous alternations in T maze test compared to control AD mice.

Ahmed et al. (2014) demonstrated the protective role of pomegranate extract (standardized) on transgenic mice (R1.40) model of AD. Pomegranate extract exposure did not enhance the working and spatial memory functions (Y-maze and Morris water maze test) in the transgenic AD mice. But it attenuated the levels of Aβ precursor protein and Aβ, along with change in the activity of γ-secretase enzyme, favouring the anti-amyloidogenic mechanism. Subash et al. (2014) indicated the role of pomegranate fruit enriched with antioxidants on brain oxidant-antioxidant imbalance in the double transgenic Swedish mouse with mutated APP (APPsw/Tg2576). Supplementation of diet containing 4% pomegranate extract to the transgenic mice attenuated the oxidative stress by diminishing the levels of lipid peroxidation products. Transgenic mice fed with pomegranate extract restored the glutathione peroxidase, superoxide dismutase, catalase, and glutathione S transferase activities and the level of glutathione. Moreover, the acetylcholinesterase and Na^+-K^+-ATPase activities were also attenuated by pomegranate extract administration in the transgenic mice.

Transgenic APPsw/Tg2576 mice expressed the memory dysfunction, enhanced anxiety-like behaviour, and impaired position discrimination learning ability and spatial learning capability and motor coordination. Administration of pomegranate enhanced the improvement in learning and memory functions and locomotion and diminished the anxiety-like behaviour (Subash et al. 2015). Braidy et al. (2015) compared the beneficiary role of dietary administration of pomegranate extract, dates, and figs on the regulations of inflammatory markers in transgenic (APPsw/Tg2576) mice model of AD. Alterations in the circulatory cytokines and neuronal Aβ1–42, ATP, and inflammatory cytokines (TNF-α,IL-1β,-2, -3, -4, -5, -6, -9, -10 and eotaxin) in transgenic mice were attenuated by diet (pomegranates, figs, or dates) administration. Moreover, their administration significantly decreased senile plaque formation and reduced the levels of brain Aβ1-40 and 1-42.

Braidy et al. (2016) studied the role of dietary administration of pomegranate extract on inflammation and synaptic plasticity in transgenic (APPsw/Tg2576) mice model of AD. The pomegranate extract attenuated the proteins (SNAP25, Munc18-1 and PSD-95, phosphorylation of cAMP-response element-binding protein and calcium-/calmodulin-dependent protein kinase IIα, synaptophysin), reduced the activity of neuroinflammation, increased autophagy, and activated the phophoinositide-3-kinase-Akt-signaling pathway. It also offered neuroprotective function by reducing β-site cleavage of Aβ precursor protein in AD mice. Esselun et al. (2021) studied the influence of urolithins against mitochondrial deficits in an in vitro model of AD. Pre-treatment of urolithins to SH-SY5Y-APP695 enhanced the mitochondrial respiration and membrane potential, increased the adenosine triphosphate synthesis, and diminished the levels of reactive oxygen species. The q-rtPCR studies indicated that exposure of urolithins regulated the gene expression involved in mitochondrial bioenergetics (peroxisome proliferator-activated receptor gamma coactivator 1-alpha) and biogenesis (oestrogen-related receptor and mitochondrial transcription factor A). But the genes involved in the autophaghy (LC3B-I & B-II, and p62) were unaltered by urolithins.

8.3 Conclusion

Pomegranate and their components are more potent in preventing the cause or inhibiting the progression of neurodegenerative changes that occurred in AD through several mechanisms. Hence, the authors recommend the continuous and daily intake/consumption of pomegranate as a supplement to maintain a healthy lifestyle against aging and its associated neurodegenerative diseases. However, a clinical validation is very much needed in the near future, which is lacking in this field of nutritional neuroscience.

References

Ahmed AH, Subaiea GM, Eid A, Li L, Seeram NP, Zawia NH (2014) Pomegranate extract modulates processing of amyloid-β precursor protein in an aged Alzheimer's disease animal model. Curr Alzheimer Res 11(9):834–843

Almuhayawi MS, Ramadan WS, Harakeh S, Al Jaouni SK, Bharali DJ, Mousa SA et al (2020) The potential role of pomegranate and its nano-formulations on cerebral neurons in aluminum chloride induced Alzheimer rat model. Saudi J Biol Sci 27(7):1710–1716

Aslam MN, Lansky EP, Varani J (2006) Pomegranate as a cosmeceutical source: pomegranate fractions promote proliferation and procollagen synthesis and inhibit matrix metalloproteinase-1 production in human skin cells. J Ethnopharmacol 103(3):311–318

Aviram M, Dornfeld L, Rosenblat M, Volkova N, Kaplan M, Coleman R et al (2000) Pomegranate juice consumption reduces oxidative stress, atherogenic modifications to LDL, and platelet aggregation: studies in humans and in atherosclerotic apolipoprotein E-deficient mice. Am J Clin Nutr 71:1062–1076

Braidy M, Essa MM, Subash S, Akbar M, Al-Adawi S, Guillemin GJ (2015) Long-term dietary supplementation of pomegranates, figs and dates alleviate neuroinflammation in a transgenic mouse model of Alzheimer's disease. PLoS One 10(3):e0120964

Braidy N, Essa MM, Poljak A, Selvaraju S, Al-Adawi S, Manivasagam T et al (2016) Consumption of pomegranates improves synaptic function in a transgenic mice model of Alzheimer's disease. Oncotarget 7(40):64589–64604

Cam M, Hısıl Y, Durmaz G (2009) Classification of eight pomegranate juices based on antioxidant capacity measured by four methods. Food Chem 112:721–726

Celik I, Temur A, Isik I (2009) Hepatoprotective role and antioxidant capacity of pomegranate (*Punica granatum*) flowers infusion against trichloroacetic acid-exposed rats. Food Chem Toxicol 47(1):145–149

DaSilva NA, Nahar PP, Ma H, Eid A, Wei Z, Meschwitz S et al (2019) Pomegranate ellagitannin-gut microbial-derived metabolites, urolithins, inhibit neuroinflammation in vitro. Nutr Neurosci 22(3):185–195

Davidson MH, Maki KC, Dicklin MR, Feinstein SB, Witchger MS, Bell M et al (2009) Effects of consumption of pomegranate juice on carotid intima-media thickness in men and women at moderate risk for coronary heart disease. Ame J Cardiol 104(7):936–942

Di Silvestro RA, Di Silvestro DJ, Di Silvestro DJ (2009) Pomegranate extract mouth rinsing effects on saliva measures relevant to gingivitis risk. Phytother Res 23:1123–1127

Duman AD, Ozgen M, Dayisoylu KS, Erbil N, Durgac C (2009) Antimicrobial activity of six pomegranate (*Punica granatum* L.) varieties and their relation to some of their pomological and phytonutrient characteristics. Molecules 14:1808–1817

Esselun C, Theyssen E, Eckert GP (2021) Effects of Urolithin a on mitochondrial parameters in a cellular model of early Alzheimer disease. Int J Mol Sci 22(15):8333

Forest CP, Padma-Nathan H, Liker HR (2007) Efficacy and safety of pomegranate juice on improvement of erectile dysfunction in male patients with mild to moderate erectile dysfunction: a randomized, placebo-controlled, double-blind crossover study. Int J Impot Res 19(6):564–567

Haidari M, Ali M, Casscells SW, Madjid M (2009) Pomegranate (*Punica granatum*) purified polyphenol extract inhibits influenza virus and has a synergistic effect with oseltamivir. Phytomedicine 16(12):1127–1136

Hamad AW, Al-Momene W (2009) Separation and purification of crude ellagic acid from white flesh of pomegranate fruits as a potent anti-carcinogenic. New Biotechnol 25(1):286

Hartman RE, Shah A, Fagan AM, Schwetye KE, Parsadanian M, Schulman RN et al (2006) Pomegranate juice decreases amyloid load and improves behavior in a mouse model of Alzheimer's disease. Neurobiol Dis 24(3):506–515

Kim YE, Hwang CJ, Lee HP, Kim CS, Son DJ, Ham YW et al (2017) Inhibitory effect of punicalagin on lipopolysaccharide-induced neuroinflammation, oxidative stress and memory impairment via inhibition of nuclear factor-kappaB. Neuropharmacology 117:21–32

Lee CJ, Chen LG, Liang WL, Wanga CC (2010) Anti-inflammatory effects of *Punicagranatum* Linne in vitro and in vivo. Food Chem 118:315–322

Liu W, Ma H, Frost L, Yuan T, Dain JA, Seeram NP (2014) Pomegranate phenolics inhibit formation of advanced glycation end products by scavenging reactive carbonyl species. Food Funct 5(11):2996–3004

Longtin R (2003) The pomegranate: nature's power fruit? J Natl Cancer Inst 95:346–348

Mazumder MK, Choudhury S, Borah A (2019) An in silico investigation on the inhibitory potential of the constituents of Pomegranate juice on antioxidant defense mechanism: Relevance to neurodegenerative diseases. IBRO Rep 6:153–159

Morzelle MC, Salgado JM, Telles M, Mourelle D, Bachiega P, Buck HS et al (2016) Neuroprotective effects of pomegranate peel extract after chronic infusion with Amyloid-β peptide in mice. PLoS One 11(11):e0166123

Noshadi B, Ercetin T, Luise C, Yuksel MY, Sippl W, Sahin MF et al (2020) Synthesis, characterization, molecular docking, and biological activities of some natural and synthetic urolithinanalogs. Chem Biodivers 17(8):e2000197

Ooi TC, Munawar MA, Mohd Rosli NH, Abdul Malek SNA, Rosli H, Ibrahim FW et al (2020) Neuroprotection of tropical fruit juice mixture via the reduction of iNOS expression and CRH level in β-amyloid-induced rats model of Alzheimer's disease. Evid Based Complement Alternat Med 2020:5126457

Rahimia HR, Arastoob M, Ostada SN (2012) A comprehensive review of Punica granatum (pomegranate) properties in toxicological, pharmacological, cellular and molecular biology researches. Iran J Pharm Sci 11(2):385–400

Rojanathammanee L, Puig KL, Combs CK (2013) Pomegranate polyphenols and extract inhibit nuclear factor of activated T-cell activity and microglial activation in vitro and in a transgenic mouse model of Alzheimer disease. J Nutr 143(5):597–605

Singh M, Arseneault M, Sanderson T, Morthy V, Ramassamy C (2008) Challenges for research on polyphenols from foods in Alzheimer's disease: bioavailability, metabolism and cellular and molecular mechanism. J Agric Food Chem 56:4855–4873

Subash S, Braidy N, Essa MM, Zayana AB, Ragini V, Al-Adawi S et al (2015) Long-term (15 mo) dietary supplementation with pomegranates from Oman attenuates cognitive and behavioral deficits in a transgenic mice model of Alzheimer's disease. Nutrition 31(1):223–229

Subash S, Essa MM, Al-Asmi A, Al-Adawi S, Vaishnav R, Braidy N et al (2014) Pomegranate from Oman alleviates the brain oxidative damage in transgenic mouse model of Alzheimer's disease. J Tradit Complement Med 4(4):232–238

Velagapudi R, Baco G, Khela S, Okorji U, Olajide O (2016) Pomegranate inhibits neuroinflammation and amyloidogenesis in IL-1β-stimulated SK-N-SH cells. Eur J Nutr 55(4):1653–1660

Vidal A, Fallarero A, Pena BR, Medina ME, Gra B, Rivera F et al (2003) Studies on the toxicity of *Punicagranatum* L.(Punicaceae) whole fruit extracts. J Ethnopharmacol 89:295–300

Viuda-Martos M, Fern'andez-L'opez J, Pérez-Álvarez JA (2010) Pomegranate and its many functional components as related to human health: a review. Compr Rev Food Sci Food Saf 9:635–653

Xu KZY, Zhu C, Kim MS, Yamahara J, Li Y (2009) Pomegranate flower ameliorates fatty liver in an animal model of type 2 diabetes and obesity. J Ethnopharmacol 123(2):280–287

Yuan T, Ma H, Liu W, Niesen DB, Shah N, Crews R et al (2016) Pomegranate's neuroprotective effects against Alzheimer's disease are mediated by urolithins, its ellagitannin-gut microbial derived metabolites. ACS Chem Neurosci 7(1):26–33

Lee CJ, Chen LG, Liang WL, Wang CC (2010) Anti-inflammatory effects of Punica granatum Linne in vitro and in vivo. Food Chem 118:315–322

Chapter 9
Nuts and Their Potential Role in Alzheimer's Disease

Ganesh Vaishnavi and Arokiasamy Justin Thenmozhi

Abstract The rising incidence of Alzheimer's disease (AD), a challenging neurological disorder, is a practically global issue. To find potential treatments for the condition, research is being done on natural products. The potential therapeutic properties of tree nuts have been analyzed in many research articles and traditional medicinal systems like Persian Medicine because of their properties to protect the brain. The purpose of this chapter is to emphasize the advantages of consuming nuts as dietary supplements and natural therapies for AD patients by providing a pharmacological evaluation of their bioactive components. The macronutrients, micronutrients, and phytochemicals found in almonds, hazelnuts, and walnuts have an effect on several pathways involved in the pathogenesis of AD, such as amyloidogenesis, tau phosphorylation, oxidative stress, cholinergic pathways, and some nontarget mechanisms, such as effects on neurogenesis and cholesterol-lowering and anti-inflammatory properties. Especially in the case of hazelnut, it reverses brain atrophy. Along with almonds, walnuts, and hazelnuts, other nuts like pistachio, pine nuts, peanuts, areca nuts, kola nuts, and pecan nuts also have the potential to ameliorate Alzheimer's disease by their bioactive components. Beyond the molecular effects associated with the phytochemicals, the utilization of these tree nuts as valuable nutrients for the prevention or perhaps management of AD may be more thoroughly investigated in scientific investigations.

Keywords Alzheimer's disease · Almonds · Hazel nuts · Walnuts · Cholesterol-lowering effect · Anti-inflammatory properties

G. Vaishnavi
Department of Molecular Microbiology, School of Biotechnology, Madurai Kamaraj University, Madurai, Tamil Nadu, India

A. Justin Thenmozhi (✉)
Department of Biochemistry, School of Biological Sciences, Madurai Kamaraj University, Madurai, Tamil Nadu, India

Department of Biochemistry and Biotechnology, Faculty of Science, Annamalai University, Chidambaram, Tamil Nadu, India

A. Justin Thenmozhi, T. Manivasagam (eds.), *Nutraceuticals for Alzheimer's Disease: A Promising Therapeutic Approach*, Nutritional Neurosciences,
https://doi.org/10.1007/978-981-99-0677-2_9

9.1 Introduction

Dementia, which mostly affects the older population and is becoming a worldwide health problem, affected 47 million individuals in 2015, and it is predicted that this figure would rise to 131 million by the year 2050 (De Lima Oliveira et al. 2018). Alzheimer's disease (AD) is a deadly neurological disease that slowly impairs cognitive abilities over the course of 5–20 years. Despite the fact that there are many different varieties of dementia, AD ranks as the most prevalent among senior individuals and is responsible for more than 60% of dementia cases. Approximately, 25 million individuals around the world, including 5.5 million Americans, are affected by AD. AD affects 10% of the overall population over the age of 65 and approximately 50% of those over the age of 85 (Chauhan and Chauha 2020). The gradual deposition of fibrillar amyloid beta protein (Aβ) as amyloid plaques and of paired helical filaments as neurofibrillary tangles in the brain is the neuropathological marker of AD (Huang and Jiang 2009).

9.2 Pathogenesis of Alzheimer's Disease

Several processes in AD pathogenesis, including amyloidogenesis, tau phosphorylation, oxidative stress, cholinergic pathways, and certain nontarget mechanisms, such as cholesterol lowering, are influenced by the macronutrients, micronutrients, and phytochemicals found in almond, hazelnut, and walnut. In scientific studies, the consumption of these nuts might be more closely examined as efficient nutrients for prevention or maybe treatment of AD, beyond the molecular actions related to the phytochemicals (Gorji et al. 2018).

The primary theories put out for the etiology of AD include the accumulation of amyloid, tau phosphorylation, impairment of intracellular signaling, oxidative stress, dysregulation of metal ions, and inflammation (Mocchegiani et al. 2014; Webber et al. 2005). The cerebrocortical cholinergic system and the somatostatin-containing neuronal systems are the most damaged; however, deficiencies in multiple neurotransmitter systems have been found in different brain regions in AD. Degeneration of basal forebrain cholinergic neurons, a key feature of AD, results in a striking impairment of cortical cholinergic neurotransmission pathways, including acetylcholine (ACh) production, release, and uptake, as well as increased choline acetyltransferase and acetylcholinesterase (AChE) activity (Perry et al. 1978, 1986). The pathophysiology of AD is thought to be influenced by a variety of environmental and genetic elements (De Jesus Moreno Moreno 2003). The main amyloid protein in AD is an amyloid beta protein (Aβ), which has 40 or 42 amino acids and can be soluble or fibrillar in nature. Before dementia is identified based on its clinical symptoms, AD's neuropathological alterations take decades to manifest.

9.3 Walnuts, Almonds, and Hazelnuts: The Three Major Nuts and Their Role in Ameliorating Alzheimer's Disease

9.3.1 Walnuts

Since ancient times, people have consumed walnuts (*Juglans regia L.*, Juglandaceae) for their nutritional needs. The edible portion of the fruit, the seed, or kernel is consumed either fresh or roasted, on its own or in other edible products, making walnuts a crop of great commercial significance to the food industry. It is well-liked and highly respected throughout the world for its dietary, wellness, and sensory qualities. The natural fresh kernels are mostly eaten whole or added to different confectioneries. They are a food that is rich in nutrients because of their protein, vitamin, and mineral profiles, as well as their fat content. Additionally, walnut kernels are an excellent source of many flavonoids, phenolic acids, and related polyphenols (Martínez et al. 2010). On this note, in a study, APP-transgenic mouse (AD-tg) was used to determine walnut's efficiency in order to treat AD. Amyloid precursor protein (APP) is broken down into amino acids to become Aβ. An animal model of AD is the APP-transgenic mouse (AD-tg), which exhibits memory impairment and Aβ deposition in the brain.

The amount of walnuts that should be consumed each day is 28–42 g or 12–18 walnut halves. Benefits of brief (14 months) nutritional supplementation with walnuts on AD-tg mouse help in relieving anxiety, improving memory, learning abilities, and motor coordination. In comparison to wild-type mice on the same diet, AD-tg mice had memory deficit, anxiety-related behavior, and impairments in motor coordination, position discrimination learning capacity, and spatial learning ability. In comparison to a control diet without walnuts, meals supplemented with walnuts (6% or 9%) in AD-tg mice exhibited an improvement in memory, learning abilities, motor development, and anxiety-related behavior. In terms of overall calorie consumption and the amounts of protein, carbohydrates, and fat in the control and experimental mice's meals, they were comparable. The Tg2576 transgenic (tg) mouse model of AD's memory, learning capacity, and anxiety were all considerably improved by walnut supplementation in the diet (Muthaiyah et al. 2014).

Aβ/ROS/NF-κB pathway modulation by walnut's peptide supplementation has also been demonstrated to successfully reduce learning and memory deficits brought on by Aβ25–35 (an active fragment of toxic Aβ) in vivo. The expression of NF-κB was decreased in the hippocampus of AD mice treated with walnut peptides compared to the control group, and walnut peptides significantly reduced the expression of pro-inflammatory cytokines such as IL-6, IL-1, and TNF-, along with the level of acetylcholinesterase (AChE), and the level of antioxidant enzymes (Zou et al. 2016).

In PC12 pheochromocytoma cells, walnut extract decreased Aβ-mediated cell death, lactate dehydrogenase release (associated with membrane damage), and DNA damage (associated with apoptosis). Aβ causes PC12 cells to produce ROS. When cells were treated with walnut extract instead of just Aβ, the amount of intracellular

ROS that resulted from Aβ treatment was substantially lower, proving that walnut extract had the ability to scavenge oxygen radicals. Numerous studies have linked the production of ROS and the subsequent oxidative stress to the mechanism of Aβ-induced cell death (Muthaiyah et al. 2011). Higher long-term nut intake, especially walnut intake, was linked to improved cognitive function in another clinical investigation involving older women (O'Brien et al. 2014). In total, 15,467 women (mean age: 74 years) who were 70 years of age or older took part in the study conducted by Arab and Ang (2015). Women who consumed five or more servings of nuts per week scored much better on cognitive tests than women who did not, and this difference was equal to 2 years of cognitive aging. Additionally, eating walnuts improved cognitive scores in the National Health and Nutrition Examination Study (NHANES) of an adult population in the United States (Arab and Ang 2015).

In a study conducted by Pribis et al., at the conclusion of the 8-week dietary intervention period, verbal and nonverbal reasoning, memory, emotional states, and other cognitive abilities were evaluated. The participants on the walnut-enriched diet showed a substantial improvement in inferential verbal thinking, but there were no significant alterations in mood, memory, or nonverbal reasoning. This can be due to the fact that in this trial, walnut supplementation only lasted for 8 weeks (Pribis et al. 2012). By defibrillating Aβ premade fibrils in vitro, methanolic extract of walnut has been demonstrated to suppress Aβ fibril production and keep Aβ in the soluble state. Walnut extract prevented synthetic Aβ from fibrillizing and dissolved Aβ fibrils. In this case, the shape of the Aβ structure was investigated using electron microscopy, and the degree of Aβ aggregation/fibrillization was evaluated using thioflavin T fluorescence spectroscopy. In fact, it has been noted that inhibiting Aβ fibrillization can lessen Aβ's neurotoxic effects. As a result, an approach to stop Aβ fibrillization can act as preventative and therapeutic implications in AD. According to the study's findings, walnut extract may both solubilize Aβ fibrils and prevent Aβ from fibrillizing (Chauhan et al. 2004).

In another study, one serving of nuts per week was reported by 187 individuals, 148 of whom mentioned eating walnuts. A considerably decreased WMH volume was associated with nut (or walnut) consumption of less than one serving per week ($p = 0.035$ for both). None of the interactions between nuts and APOE-4 were statistically significant. Participants who consumed less than one serving of walnuts per week had substantially higher GMv (grey matter volume) in the anterior/middle cingulate cortex, a region important for cognition and linked to healthy aging. Consumption of nuts, especially walnuts, is associated with favorable GMv and cerebral vasculature phenotypes (Sala-Vila et al. 2020).

MCI, dementia, and many other age-related disorders are all influenced by oxidative stress and inflammation. Walnuts include a variety of antioxidant and anti-inflammatory compounds that may work in concert to reduce inflammation and oxidative damage. Walnuts have the potential to reduce oxidative stress by enhancing antioxidant defense as well as lowering free radical levels, which prevents oxidative damage to lipids and proteins (Chauhan and Chauha 2020). When compared to a control diet, including walnuts in the diet has no negative effects on weight reduction efforts or causes weight gain (Neale et al. 2017).

9.3.2 *Almonds*

Almonds (*Prunus amygdalus*), the most widely consumed tree nut in the world and the leading producer of tree nuts, are members of the Rosaceae family, which also includes apples, pears, prunes, and raspberries. Typically, they are used as snacks and as ingredients in a wide range of processed meals, particularly in bakery and confectionery goods. The edible almond fruit (*P. amygdalus*), which resembles a peach, is divided into three distinct parts: the inside kernel or meat, the center region of the shell, and the outside green shell cover. Hard-shelled and soft-shelled almond varietals are distinguished by the texture of their shells (Esfahlan et al. 2010). The United States produces the most almonds in the world, and the majority of those are farmed in California (Esfahlan et al. 2010).

In a report, it was shown that adding almonds to the food of rats for 7 and 14 consecutive days at doses of 150, 300, and 600 mg/kg per kg substantially cured scopolamine-induced amnesia. They demonstrated that almonds decreased the activity of the enzyme cholinesterase in the rats' brains, as well as lowered cholesterol and triglycerides and slightly increased glucose, suggesting that almonds may be useful in the treatment of AD (Kulkarni et al. 2010). Rats fed with almond paste for 28 days experienced substantial improvements in learning and memory compared to the control group while also eating less and having lower plasma cholesterol levels. Following oral ingestion of almonds, increased brain tryptophan (TRP) monoamine levels and serotonergic turnover were also seen in rats (Haider et al. 2012). Another study found that rats' memory retention was greatly better after receiving almond treatment for 28 days. The scopolamine-induced amnesia paradigm likewise showed almond's memory-improving effects. According to that study, acetylcholine played a part in how almonds reduced the effects of scopolamine on amnesia (Batool et al. 2016).

9.3.3 *Hazelnuts*

Hazelnut (*Corylus avellana*), known for its nutritional and nutraceutical benefits, has made its way into nontraditional meals. Because of its unique combination of fat (mostly oleic acid), protein, carbohydrate, dietary fiber, vitamins (vitamin E), minerals, phytosterols (primarily α-sitosterol), and antioxidant phenolics, hazelnuts play a significant role in human nutrition and health. The *Corylus* genus and *Betulaceae* family are the home of the hazelnut. Hazelnuts have a carbohydrate content between 10% and 22%, 1–3% cellulose and pectin, 10–24% protein, 70–73% fat, and 2.4–2.8% ash. It is a good source of phytochemicals, fats, notably monounsaturated fats, and bioactive substances that can be dissolved in fat (tocopherol, phytosterols). Additionally, it includes dietary fiber, minerals, vitamins, antioxidant phenolics, vital amino acids, and other compounds. These can lower cholesterol levels, avoid metabolic syndrome, and lower the risk of cardiovascular illnesses. Catechin, epicatechin, epicatechin gallate, and gallic acid are abundant in the main by-products of hazelnuts (hard shells). Both the raw and roasted varieties of hazelnut can be eaten. Depending on the temperature and duration, roasting enhances the product's flavor, color, or texture (Wani et al. 2020). Around 650,000 MT of hazelnuts were produced in Turkey annually in 2001, accounting for roughly 70% of all hazelnuts produced worldwide. Turkey is followed in production by Italy (130,000 MT annually), the United States (45,000 MT annually), and Spain (around 25,000 MT annually). Additionally, Turkey accounts for more than 80% of the global hazelnut trade, which is worth about $1 billion annually (Alasalvar et al. 2003).

Hazelnuts are a great source of manganese, copper, and selenium as well as potassium, chromium, iron, magnesium, phosphorus, calcium, and zinc. Vitamins (K, E, and A) and water-soluble vitamins are both present in hazelnuts. Additionally, it includes proanthocyanidins, free amino acids, proteins, sugars, and organic acids. Significant concentrations of tocopherols, phytosterols, squalene, and polyphenols have all been found in hazelnut oil, according to reports (Alasalvar et al. 2010). According to one study, hazelnut kernel supplementation (800 mg/kg/day) in rats' diets improved memory and decreased anxiety while also reducing levels of cyclooxygenase-2, interleukin-1, tumor necrosis factor, B-cell lymphoma 2 (Bcl-2), Bcl-2-associated X protein, and caspase 3 that are associated with the neuroinflammation and apoptosis caused by amyloid-β (Aβ).

Hazelnut reduced body weight increase, making it potentially useful for disorders like weight management. Additionally, rats given hazelnuts orally showed better memory, decreased anxiety-related behavior, and showed protective efficacy by reducing apoptosis and inflammation in response to intra-hippocampal injection of Aβ. In addition, hazelnut, a nutrient-rich food, may be used in novel products to promote healthy aging and may be an effective diet for the treatment of AD (Bahaeddin et al. 2017). Additionally, the hazelnut-supplemented food boosted grooming and horizontal movement while decreasing rearing activities. While anxiety markers decreased at lower concentrations of Italian and larger concentrations of Turkish hazelnut, working memory also improved significantly with both cultivars of hazelnut. Both types of hazelnuts were linked to lower levels of acetylcholinesterase activity, superoxide dismutase activity, nitric oxide, and caspase-3, but higher amounts of dopamine (Olofinnade et al. 2021).

9.4 Beneficial Effect of the Nuts when Taken Together

When eaten alone, especially before bed, all of these nuts mentioned above are regarded as "brain food," contributing to mental alertness, focus, memory, and recall abilities. This could enhance the quality of sleep (Khorasani 2001). Almonds and hazelnuts are especially mentioned in several Persian medicinal literatures as being good for memory, protecting brain tissue, and preventing brain atrophy (Vanmierlo et al. 2015).

For those who are regularly exposed to these neurotoxins from heavy metals, neurological illnesses require special consideration. Therefore, one study was performed to ascertain if almond and walnut consumption can prevent cadmium-related cognitive impairment and oxidative stress. In MWM (Morris water maze test), OFT (open-field test), and NOR (novel object recognition test) paradigms, weekly cadmium induction for 1 month resulted in reduced spatial, habituation, and recognition memory, respectively. Following cadmium injection, a neurochemical investigation showed decreased acetylcholine and increased acetylcholinesterase activity, indicating cholinergic dysfunction. The treatment of almonds and walnuts for 28 days dramatically reduced these behavioral and neurochemical abnormalities. According to Batool et al., 400 mg/kg/day of almond and walnut supplementation for 28 days prevented cadmium-induced memory impairment in rats through cholinergic and antioxidant actions. Antioxidant elements in these nuts may increase the antioxidant capacity of neurons, lowering oxidative stress brought on by cadmium and enhancing memory. For the normal community exposed to heavy metals through employment or daily activities, the findings acquire considerable relevance. Consuming nuts regularly may protect us against the toxicity caused by heavy metals as a result of environmental pollution (Batool et al. 2017).

In two PREDIMED (Prevención con Dieta Mediterránea) clinical studies from Spain, healthy adult volunteers on the Mediterranean diet supplemented with 30 g mixed nuts/day (15 g walnuts, 7.5 g hazelnuts, and 7.5 g almonds) demonstrated higher cognitive function than the control group on a low-fat diet, and memory was considerably enhanced when compared to baseline scores in that group. A total of 522 individuals with a median age of 74.6 years or 447 adults with a median age of 66.9 years who were at high cardiovascular risk but did not have either CVD or cognitive impairment participated in these investigations. At the conclusion of the clinical study, 6.5 years after the start of the food intervention, a baseline examination was not conducted, but global cognitive function was evaluated using the Mini-Mental State Examination (MMSE) and Clock Drawing Test (CDT). The second research compared results from six distinct neuropsychological tests administered at baseline and 4.1 years following dietary intervention to determine rates of cognitive change over time (Martinez-Lapiscina et al. 2013; Valls-Pedret et al. 2015).

9.5 Bioactive Ingredients in Almond, Hazelnut, and Walnut and Their Neuropharmacological Mechanisms Underlying Their Neuroprotective Effects

Potential compounds like phenolic compounds, omega-3 fatty acids, fat-soluble vitamins, isothiocyanates, and carotenoids help in the prevention of AD. In addition to acting as antioxidants and anti-inflammatory agents, they also actively modulate the pathogenic molecular pathways involved in the development of AD, such as the creation of amyloid plaques and tau tangles, which are the two main pathological hallmarks of AD. Studies using in vivo animal models and clinical and epidemiological data reveal that nutritional intervention benefits older adults' health and may halt the age-related cognitive decline, particularly when the diet contains many bioactive nutrients (Grodzicki and Dziendzikowska 2020).

The enzyme α-secretase, which cleaves APP into a soluble sAPPα protein and another fragment that is subsequently proteolyzed by γ-secretase, catalyzes the process. An intracellular APP domain (AICD) and a p3 fragment are produced by this mechanism. However, in pathological circumstances, APP might enter a different endosomal-lysosomal route for APP proteolytic processing. Additionally, APP biochemical changes are influenced by β-secretase, which functions closer to the APP N-terminus, close to the lumen of the cellular organelles, whereas γ-secretase operates closer to the APP C-terminus, which is submerged in the cytoplasm.

A sAPPβ protein is produced by the catalytic activity of the β-secretase enzyme. Along with AICD, amyloid beta is another insoluble peptide produced by this mechanism. The alternate APP cleavage by β-secretase occurs 50% more frequently in AD patients compared to healthy people. Aβ is therefore highly concentrated in the extracellular space, where it binds to apolipoprotein E (APOE), degenerating axons, microglia, and astrocytes that have been activated by pro-inflammatory cytokines (Sharma et al. 2018a, b; Kinney et al. 2018; Yin et al. 2015; Kumar et al. 2018).

9.5.1 Fatty Acids

Considering the two unsaturated fatty acids that are most prevalent in almonds and hazelnuts are oleic acid (number 9) and linoleic acid (number 6). It has been shown that mice on a cholesterol-free diet may lower their Aβ levels by supplementing with oleic acid. The primary β-secretase in neurons responsible for Aβ production in the brain is β-secretase 1 (BACE1); oleic acid and linoleic acid have been shown to inhibit BACE1 in vitro. BACE1 is the enzyme that cleaves the APP (Youn et al. 2014). It has been demonstrated that oleic acid greatly reduces prolyl endopeptidase activity, which is elevated in AD brains (Park et al. 2006). Oligomerization and fibrillation were observed to be inhibited by conjugated linoleic acid (Lee et al. 2013).

Another essential PUFA, alpha-linolenic acid (ALA, 18:3 *n*-3), is the precursor of the PUFAs docosahexaenoic acid (DHA, 22:6 *n*-3) and eicosapentaenoic acid (EPA, 20:5 *n*-3), and it has been shown to have antioxidative and neuroprotective properties and the ability to speed up brain development in preterm and neonates (Kim et al. 2014). There is evidence that long-term ALA supplementation in rats prevents cognitive decline, raises DHA and EPA levels in the cortex and hippocampus, and lowers Aβ levels and tau phosphorylation. Additionally, by preventing BACE1 activity, ALA reduced the amyloidogenic processing of APP and avoided CREB (cAMP response element-binding protein) malfunction in the cortex and hippocampus. In older rats eating a high-fat diet, ALA therapy also reduced the phosphorylation of eIF2 (eukaryotic initiation factor 2), which was PERK (protein kinase RNA-like endoplasmic reticulum kinase) dependent. This neutralized the rise of ATF4 (activating transcription factor 4) (Gao et al. 2016).

Tree nuts are very nutrient dense and contain macronutrients (fat, protein, and carbohydrates), micronutrients (minerals and vitamins), fat-soluble bioactives (monounsaturated fatty acids [MUFA], polyunsaturated fatty acids [PUFA], monoacylglycerols [MAG], diacylglycerols [DAG], triacylglycerols [TAG], and phospholipids. In particular, beta-sitosterol, campesterol, and stigmasterol are found in large quantities in almonds, hazelnuts, and walnuts (Alasalvar et al. 2003; Maguire et al. 2004).

9.5.2 Vitamins

Vitamin E can also aid in the prevention of AD. Eight different compounds make up this mixture, including four tocopherols and four tocotrienols, each of which can take one of the following four chemical forms: α, β, γ, or δ. Their main sources are nuts, seeds, and vegetable oils (Boccardi et al. 2016).

9.5.3 Phytosterols

According to various in vitro and in vivo investigations, phytosterols may impact the expression, activity, and availability of β-secretases as well as other molecular processes that contribute to AD, such as the creation of Aβ through the processing of APP (Vanmierlo et al. 2015).

9.5.4 Folic Acid

During embryogenesis, folate insufficiency is linked to cognitive decline and AD, as well as neural tube abnormalities. Through a number of methods, including the reduction of inflammatory markers and oxidative stress, the prevention of excessive Ca^{++} influx, and the buildup of hyperphosphorylated tau and Aβ, folic acid may function as a protective factor and enhance cognition in AD (Hinterberger and Fischer 2013).

9.5.5 Thiamine

In the entire brain as well as in particularly damaged areas, thiamine-dependent enzymes transketolase, pyruvate dehydrogenase complex (PDHC), and

2-ketoglutarate dehydrogenase complex (KGDHC) are less effective in mice with thiamine deficiency. In addition, it was shown that thiamine deficiency caused oxidative stress, inflammation, and the development of amyloid plaque in the hippocampus of Tg19959 AD mice (Gibson et al. 1988). The Aβ buildup in the brains of AD model mouse models has been shown to be reversed by thiamine administration (Zhang et al. 2011).

9.5.6 *Niacin*

Pretreatment with nicotinamide in Aβ-injected mice significantly decreased the gene expression of presenilin 1 and amyloid precursor protein, increased the expression of sirtuin 1 (a conserved NAD+-dependent enzyme beneficial against AD), and decreased nuclear factor-KB (NF-kB) in the brain tissue (Kim and Yang 2017).

9.5.7 *Phenolic Compounds*

Walnuts are the main source of gallic acid. Gallic acid pretreatment of microglia cells prevented NF-kB acetylation, decreased cytokine production, and protected neuronal cells against Aβ-induced neurotoxicity and cell death. Gallic acid restored cognitive impairment brought on by Aβ in mice, as seen in tests using the Y-maze and passive avoidance (Kim et al. 2011). Ellagic acid (EA) was discovered to be a powerful inhibitor of the advanced glycation end products in all three of these nuts, notably in walnut (AGEs). Compared to aminoguanidines, it was more powerful. Numerous chronic human illnesses, including AD, are largely attributed to AGEs, and glycation of Aβ promotes its aggregation (Xia and Mo 2016).

To lessen or delay the onset of neurodegenerative illnesses, ellagic acid may activate a variety of cell signaling pathways. Ellagic acid has strong neuroprotective benefits due to its ability to scavenge free radicals, chelate iron, activate several cell signaling pathways, and reduce mitochondrial abnormalities. Additionally, S-nitrosylation of protein disulfide isomerase (PDI), which is thought to play a role in the pathogenesis of AD, was directly hampered by EA. Additionally, it has been proven that EA therapy enhanced memory performance in an animal model of AD and showed inhibitory effects on β-secretase (BACE1) (Ahmed et al. 2016). It has been demonstrated that caffeic acid protects PC12 cells against A-induced toxicity by reducing intracellular calcium influx and tau phosphorylation by reducing GSK-3 activation (Sul et al. 2009). Comparatively to mice that had received Aβ25–35 injections, caffeic acid enhanced spatial cognition and memory in the T-maze and object recognition tests (Kim et al. 2015).

9.5.8 Flavonoids

Flavonoids are a class of naturally occurring chemicals that have the capacity to regulate intracellular signals that support cellular survival, providing free radical scavenging characteristics and neuroprotection from oxidative stress (Mercer et al. 2005; Harnly et al. 2006). Chemicals known as flavonoids have anti-inflammatory, antibacterial, anticarcinogenic, anti-allergic, immune-stimulating, and antiviral properties (Brown 1980). Flavonoids also possess the vasodilating ability (Duarte et al. 1993). Since several flavonoids, including quercetin, luteolin, and catechins, may be superior antioxidants to vitamin C, vitamin E, and carotene, the antioxidant activity of flavonoids has received a lot of interest in the pharmaceutical field (Rice-evans et al. 1995).

In addition to its many other biological advantages, quercetin is seen to have a vital role in protecting neuronal cells against oxidative stress-induced neurotoxicity. In primary hippocampal cells, the antioxidant quercetin, which is mostly present in almonds, decreased Aβ-induced cytotoxicity, protein oxidation, lipid peroxidation, and apoptosis when used at lower levels. Quercetin has also been associated with improved neurogenesis, synaptogenesis, and higher cell proliferation in the hippocampal neurons, all of which are dose dependent (Ansari et al. 2009). Transgenic mice treated with catechin showed a substantial reduction in behavioral impairment, Aβ-42 production, APP-C99/89 expression, Wnt protein levels, γ-secretase activity, and MAPK activation (Lim et al. 2013).

Another study found that catechins altered the shape of the fibrillar form of Aβ by mechanistically inhibiting the late stages of Aβ soluble aggregation development. The aggregation of Aβ was mechanistically inhibited by catechins and theaflavins. Catechins appear to only be able to bind big Aβ-aggregate conformations because they only impede the last stages of Aβ-soluble aggregate development. Theaflavins demonstrated a wider range of activity, exhibiting a subtle to pronounced inhibition across the mechanistic stages of Aβ aggregation. The morphology of fAβ produced in their presence might be changed by catechins and theaflavins both. The morphology, or structural conformation, of fAβ has been linked to cytotoxic effects (Chastain 2016).

Kaempferol, which is present in the skin of almond and hazelnut trees, has the potential to greatly improve Aβ-induced performance. Additionally, cell pretreatment with kaempferol had higher cell viability than cells pretreated with vitamin C (Kim et al. 2010). Another study showed that kaempferol has a protective impact on PC12 neuroblastoma β-amyloid peptide-induced damage when compared to α- and β-estradiol (Roth et al. 1999). Treatment with epicatechin in female C57BL/6 mice increased angiogenesis and neuronal spine density, but not neonatal cell survival, which were linked to better spatial memory recall in the dentate gyrus of the hippocampal brain. While decreasing the expression of TNF, Lypla3, IL-20, and Casp3 and increasing the expression of learning-related genes such as SNAP-25 and Kif17, epicatechin consumption especially when combined with exercise also boosted the expression of genes involved in learning (Van Praag et al. 2007).

Due to their antioxidant and anti-aggregation properties, epigallocatechin and epigallocatechin gallate greatly increased the RNA production of Aβ disintegrin and metalloproteinase domain-containing protein 10 (ADAM10), which cleaves APP in a way that precludes Aβ release. According to one study, catechins have anti-aggregation properties that can reduce the amount of BACE1 mRNA that is expressed when Aβ is present (Chastain 2016). Almonds contain flavonol aglycone isorhamnetin. In cultured PC12 cells stimulated with the nerve growth factor, isorhamnetin has been demonstrated to promote the expression of the protein marker for neurite outgrowth and neurofilament (NGF) (Xu et al. 2012).

In addition to reducing Aβ25–35-related abnormalities such as decreased choline acetyltransferase activity, increased H_2O_2, monoamine oxidase activity, and increased iNOS and IL-β, isorhamnetin was also shown to have neuroprotective effects against Aβ25–35-induced memory impairment and oxidative damage in rats (Asha and Sumathi 2016). Myricetin was able to control metal-induced Aβ aggregation and neurotoxicity in human neuroblastoma cells in vitro and has been demonstrated to be particularly effective against metal-associated Aβ over metal-free Aβ species (DeToma et al. 2011).

Myricetin pre- and simultaneous treatment decreased Aβ neurotoxicity in a concentration-dependent manner in primary cortical neuron cultures from rats. Additionally, it decreased Casp3 activation and apoptosis brought on by Aβ1–42. This study also demonstrated that administering myricetin dramatically reduced the levels of Aβ-1–40 and Aβ-1–42 in culture media. This work introduces two mechanisms: direct binding and inhibition of BACE-1 and activation and upregulation of secretase protein levels (Shimmyo et al. 2008). According to Alasalvar et al. almond, hazelnut, and walnut contain proanthocyanidin and anthocyanin, two additional significant polyphenols; hazelnut has the highest proanthocyanidin concentration (Alasalvar et al. 2010). Anthocyanins have been found to protect Neuro2a cells against the damage caused by Aβ and to lessen the cognitive decline in an AD mice model. Additionally, anthocyanins have demonstrated a protective effect against memory loss through interacting with the GABA receptor (Yamakawa et al. 2016).

9.5.9 Lignans

Tree nuts possess lignans. Lignans from various plants have lately been investigated for their anti-AD properties in numerous studies, and it has been demonstrated that they can reduce AD-induced neurodegeneration by reducing oxidative stress, lowering cholinesterase activity, and blocking inflammatory signaling pathways (Zhao et al. 2016).

9.5.10 Tannins

According to previous studies, hazelnuts and other nut skin (Table 9.1) contain a lot of tannin compounds (Contini et al. 2008). Tannic acid suppressed fAβ production from Aβ (1–40) and Aβ (1–42) as well as their extension in a dose-dependent manner. It also caused preformed fAβs to destabilize in a dose-dependent way (Ono et al. 2004).

List of phenolics, vitamins, fatty acids, amino acids, choline, sphingolipids, and phytosterol bioactive molecules of three important Persian medicinal nuts (walnut, almond, and hazelnut) and their Alzheimer's disease-associated mechanisms are tabulated and reviewed in detail by Gorji et al. (2018).

9.5.11 Choline

Choline, a precursor of phosphatidylcholine and sphingomyelin that are necessary for the structural integrity of cell membranes, can be found in tree nuts. Acetylcholine production and cholinergic neurotransmission in people depend on choline (Zeisel 2006). Supplementing with dietary choline increases acetylcholine production and improves memory performance. Long-term treatment of almonds has been shown to raise the amount of acetylcholine in the hippocampus and frontal cortex of

Table 9.1 Classification of major bioactive polyphenolic compounds present in nuts (modified from Gorji et al. 2018)

Phenolic compounds types	Classes	Bioactive compounds
Phenolic acid	Hydroxybenzoic acids	Gallic acid, ellagic acid
	Hydroxycinnamic acids	Caffeic acid
Flavonoids	Flavonols	Isorhamnetin, quercetin, quercetin 3-O-glucuronide, kaempferol, myricetin
	Flavones	Apigenin, luteolin
	Flavanones	Hesperetin, hesperidin, naringenin, naringin
	Isoflavones	Genistein
	Flavanols	Catechin, epigallocatechin, epigallocatechin gallate, epicatechin
	Anthocyanins	Cyanidin
Non-flavonoids	Stilbenes	Polydatin, piceatannol, oxyresveratrol
	Tannins	Pedunculagin, gallotannins, ellagitannins (tellimagrandin, casuarictin)
	Lignans	Secoisolariciresinol, (+)-lariciresin

the brain, improve memory function in healthy rats, and lessen memory deficits in an animal model of amnesia (Batool et al. 2016).

9.5.12 Amino Acids

Different cultivars of almond, hazelnut, and walnut have been shown to contain different amounts of protein. The two main protein fractions in various cultivars of almond and walnut are globulins and albumin (Sze-Tao and Sathe 2000). Arginine, histidine, lysine, phenylalanine, leucine, valine, tryptophan, methionine, and cysteine are among the amino acids found in almond, hazelnut, and walnut. The relative levels of arginine, aspartic acid, and glutamic acid are the highest (Venkatachalam and Sathe 2006; Alasalvar et al. 2010).

The semi-essential amino acid arginine (Arg) has an impact on a variety of physiological and pathological processes. Due to Arg's function as a precursor to nitric oxide (NO), it might be beneficial for treating AD (Rath et al. 2014). Numerous studies were examined by Yi et al. (2009) to determine whether physiological NO levels may play a part in AD. They provided a thorough explanation of the function of the Arg/NO pathway through a variety of processes, such as regulation of cerebral blood flow, modulation of oxidant-mediated neuroinflammation, modulation of atherosclerosis, antioxidative stress, glucose metabolism, and insulin action (Yi et al. 2009).

9.5.13 Sphingolipids

The amount of sphingolipids in certain nuts, including almond, hazelnut, and walnut, was measured by Fang et al. They discovered that the concentration of cerebroside (d18:2-C16:0 h-Glu) was highest in almond and lowest in hazelnut (Fang et al. 2005). In one study, it was discovered that dietary glucocerebrosides had some ameliorative effects on the aberrant sphingolipid metabolism in mice with AD that significantly differed from normal mice's sphingolipid profile (Song et al. 2017). Additionally, cerebrosides reduced Aβ accumulation, learning and memory impairments and oxidative stress in demented mice. Additionally, they demonstrated that the mitochondria-dependent apoptotic pathway controlled the in vitro protective effect (Che et al. 2017).

Thus, a number of pathways involved in the pathogenesis of AD (Fig. 9.1), such as amyloidogenesis, tau phosphorylation, oxidative stress, cholinergic pathways, and some nontarget mechanisms like effects on neurogenesis and cholesterol lowering, can be mitigated by almonds, hazelnuts, and walnuts which contain the macronutrients, micronutrients, phytochemicals and other bioactive components mentioned above.

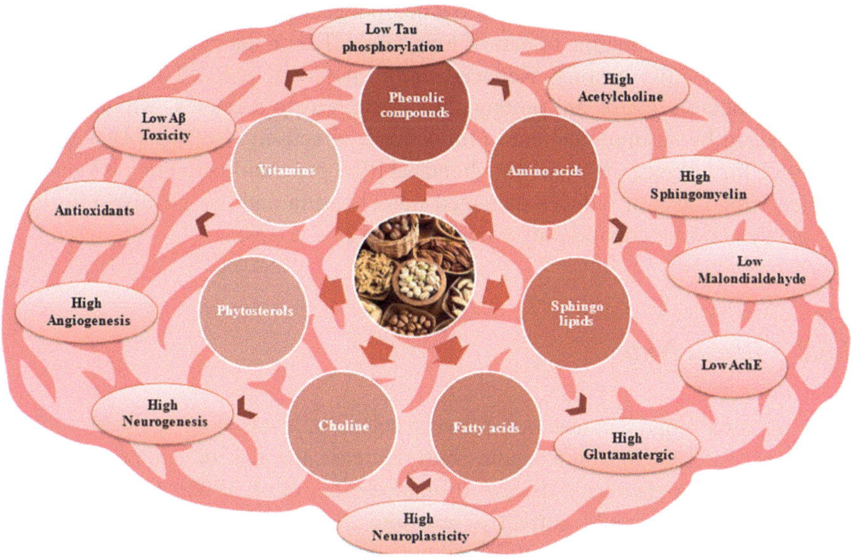

Fig. 9.1 Presence of major bioactive compounds in nuts and their neuroprotective effects that help fight against Alzheimer's disease (figure adopted and modified from Gorji et al. 2018)

9.6 Other Nuts that Helps in Preventing AD

9.6.1 Pistachio

Pistachio (*Pistacia vera*), a tiny tree in the cashew family (Anacardiaceae), is grown in dry regions with warm or temperate climates for its delicious seeds. Flavonoids, antioxidants, carotenoids, vitamins E and B, and other nutrients can all be found in pistachios. As compared to the vehicle-treated group, the study's findings showed a significantly longer latency to enter the dark room in the groups receiving various doses of pistachio extract. Additionally, compared to the vehicle group, the treated groups spent less time in the dark room. The effectiveness of the pistachio extract at different doses was compared, and it was shown that 100 mg/kg was superior to 10 and 50 mg/kg. According to the study's findings, giving rats pistachio extract, which is high in vitamins, flavonoids, and antioxidant components, can help them learn and remember new things (Salari et al. 2014).

Pistachios have a better nutritional profile than other nuts since they are low in fat and high in polyunsaturated fatty acids (13.3 g/100 g) and monounsaturated fatty acids (24.5 g/100 g), minerals (potassium, phosphorus, magnesium, and calcium), and vitamins (vitamins A, E, C, and B). Pistachio phytochemicals exhibit high bioavailability, which supports the positive association between pistachio consumption and health-related outcomes (Mandalari et al. 2013). Pistachios can be a beneficial addition to a diet, especially when combined with exercise training, as a way to improve dysmetabolism and slow neurodegeneration. Pistachio intake effectively nullifies the damaging effects of long-term HFD on the mouse brain by acting in a neuroprotective manner. Reduced brain apoptosis, reduced brain lipid, and reduced oxidative stress are among the neuroprotective effects, which are also accompanied by an improvement in mitochondrial activity. Regular pistachio consumption, for instance, reduces mitochondrial ROS generation brought on by a high-fat diet, which in turn reduces damaged mitochondria and has positive effects on mitophagy and mitochondrial dynamics (Nuzzo et al. 2020).

9.6.2 *Brazil Nuts*

Some research found a connection between selenium levels and cognitive function, indicating that selenium deficiencies may increase the risk of neurodegenerative conditions like Alzheimer's disease (Ceballos-Picot et al. 1996). Consuming one Brazil nut per day could correct a selenium deficiency. We have found some evidence that supplementing with Brazil nuts may benefit several cognitive abilities in older people with MCI, particularly semantic verbal fluency and constructional praxis (Rita Cardoso et al. 2016). One nut consumption, the highest food source of selenium, contributed 288.75 μg of selenium per day, above the recommended intake level of selenium (55 μg/day), while staying below the acceptable upper intake limit (400 μg/day) (IOM 2002).

Patients who received one Brazil nut each day for 3 months were able to make up for a deficiency they had in terms of their nutritional condition. A further indication that the Brazil nut can significantly alleviate oxidative stress and inflammation in hemodialysis patients is the considerable increase in GPx levels and the significant drop in levels of 8-isoprostane, 8-OHdG, and cytokines following supplementation (Stockler-Pinto et al. 2014).

9.6.3 *Pine Nuts*

Pine nut albumin hydrolysate (fraction <3 kDa) and its short peptide derivative Trp-Tyr-Pro-Gly-Lys (WYPGK) increased synaptic plasticity and boosted learning and memory performance in the scopolamine-induced mouse model. Analysis using H & E and Nissl staining revealed that the degree of damage to hippocampus neurons had decreased. Treatment with mitochondrial sirtuin 3 (SIRT3) inducer and inhibitor molecules in an H_2O_2-induced PC12 cell model demonstrated that the <3 kDa fraction and WYPGK activated SIRT3, which reduced Ace-SOD2 acetylation and increased the expression of SYP, SYN-1, SNAP25, and PSD95, thereby enhancing synaptic plasticity and increased the expression of brain-derived neurotrophic factor and the ERK/CREB pathway. Thus, WYPGK and 3 kDa enhance learning and memory function in vitro and in vivo by inducing SIRT3-induced synaptic plasticity (Lu et al. 2021).

In general, pine nuts are regarded as a storehouse of various nutrients, including oil-containing compounds, proteins, carbs, fibers, and minerals (Sharma et al. 2018a, b). When consumed in large quantities together with food, chehelghoza pine nuts have been shown to increase brain weight, reduce anxiety, and enhance both short- and long-term memory. The chehelghoza pine nut mixed with the diet dosage that was most beneficial was 25% (Alami and Mousavi 2020).

9.6.4 Peanuts

Because the seed develops underground, *Arachis hypogaea L.* can also be considered a nut which is also known as peanut, groundnut, monkey nut, goober, or earth nut. It belongs to the Leguminosae family's division Papilionaceae (Caballero et al. 2003). Monounsaturated fats, which are prioritized in the heart-healthy diet, are abundant in peanuts. Peanuts include a variety of additional nutrients, in addition to monounsaturated fats, that have been demonstrated in several studies to support heart health. Peanuts are a good source of protein, manganese, niacin, folate, vitamin E, and flavonoid resveratrol (Ansari et al. 2015).

Peanuts have high concentrations of niacin (13.52×10^{-3} g) and vitamin E (69.3×10^{-4} g), both of which are useful in preventing the onset of Alzheimer's disease and reducing the rate of aging-related cognitive deterioration (La Fata et al. 2014; Morris et al. 2004). Niacin and vitamin E-rich meals have a protective effect and be beneficial in preventing cognitive decline (Morris et al. 2004). By reducing nerve degeneration, the resveratrol in peanuts demonstrates preventive qualities against AD (Chen et al. 2005). Consuming peanuts can consequently reduce oxidative stress in the neurons, which in turn can slow down age-related neuronal degeneration (Butterfield et al. 2002).

9.6.5 Areca Nut

Areca catechu, a member of the Arecaceae family, is found in various parts of East Africa, Asia, and the tropical Pacific (Lechner et al. 2019). The plant is well known for its seeds, sometimes known as areca or betel nuts. It includes a variety of phytochemicals, including alkaloids, tannins, and polyphenols (Peng et al. 2015), and Bhat et al. have explored its effectiveness in reducing AD symptoms (Bhat et al. 2017). Numerous studies on improving memory in both lab animals and people have been conducted using arecoline, the main alkaloid of the areca nut.

According to reports, subcutaneous injections of cholinergic medicines including arecoline, edrophonium, oxotremorine, and tacrine, either alone or in combination, considerably improved the memory retention ability in mice (Flood et al. 1985).

Further observation revealed that arecoline and tacrine provided together were substantially more effective than either medicine administered alone (Flood and Cherkin 1988). The biological activity of *A. catechu L.*'s aqueous extract against AD was good and excellent. Both AChE and BChE (butyrylcholinesterase) were inhibited by it. Amyloid beta and BACE1 accumulation inhibitory activity was also assessed, and it demonstrated 82% (100 g/mL). Additionally, PC12 neurons showed strong neuroprotectivity against H_2O_2-induced cell death. The water maze test findings showed that the extract, at dosages of 1.5 and 3 mg/kg, may repair the memory impairment that scopolamine caused in rats (Bozorgi et al. 2021).

9.6.6 Kola Nuts

Cola nitida and *Cola acuminata* alkaloid extracts have antioxidant and potent inhibitory effects on AChE, BChE, and MAO (monoamine oxidase) activities. Because the extracts target numerous pathways, including the control of neurotransmitter concentrations via inhibition of AChE and MAO, as well as the reduction of oxidative stress, they may have positive neuroprotective effects. These pathways may be affected, and the synergism that results might increase the neuroprotective effect's efficacy. *Cola acuminata* extract may prove to be more effective as a neuroprotective agent given the increased efficacy in MAO inhibition and Fe^{2+}-induced lipid-peroxidation reported in the former as compared to *Cola nitida* extract. However, to further understand the therapeutic potentials of these extracts, we advise more in vivo research and potential clinical trials (Oboh et al. 2019).

9.6.7 Pecan Nuts

Among the five varieties of nuts, like Brazil, pecan, pine, pistachio, and cashew nuts, Pecan nuts were shown to have the highest ratio of unsaturated/saturated fatty acids (13.54), with a total unsaturated fatty acid content in the oil produced from pecan nuts that reached as high as 93%u (Atanasov et al. 2018).

9.7 Other Benefits of Nuts

Due to their rich nutritional profile and numerous health benefits, nuts are of utmost importance around the world. Numerous scientific studies have shown how they work to fight dementia or memory loss, oxidative damage, inflammation, and the aging process. The nutritious profile of nuts contains essential fats (mostly mono- and polyunsaturated fatty acids), proteins (source for arginine, lysine, and tryptophan), vitamins (riboflavin, folate, and various tocopherols), fibers, minerals (calcium, sodium, magnesium, phosphorus, and potassium), and trace elements. Different nuts like almond, walnut, pistachio, Brazil nut, peanut, pecans, and cashew contain important minerals for brain health such as copper, zinc, and selenium. Nuts also have the potential to mitigate negative effects (Arslan et al. 2020).

Nuts like almonds, hazelnut, and walnut play a role in other health benefits, it appears that almonds may be useful for controlling total cholesterol and elevating HDL. By lowering endothelial nitric oxide (NO) synthase inhibition and encouraging NO release, almonds were able to partially restore the vascular reactivity of isolated aortas and prevent high-fat diet-induced endothelial deterioration (Jamshed and Gilani 2014).

9.8 Side Effects

Allergy is the most well-known adverse impact of eating these nuts. Regional variations in the occurrence of tree nut allergies have been noted. According to reports of allergic reactions, hazelnuts are the most allergenic nut in Europe, followed by almonds and walnut in the UK and walnut in the USA (McWilliam et al. 2015). In three cases of kids who had hematuria, dysuria, kidney stones, and hyperoxaluria, it was hypothesized that consuming too many almond milk products could be one of the contributing factors (Ellis and Lieb 2015). Another case report study described a patient who had renal failure after consuming daily amounts of 50–100 g of marzipan containing almonds and 150–200 g of almonds. This investigation suggested that higher intestinal oxalate absorption in this patient was likely caused by intestinal dysbiosis and the absence of the oxalate-degrading bacteria *Oxalobacter formigenes* (Haaskjold et al. 2015).

9.9 Conclusion

Many studies have been conducted to identify the precise pathomechanisms of AD and subsequently develop efficient and/or preventive therapies. Finding a single pure substance to treat AD seems impossible due to the intricate pathways that are involved in the disease. In addition to the ongoing drug discovery research, food nutrients and nutraceuticals may be employed to improve AD patients' health more sustainably. By reviewing their AD-related bioactive components, it can be concluded that these nuts and their phytochemicals may be involved in a number of mechanisms (Fig. 9.2), including those that affect amyloidogenesis, tau phosphorylation, oxidative stress, and cholinergic pathways, as well as some nontarget mechanisms, such as those that have the ability to lower cholesterol and have anti-inflammatory properties, as well as those that have an impact on neurogenesis. However, more research is required for more clinical outcomes on their proper

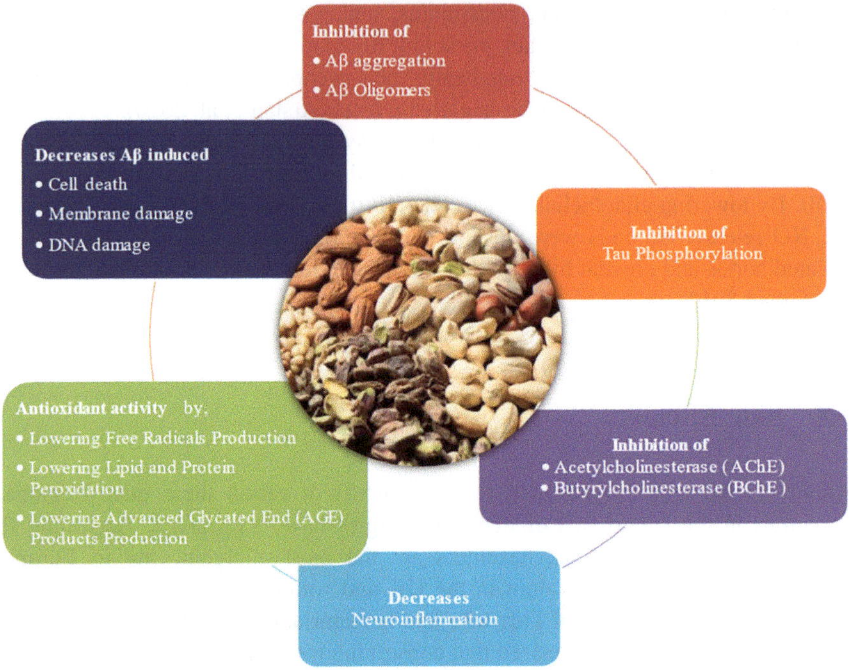

Fig. 9.2 Representation of the potential AD-related mechanisms of nuts that help to overcome neurological impairments

use. Thus, consuming these nuts as nutritional supplements help to prevent, overcome, and tackle the Alzheimer's disease.

References

Ahmed T, Setzer WN, Nabavi SF, Orhan IE, Braidy N, Sobarzo-Sanchez E et al (2016) Insights into effects of Ellagic acid on the nervous system: a mini review. Curr Pharm Des 22(10):1350–1360

Alami K, Mousavi SY (2020) Afghan Chehelghoza (*Pinus gerardiana L.*) pine nut diet enhances the learning and memory in male rats. Nutr Diet Suppl 12:277–288

Alasalvar C, Shahidi F, Liyanapathirana CM, Ohshima T (2003) Turkish tombul hazelnut (Corylus avellana L.). 1. Compositional characteristics. J Agric Food Chem 51(13):3790–3796

Alasalvar C, Shahidi F, Raton B (2010) Tree nuts: composition, phytochemicals, and health effects. Chromatographia 72(5):589–594

Ansari MA, Abdul HM, Joshi G, Opii WO, Butterfield DA (2009) Protective effect of quercetin in primary neurons against Abeta(1-42): relevance to Alzheimer's disease. J Nutr Biochem 20(4): 269–275

Ansari MA, Prakash N, Punitha P, Baishya LK (2015) Post-harvest management and value addition of groundnut. ICAR research Complex NEH Region Manipur centre, Lamphelpat

Arab L, Ang A (2015) A cross sectional study of the association between walnut consumption and cognitive function among adult US populations represented in NHANES. J Nutr Health Aging 19:284–290

Arslan J, Gilani AUH, Jamshed H, Khan SF, Kamal MA (2020) Edible nuts for memory. Curr Pharm Des 26(37):4712–4720

Asha D, Sumathi T (2016) Nootropic activity of isorhamnetin in amyloid beta 25–35 induced cognitive dysfunction and its related mRNA expressions in Alzheimer's disease. Int J Pharm Sci Res 7(8):3233

Atanasov AG, Sabharanjak SM, Zengin G, Mollica A, Szostak A, Simirgiotis M et al (2018) Pecan nuts: a review of reported bioactivities and health effects. Trends Food Sci Technol 71: 246–257

Bahaeddin Z, Yans A, Khodagholi F, Hajimehdipoor H, Sahranavard S (2017) Hazelnut and neuroprotection: improved memory and hindered anxiety in response to intra-hippocampal Aβ injection. Nutr Neurosci 20(6):317–326

Batool Z, Sadir S, Liaquat L, Tabassum S, Madiha S, Rafiq S et al (2016) Repeated administration of almonds increases brain acetylcholine levels and enhances memory function in healthy rats while attenuates memory deficits in animal model of amnesia. Brain Res Bull 120:63–74

Batool Z, Agha F, Ahmad S, Liaquat L, Tabassum S, Khaliq S et al (2017) Attenuation of cadmium-induced decline in spatial, habituation and recognition memory by long-term administration of almond and walnut supplementation: role of cholinergic function. Pak J Pharm Sci 30(1):273

Bhat SK, Ashwin D, Mythri S, Bhat S (2017) Areca nut (Areca catechu L) decreases Alzheimer's disease symptoms: compilation of research works. J Med Plants Stud 5(5):4–9

Boccardi V, Baroni M, Mangialasche F, Mecocci P (2016) Vitamin E family: role in the pathogenesis and treatment of Alzheimer's disease. Alzheimers Dement (N Y) 2(3):182–191

Bozorgi M, Najafi Z, Omidpanah S, Sadri A, Narimani Z, Moghadam FH et al (2021) Investigation of anti-Alzheimer's activity of aqueous extract of areca nuts (Areca catechu L.): in vitro and in vivo studies. Bol Latinoam Caribe Plantas Med Aromát 20(4)

Brown JP (1980) A review of the genetic effects of naturally occurring flavonoids, anthraquinones and related compounds. Mutat Res 75(3):243–277

Butterfield DA, Castegna A, Pocernich CB, Drake J, Scapagnini G, Calabrese V (2002) Nutritional approaches to combat oxidative stress in Alzheimer's disease. J Nutr Biochem 13(8):444–461

Caballero B, Trugo LC, Finglas PM (2003) Encyclopedia of food sciences and nutrition. Academic, New York

Ceballos-Picot I, Merad-Boudia M, Nicole A, Thevenin M, Hellier G, Legrain S et al (1996) Peripheral antioxidant enzyme activities and selenium in elderly subjects and in dementia of Alzheimer's type—place of the extracellular glutathione peroxidase. Free Radic Biol Med 20(4): 579–587

Chastain SE (2016) Therapeutic potential of catechins and derivatives for the prevention of Alzheimer's disease. University of South Carolina, Columbia

Chauhan A, Chauha V (2020) Beneficial effects of walnuts on cognition and brain health. Nutrients 12(2):550

Chauhan N, Wang KC, Wegiel J, Malik MN (2004) Walnut extract inhibits the fibrillization of amyloid beta-protein, and also defibrillizes its preformed fibrils. Curr Alzheimer Res 1:183–188

Che H, Du L, Cong P, Tao S, Ding N, Wu F et al (2017) Cerebrosides from sea cucumber protect against oxidative stress in SAMP8 mice and PC12 cells. J Med Food 20(4):392–402

Chen J, Zhou Y, Mueller-Steiner S, Chen LF, Kwon H, Yi S et al (2005) SIRT1 protects against microglia-dependent amyloid-β toxicity through inhibiting NF-κB signaling. J Biol Chem 280(48):40364–40374

Contini M, Baccelloni S, Massantini R, Anelli G (2008) Extraction of natural antioxidants from hazelnut (*Corylus avellana L.*) shell and skin wastes by long maceration at room temperature. Food Chem 110(3):659–669

De Jesus Moreno Moreno M (2003) Cognitive improvement in mild to moderate Alzheimer's dementia after treatment with the acetylcholine precursor choline alfoscerate: a multicenter, doubleblind, randomized, placebo-controlled trial. Clin Ther 25(1):178–193

De Lima Oliveira BC, Bellozi PMQ, Reis HJ, De Oliveira ACP (2018) Inflammation as a possible link between dyslipidemia and Alzheimer's disease. Neuroscience 376:127–141

DeToma AS, Choi JS, Braymer JJ, Lim MH (2011) Myricetin: a naturally occurring regulator of metal-induced amyloid-β aggregation and neurotoxicity. Chembiochem 12(8):1198–1201
Duarte J, Pérez-Vizcaíno F, Zarzuelo A, Jiménez J, Tamargo J (1993) Vasodilator effects of quercetin in isolated rat vascular smooth muscle. Eur J Pharmacol 239(1–3):1–7
Ellis D, Lieb J (2015) Hyperoxaluria and genitourinary disorders in children ingesting almond milk products. J Pediatr 167(5):1155–1158
Esfahlan AJ, Jamei R, Esfahlan RJ (2010) The importance of almond (*Prunus amygdalus L.*) and its by-products. Food Chem 120(2):349–360
Fang F, Ho CT, Sang S, Rosen RT (2005) Determination of sphingolipids in nuts and seeds by a single quadrupole liquid chromatography–mass spectrometry method. J Food Lipids 12(4): 327–343
Flood JF, Cherkin A (1988) Effect of acute arecoline, tacrine and arecoline+ tacrine post-training administration on retention in old mice. Neurobiol Aging 9:5–8
Flood JF, Smith GE, Cherkin A (1985) Memory enhancement: supra-additive effect of subcutaneous cholinergic drug combinations in mice. Psychopharmacology 86(1):61–67
Gao H, Yan P, Zhang S, Nie S, Huang F, Han H et al (2016) Chronic alpha-linolenic acid treatment alleviates age-associated neuropathology: roles of PERK/eIF2alpha signaling pathway. Brain Behav Immun 57:314–325
Gibson GE, Sheu KF, Blass JP, Baker A, Carlson KC, Harding B et al (1988) Reduced activities of thiamine-dependent enzymes in the brains and peripheral tissues of patients with Alzheimer's disease. Arch Neurol 45(8):836–840
Gorji N, Moeini R, Memariani Z (2018) Almond, hazelnut and walnut, three nuts for neuroprotection in Alzheimer’s disease: a neuropharmacological review of their bioactive constituents. Pharmacol Res 129:115–127
Grodzicki W, Dziendzikowska K (2020) The role of selected bioactive compounds in the prevention of Alzheimer’s disease. Antioxidants 9(3):229
Haaskjold YL, Drotningsvik A, Leh S, Marti HP, Svarstad E (2015) Renal failure due to excessive intake of almonds in the absence of Oxalobacter formigenes. Am J Med 128(12):e29–e30
Haider S, Batool Z, Haleem DJ (2012) Nootropic and hypophagic effects following long term intake of almonds (Prunus amygdalus) in rats. Forum Nutr 27(6):2109–2115
Harnly JM, Doherty RF, Beecher GR, Holden JM, Haytowitz DB, Bhagwat S et al (2006) Flavonoid content of US fruits, vegetables, and nuts. J Agric Food Chem 54(26):9966–9977
Hinterberger M, Fischer P (2013) Folate and Alzheimer: when time matters. J Neural Transm 120(1):211–224
Huang HC, Jiang ZF (2009) Accumulated amyloid-β peptide and hyperphosphorylated tau protein: relationship and links in Alzheimer’s disease. J Alzheimers Dis 16:15–27
IOM—Institute of Medicine (2002) Dietary reference intakes for vitamin C, vitamin E, selenium, and carotenoids. Academy, Washington
Jamshed H, Gilani AH (2014) Almonds inhibit dyslipidemia and vascular dysfunction in rats through multiple pathways. J Nutr 144(11):1768–1774
Khorasani MA (2001) Makhzan al Advieh (The storehouse of medicaments). Research institute for Islamic and Complementary Medicine, Iran University of Medical Sciences. Bavardaran Press (In Persian), Tehran, Iran, p 351
Kim EJ, Yang SJ (2017) Nicotinamide reduces amyloid precursor protein and Presenilin 1 in brain tissues of amyloid Beta-tail vein injected mice. Clin Nutr Res 6(2):130–135
Kim JK, Choi SJ, Cho HY, Hwang HJ, Kim YJ, Lim ST et al (2010) Protective effects of kaempferol (3,4′,5,7-tetrahydroxyflavone) against amyloid beta peptide (Abeta)-induced neurotoxicity in ICR mice. Biosci Biotechnol Biochem 74(2):397–401
Kim MJ, Seong AR, Yoo JY, Jin CH, Lee YH, Kim YJ et al (2011) Gallic acid, a histone acetyltransferase inhibitor, suppresses beta-amyloid neurotoxicity by inhibiting microglial-mediated neuroinflammation. Mol Nutr Food Res 55(12):1798–1808
Kim KB, Nam YA, Kim HS, Hayes AW, Lee BM (2014) Alpha-linolenic acid: nutraceutical, pharmacological and toxicological evaluation. Food Chem Toxicol 70:163–178

Kim JH, Wang Q, Choi JM, Lee S, Cho EJ (2015) Protective role of caffeic acid in an Aβ25-35-induced Alzheimer's disease model. Nutr Res Pract 9(5):480–488

Kinney JW, Bemiller SM, Murtishaw AS, Leisgang AM, Salazar AM, Lamb BT (2018) Inflammation as a central mechanism in Alzheimer's disease. Alzheimers Dement (N Y) 4:575–590

Kulkarni KS, Kasture S, Mengi S (2010) Efficacy study of Prunus amygdalus (almond) nuts in scopolamine-induced amnesia in rats. Ind J Pharmacol 42(3):168

Kumar K, Kumar A, Keegan RM, Deshmukh R (2018) Recent advances in the neurobiology and neuropharmacology of Alzheimer's disease. Biomed Pharmacother 98:297–307

La Fata G, Weber P, Mohajeri MH (2014) Effects of vitamin E on cognitive performance during ageing and in Alzheimer's disease. Nutrients 6(12):5453–5472

Lechner M, Breeze CE, Vaz F, Lund VJ, Kotech B (2019) Betel nut chewing in high-income countries—lack of awareness and regulation. Lancet Oncol 20(2):181–183

Lee E, Eom J-E, Kim H-L, Baek KH, Jun K-Y, Kim H-J et al (2013) Effect of conjugated linoleic acid, μ-calpain inhibitor, on pathogenesis of Alzheimer's disease. Biochim Biophys Acta 1831(4):709–718

Lim HJ, Shim SB, Jee SW, Lee SH, Lim CJ, Hong JT et al (2013) Green tea catechin leads to global improvement among Alzheimer's disease-related phenotypes in NSE/hAPP-C105 Tg mice. J Nutr Biochem 24(7):1302–1313

Lu H, Fang L, Wang J, Zhao F, Liu C, Gao Y, Min W (2021) Pine nut antioxidant peptides ameliorate the memory impairment in a scopolamine-induced mouse model via SIRT3-induced synaptic plasticity. Food Funct 12(17):8026–8036

Maguire LS, O'Sullivan SM, Galvin K, O'Connor TP, O'Brien NM (2004) Fatty acid profile, tocopherol, squalene and phytosterol content of walnuts, almonds, peanuts, hazelnuts and the macadamia nut. Int J Food Sci Nutr 55(3):171–178

Mandalari G, Bisignano C, Filocamo A, Chessa S, Sarò M, Torre G et al (2013) Bioaccessibility of pistachio polyphenols, xanthophylls, and tocopherols during simulated human digestion. Nutrition 29(1):338–344

Martínez ML, Labuckas DO, Lamarque AL, Maestri DM (2010) Walnut (Juglans regia L.): genetic resources, chemistry, by-products. J Sci Food Agric 90(12):1959–1967

Martinez-Lapiscina EH, Clavero P, Toledo E, Estruch R, Salas-Salvado J, San Julian B et al (2013) Mediterranean diet improves cognition: the PREDIMED-NAVARRA randomised trial. J Neurol Neurosurg Psychiatry 84:1318–1325

McWilliam V, Koplin J, Lodge C, Tang M, Dharmage S, Allen K (2015) The prevalence of tree nut allergy: a systematic review. Curr Allergy Asthma Rep 15(9):54

Mercer LD, Kelly BL, Horne MK, Beart PM (2005) Dietary polyphenols protect dopamine neurons from oxidative insults and apoptosis: investigations in primary rat mesencephalic cultures. Biochem Pharmacol 69(2):339–345

Mocchegiani E, Costarelli L, Giacconi R, Malavolta M, Bass A, Piacenza F et al (2014) Vitamin E–gene interactions in aging and inflammatory age-related diseases: implications for treatment. Ageing Res Rev 14:81–101

Morris MC, Evan DA, Bienias JL, Scherr PA, Tangney CC, Hebert LE et al (2004) Dietary niacin and the risk of incident Alzheimer's disease and of cognitive decline. J Neurol Neurosurg Psychiatry 75(8):1093–1099

Muthaiyah B, Essa MM, Chauhan V, Chauhan A (2011) Protective effects of walnut extract against amyloid beta peptide-induced cell death and oxidative stress in PC12 cells. Neurochem Res 36(11):2096–2103

Muthaiyah B, Essa MM, Lee M, Chauhan V, Kaur K, Chauhan A (2014) Dietary supplementation of walnuts improves memory deficits and learning skills in transgenic mouse model of Alzheimer's disease. J Alzheimers Dis 42(4):1397–1405

Neale EP, Tapsell LC, Martin A, Batterham MJ, Wibisono C, Probst YC (2017) Impact of providing walnut samples in a lifestyle intervention for weight loss: a secondary analysis of the HealthTrack trial. Food Nutr Res 61:1344522

Nuzzo D, Galizzi G, Amato A, Terzo S, Picone P, Cristaldi L et al (2020) Regular intake of pistachio mitigates the deleterious effects of a high fat-diet in the brain of obese mice. Antioxidants 9(4):317

O'Brien J, Okereke O, Devore E, Rosner B, Breteler M, Grodstein F (2014) Long-term intake of nuts in relation to cognitive function in older women. J Nutr Health Aging 18:496–502

Oboh G, Ademosun AO, Ogunsuyi OB, Oyedola ET, Olasehinde TA, Oyeleye SI et al (2019) In vitro anticholinesterase, antimonoamine oxidase and antioxidant properties of alkaloid extracts from kola nuts (Cola acuminata and Cola nitida). J Complement Integr Med 16(1). https://doi.org/10.1515/jcim-2016-0155

Olofinnade AT, Onaolapo AY, Onaolapo OJ, Olowe OA (2021) Hazelnut modulates Neurobehaviour and ameliorates ageing-induced oxidative stress, and Caspase-3-mediated apoptosis in mice. Curr Aging Sci 14(2):154–162

Ono K, Hasegawa K, Naiki H, Yamada M (2004) Anti-amyloidogenic activity of tannic acid and its activity to destabilize Alzheimer's β-amyloid fibrils in vitro. Biochim Biophys Acta 1690(3): 193–202

Park YS, Jang HJ, Lee KH, Hahn TR, Paik YS (2006) Prolyl endopeptidase inhibitory activity of unsaturated fatty acids. J Agric Food Chem 54(4):1238–1242

Peng W, Liu YJ, Wu N, Sun T, He XY, Gao YX et al (2015) Areca catechu L.(Arecaceae): a review of its traditional uses, botany, phytochemistry, pharmacology and toxicology. J Ethnopharmacol 164:340–356

Perry EK (1986) The cholinergic hypothesis—ten years on. Br Med Bull 42(1):63–69

Perry EK, Tomlinson BE, Blessed G, Bergmann K, Gibson PH, Perry RH (1978) Correlation of cholinergic abnormalities with senile plaques and mental test scores in senile dementia. Br Med J 2(6150):1457–1459

Pribis P, Bailey RN, Russell AA, Kilsby MA, Hernandez M, Craig WJ et al (2012) Effects of walnut consumption on cognitive performance in young adults. Br J Nutr 107(9):1393–1401

Rath M, Müller I, Kropf P, Closs EI, Munder M (2014) Metabolism via arginase or nitric oxide synthase: two competing arginine pathways in macrophages. Front Immunol 5:532

Rice-evans CA, Miller NJ, Bolwell PG, Bramley PM, Pridham JB (1995) The relative antioxidant activities of plant-derived polyphenolic flavonoids. Free Radic Res 22(4):375–383

Rita Cardoso B, Apolinário D, Da Silva BV, Busse AL, Magaldi RM, Jacob-Filho W et al (2016) Effects of Brazil nut consumption on selenium status and cognitive performance in older adults with mild cognitive impairment: a randomized controlled pilot trial. Eur J Nutr 55(1):107–116

Roth A, Schaffner W, Hertel C (1999) Phytoestrogen kaempferol (3,4′,5,7-tetrahydroxyflavone) protects PC12 and T47D cells from beta-amyloid-induced toxicity. J Neurosci Res 57(3): 399–404

Salari E, Baloochi M, Shamsizedeh A, Ayoobi F, Allahtavakoli M, Taghavi Y et al (2014) Effect of the hydroalcoholic extract of pistachio on avoidance learning in male Wistar rats. 180–187

Sala-Vila A, Crous-Bou M, Sánchez-Benavides G, De Arenaza-Urquijo EM, Suárez-Calvet M, Milà-Alomà M et al (2020) Eating a weekly serving of walnuts relates to beneficial brain imaging phenotypes in a cohort at increased risk of Alzheimer's disease. Curr Dev Nutr 4 (Suppl_2):1234–1234

Sharma A, Sharma L, Goyal R (2018a) A review on himalayan pine species: ethnopharmacological, phytochemical and pharmacological aspects. Pharm J 10(4):611

Sharma P, Srivastava P, Seth A, Tripathi PN, Banerjee AG, Shrivastava SK (2018b) Comprehensive review of mechanisms of pathogenesis involved in Alzheimer's disease and potential therapeutic strategies. Prog Neurobiol 174:53–89

Shimmyo Y, Kihara T, Akaike A, Niidome T, Sugimoto H (2008) Multifunction of myricetin on Aβ: neuroprotection via a conformational change of Aβ and reduction of Aβ via the interference of secretases. J Neurosci Res 86(2):368–377

Song Y, Cong P, Lu L, Wan Y, Tang Q, Zhang H et al (2017) Effects of dietary glucocerebrosides from sea cucumber on the brain sphingolipid profiles of mouse models of Alzheimer's disease. Food Funct 8(3):1271–1281

Stockler-Pinto MB, Mafra D, Moraes C, Lobo J, Boaventura GT, Farage NE et al (2014) Brazil nut (Bertholletia excelsa, HBK) improves oxidative stress and inflammation biomarkers in hemodialysis patients. Biol Trace Elem Res 158(1):105–112

Sul D, Kim HS, Lee D, Joo SS, Hwang KW, Park SY (2009) Protective effect of caffeic acid against beta-amyloid-induced neurotoxicity by the inhibition of calcium influx and tau phosphorylation. Life Sci 84(9–10):257–262

Sze-Tao KWC, Sathe SK (2000) Walnuts (Juglans regia L): proximate composition, protein solubility, protein amino acid composition and protein in vitro digestibility. J Sci Food Agric 80(9):1393–1401

Valls-Pedret C, Sala-Vila A, Serra-Mir M, Corella D, de la Torre R, Martinez-Gonzalez MA et al (2015) Mediterranean diet and age-related cognitive decline: a randomized clinical trial. JAMA Intern Med 175:1094–1103

Van Praag H, Lucero MJ, Yeo GW, Stecker K, Heivand N, Zhao C et al (2007) Plant-derived flavanol (−) epicatechin enhances angiogenesis and retention of spatial memory in mice. J Neurosci 27(22):5869–5878

Vanmierlo T, Bogie JF, Mailleux J, Vanmol J, Lütjohann D, Mulder M et al (2015) Plant sterols: friend or foe in CNS disorders? Prog Lipid Res 58:26–39

Venkatachalam M, Sathe SK (2006) Chemical composition of selected edible nut seeds. J Agric Food Chem 54(13):4705–4714

Wani IA, Ayoub A, Bhat NA, Dar AH, Gull A (2020) Hazelnut. In: Nayik GA, Gull A (eds) Antioxidants in vegetables and nuts-properties and health benefits. Springer, Singapore, pp 559–572

Webber KM, Raina AK, Marlatt MW, Zhu X, Prat MI, Morelli L et al (2005) The cell cycle in Alzheimer disease: a unique target for neuropharmacology. Mech Ageing Dev 126(10):1019

Xia W, Mo H (2016) Potential of tocotrienols in the prevention and therapy of Alzheimer's disease. J Nutr Biochem 31:1–9

Xu SL, Choi RC, Zhu KY, Leung KW, Guo AJ, Bi D et al (2012) Isorhamnetin, a flavonol aglycone from Ginkgo biloba L., induces neuronal differentiation of cultured PC12 cells: potentiating the effect of nerve growth factor. Evidence-based complementary and alternative medicine. Evid Based Complement Alternat Med 2012:278273

Yamakawa MY, Uchino K, Watanabe Y, Adachi T, Nakanishi M, Ichino H et al (2016) Anthocyanin suppresses the toxicity of Aβ deposits through diversion of molecular forms in in vitro and in vivo models of Alzheimer's disease. Nutr Neurosci 19(1):32–42

Yi J, Horky LL, Friedlich AL, Shi Y, Rogers JT, Huang X (2009) L-arginine and Alzheimer's disease. Int J Clin Exp Pathol 2(3):211

Yin RH, Yu JT, Tan L (2015) The role of SORL1 in Alzheimer's disease. Mol Neurobiol 51:909–918

Youn K, Yun EY, Lee J, Kim JY, Hwang JS, Jeong WS et al (2014) Oleic acid and linoleic acid from Tenebrio molitor larvae inhibit BACE1 activity in vitro: molecular docking studies. J Med Food 17(2):284–289

Zeisel SH (2006) Choline: critical role during fetal development and dietary requirements in adults. Annu Rev Nutr 26:229

Zhang Q, Yang G, Li W, Fan Z, Sun A, Luo J et al (2011) Thiamine deficiency increases betasecretase activity and accumulation of beta-amyloid peptides. Neurobiol Aging 32(1): 42–53

Zhao X, Liu C, Xu M, Li X, Bi K, Jia Y (2016) Total lignans of Schisandra chinensis ameliorates Aβ1-42-induced neurodegeneration with cognitive impairment in mice and primary mouse neuronal cells. PLoS One 11(4):e0152772

Zou J, Cai PS, Xiong CM, Ruan JL (2016) Neuroprotective effect of peptides extracted from walnut (Juglans Sigilata Dode) proteins on Abeta25-35-induced memory impairment in mice. J Huazhong Univ Sci Technolog Med Sci 36(1):21–30

Chapter 10
Anti-Alzheimeric Role of Ginger

Ponram Ramalakshmi, Ponnambalam Annapoorani, Thamilarasan Manivasagam, Arokiasamy Justin Thenmozhi, and Selvaraju Aruna Devi

Abstract The rhizome of the *Zingiber officinale* is known as ginger with prevalence in Asia and is mostly grown in India. Ginger is demonstrated to have several pharmacological functions like antioxidant, anti-inflammatory, and antiapoptotic properties due to the occurrence of numerous active compounds. It influences the nervous system that aids to improve cognitive functions during initial stage of dementia. In this chapter, we discuss about the various functional properties of ginger on the prevention of Alzheimer's disease (AD).

Keywords Ginger · Active components · Antioxidants · Anti-inflammation · Anti-Alzheimeric role

10.1 Introduction

*10.1.1 Ginger (***Zingiber Officinale***)*

Ginger is a rhizome originating in India or Southeast Asia. Now it is commonly planted in West Indies, China, Mexico, and in other countries (Banerjee et al. 2011). It is classified under "Generally Recognized as Safe" by the FDA, USA (Mahdy

P. Ramalakshmi · P. Annapoorani (✉)
Department of Biochemistry, V.V. Vanniaperumal College for Women, Virudhunagar, Tamil Nadu, India
e-mail: annapoorani@vvvcollege.org

T. Manivasagam · S. Aruna Devi
Department of Biochemistry and Biotechnology, Faculty of Science, Annamalai University, Chidambaram, Tamil Nadu, India

A. Justin Thenmozhi
Department of Biochemistry, School of Biological Sciences, Madurai Kamaraj University, Madurai, Tamil Nadu, India

Department of Biochemistry and Biotechnology, Faculty of Science, Annamalai University, Chidambaram, Tamil Nadu, India

A. Justin Thenmozhi, T. Manivasagam (eds.), *Nutraceuticals for Alzheimer's Disease: A Promising Therapeutic Approach*, Nutritional Neurosciences,
https://doi.org/10.1007/978-981-99-0677-2_10

et al. 2014). The rhizome consists of two main kinds of compounds, (1) volatile materials including essential oil and (2) nonvolatile substances like oleoresin (a pungent compound), organic and inorganic compounds (Vernin and Parkanyi 2005). The main compounds found in volatile oil are the sesquiterpenes, like zingiberene, curcumene, and farnesene (36%, 18%, and 10%, respectively), and monoterpenes, like linalool, cineole, borneol (1.3%;1.3% and 2.2% respectively), geranial or citral A, and neral or citral B (1.4%;0.8% respectively) and non-volatiles such as shogaols, gingers, zingerone and paradols. The minor constituents found in ginger rhizome are lipids, polysaccharides, organic acids, and fibrous compounds (Vernin and Parkanyi 2005).

10.1.2 Phytochemical Properties of Ginger

The ginger rhizome is used as a food adjunct that improves the taste and flavor of the food substances. Although globally it is used as a food additive and spice, it is reported to offer several pharmacological functions (Fig. 10.1) due to the presence of numerous potent active substances (Saha et al. 2014; Pashaei-Asl et al. 2017).

The alcoholic extract of ginger contains enormous amount of antioxidants and offered potent free radical scavenging function. The methanolic and ethanolic ginger extracts possess enhanced antioxidant effects than aqueous extracts. The hydroxyl groups and solubilizing side chains found in the active compounds are responsible for this antioxidant activity (Adel and Prakash 2010).

Manju and Nalini (2005) indicated that exposure of ginger extract to experimental model of colon carcinogenesis diminished the activities of enzymatic antioxidants like SOD, GPx, and catalase and enhanced the levels of superoxide radicals and hydrogen peroxide as compared to control rats. Another experiment indicated that ginger administration offered an antidiabetic effect by reducing the levels of circulatory glucose, cholesterol, and triacylglycerol (Khandouzi et al. 2015). Due to the

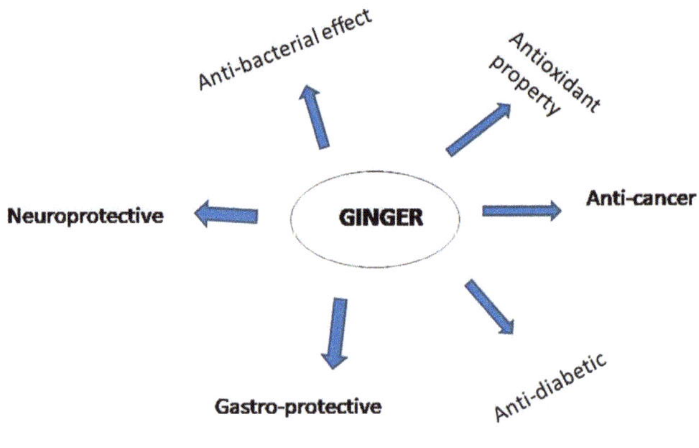

Fig. 10.1 Pharmacological properties of ginger

antiapoptotic and anti-inflammatory effect of ginger, the administration of ginger extract reduced the activation of Bax/Bcl-2 ratio and caspase-3 by inhibiting the levels of IL-1β. Guillemin and Brew (2002) studied the neuroprotective effect of ginger against monosodium glutamate induced neurotoxicity by regulating neurotransmitter levels, reducing 8-hydroxy-2′- deoxy guanosine (8-OHdg) and accumulation of amyloid proteins due to the presence of various polyphenols. The aqueous extract of ginger enhanced the activity of cholinergic neurons, reduced the activity of AChE, and enhanced the superoxide dismutase/malondialdehyde ratio and diminished MDA levels in the brain.

10.1.3 Ginger and Alzheimer's Disease

Grzanna et al. (2004) performed an in vivo experiment and assessed the anti-inflammatory role of ginger in in vitro model of AD (THP-1 monocytic cells). Pretreatment of mixture of solution consisting of ginger extracts and *Alpinia galangal* to lipopolysaccharide and $A\beta_{1-42}$ inhibited the expressions of chemokine ligand 2, chemokine monocyte chemoattractant protein-1/C-C motif, CXC ligand 10, IL-1β, TNF-α, COX-2, and macrophage inflammatory protein-1α/C-C motif chemokine ligand 3 as compared to control cells alone treated with toxins. The neuroprotective effect of aqueous methanol and chloroform extract of several herbs including ginger by analyzing the cell viability in $A\beta_{1-42}$ induced toxicity in PC12 and primary neuronal cells was studied. Although aqueous methanol and chloroform extracts of many herbs showed toxicity or no beneficial effect in high concentrations, only seven herbs including ginger extract offered the neuroprotective function (Kim et al. 2007).

The ginger root extract administration to the rat injected with the intracerebroventricular injection of Aβ-protein and subsequent administration of aluminum chloride (AD model) offered neuroprotection by improving the memory and learning deficits, histopathological changes in various brain regions, and expression of enzymatic antioxidants (catalase and superoxide dismutase) and reduced the levels of interleukin-1β, nuclear factor-κB, and malondialdehyde (Zeng et al. 2013). Mathew and Subramanian (2014) studied the protective role of dry ginger in the primary hippocampal cells treated with monomeric Aβ. Pretreatment of ginger enhanced the cell survival by preventing the Aβ oligomer aggregation and dissociation.

Lim et al. (2016) evaluated the neuroprotection of ginger on learning and memory dysfunctions in double-transgenic (APP/PSEN1) models of mice. Oral administration of peony root and ginger diminished the Aβ accumulation and cognitive dysfunctions. Administration of ginger extract diminished the inflammatory marker expression of NF-κB, GFAP, TNF-α, IL-1β, COX-2, and IL-6 in transgenic AD mice. Peony root and ginger treatment reduced the levels of GFAP andCOX-2 expression in astrocytes.

Sutalangka and Wattanathorn (2017) investigated the neuroprotective effect of *Cyperus rotundus* and *Zingiber officinale* mixture against the rat injected with intracerebroventricular injection of AF64A (model with cholinergic deficiency). The combined extract enhanced the spatial memory (Morris water maze) and diminished the oxidative imbalance (enhanced the SOD and catalase activity) in the hippocampus and enhanced the cholinergic activities (diminishing AChE activity). Oral administration of combined extract enhanced the neuronal mass in the dentate gyrus and enhanced the expression of ERK1/2 signaling molecules. Huh et al. (2018) indicated the protective effect of fermented ginger containing 6-paradol against ICR mice injected with the intraperitoneal injection of scopolamine (AD model). Oral treatment of 6-paradol enhanced the protective role of ginger when compared to control mice during the analysis of recognition memory. In addition, the protective role of ginger was demonstrated in AD mice infused with $A\beta_{1-42}$ in the hippocampus. The results suggested that fermented ginger enriched with 6-paradol enhanced the memory and learning impairments by improving the neuronal density in hippocampal neurons induced by $A\beta_{1-42}$.

10.1.4 6-Gingerol and AD

Exposure of [6]-gingerol to SH-SY 5Y human neuroblastoma cells treated with $A\beta_{25-35}$ enhanced the cell viability and reduced cell death by improving mitochondrial membrane potential and decreasing DNA fragmentation and regulating BCl2/Bax ratio (Lee et al. 2011). Moreover, neuroprotective effect of this compound is partially due to its antioxidant effect (enhanced the activities of enzymatic antioxidant such as heme oxygenase-1 and c-glutamylcysteine ligase via the regulation of NRF2). Zeng et al. (2015) analyzed the 6-gingerol in $A\beta_{1-42}$ induced rat PC12 cells to prove its neuroprotective effect. Exposure to 6-gingerol significantly enhanced the cell viability and diminished the oxidant-antioxidant imbalance by reducing the levels of ROS (nitric oxide and MDA) and also by enhancing the activity of SOD activity as compared to $A\beta_{1-42}$ alone treated group. Moreover, it offered the neuroprotective role by activating AKT and inhibiting of GSK-3β signaling pathways, thereby decreasing the breakdown of APP through NF-κB inhibition, evading Aβ synthesis.

El Halawany et al. (2017) indicated the neuroprotective effects of 6-gingerol, an active compound of ginger, in the mice model of AD injected intracerebroventricularly with streptozotocin. Intraperitoneal injection of 6-gingerol decreased the levels of $A\beta_{1-42}$, β- and γ-secretase (enzymes involved in amyloidogenesis pathway), and COX-2, with the enhanced activity of α-secretase (enzyme involved in non-amyloidogenesis pathway). It also normalized the behavioral and memory impairment as indicated by the scores observed in Morris water maze and Y-maze test. The neuroprotective effects of ginger (fresh and dried) extract and 6-gingerol on memory deficits were manifested by an intraperitoneal injection of scopolamine in mice as analyzed by Kim et al. (2018). The learning and memory assessment were carried out by performing Morris water maze, contextual fear conditioning, Y-maze, and passive avoidance tests. Mice pretreated with both the ginger extracts offered good scores in behavioral test as compared to scopolamine alone treated animals. Pretreatment of 6-gingerols offered more neuroprotection by alleviating memory deficits caused by scopolamine by enhancing the expression of BDNF that is reported to play a key role in supporting the survival, plasticity, and function of cholinergic neurons (Kim et al. 2018).

10.1.5 6-Shogaol and AD

The protective role of 6-shogaol, an active component of ginger, on memory and learning deficits and inflammation in unilateral injection of hippocampal $A\beta_{1-42}$ oligomers was demonstrated by Moon et al. (2014). Oral treatment of 6-shogaol reduced the apoptotic markers and attenuated the learning and memory deficits. It also diminished the astrocyte and microglia activation in the hippocampus indicated by the decline in the expression of GFAP and macrophage-1 antigen. Na et al. (2016) showed the protective role of 6-shogaol in the cellular and animal models of AD. Exposure of 6-shogaol on the $A\beta_{1-42}$ induced mice model of AD improved the behavioral deficits and inhibited the expression of cathepsin B (apoptotic marker) and cysteinyl leukotriene 1 receptor (inflammatory marker). In the cellular model of AD (hippocampal HT22 cells treated with $A\beta_{1-42}$), pre-exposure of 6-shogaol enhanced the cell viability and inhibited the cytotoxicity as compared to $A\beta_{1-42}$ alone exposed cells. Na et al. (2017) also indicated the protective role of 6-shogaol by activation of sortilin-related receptor 1 (involved APP processing and Aβ secretion) and APP cleaving enzyme 1 and also the suppression of Aβ aggregation in HT22 cells induced with $A\beta_{1-42}$. The researchers carried out the experiment in transgenic APP/PSEN1 mice in support of in vitro studies (Na et al. 2016).

10.1.6 Zerumbone and AD

Jafarian et al. (2019) indicated the neuroprotective effect of zerumbone, an active compound of ginger in in vivo model of AD. Intraperitoneal injection of zerumbone ameliorated scopolamine induced memory dysfunctions, hypoactivity, and anxiety as studied by elevated plus maze, Morris water maze, and open field tests. Li et al. (2020) investigated the protective effect of zerumbone in cellular and animal models of AD. In the transgenic APP/PSEN1 mouse model, oral administration of zerumbone significantly enhanced the nesting and learning skill, memory status, and also social relation. It also diminished the accumulation of hippocampal and cortical Aβ plaques and inflammation, thereby attenuating the behavioral symptoms resembling AD. To induce the in vitro model of AD, cortical microglial cells obtained from C57BL/6 J mice were induced with Aβ1–42 or lipopolysaccharide.

Exposure of zerumbone suppressed the neuroinflammation through attenuating MAPK/NF-κB signaling pathway.

10.1.7 Zingerone and AD

Kim et al. (2010) proved the antioxidant and anti-inflammatory role of zingerone by reducing the levels of reactive oxygen species and NF-kB. Oral administration of zingerone inhibited the nuclear NF-kB translocation and the expression of pro-inflammatory markers like iNOS and COX-2 by downregulating several age-related signaling pathways such as JNK, P38, MAPK, and ERK.

10.2 Conclusion

Globally, people now prefer the natural substances for health maintenance than the synthetic compounds because of their lowered side effects. Ginger is a potent antioxidant and anti-inflammatory agent, with minimal side effects. Many cellular and animal experiments demonstrated that ginger is more effective against progression of AD. The phytochemical constituents found in ginger are responsible for protective effect by improving the learning and memory dysfunction. It is suggested that consuming right proportion of ginger in any form or as a part of food may improve the mental health.

References

Adel PRS, Prakash J (2010) Chemical composition and antioxidant properties of ginger root (Zingiber officinale). J Med Plant Res 4:2674–2679

Banerjee S, Mullick H, Banerjee J, Ghosh A (2011) Zingiber officinale: 'a natural gold'. Int J Pharm Bio-Sci 2(1):283–294

Grzanna R, Phan P, Polotsky A, Lindmark L, Frondoza CG (2004) Ginger extract inhibits beta-amyloid peptide-induced cytokine and chemokine expression in cultured THP-1 monocytes. J Altern Complement Med 10:1009–1013

Guillemin GJ, Brew BJ (2002) Implications of the kynurenine pathway and quinolinic acid in Alzheimer's disease. Redox Rep 7:199–206

Halawany AME, Sayed NS, Abdallah HM, Dine RSE (2017) Protective effects of gingerol on streptozotocin-induced sporadic Alzheimer's disease: emphasis on inhibition of beta-amyloid, COX-2, alpha-, beta-secretases and APH1a. Sci Rep 7:2902

Huh E, Lim S, Kim HG, Ha SK, Park HY, Huh Y et al (2018) Ginger fermented with Schizosaccharomycespombe alleviates memory impairment via protecting hippocampal neuronal cells in amyloid beta1-42 plaque injected mice. Food Funct 9:171–178

Jafarian S, Ling KH, Hassan Z, Perimal-Lewis L, Sulaiman MR, Perimal EK (2019) Effect of zerumbone on scopolamine-induced memory impairment and anxiety-like behaviours in rats. Alzheimers Dement 5:637–643

Khandouzi N, Shidfar F, Rajab A, Rahideh T, Hosseini P, Taherif MM (2015) The effects of ginger on fasting blood sugar, hemoglobina 1c, apolipoprotein b, apolipoprotein a-i and malondialdehyde in type 2 diabetic patients. Iran J Pharm Res 14(1):131–140

Kim DS, Kim JY, Han YS (2007) Alzheimer's disease drug discovery from herbs: Neuroprotectivity from beta-amyloid (1-42) insult. J Altern Complement Med 13:333–340

Kim MK, Chung SW, Kim DH, Kim JM, Lee EK, Kim JY et al (2010) Modulation of age-related NF-kappa B activation by dietary zingerone via MAPK pathway. Exp Gerontol 45:419–426

Kim CY, Seo Y, Lee C, Park GH, Jang JH (2018) Neuroprotective effect and molecular mechanism of [6]-gingerol against scopolamine-induced amnesia in C57BL/6 mice. Evidence based complementary. Altern Med Ecam 2018:8941564

Lee C, Park GH, Kim CY, Jang JH (2011) [6]-Gingerol attenuates β-amyloid-induced oxidative cell death via fortifying cellular antioxidant defense system. Food Chem Toxicol 49:1261–1269

Li L, Wu XH, Zhao XJ, Xu L, Pan CL, Zhang ZY (2020) Zerumbone ameliorates behavioral impairments and neuropathology in transgenic APP/PS1 mice by suppressing MAPK signaling. J Neuroinflammation 17:61

Lim S, Choi JG, Moon M, Kim HG, Lee W, Bak HR et al (2016) An optimized combination of ginger and peony root effectively inhibits amyloid-beta accumulation and amyloid-beta mediated pathology in abetaPP/PS1 double-transgenic mice. J Alzheimers Dis 50:189–200

Mahdy K, Gouda N, El-Fattah A, Yassin N, El-Shenawy S, Farrag AR et al (2014) Protective effect of ginger (Zingiber officinale) on Alzheimer's disease induced in rats. J Neuroinfect Dis 5(2): 159

Manju V, Nalini N (2005) Chemopreventive efficacy of ginger, a naturally occurring anticarcinogen during the initiation, postinitiation stages of 1, 2 dimethylhydrazine-induced colon cancer. Clin Chim Acta 358(1):60–67

Mathew M, Subramanian S (2014) In vitro evaluation of anti-Alzheimer effects of dry ginger (Zingiber officinale roscoe) extract. Indian J Exp Biol 52:606–612

Moon M, Kim HG, Choi JG, Oh H, Lee PK, Ha SK et al (2014) An active constituent of ginger, attenuates neuroinflammation and cognitive deficits in animal models of dementia. Biochem Biophys Res Commun 449:8–13

Na JY, Song K, Lee JW, Kim S, Kwon J (2016) 6-Shogaol has anti-amyloidogenic activity and ameliorates Alzheimer's disease via CysLT1R-mediated inhibition of cathepsin B. Biochem Biophys Res Commun 477:96–102

Na JY, Song K, Lee JW, Kim S, Kwon J (2017) Sortilin-related receptor 1 interacts with amyloid precursor protein and is activated by 6-shogaol, leading to inhibition of the amyloidogenic pathway. Biochem Biophys Res Commun 484:890–895

Pashaei-Asl R, Pashaei-Asl F, Gharabaghi PM et al (2017) The inhibitory effect of ginger extract on ovarian cancer cell line; application of systems biology. Adv Pharm Bull 7:241–249

Saha A, Blando J, Silver E, Beltran L, Sessler J, DiGiovanni J (2014) 6-Shogaol from dried ginger inhibits growth of prostate cancer cells both in vitro and in vivo through inhibition of STAT3 and NF-κB signaling. J Cancer Prev 7:627–638

Sutalangka C, Wattanathorn J (2017) Neuroprotective and cognitive-enhancing effects of the combined extract of cyperusrotundus and zingiberofficinale. BMC Complement Altern Med 17:135

Vernin G, Parkanyi C (2005) Chemistry of ginger. In: Ravindran PN, Nirmal Babu K (eds) Ginger: the genus Zingiber. CRC Press, Boca Raton, pp 87–180

Zeng GF, Zhang ZY, Lu L, Xiao DQ, Zong SH, He JM (2013) Protective effects of ginger root extract on Alzheimer disease-induced behavioral dysfunction in rats. Rejuvenation Res 16:124–133

Zeng GF, Zong SH, Zhang ZY, Fu SW, Li KK, Fang Y et al (2015) The role of 6-gingerol on inhibiting amyloid beta protein-induced apoptosis in PC12 cells. Rejuvenation Res 18:413–421

Chapter 11
Effect of Pepper and Its Components on Alzheimer's Disease

Anna Kuppusamy Iyer Venkatachari Sama, Ponnambalam Annapoorani, Arokiasamy Justin Thenmozhi, Thamilarasan Manivasagam, and Tamiloli Hemalatha

Abstract Alzheimer's disease is a progressive neurodegenerative disease that affects approximately 47 million people globally. Although the impaired levels of acetylcholine are responsible for memory loss in AD, there is no drug yet discovered to cure or treat completely while relieving the symptoms. Black pepper (*Piper nigrum*, Piperaceae) considered as the 'king of the spices' is reported to have medicinal properties due to the presence of potent components. In this chapter, we discuss the protective role of extracts of various types of pepper, in combination with other medicinal plant extracts and its potent components by diminishing the learning and memory loss, oxidative stress, inflammation and apoptosis and maintain the normal architecture of brain organs in several in vitro and in vivo models of AD.

Keywords Pepper · Learning and memory improvement · Antioxidant · Anti-inflammatory role

A. K. I. V. Sama · P. Annapoorani (✉)
Department of Biochemistry, V.V. Vanniaperumal College for Women (Affilated to Madurai Kamaraj University, Madurai), Virudhunagar, Tamil Nadu, India
e-mail: annapoorani@vvvcollege.org

A. Justin Thenmozhi
Department of Biochemistry, School of Biological Sciences, Madurai Kamaraj University, Madurai, Tamil Nadu, India

Department of Biochemistry and Biotechnology, Faculty of Science, Annamalai University, Chidambaram, Tamil Nadu, India

T. Manivasagam · T. Hemalatha
Department of Biochemistry and Biotechnology, Faculty of Science, Annamalai University, Chidambaram, Tamil Nadu, India

A. Justin Thenmozhi, T. Manivasagam (eds.), *Nutraceuticals for Alzheimer's Disease: A Promising Therapeutic Approach*, Nutritional Neurosciences,
https://doi.org/10.1007/978-981-99-0677-2_11

11.1 Introduction

Spices are the dry, aromatic and pleasant components, obtained from the plants. It is acquired from various parts of the plants (Table 11.1) in a dried form including the bark (cinnamon), berry/fruits (cloves, black pepper, chili), seeds (cumin), flowers/buds (saffron, cloves) and root (turmeric/ginger) (Embuscado 2015). The origin of many spices is the Mediterranean and Middle East Asian countries, and most of them were utilised from thousands of years from the old Roman and Egyptian times (Peter 2004). Although they are mainly used to improve the aroma, flavour and colour of beverages and food substances, they are reported to offer protection against several acute and chronic, communicable and non-communicable diseases and help to maintain the health of the people. The prolonged historical use of spices and growing research proved the beneficiary role of spices against several chronic diseases including cancer, neurodegenerative diseases, type 2 diabetes and cardiovascular diseases (Tapsell et al. 2006; Kaefer and Milner 2008; Iriti et al. 2010).

11.2 Average Dietary Intake of Spices

About 0.5 g of spices/person/day was consumed by European peoples, while 1.3–1.9 g of spices/person/day was taken by the Australians and New Zealand residents, and about 1.8 g of spices/person/day was consumed by peoples of Africa. Moderate usage of spices was found in East and Middle East Asian peoples (2.6–3.1 g of spices/person/day). Indian, Latin American and South African peoples are the highest consumers of spices (~4.4 g/ day) (Vázquez-Fresno et al. 2019). The consumption of turmeric alone by Indian peoples was found to be about 1.5 g/person/day (Sharma et al. 2001). Although the intake of spices and herbs were usually more in the countries like India, Peru, Mexico, China and Thailand, due to the change in food habits and increasing choice for spicy food substances, their consumption were enhanced in several many developed nations in Northern Europe and America also (Williams 2006).

11.3 Pepper

Pepper (Table 11.2) is considered as a chief spice, containing more amounts of aromatic medicinal substances with significant amounts of numerous other functional compounds obligating health-enhancing properties. Its trade accounts for approximately 1/5th of the world's spice trade. Its uses are enhancing progressively in several fields like pharmaceutical industry, food processing, etc, as it is reported as a vital source of antioxidants with anti-carcinogenic activities.

Table 11.1 Types of spices with their main constituents

S. No	Spice	Scientific name	Components	Part of the plant
1	Basil	*Ocimum basilicum*	Linalool, methyl cinnamate, camphor and β-elemene	Leaf
2	Marjoram	*Origanum majorana*	Sinapic acid, ferulic acid, vanillic acid, caffeic acid, rosmarinic acid, syringic acid, p- and m-hydroxybenzoic acid, and coumarinic acid	Leaf
3	Oregano	*Origanum vulgare*	β-Fenchyl alcohol, thymol, carvacrol and γ-terpinene	Leaf
4	Lemongrass	*Cymbopogon*	β-Myrcene, geranial and neraland	Leaf
5	Peppermint	*Mentha* x *piperita*	Menthol and menthone	Leaf
6	Rosemary	*Rosmarinus officinalis*	Camphene, camphor, 1,8-cineol, α-pinene, limonene and linalool	Leaf
7	Sage	*Salvia officinalis*	Camphene, 1,8-cineole, α-thujone, camphor,β-thujone and sesquiterpene α-humulene	Leaf
8	Parsley	*Petroselinum crispum*	Falcarinol, myristicin, apiol, 1-allyl-2,3,4,5-tetramethoxybenzene, b-phellandrene, 1,3,8-p-menthatriene, oxypeucedanin, b-pinene, terpinolene and apiin	Leaf
9	Spearmint	*Mentha spicata*	Limonene, carvone, cis-dihydrocarvone and 1,8-cineole	Leaf
10	Tarragon	*Artemisia dracunculus*	Myrcene, methyl ethers, limonene, ocimene, α-pinene, β-pinene, camphene and linalool	Leaf
11	Thyme	*Thymus vulgaris*	Thymol, p-cymene and γ-terpinene	Leaf
12	Dill	*Anethum graveolens*	Carvone, limonene, α-phellandrene, dill ether and myristicin	Leaf/ seed
13	Fennel	*Foeniculum vulgare*	Estragole, trans-anethole and fenchone	Leaf/ seed
14	Black pepper	*Piper nigrum*	Sabinene, piperine, myrcene, α- and β-pinene, limonene, linalool, germacrene D α-phellandrene and β-caryophyllene	Berry
15	Caraway	*Carum carvi*	Oleic acids, petroselinic and linoleic	Fruit
16	Chili pepper	*Capsicum annuum, C. baccatum, C. chinense, C. frutescens, C. pubescens*	Capsaicin	Fruit

(continued)

Table 11.1 (continued)

S. No	Spice	Scientific name	Components	Part of the plant
17	Anise	*Pimpinella anisum*	Palmitic, trans-anethole and oleic acids	Seed/ fruit
18	Cinnamon	*Cinnamomum*	Cinnamate, cinnamaldehyde and cinnamic acid	Bark
19	Clove	*Syzygium aromaticum*	Eugenyl acetate, β-caryophyllene, α-humulene and eugenol	Bud
20	Cumin	*Cuminum cyminum*	Cuminaldehyde, cymene and terpenoids	Seed
21	Fenugreek	*Trigonella foenum-graecum*	Diosgenin, fenugreekine, trigonelline, carpaine, gentianine, apigenin, luteolin, orientin, quercetin, vitexin, isovitexin, 4-hydroxyisoleucine, yamogenin, tigogenin and neotigogenin	Seed
22	Nutmeg	*Myristica fragrans*	Myristicin, 4-terpineol and sabinene	Seed
23	Saffron	*Crocus sativus*	Safranal, picrocrocin and crocin	Stigma (flower)
24	Ginger	*Zingiber officinale*	Paradols, shogaols and gingerols	Root
25	Turmeric (curcumin)	*Curcuma longa*	Curcumin demethoxycurcumin, 5′-methoxycurcumin, turmerone and dihydrocurcumin	Root

Table 11.2 Taxonomical classification of black pepper

Kingdom	Plantae
Division	Magnoliophyta
Class	Magnoliopsida
Order	Piperales
Family	Piperaceae
Genus	*Piper*
Species	*Nigrum*

11.3.1 Classifications

It is classified based on diverse types and varieties, range of colours, grades and qualities. Few peppers were named according to the regions were they are grown or ports where they are exported (Tellicherry and Malabar pepper from Malabar regions of Kerala, Lampong pepper from Indonesia, Sarawak pepper (white) from Sarawak state of Malaysia, Brazilian pepper from Brazil); few are based on their colour: black, white, yellow, pink and green peppers. The cubeb and long pepper belong to another two species of piper family, which were more famous during ancient days but with little applications today. However, the term "pepper" globally refers to the dried fruit from the climbing shrub, *Piper nigrum*, belonging to India. On average, each American consumes about 4 ounces of black pepper annually.

11.3.2 Components of Pepper

Black pepper contains essential oils, lipase enzyme, piperine, fibre, eugenol and minerals. The main constituents of essential oil are limonene, β-caryophyllene and α- and β-pinene (Table 11.3). The pungent and irritant nature of black pepper is due to the presence of piperine and its isomers. Chemicals like piperine and 1-piperoylpiperidine are the major bioactive components present in both black and white peppers (Vázquez-Fresno et al. 1992).

11.3.3 Pepper and Ayurveda

The maricam (*Piper nigrum*) plant were called by about 20 names in Ayurvedic text (Table 11.4).

Table 11.3 Major chemical compounds found in black pepper

Chemical compound	Type of odour
α–terpineol	Floral
Nerol	Fresh, floral, herbal
Hexanol	Green apple
α-Pinene	Terperic, oxidised
Nerolidol	Mild spicy, rooty
Dihydrocarveol	Warm, woody
Acetophenone	Irritant, sharp
Citral	Citrussy
1, 8–Cineol	Camphory
Piperonal	Sweet, floral

Table 11.4 Other names of pepper in Ayurveda and their meaning

Names	Meaning
Marica	Destroys germs and worms
Krsna	Black
Katu	Pungent
Laghu	Light to digest
Teekshna	Piercing
Vrttaphalam	Globose fruit
Vallijam	Sprouts on a creeper
Sirovamtam	Stalk located on the top
Dharmapattana	Grows in the Travancore region of Kerala
Yavanesta	Liked by Europeans
Kaphavirodhi	Destroyer of kapha

P. nigrum, P. betle, P. longum, P. brachystachyum and P. cubeba belonging to the family of Piperaceae were mentioned extensively in Ayurveda. This herb was mentioned in ayurvedic texts like Materia Medica of the Ayurveda, Sushruta Samhita and Caraka Samhita. From the ancient period, the name, morphology and applications were found in several Ayurvedic texts. Although this medical system indicates seven Piperaceae herbs, the marica (*P. nigrum*) or black pepper is considered as an important spice due to their medicinal properties (Ram Manohar 2008). For the treatment of several digestive disorders, the mixer containing pepper and butter milk were used. It is also suggested for controlling the symptoms and pathology of common cold but is not desirable if patient has fever. Trikatu (trio of pungents) containing black pepper, long pepper and dry ginger is used for common ailments in various Ayurvedic pharmaceuticals and also a component of several main Ayurvedic formulations (Ram Manohar 2008). It is utilized in the management of earache, oral abscesses, indigestion, sunburn, insect bites, gangrene, constipation, lung disease, diarrhoea, hernia, tooth decay, heart disease, joint pain, hoarseness, liver problems, insomnia and toothaches (Kuete et al. 2013).

11.3.4 Pharmacological Applications of Black Pepper

Pepper is considered as the 'king of spices' that can improve the digestive capacity; enhance appetite; treat bellyache, dysentery, cough, common cold, throat infections and fever; and kills worms and smoothens the piles. It induces blood flow and holds a broad pharmacological properties such as antimicrobial, antipyretic, analgesic and anti-inflammatory. It also possesses the protective effect against hepatotoxicity and has anti-mutagenic and anti-carcinogenic properties. Due to the presence of the phenolic amides, pepper showed a potent antioxidant capacity than the other synthetic compounds such as butylated hydroxyanisole, butylated hydroxytoluene, etc. (Meghwal and Goswami 2012).

11.3.5 Pepper and Alzheimer's Disease

Various in silico, in vitro and in vivo experiments indicated the anti-alzheimeric role of pepper (Table 11.5).

11.4 Conclusion

In olden days, people considered pepper as having the beneficiary effect only on the digestive system and enhancing energy expenditure. But now, there are several in vitro and in vivo experiments indicating the neuroprotective effect of pepper

Table 11.5 Role of pepper in Alzheimer's disease

Pepper/constituents	AD models	Parameters analysed	Author(s)/year published
Piperine	Bilateral and intracerebroventricular injection of ethylcholineaziridinium ion	Enhanced memory deficits and reduced neurodegeneration in the hippocampus by diminished lipid peroxidation and acetylcholinesterase enzyme	Chonpathompikunlertet al. (2010)
Total plant extracts of Salvia triloba and *P. nigrum*	$AlCl_3$-induced rat model	Ameliorated the oxidative stress	Mahdy et al. (2012)
Synergestic effect of *S. triloba* L. and *Piper nigrum* extracts	$AlCl_3$-induced rat model	Regulation of acetylcholine, acetylcholinesterase, C-reactive protein, NF-kB, and monocyte chemoattractant protein-1	Ahmed et al. (2013)
N-[2-(3,4-dimethoxyphenyl)ethyl]-3-phenyl-acrylamide isolated from Sichuan pepper	AβPP-transgenic mice	Memory enhanced in Morris water maze test, disassociation of Aβ oligomers, inhibition of Aβ-mediated apoptosis and their gene expression, and reduction in calcium toxicity	Tang et al. (2013)
Piperine solid lipid nanoparticles with polysorbate-80 coating	Experimental model induced by ibotenic acid	Improve the memory in force swimming test, superoxide dismutase, acetylcholinesterase and histopathology in cortex	Yusuf et al. (2013)
Piper nigrum fruits (methanolic extract)	$A\beta_{1-42}$ infusion in rat hippocampus	Enhanced learning in Y-maze and radial arm-maze test Attenuation of the malonaldehyde levels and activities of enzymatic antioxidants	Hritcu et al. (2014)
Piper nigrum fruits (methanolic extract)	$A\beta_{1-42}$ infusion in rat hippocampus	Elevated memory and learning in plus-maze test and forced swimming test Attenuated the activities of glutathione peroxidase, superoxide dismutase and catalase, reduced amount of glutathione, protein carbonyl and malondialdehyde	Hritcu et al. (2015)
Various red pepper extracts	Hippocampal injections of β-amyloid (25–35) or (35–25) in rats	Improve memory in passive avoidance test and water maze test Induced the phosphorylation of CREB and GSK and prohibited the phosphorylation of tau and accumulation of Aβ	Yang et al. (2015)

(continued)

Table 11.5 (continued)

Pepper/constituents	AD models	Parameters analysed	Author(s)/year published
Alkamides of pepper	In vitro experiments	Piperine, piperettine, feruperine and piperettyline inhibited the acetylcholine and butylcholine esterase	
Polyphenol-enriched extracts of bell pepper (ripe and unripe)	In vitro studies on amyloid production and aggregation	Diminished the expression of β-secretase and Aβ1–40 aggregation	Ogunruku et al. (2017)
Piperlongumine B and analogues	Ellman's assays	Reduced activity of acetylcholinesterase	Wiemann et al. (2017)
N-phenethyl cinnamide derivatives of pepper	Aβ-induced and transgenic drosophila model	Antioxidant and Aβ disaggregation	Chai et al. (2018)
Piperlongumine	Lipopolysaccharide-induced amyloidogenesis	Reduced Aβ accumulation and suppressed the β- and γ-secretase activities Diminished the expression of inflammatory proteins in in vitro and in vivo models of AD	Gu et al. (2018)
Piperine	Streptozotocin was infused bilaterally in the hippocampus	Improvement of locomotion and memory in the open field and Morris water maze test Attenuated acetyl choline esterase, oxidative-antioxidative indices and inflammatory markers	Wang et al. (2019)
Pepper oleoresin	Scopolamine-induced rat model of AD	Elevated memory and locomotion in Morris water maze test and open-field test Attenuated the neurotoxin-induced levels and activities of AChE, malondialdehyde, reduced glutathione, glutathione peroxidase, superoxide dismutase and catalase	Rajashri et al. (2020)
Capsaicin, the pungent compound in pepper	APP/PS1 transgenic mice	Amyloid precursor protein processing shifted to α-cleavage, diminished neuroinflammation, Aβ generation, tau hyperphosphorylation and neurodegeneration	Wang et al. (2020)
Combination of aqueous and hydroalcoholic extract of cumin and black pepper	Immobility stress-induced memory loss	Improved motor coordination and learning ability (novel object detection, shuttle box and rotarod test) Regulated the levels of reduced glutathione and malondialdehyde and activities of acetylcholinesterase, superoxide dismutase and catalase	Rashedinia et al. (2021)

Black pepper cold-pressed oil	Scopolamine-induced Rat model of AD	Elevation of memory in passive avoidance and Morris water maze test. Increased activity of catalase and superoxide dismutase and reduced malondialdehyde acetylcholinesterase levels and histopathological changes in the hippocampi	Mostafa et al. (2021)
Synergistic role of *Piper nigrum* fruit and Cinnamumzeylanicum bark extracts	Scopolamine-induced mice model of AD	Improvement of memory in passive avoidance test and object recognition test	Teymuori et al. (2021)
Essential oils from black and white pepper	Aluminium trichloride-induced zebra fish model of AD	Acetylcholinesterase inhibitory activity	Chen et al. (2022)

and its constituents against various models of AD through its potent acetylcholine inhibitory, antioxidant, anti-inflammatory and anti-apoptotic effect. Further clinical trials involving several dietary doses of black pepper and their components are very much needed to confirm this beneficial role.

References

Ahmed HH, Salem AM, Sabry GM, Husein AA, Kotob SE (2013) Possible therapeutic uses of Salvia triloba and Piper nigrum in Alzheimer's disease-induced rats. J Med Food 16(5):437–446

Chai T, Zhao XB, Wang WF, Qiang Y, Zhang XY, Yang JL (2018) Design, synthesis of n-phenethylcinnamide derivatives and their biological activities for the treatment of Alzheimer's disease: antioxidant, beta-amyloid disaggregating and rescue effects on memory loss. Molecules 23(10):2663

Chen SX, Xiang JY, Han JX, Yang-Feng LHZ, Chen H et al (2022) Essential oils from spices inhibit cholinesterase activity and improve behavioral disorder in $AlCl_3$ induced dementia. Chem Biodivers 19(1):e202100443

Chonpathompikunlert P, Wattanathorn J, Muchimapura S (2010) Piperine, the main alkaloid of thai black pepper, protects against neurodegeneration and cognitive impairment in animal model of cognitive deficit like condition of Alzheimer's disease. Food Chem Toxicol 48(3):798–802

Embuscado ME (2015) Spices and herbs: natural sources of antioxidants—a mini review. J Funct Foods 18:811–819

Gu SM, Lee HP, Ham YW, Son DJ, Kim HY, Oh KW et al (2018) piperlongumine improves lipopolysaccharide-induced amyloidogenesis by suppressing NF-KappaB pathway. Neuro Molecular Med 20(3):312–327

Hritcu L, Noumedem JA, Cioanca O, Hancianu M, Kuete V, Mihasan M (2014) Methanolic extract of Piper nigrum fruits improves memory impairment by decreasing brain oxidative stress in amyloid beta(1-42) rat model of Alzheimer's disease. Cell Mol Neurobiol 34(3):437–449

Hritcu L, Noumedem JA, Cioanca O, Hancianu M, Postu P, Mihasan M (2015) Anxiolytic and antidepressant profile of the methanolic extract of piper nigrum fruits in beta-amyloid (1-42) rat model of Alzheimer's disease. Behav Brain Funct 29(11):13

Iriti M, Vitalini S, Fico G, Faoro F (2010) Neuroprotective herbs and foods from different traditional medicines and diets. Molecules 15:3517–3555

Kaefer CM, Milner JA (2008) The role of herbs and spices in cancer prevention. J Nutr Biochem 19: 347–361

Kuete V, Viertel K, Efferth T (2013) Antiproliferative potential of African medicinal plants. In: Medicinal plant research in Africa (Pharmacology and Chemistry). Elsevier, Amsterdam, pp 711–724

Mahdy K, Shaker O, Wafay H, Nassar Y, Hassan H, Hussein A (2012) Effect of some medicinal plant extracts on the oxidative stress status in Alzheimer's disease induced in rats. Eur Rev Med Pharmacol Sci 16(3):31–42

Meghwal M, Goswami TK (2012) Chemical composition, nutritional, medicinal and functional properties of black pepper: a review. Sci Rep 1:172

Mostafa NM, Mostafa AM, Ashour ML, Elhady SS (2021) Neuroprotective effects of black pepper cold-pressed oil on scopolamine-induced oxidative stress and memory impairment in rats. Antioxidants 10(12):1993

Ogunruku OO, Oboh G, Passamonti S, Tramer F, Boligon AA (2017) Capsicum annuum var. grossum (bell pepper) inhibits β-secretase activity and β-amyloid 1-40 aggregation. J Med Food 20(2):124–130

Peter KV (2004) Handbook of herbs and spices. CRC Press, Boca Raton, p 2004

Rajashri K, Mudhol S, Peddha MS, Borse BB (2020) Neuroprotective effect of spice oleoresins on memory and cognitive impairment associated with scopolamine-induced Alzheimer's disease in rats. ACS Omega 5(48):30898–30905

Ram Manohar P (2008) Piperaceae crops-technologies and future perspectives. National Seminar on Piperaceae-harnessing agro-technologies for accelerated production of economically important piper species. Indian Institute of spices research, Calicut. 302–310

Rashedinia M, Mojarad M, Khodaei F, Sahragard A, Khoshnoud MJ, Zarshenas MM (2021) The effect of a traditional preparation containing piper *Nigrum l.* and *Bunium persicum* (boiss.) B. fedtsch. on immobility stress-induced memory loss in mice. Biomed Res Int 2021:5577594

Sharma RA, McLelland HR, Hill KA, Ireson CR, Euden SA, Manson MM (2001) Pharmacodynamic and pharmacokinetic study of oral curcuma extract in patients with colorectal cancer. Clin Cancer Res 77:1894–1900

Tang M, Wang Z, Zhou Y, Xu W, Li S, Wang L et al (2013) A novel drug candidate for Alzheimer's disease treatment: gx-50 derived from zanthoxylumbungeanum. Alzheimer Dis 34(1):203–213

Tapsell LC, Hemphill I, Cobiac L, Patch CS, Sullivan DR, French M (2006) Health benefits of herbs and spices: the past, the present, the future. Med J Aust 185:S4–S24

Teymuori M, Yegdaneh A, Rabbani M (2021) Effects of piper nigrum fruit and Cinnamumzeylanicum bark alcoholic extracts, alone and in combination, on scopolamine-induced memory impairment in mice. Res Pharm Sci 16(5):474–481

Vázquez-Fresno R, Rosana AR, Sajed T, Onookome-Okome T, Wishart NA, Wishart DS (2019) Herbs and spices-biomarkers of intake based on human intervention studies – a systematic review. Genes Nutr 14:18

Wang C, Cai Z, Wang W, Wei M, Kou D, Li T et al (2019) Piperine attenuates cognitive impairment in an experimental mouse model of sporadic Alzheimer's disease. J NutrBiochem 70:147–155

Wang J, Sun BL, Xiang Y, Tian DY, Zhu C et al (2020) Capsaicin consumption reduces brain amyloid-beta generation and attenuates Alzheimer's disease-type pathology and cognitive deficits in APP/PS1 mice. TranslPsychiatry 10(1):230

Wiemann J, Karasch J, Loesche A, Heller L, Brandt W, Csuk R (2017) Piperlongumine B and analogs are promising and selective inhibitors for acetylcholinesterase. Eur J Med Chem 139: 222–231

Williams PG (2006) Health benefits of herbs and spices. Publ Health Med J Aust 1854:S17–S18

Yang HJ, Kwon DY, Kim MJ, Kang S, Moon NR, Daily JW et al (2015) Red peppers with moderate and severe pungency prevent the memory deficit and hepatic insulin resistance in diabetic rats with Alzheimer's disease. Nutr Metab (Lond) 12:9

Yusuf M, Khan M, Khan RA, Ahmed B (2013) Preparation, characterization, in vivo and biochemical evaluation of brain targeted Piperine solid lipid nanoparticles in an experimentally induced Alzheimer's disease model. J Drug Target 21(3):300–311

Rajashri K, Mudhol S, Peddha MS, Borse BB (2020) Neuroprotective effect of spice oleoresins on memory and cognitive impairment associated with scopolamine-induced Alzheimer's disease in rats. ACS Omega [illegible]

Chapter 12
Potential Therapeutics from Ayurveda, Siddha, and Homeopathic Medical System for Alzheimer's Disease

Samadder Chandrani and Arokiasamy Justin Thenmozhi

Abstract As a progressive and cognitively debilitating disease, Alzheimer's disease ranks the third cause of death for older people, right after cancer and heart disease. Despite the far-reaching understanding of its pathogenicity, the current FDA-approved drugs can only partially inhibit the disease progression. Options from alternative therapies target overcoming present therapies' inadequate efficacy, adverse effects, and poor patient compliance. Significant leads have emerged from scientific studies of phytoconstituents of Ayurvedic herbs, Siddha medicines, and homeopathic formulations. These therapeutic interventions aim toward the prevention and postponement of Alzheimer's disease onset, and many are already in different phases of clinical trials. Traditional medicine has been practiced since time immemorial, despite little understanding of the mechanism of their action. However, current scientific leads are consistently reporting anti-inflammatory, anti-amyloidogenic, anti-cholinesterase, hypolipidemic, and antioxidant activities in the herbal phytoconstituents, i.e., lignans, flavonoids, tannins, polyphenols, triterpenes, sterols, and alkaloids. This review aims to collectively present the scientific evidence of the vast potential of Ayurvedic medicinal plants; medicinal plants from the Siddha system; and homeopathic plant remedies in the prevention and treatment of Alzheimer's disease. Information was collected from scientific databases like PubMed, Semantic Scholar, and Google Scholar, along with reports by government bodies and documentation. Following a comprehensive literature search, the author provides a review encompassing (1) Ayurvedic, Siddha, and homeopathy holistic approaches to manage dementia and AD and (2) phytoconstituents of traditional herbs in Siddha, Ayurveda, and homeopathic

S. Chandrani
Department of Molecular Microbiology, School of Biotechnology, Madurai Kamaraj University, Madurai, Tamil Nadu, India

A. Justin Thenmozhi (✉)
Department of Biochemistry, School of Biological Sciences, Madurai Kamaraj University, Madurai, Tamil Nadu, India

Department of Biochemistry and Biotechnology, Faculty of Science, Annamalai University, Chidambaram, Tamil Nadu, India

A. Justin Thenmozhi, T. Manivasagam (eds.), *Nutraceuticals for Alzheimer's Disease: A Promising Therapeutic Approach*, Nutritional Neurosciences,
https://doi.org/10.1007/978-981-99-0677-2_12

formulations effective in AD with references to their current clinical uses and future prospects.

Keywords Alzheimer's disease · Ayurveda drugs · Siddha herbs · Homeopathy approaches · Plant extracts and active components

12.1 Introduction

Alzheimer's disease is a progressive and debilitating neurodegenerative disease, affecting 5% of men and 6% of women, above the age of 60, annually, worldwide. The overall prevalence of dementia in India is 33.6% per 1000 persons, with the disease accounting for around 54% and vascular dementia for about 39%. In the United States, it affects over 4.5 million people, or about 10% of the population over the age of 65, and it is predicted that this number will increase fourfold by 2050. Once a person reaches the age of 80, the frequency rises to 50%. Characterized by progressive cognitive deterioration and declining activity, the pathology of this disease is highly complex, involving the influence of genetic, familial, and environmental factors; age; lifestyle; diet, etc. Alzheimer's is a common form of presenile and senile dementia.

The primary neuropathology of Alzheimer's disease is neurofibrillary tangles, which are an aberrant collection of protein pieces, and atrophying neurons, mostly affecting the temporoparietal and frontal cortex. Dystrophic neuritis, reactive astrocytes, and activated microglia surround the comparatively intractable dense cores of neurotic plaques, which are 510 nm thick amyloid fibrils. The cerebral artery wall can have a high concentration of monocytes/macrophages, while the nearby parenchyma is highly reactive surrounded by activated microglial cells. The 39–42-amino acid (beta) amyloid peptide is the primary protein component of amyloid in Alzheimer's brain (Bamberger and Landreth 2002; Di Patre et al. 1999).

12.2 Siddha System

Plants and conventional herbal remedies have been an important source of medication since more than 5000 years ago, with Rigveda having the first evidence between 4500 and 1600 B.C. Even in modern times, more than 70% of individuals living in rural areas receive their health treatment through complementary medical practices, particularly Ayurvedic, Unani, and Siddha (Bhattacharjee 2001). The Siddha system, the most revered of all medical traditions, is practiced in South India and most commonly in Tamil Nadu. According to Rule 3a of the Drugs and Cosmetics Act of 1940, an "Ayurvedic, Siddha, or Unani drug" is any medication made solely in accordance with the formulae described in the authoritative books of [Ayurveda, Siddha, Unani Tibb systems of medicine] and intended for internal or external use in the diagnosis, treatment, mitigation, or prevention of diseases or disorders.

Attributed to its effectiveness with minimal side effects (Sudarsanam and Moonandi 2014), this Dravidian system of medicine was thought to have been founded by Siddhic powers. Its influence ranges across various age-old practices of traditional medicine such as alchemy, philosophy, yoga, varma, and other external therapies.

Traditional medicine is defined by the World Health Organization (WHO) as the body of knowledge, abilities, and practices based on theories that are inherent to various cultures and used in the preservation and improvement of health against physical and mental illness. Several molecular mechanisms have been suggested to explain the actions of the herbs, with the most significant ones being the ability to act on several molecular targets in an additive or even synergistic manner as well as free radical scavenging activity and a neuroprotective effect. Herbs such as *Glycyrrhiza*, *Ginkgo biloba L.*, *Camellia sinensis* Kuntze (green tea), *Withania somnifera* (Ashwagandha), *Achyranthes aspera* L., *Nardostachys jatamansi*, *Tinospora cordifolia*, *Centella asiatica*, *Allium sativum*, and curcumin are most commonly suggested in neuroprotective treatments.

In the Siddha system of medicine, drugs of plant, animal, and mineral origin are employed in their natural or so-called crude forms separately or in their mixture or combination to make a compound preparation. Plant drugs, particularly herbaceous drugs, are commonly employed as complete plants or their parts are used such as root, stem, leaf, flower, seed, fruit variations of stem and root, bark of a stem or root, wood, and their exudates or gums. These formulations are either dried or consumed fresh or in juice form. Siddha system also has dosage forms of churnas, bhasma, kwatha, taila, tablets, and ointments.

In the Siddha system, vatham (biological air humor) that travels through the brain and the nerves, when out of balance, is said to cause nervous system illnesses. Alzheimer's disease is one of the 80 varieties of vatha illness that are categorized by this system (Rao et al. 2012; Eugenie et al. 2019). It is a common practice in Siddha system to view the nervous system in a holistic manner. It considers the nerve impulse to be the wind/air (vatham) that controls both voluntary and involuntary functions of the body (Siddha Noi Nadal 2023). The bioenergies Tridhosham and Thrigunam control the bodily physiological and psychological processes. Thus, in Siddha, any disruptions in these energies can negatively affect Buddhi (the psychological mind), Indriyam (the cognitive and motor organs), and Manam (intellectual mind). As a result, maintaining a trihumoral balance in the environment is crucial for managing Alzheimer's disease effectively (Murugesa Mudaliar 1998).

The important component of every Siddha formulation is the selection of suitable herbs, with necessary tastes, that balance humors as a basic treatment. Vatham is the driving force behind the other doshas out of the three (pitham and kapham). An increase in the vatham dosha causes tissue loss of brain cells, which results in a degenerative condition resembling AD. The Siddha method states that the selection of herbs that have the ability to rectify deranged humors by their nature of opposite action due to their suvai (taste), veeryam (potency), and taste after digestion is the basis for treating ailments (deranged humors) (vipaakam). Each flavor is a fusion of two of the panchabootham elements, which include earth, water, fire, air, and space. These comparable panchabootham components also make up our human body.

Most Siddha pharmacokinetics aimed at neuroprotection focus greatly on these selected herbs that can act on trihumors, i.e., *Bacopa monnieri* (Brahmi), *Withania somnifera* (Amukkara), *Clitoria ternatea* (Kaakataan), *Centella asiatica* (Vallaarai), *Curcuma longa* (Manjal), *Glycyrrhiza glabra* (Athimadhuram), and *Tinospora cordifolia* (Seenthil).

Siddha systems of medicine depend on siddhi formulations like Vallarai Nei and Brahmi Nei to target oxidative stress-induced neurodegenerative disorders and memory loss. And most commonly, these formulations are prepared in milk and ghee. The most common formulations are milk- and ghee-treated *Curcuma aromatica*, *Acorus calamus*, and *Zingiber officinale*. In one study, Kausalya et al. (2017) investigated the scientific rationale behind the practice by comparing the preparations of herb mixtures with milk and ghee independently as well as in combination. They could find that the free radical scavenging activity (84.88%) of milk-treated sample of *Acorus calamus* and ghee-treated sample of herbal mixture exhibited lesser malondialdehyde (MDA) content (0.04 nM/100 g) and showed higher inhibition of lipid peroxidation (99.65%) on goat brain homogenate model. They found 3-hydroxy-3,4-dimethoxyflavone in *Z. officinale*, cinnamic acid in *C. aromatica*, and taxifolin-3-glucopyranoside, velutin, and methyl digallate in *A. calamus* in their LC-MS/MS studies; the herbal mixture treated with ghee was more stable, having smaller nanoparticles. All the while milk promoted the release of antioxidants in the selected herbs.

12.3 Ayurveda

Ayurveda is a traditional medicinal system indigenous to India and the Indian subcontinent. One of the oldest references about Alzheimer's in Ayurvedic literature was possibly a condition termed Prana Vritti Samana Vatavyadhi under the category of Vatavyadhi (neurodegenerative diseases). Its pathological descriptions and symptomology have been shown to match the western symptomology of Alzheimer's (Rao et al. 2012; Shantala and Nagarjun 2013). Bredesen and Rammohan et al., based on their multiple studies (Bredesen 2014, 2015, 2016; Bredesen et al. 2016), have proposed three distinct subtypes of Ayurvedic interpretation of Alzheimer's. They are Vata, Pitta, and Krimi, identification of which can help distinguish patients of different severity of cognitive decline and hence categorize their therapeutic recommendations.

In the Vata type, Alzheimer's disease is associated with an age-associated neurodegenerative condition arising from Vata-provoking lifestyle that involves practicing unstable routines (Rao et al. 2012; Gokani 2014; Manyam and Kumar 2013). It was also noted in the metabolomic profile of Vata-type AD that a lifestyle consisting of overwhelming stress and eating a dry, cold, light diet (Rao et al. 2012; Ninivaggi 2010) can trigger the withdrawal of trophic state and reduce hormonal support (Bredesen 2015). Symptomology of these patients is shown to associate with memory decline, mood swings, insomnia, confusion, and emotional vulnerability.

Patients above the age of 80 experiencing these symptoms fall under this category (Bredesen 2015).

Ayurvedic rejuvenation therapies (Rasayana and Vajikarana) are commonly recommended for Vata-type Alzheimer's patients, which requires them to have well-monitored long-term care that helps increase their body's inherent resistance and immunity. A diet that balances Vata's light, chilly, and dry traits is advised (Ninivaggi 2010). As a result, foods must be wet, freshly made, steaming, and gently seasoned. Foods with sweet, sour, and salty flavors aid in the relief of Vata imbalance. It is preferable to eat whole grains, fresh veggies, nuts, ghee, coconut, or sesame oil. Consuming fruits and vegetables of all colors can help to increase your intake of antioxidants, minerals, and vitamins. Fruits and vegetables are also high in fiber and can assist with constipation and other digestive difficulties. Fruits and green and leafy vegetables should be consumed on a daily basis as part of a brain-protective diet. Meals should be eaten on a regular basis, at the same time every day. Ayurveda also prescribes a minimum of 12 h of fasting between the previous day's last meal (Murphy et al. 2014) and the first meal the next day, which supports the research that fasting induces hypoglycemia state in the brain, resulting in autophagy, lipid catabolism, and release of brain-protective ketone bodies, all of which improves brain plasticity and cognition (Murphy et al. 2014; McCarty et al. 2015; Hashim and VanItallie 2014).

Ayurvedic herbs, practiced since ancient times, promise the stimulant and sedative properties of the nervines as they address nervous system excesses or deficiencies (Rao et al. 2012). Also, combinations (or Rasayana) of multiple herbs' antioxidant activity, free radical scavenging activity, ability to increase cholinergic activity, neurogenesis, and memory-enhancing properties (Rao et al. 2012) provide both neuroceutical and cogniceutical qualities. One such herb formulation is that of Ashwagandha, Brahmi, Gotu Kola, Guduchi, and Shankhpushpi. They can be consumed orally, as herbal teas, or made into a paste with suitable carrier oil and administered intranasally (Rao 2012; Rao et al. 2012; Pires et al. 2009). Such Ayurvedic formulations carry the benefit of specificity and energetics of individual herb matching with the patient's subtype. Because it is quickly delivered, avoids the blood–brain barrier, and directly targets the central nervous system, intranasal intervention is more beneficial than the other forms of administration (Illum 2003; Jones et al. 1997). Most often, a combination of therapies stimulates enhanced lymphatic drainage. Patients with Vata-type AD are advised to receive palliative care (Shamana Chikitsa). Oleation and fomentation therapies are crucial because they counteract the dryness and coldness that Vata causes, boosting cerebral and systemic blood flow (Buckle et al. 2008; Ouchi et al. 2006). It is advised to employ medicinal oils with plants specific to the Vata subtype in transcranial and massage therapies (Abhyanga) (Rao 2012; Rao et al. 2012). Medicated oils are well known for their ability to nourish the neurological system and promote cellular renewal (Lindgren et al. 2012). To aid in the internal absorption of the medicinal oil, the patient receives a fomentation-herbal steam bath (Swedana) after receiving a herbal oil massage (Rao 2012; Rao et al. 2012; Wang et al. 2011).

Walking in a park, on the beach, or in a garden is a fantastic form of therapy, especially for Vata-type AD patients, as these activities introduce the earth element's slow, steady, and heavy properties to the nervous system (Colcombe and Kramer 2003; Kramer et al. 1999; Ngandu et al. 2015; Tolppanen et al. 2015). Yoga, dhyana, and relaxing breathing exercises (Pranayama) are recommended for patients dealing with mild or early Alzheimer's in order to maintain a calm and balanced nervous system (Eyre et al. 2016; Lavretsky et al. 2013; Luders et al. 2011; Newberg et al. 2012). Studies on healthy volunteers have shown that using diaphragmatic breathing techniques can enhance several elements of cognitive performance (Ferreira et al. 2015). Practicing mindfulness also lessens excess Vata characteristics linked to AD by lowering stress proteins, improving the lipid profile, reducing oxidative stress, strengthening neural circuits, and increasing cognitive reserve capacity (Tang et al. 2015).

In the second type, people with Pitta nature, who naturally exhibit excess heat, have been shown to exhibit the metabolic profile of type 1 Alzheimer's (inflammatory AD) (Bredesen 2015). Degeneration of hippocampal/cortical cells is most frequently observed in Alzheimer's patients having a Pitta-vitiating lifestyle (Mishra et al. 2001; Purvya and Meena 2011; Ninivaggi 2010). One of the key pathologies of this form of Alzheimer's is the enhanced proinflammatory cytokines, inflammatory microglia and activated astroglia, chemokines, and acute-phase reactants in patient brains (Theendakara et al. 2016; Akiyama et al. 2000; Bredesen 2015). Risk is associated with regular intake of hot, sour, salty, strongly acidic, and fermented foods, such as alcohol. Anger and chronic exposure to sun and heat are also notable risk factors (Ninivaggi 2010; Mishra et al. 2001).

In such cases, a preventive lifestyle and taking cool herbs and oils are majorly recommended. Patients are suggested to consume only moist, warm, fresh meals with cooling spices that incorporate the sweet, bitter, and astringent properties. Similarly, food items that trigger heat and inflammation, i.e., salty, sour, and pungent flavors, can be avoided which can include pickled and fermented dishes (Ninivaggi 2010).

Pungent, sour, and salty diets need to be avoided as they can trigger heat and underlying inflammation. It is also advisable to stay away from pickled and heavily fermented foods. A diet rich in ghee, coconut and olive oils, dairy products, bitter vegetables, bitter greens, and raw salads and ripened fruits enhances the intake of inflammatory-suppressing antioxidants. Fasting in between two meals and intake of cooling nervines can also help to reduce inflammation or excess heat. A well-known powerful anti-inflammatory spice is turmeric. Guggulu is included in the recipe to (1) cure inflammatory conditions and problems with lipid metabolism, (2) function as an antioxidant, and (3) prevent the action of the enzyme acetylcholinesterase (Rao 2012; Rao et al. 2012).

In the third type, people with the Kapha lifestyle have rarely been observed to have Alzheimer's disease. Traditionally, Kapha refers to having symptoms of parasitic infection or toxin poisoning. Therefore, Alzheimer's disease patients who did not fall under either of the above categories were categorized in this type. Risk factors would include (1) exposure to mycotoxins, aquatoxins, and other pathogens and/or (2) exposure to toxic chemicals and metals. Patients falling under this category generally have early-onset Alzheimer's and are highly stressed individuals, with insomnia and depression, and can carry the ApoE genotype (Lad 2006; Ninivaggi 2010). Patients may have been exposed to biotoxins/mycotoxins, tick-borne toxins, aquatoxins that trigger Lyme's disease, and high levels of arsenic, lead, mercury, or other metal pollutants. Widespread cerebral atrophy and frontal–temporal–parietal abnormalities cauhght by FDG-PET also confirm the underlying cognitive dysregulation in the patients. Lower cognitive performance can also be due to low zinc deficiency in these patients (Bredesen 2015).

In Ayurveda, the application of various herbal formulations, termed as Rasayana, increases cellular oxygenation and promotes neurogenesis by restoring homeostasis. These compounds are studied to be more bioavailable and less toxic, while exhibiting antioxidant, anti-amyloidogenic, anti-inflammatory, neuroprotective, and immunomodulatory effects. Most phytodrugs and their standardized extracts have been shown to regulate APP metabolism toward the α-secretase pathway while restricting the stabilization of Aβ fibrils (Joy 2015; Sharma and Amin 2015; Sharma et al. 2019). In the clinical scenario, the well-studied herb *Ginkgo biloba* is in prevention trials for Alzheimer's disease (Wang et al. 2016; Singh et al. 2019), while there is an ongoing exploration of *Buchanania axillaris* Desr. (Anacardiaceae), *Hemidesmus indicus* Linn. (Apocynaceae), and *Rhus mysorensis* Heyne (Anacardiaceae) in Alzheimer's disease treatment (Penumala et al. 2018). One of the most well-explored herbs, Brahmi Ghrita, has shown to have anti-inflammatory properties, clearance of small channels, and detoxification of the brain's toxic metabolic byproducts, thereby promoting neuroprotection (Chaudhari et al. 2017). Ayurvedic therapies have shown to have multitargeted approach and are at their best when used as a combination of complementary herbs. This becomes favorable for multifactorial pathologies such as neurodegenerative diseases that have complex etiologies, varying from patient to patient.

The herbs can unofficially be categorized as excelling in certain beneficiary factors. Herbs such as *Terminalia chebula* (Afshari et al. 2016), *Passiflora incarnata* (Ingale and Kasture 2017), *Typhonium trilobatum* (Lopa et al. 2019), *Satureja cuneifolia* (Taslimi et al. 2020), *Anisomeles indica* (Uddin et al. 2016), *Curcuma longa* (Thakur et al. 2019), *Bacopa monnieri* (Dubey and Chinnathambi 2019), *Crocus sativus* L. (Hatziagapiou et al. 2019; Finley and Gao 2017), *Macrosphyra longistyla* (Elufioye et al. 2019), *Cinnamomum zeylanicum* (Tepe and Ozaslan 2020), *Melissa officinalis* (Liou et al. 2003; Mahboubi 2019), *Caesalpinia crista* (Ravi et al. 2018), *Camellia sinensis* (Kleinrichert and Alappat 2019), and *Scoparia dulcis* (Kleinrichert and Alappat 2019) present with potent free radical scavenging activity and neuroprotective properties and thus constitute an important part of treatment. Some herbs have a natural anti-amyloidogenic property in their polyphenolic, alkaloid, and cannabinoid extracts. Plants such as *Grewia tiliaefolia* (Sheeja Malar et al. 2017), *Cassia tora* (Chethana et al. 2017), *Elettaria cardamomum* (Chowdhury and Kumar 2020), *Caesalpinia crista* (Chethana et al. 2018), *Perilla frutescens* (Lee et al. 2019; Kim et al. 2017), *Guettarda speciosa* (Tan et al. 2019), *Dryopteris crassirhizoma* (Joo et al. 2016), *Dracocephalum moldavica* L. (Liu et al. 2018), *Bacopa monnieri* (L.) Wettst (Eze et al. 2019), *Lawsonia inermis* (Dhouafli et al. 2019), and *Sargassum horridum* (Castro-Silva et al. 2020) can be used for Alzheimer's without any side effects. Some plant extracts also have anti-inflammatory property that can be effective in the treatment of Alzheimer's of inflammatory origin. Multiple in vitro and in vivo experiments assessing the anti-inflammatory properties of *Terminalia chebula* (Afshari et al. 2016), *Crocus sativus* L. (Hatziagapiou et al. 2019; Finley and Gao 2017), *Lagerstroemia indica* (Al-Snafi 2019), *Limonium spathulatum* (Mazouz et al. 2020), *Okinawa propolis* (Shahinozzaman et al. 2018), *Corydalis dubia* (Wangchuk et al. 2016), and *Pancratium parvum* (Patil et al. 2020) extracts have confirmed their non-toxicity in brain tissues. Despite being already in use in multiple clinical causes, numerous plant secondary metabolites with neuroprotective properties are still undergoing clinical trials for the treatment of Alzheimer's disease. These include nerve growth factor, valproate (and other GSK inhibitors), nicotinic agonists, stress kinase inhibitor CEP-1347, minocycline, and metal chelators (Longo and Massa 2004). Plants like *Bacopa monnieri* (L.) Wettst (Eze et al. 2019), *Grewia tiliaefolia* (Sheeja Malar et al. 2017), *Vernonia amygdalina* (Oboh et al. 2022), *Levisticum officinale* (Amraie et al. 2020), *Schisandra chinensis* (Sowndhararajan et al. 2018), *Withania somnifera* (Singh and Ramassamy 2017), *Ginkgo biloba* (Singh et al. 2019), and *Kigelia africana* (Falode et al. 2017) have neuroprotective properties, improve neuron cell regeneration, and prevent neurodegeneration.

12.4 Homeopathy

Homeopathy, as compared to others, has been well practiced and was considered a popular form of "contemporary" treatment since the 1990s. In fact, over 8.6% of the population in the UK were regular users of homeopathy, while around 20% of the Americans considered themselves homeopaths (Thomas et al. 2001). The decline in its practice was seen following Flexner Report of 1910, which has commentary on modern medical education. Originally, homeopathy was designed and standardized in 1796 by the German physician Samuel Hahnemann. He believed in "similia similibus curentur," which literally means "let like be cured by like." This was proved by quinine treatment of malaria, the overdosage of which can produce malaria-like symptoms in patients. With the following concept, Hahnemann adopted the idea of "potentization" in controlling the dosage of homeopathic medicines. The core of homeopathic remedy is the "mother tincture." It is made by soaking the necessary substance for a number of weeks in an ethanol-water mixture. Insoluble materials are first mixed with lactose and then suspended in the same liquid. The mother tincture is then serially diluted, commonly called "potentized" with forceful shaking (succussion) at each stage. The most popular dilution scales are centesimal scales (c, C, or CH), with dilutions in stages of 1:100, and decimal scales, with consecutive dilutions of 1:10 (often denoted by x in English-speaking countries, D or DH in most other countries). There are also additional dilution techniques in use, such as the 50 Millesimal (LM) and Korsakov techniques (Adler and Adler 2006). However, a lot of controversy surrounds the claims made for homeopathy. Despite finding multiple pieces of evidences of its efficacy, the absence of a concrete molecular mechanism of its action in ultramolecular dilutions against neurodegenerative diseases that aligns with current scientific concepts decreases the prospects of future studies. A systematic review of its safety and effectivity in controlled clinical trials is a necessity (Linde et al. 1997).

A popular homeopathic remedy *Lycopodium clavatum* was used for liver, urinary, and digestive problems until lately. Its Lyc alkaloid extract was shown to have acetylcholinesterase (AchE) inhibitory activity (Calderon et al. 2013). Also known as club moss, *Lycopodium clavatum* is the major genus of lycopodium and family Lycopodiaceae, and it has also found its use in rheumatism, epilepsy, and pulmonary disorder. Hanif et al. (2015) conducted a study to explore the Lyc activity on learning, memory, and cerebral blood flow in Wistar rats with intracerebroventricularly (ICV) administered streptozotocin (STZ)-induced memory impairment. In the experimental setup, the induced rats were orally administered with Lyc mother tincture (MT) (with absolute alcohol as the vehicle of dilution) 30, 200, and 1000 for 17 days and the learning and memory activity was evaluated using the Morris water maze test on 14th, 15th, and 16th days. Also, the cerebral blood flow was measured using a laser Doppler flow meter on the 17th day. They also noted that the STZ reduces cerebral blood flow through vascular amyloidosis, oxidative stress, and endothelial dysfunction. Naturally, administration of Lyc MT 30 and 200 could effectively reduce the effects of STZ on learning and memory impairment; however, with the increase of potency, improvement in learning was not statistically significant. Cerebral blood flow was restored in all the potencies of Lyc MT.

Despite multiple evidences published in international journals, controversies and disbelief continue to surround the biological and pharmacological effectivity of "ultralow doses" and "ultrahigh dilutions" in homeopathic remedies. The mechanism of action of ultrahigh dilutions continues to be unknown, despite high levels of patient satisfaction and a significant rise of technological advancement in physics, chemistry, biology, agriculture, and environmental science.

12.5 Neuroprotective Effect of Siddha and Ayurvedic Herbs

Well-known herbs associated with Siddha and Ayurveda and their applications are mentioned in Fig. 12.1.

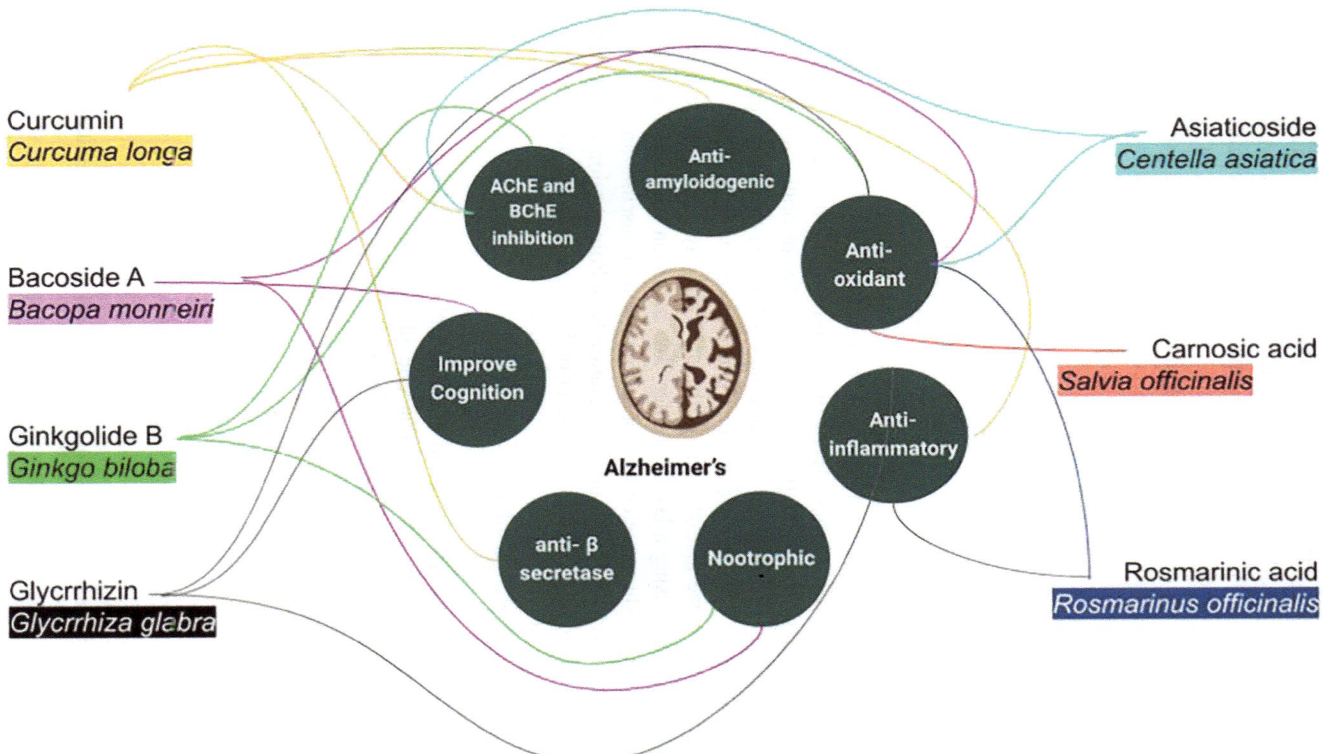

Fig. 12.1 The multifaceted potential of herbal compounds, which are integral to Ayurvedic and Siddha-based treatment against neurodegeneration and Alzheimer's disease. In the figure, different neuroprotective properties of each herbal compound and the scientific name of the herb sources have been color coded to distinguish among themselves. From the figure, it can be interpreted that all the compounds have multitargeted, neuroprotective effects against Alzheimer's disease. Neuroprotective activities of Ayurvedic herbal compounds can complement each other, which supports the existing theories of herbal mixture-based treatment in Ayurveda

12.5.1 Glycyrrhiza glabra

G. glabra is a perennial herb that can grow 1 m in height and has pinnate leaves of length 7–15 cm. Their flowers are purple to pale whitish blue in color and are arranged in a hermaphrodite inflorescence (Dhingra and Sharma 2006). Among the many bioactive substances found in *Glycyrrhiza glabra* are linalool oxide, geraniol, benzoic acid, terpinen, tetramethylpyrazine, propionic acid, ethyl linolenate, butanediol, furfuraldehyde, methyl ethyl ketone, furfuryl formate, and trimethylpyrazine. In studies with scopolamine-induced dementia, this plant was found to have memory-improving properties. According to Dhingra et al., *Glycyrrhiza glabra* improves memory in mice. Three dose levels of plant extracts (75, 150, and 300 mg/kg) were given to mice over the course of 7 days, and the dose at 150 mg/kg was found to be the most beneficial (Arpita 2018). It is mostly used in Siddha and Ayurveda to treat "Vata" (which governs all movements and activities; Pitta regulates heat and energy levels, and Kapha controls growth, structural changes, and lubrication) and "Kapha" (which regulates numerous transformations) inflammations (Agarwal et al. 2002). In the first study to demonstrate licorice's ability to boost learning and memory, mice's performance on learning and memory tests was dramatically improved after receiving 150 mg/kg of licorice extract (equal to 5.19 g of dried plant material) orally for 7 days. With a pretreatment of 150 mg/kg for 7 days, licorice extract additionally shielded the mice against the learning and memory deficits caused by interoceptive stimuli (scopolamine and diazepam). Licorice extract may have the antioxidant characteristic that causes brain cells to be subjected to less oxidative stress, resulting in less brain damage and increased neuronal function, hence improving memory. Thus, a combination of anti-inflammatory, antioxidant, and neuroprotective roles may all be contributing to the licorice extract's overall memory-improving impact (Agarwal et al. 2002).

12.5.2 Curcuma longa

An important ingredient of Indian cuisine, *Curcuma longa* (or turmeric, Haldi, Indian saffron) is a rhizome of the Zingiberaceae family. They contain curcuminoids, of which curcumin (50–60%) has antioxidant, anti-inflammatory, and lipophilic action that improves the cognitive functions in patients with AD. A rising body of research suggests that the key event in Alzheimer's disease pathogenesis is influenced by oxidative stress, free radicals, beta-amyloid, brain dysregulation brought on by bio-metal toxicity, and aberrant inflammatory responses. The overall memory in AD patients has improved as a result of curcumin's several benefits, including diminished beta-amyloid plaques, postponed neuronal degeneration, metal chelation, anti-inflammatory and antioxidant activities, and reduced microglia production (Mishra and Palanivelu 2008). Studies and research findings (Pandav et al. 2000; Ng et al. 2006) suggest that AD incidence and prevalence are lower in India. In India, the prevalence of AD among persons aged 70–79 is 4.4 times less than it is in the United States (Pandav et al. 2000). Researchers looked into the relationship between curry consumption and cognitive function in 1010 Asians between the ages of 60 and 93. The results of the study showed that people who had curry occasionally (less than once per month) and frequently (more than once per month) outperformed people who consumed it never or infrequently (Ng et al. 2006). According to a UCLA study, curcumin may aid macrophages in removing the amyloid plaques linked to Alzheimer's disease. Macrophages are crucial components of the immune system. Blood samples from nine volunteers—six AD patients and three healthy controls—were treated with macrophages and curcumin. Then beta-amyloid was added. When compared to AD patients whose macrophages were not treated with curcumin, those whose macrophages were treated demonstrated improved absorption and swallowing of the plaques. Curcumin could therefore aid the immune system in eliminating the amyloid protein (Zhang et al. 2006).

An essential component in the etiology of AD is neuroglia. On microglia, curcumin has antiproliferative effects. Curcumin affects neuroglial differentiation and proliferation at low doses. The University of Southern California Los Angeles explored and investigated its suppression of microglial proliferation and differentiation. Researchers (Ambegaokar et al. 2003) demonstrated that curcumin dose-dependently inhibits the proliferation of neuroglial cells by causing them to differentiate into mature cells or suffer apoptosis in C6 rat glioma 2B clone cells, a mixed colony of both neuroglial cells. The same study demonstrated that curcumin increased CNP, a marker enzyme for oligodendrocytes (2′3′-cyclic nucleotide 3′- -phosphohydrolase). In THP-1 monocytic cells, curcumin prevents from causing the production of the Egr-1 protein and Egr-1 DNA-binding activity. Studies have demonstrated Egr-1's contribution to the amyloid peptide-induced induction of the cytochemokine gene in monocytes. Curcumin lowers the inflammation via inhibiting Egr-1's DNA-binding function. Curcumin can lessen the chemotaxis of monocytes, which can happen in response to chemokines from active microglia and astrocytes in the brain (Pendurthi and Rao 2000; Park and Kim 2002).

One injection of curcumin (1 and 2 mg/kg, intravenously) after focal cerebral ischemia/reperfusion in rats reduced the infarct volume, improved neurological function, decreased mortality, and decreased the water content in the brain, according to a Nanjing Medical University (China) study (Jiang et al. 2007). In comparison to AD mice which did not get curcumin treatment, those that received low dosages of the herb saw a 40% reduction in beta-amyloid levels. Low doses of curcumin also led to a 43% reduction in the "plaque burden" that this beta-amyloid has on the AD mice's brains. Unexpectedly, low amounts of curcumin administered over a longer period of time were actually more effective than large doses in halting AD's neurodegenerative process (Yang et al. 2005). Curcumin binds to amyloid beta and prevents its self-assembly at greater concentrations. Curcumin has been shown to inhibit aggregation and disaggregates to generate fibrillar A-beta 40 in in vitro tests. A Japanese study showed that curcumin destabilizes the fA-beta (1–40) and fA-beta (1–42) as well as their extension utilizing fluorescence spectroscopic examination with thioflavin T and electron microscopic analyses (Ono et al. 2004). Isoxazoles and pyrazoles produced from curcumin bind to the amyloid beta peptide (Aβ) and prevent the metabolism of amyloid precursor proteins (APP) (Narlawar et al. 2008).

Multiphoton microscopy has shown that curcumin passes the blood–brain barrier and shrinks the senile plaques that already exist in APPswe/PS1dE9 mice (Garcia-Alloza et al. 2007). In a different investigation, it was discovered that curcumin increased the removal of amyloid beta from AD patients' brains by phagocytosis (Fiala et al. 2007). When curcumin is consumed as it is, 38–75% of it is eliminated in the feces. However, with food, absorption seems to be better. There have been a few reports of adverse contact dermatitis caused by curcumin (Liddle et al. 2006). But heavy drinkers with a history of liver illness, and those who take prescription drugs that are metabolized by the liver, should avoid using curcumin on a regular basis because it can be hazardous to the liver. In human clinical trials, curcumin was demonstrated to be pharmacologically secure at doses up to 10 g/day. There was no harm from curcumin in phase 1 human trial with 25 individuals using up to 8000 mg/day for 3 months (Chainani 2020).

12.5.3 Bacopa monnieri

Bacopa monnieri (L.) of Scrophulariaceae family, commonly known as Brahmi and Aindri in Sanskrit, is an important medicinal plant in Indian traditional medicine. It most commonly acts as a nerve tonic and acts as a therapeutic against insomnia, epilepsy, asthma, and rheumatism. Bacoside A is an aglycone unit, with pseudo-jujubogenin moieties that are assumed to carry the neuropharmacological property of Brahmi. It also carries bacopaside III, bacopaside X, bacoside A3, and bacopasaponin C, and they have been found to prevent Aβ aggregation and provide neuroprotection against Aβ-induced toxicity (Chaudhari et al. 2017).

In one study, they showed that the bioactive constituent, bacoside A, present in the *B. monnieri* extract (BME) when treated with rat serum, could directly or indirectly interact with the neurotransmitter systems to improve memory and learning ability. In one study, BM was shown to enhance protein kinase activity in the hippocampus along with inhibiting cholinergic degeneration in a rat model of Alzheimer's disease (Sireeratawong et al. 2016). In a different study, cognitive imbalance in rat brain was induced by administering colchicines and ibotenic acid (intracerebroventricularly) into nucleus basalis magnocellularis. The associated effects of acetylcholine depletion, acetylcholinesterase activity, and decreased muscarinic cholinergic receptor activity were quite effectively reversed by the standardized extract of BM. Such BM-treated neurons were seen to be expressing a low level of ROS (Arpita 2018). In order to determine the minimum dose that was nontoxic as well as neuroprotective, Sireeratawong et al. studied the presence of any histopathological abnormality in internal organs (liver, kidney), wherein 5000 mg/kg dose of *B. monnieri* extract (BME) was safe and did not cause any observable signs or symptoms of toxicity, and also neither gross nor histopathological abnormalities were observed in any of the internal organs, including the liver and kidney of *B. monnieri*-treated rats. Although the dose used in the study was 1000 times higher than the normally allowed dose in humans (5 mg/kg/day), there was no observable signs of toxicity, indicating the safety of *B. monnieri* for human use.

Multiple evidence of safety of *B. monnieri* in in vitro and in vivo studies has encouraged the research of its prospects of stimulating neuroprotectivity on human subjects. In some clinical studies, a standardized extract of the herb, Bacognize, in fixed doses (300 mg twice daily) was given to geriatric Alzheimer's patients in a 6-month trial (Goswami et al. 2011). At the end of the trial period, the patients showed statistically significant improvement in multiple criteria of Mini-Mental State Examination Scale (MMSES), including orientation of time, place, and person; attention; and their language ability in terms of reading, writing, and comprehension. In a similar study conducted in India, significant cognitive enhancement was observed in a group of medical students from the Government Medical College, Nagpur, India, who took 150 mg of the standardized extract, over a period of 15 days (Kumar et al. 2016). Upon biochemical analysis, it was revealed that most of the individuals had elevated serum calcium levels and enhanced memory. Thakkar et al. (2017) used the inclusion complex of Bacognize (containing 16% bacosides) and cyclodextrin to produce different molar ratios of *B. monnieri* via coprecipitation technique, improving the low solubility of Bacognize. The findings showed that the inclusion of complex at a molar ratio of 1:4 can increase *B. monnieri*'s solubility and stability in the inclusion complex three times. In accordance with another study, the d2 concentration test performance of ten German patients with mild cognitive impairment (mean age: 61.88–6.69 years) improved after receiving individual doses of *B. monnieri* and *Sideritis scardica* extracts. However, a study from Swinburne University in Australia found that giving *B. monnieri* (2150 mg) for 90 days to 107 participants (aged 18–60) improved their performance on a task requiring structural working memory in healthy individuals without a history of neurological disorders, gastrointestinal disorders, or chronic infections. Not only that, but also none of the healthy participants took any cognitive-enhancing medications.

According to a study, *B. monnieri* consumption for 3 months in 76 adult volunteers between the ages of 40 and 65 at the University of Wollongong in Australia had a substantial impact on memory recall (Roodenrys et al. 2002). In a similar incident, subjects without dementia aged 65 and older who consistently took BME (300 mg/day) for 84 days at the University of Catania in Italy performed better on the Stroop task, which measures the ability to ignore unneeded input, and restrained recall tests (Calabrese et al. 2008). Furthermore, administration of *B. monnieri* (300 mg/day) to healthy volunteers over 55 demonstrated improvement in their oral learning, memory attainment, and inhibited recall in Lismore, New South Wales, Australia (Morgan and Stevens 2010). In another study conducted at the Swinburne University of Technology in Melbourne, Australia, healthy volunteers between the ages of 18 and 44 showed increased and preserved cognitive performance when given a larger single dose in a double-blind, placebo-controlled trial (Downey et al. 2013). When Bacopa was given to 40 schoolchildren from rural India, ages 6–8, for 90 days as syrup (proportionate to 10 g dried Bacopa daily), improvements in rapid memory and response performance were noted (Sharma et al. 1987). The combined extracts of *B. monnieri* and *Ginkgo biloba* were used to test anti-dementia and anticholinesterase effects in adult male Swiss mice, which were induced by scopolamine (3 mg/kg BW). Passive avoidance (PA) test was used to determine the anti-dementia activity. After administering the combination of extracts of *B. monnieri* at 30 mg/kg and *G. biloba* at 15, 30, and 60 mg/kg for 7 days, their findings showed a significantly longer transfer latency time (TLT) and no transfer response (NTR) (Das et al. 2002).

12.5.4 Centella asiatica

Centella asiatica belongs to Apiaceae family and is found throughout India, Sri Lanka, and Bangladesh. Its extracts have been shown to decrease beta-amyloid levels and oxidative stress, prevent the shrinkage of neuronal processes, and protect against beta-amyloid-associated toxicity and behavioral abnormalities. Treatment with *Centella asiatica* has been shown to greatly raise antioxidant defense, reduce cell death signals, and restore mitochondrial deficiencies. Of the components, asiatic acid has gained the most attention in preclinical models. Asiatic acid has antioxidant and neuroprotective activities and can pass the blood–brain barrier (Soumyanath 2012).

Triterpenes, asiatic acid, asiaticoside, madecassoside, sapogenins, glycosides, madecassic acid, vellarin, and centelloside are few of many bioactive substances it contains (Dhanasekaran et al. 2009). According to an in vitro investigation, asiatic acid and asiaticoside may play a role in the treatment of Alzheimer's disease and the prevention of beta-amyloid toxicity by reducing hydrogen peroxide-induced cell death, lowering free radical concentrations, and inhibiting beta-amyloid cell death. *Centella asiatica* extract is an important plant for nerve and brain cells capable of improving intelligence, memory, and longevity. It corrected the beta-amyloid pathology in mouse brains and altered oxidative stress response components (Arpita 2018). In a study, Orhan et al. used GC-MS for the first time to analyze the essential oil content of *C. asiatica* grown in Turkey and found that beta-copaene was the main constituent (Mishra et al. 2011). A potential role for *Centella asiatica* (Vallarai) in the treatment and prevention of AD and beta-amyloid toxicity has been suggested by the asiaticoside derivatives as asiatic acid and asiaticoside, which reduced hydrogen peroxide-induced cell death, decreased free radical concentrations, and inhibited beta-amyloid cell death in vitro. As mentioned in earlier investigations, the beta-amyloid pathology in the brains might be reversed with Vallarai extracts (Begum et al. 2008). When given to mice for 15 days at doses of 200, 500, 700, and 1000 mg/kg (b.w.), *C. asiatica* extract improved learning and memory as measured by the radial-armed labyrinth test; however, it had no effect on locomotor activity. On the other hand, the extract was found to increase AChE activity and dendritic arborization in CA3 neurons in the hippocampus. In a related study, the fresh leaf extract of

Centella asiatica was administered to adult mice at doses of 2, 4, and 6 mL/kg for 2, 4, and 6 weeks, respectively. When the mice's removed brains were examined under a microscope, it was clear that the extract given at the 6 mL/kg dose for 6 weeks had significantly increased dendritic arborization in the neurons. These authors also arrived at a similar conclusion, noting that mice given *C. asiatica* juice derived by crushing fresh leaves showed improved dendritic arborization (Mishra et al. 2011).

12.5.5 Ginkgo biloba

Belonging to the Ginkgoaceae family, the leaves of *Ginkgo biloba* are generally used in most preparations, and they are rich in tocopherols, phenolics, flavonoids, α-linolenic, etc., with α-tocopherol being the most abundant vitamer. The inherent properties of such bioactives have shown to provide the neuroprotective, free radical scavenging, and membrane-stabilizing activity from ginkgo. The ginkgolide B also acts as a platelet-activating factor inhibitor. Suppression of 3′,5′-cyclic guanosine monophosphate (GMP) phosphodiesterase, prevention of age-related loss of muscarinic cholinoceptors and adrenoceptors, and augmentation of choline uptake in the hippocampus are additional pharmacologic effects. Additionally, ginkgo extract has been discovered to delay the buildup of beta-amyloid (Sierpina et al. 2003).

12.5.6 Salvia officinalis

Commonly called as sage, *Salvia officinalis* L. belongs to the Labiatae family. They are frequently referred for Alzheimer's treatment as they are rich in antioxidants carnosic acid and rosmarinic acid. Antioxidants can normally help to protect against oxidative damage in brain (Akhondzadeh 2003). Sage is also rich in phenolic compounds (e.g., coumarins, flavonoids, tannins).

12.5.7 Rosmarinus officinalis

Commonly called Satapatrika, this herb belongs to the family Labiatae. The therapeutic action of Satapatrika is by virtue of its chemical constituents apigenin, carvacrol, eugenol, oleanolic acid, thymol, and ursolic acid that act as COX-2 inhibitors, containing around two dozen antioxidants, such as carnosic acid and ferulic acid and a dozen anti-inflammatory compounds. Some of its constituents, i.e., butylated hydroxytoluene (BHT) and butylated hydroxyanisole (BHA), have even greater antioxidant activity than their synthetic counterparts (Mohaddese et al. 2019).

Apart from these Ayurvedic plants, various other plants and their components also have anti-Alzheimer's effect (Table 12.1).

Table 12.1 Some other commonly known herbs and plants used in Ayurvedic treatment of Alzheimer's disease

Bioactive compound	Herb name	Ayurvedic term	Neuroprotective activity	References
Anthocyanin	*Punica granatum*	Dadima	Antioxidant activity, improves learning abilities and memory retention	Cambay et al. (2011)
Inositol, cyclohexene, 1 methyl-4-(1-methylethylidene)	*Clitoria ternatea*	Aparajita	Increases ACh levels and nootropic	Rai et al. (2002)
Pterocarpan, pterocarpanoids gangetin, gangetinin, desmodin	*Desmodium gangeticum*	Shalparni	AChE inhibition, nootropic	Joshi and Parle (2006)
β-Asarone, α-asarone	*Acorus calamus*	Vacha	Sedative, neuroprotective, nootropic	Vohora et al. (1990)
1,8-Cineole, α-pinene, β-pinene	*Salvia lavandulaefolia*	Shati	AChE inhibition	Perry et al. (2000)
Hayatine, hayatinine, berberine	*Cissampelos pareira*	Paatha	AChE inhibition, antioxidant, anti-inflammatory	Pramodinee et al. (2011)
Sitoindoside IX, sitoindoside X, withanolides, withanols	*Withania somnifera*	Ashwagandha	Anti-inflammatory, antioxidant, Aβ-inhibition, AChE inhibition, regenerates damaged axons, dendrites, and synapses	Kulkarni and Dhir (2008)
Triterpenoids, flavonol glycosides, anthocyanins	*Convolvulus pluricaulis*	Shankhpushpi	Antidementia, AChE inhibition, nootropic	Gujran et al. (2007)
Guggulsterones, manusumbionic acid	*Commiphora wightii*	Guggulu	Antidementia, AChE inhibition, nootropic	Gharibi et al. (2013)
Piperidine alkaloids	*Piper nigrum* and *Piper longum*	Piperine	Neuroprotective, inhibits AChE and β-secretase enzymes, attenuates oxidative stress and cognitive deficits	Hritcu et al. (2014)
Lycopodium alkaloid	*Huperzia serrata*	Huperzine A	Neuroprotective role, mitochondrial protection from Aβ aggregation-induced toxicity, inhibitor of Aβ, also promotes the production of BDNF	Choudhury et al. (2014)

12.6 Conclusion

The present pharmaceutical methods for reducing Alzheimer's disease are only marginally effective. As a result, several novel treatments have been tried and tested, while exploring Indian traditional practices of Ayurveda and Siddha as well as the globally practiced homeopathy. These systems can holistically manage such neurodegenerative diseases via special modes: pharmacological (Rasayana and nootropic herbs, Panchakarma bio-purification) and non-pharmacological (spiritual healing/ mantra therapy, psychotherapy/yoga/meditation). The current research environment should shift toward the combinatorial approach of these systems in the hunt for appropriate therapeutic options for the treatment of Alzheimer's disease. The well-known herb for neuroprotection is Brahmi (*Bacopa monnieri*). Brahmi's impact on many facets of neurodegeneration is well understood. Brahmi may have anti-inflammatory, viability, and antiproliferative properties, based on previous studies. Also, based on the key findings mentioned above, curcumin will result in a promising Alzheimer's disease treatment. The chemical characteristics of curcumin have been explored in clinical settings, and its different effects on Alzheimer's disease point to the possibility of doing additional research and developing better medications based on its multitargeted potential. Curcumin, i.e., Longvida® (Verdure Sciences, Noblesville, IN, USA), is currently in phase II clinical trials for anti-amyloidogenic, anti-β-secretase, anti-inflammatory, and anti-ChE activity. Because of their natural and safer approach, herbal and traditional systems like Ayurveda, Siddha, and homeopathy are being looked into as alternatives to the existing treatment techniques for Alzheimer's disease due to their limited efficacy and adverse effects. These bioavailable formulations are now being utilized to change how Alzheimer's disease is treated.

According to homoeopathic doctrine, the best kind of treatment is the quick, gentle, and permanent restoration of health or the quick, safe, and harmless eradication and annihilation of the disease in its entirety. Homeopathy provides reasonably effective control, alleviation, and therapy (not a cure). The holistic approach of homeopathy emphasizes treating the entire patient, not just the affected areas. Each person is distinct, and because everyone is affected by dementia in a different way, no two cases will unfold exactly the same way. Each case of Alzheimer's is studied separately by homoeopathy. A thorough assessment of the mind and emotions is also part of the investigation, following which a constitutional medicine is administered if it is appropriate. Inhibiting AChE, MAO, secretase enzymes, synaptic damage, and tau protein hyperphosphorylation; attenuating A, oxidative stress, neurodegeneration, and cognitive deficits, and improving memory are just a few of the multifaceted and multitargeted actions that various traditional herbs, phytoconstituents, and alkaloids use to mediate their anti-AD activity. Ayurveda and herbal treatments aid in reducing morbidity; controlling AD at the preventative, promotive, and curative levels; and enhancing the quality of life. The many degenerative cascades of AD can be treated using traditional treatments. This chapter offers a variety of revitalizing herbs and details on how they work to cure

AD. According to Siddha's pathologic theory for all human illnesses and scientific study on phytocompounds to boost memory and cognition, revitalize brain processes, and improve quality of life in patients with AD, these herbs have been validated in terms of the trihumoral foundations of AD. Therefore, a solid understanding of the fundamentals of conventional systems combined with modern science may generate innovative ideas for preventive and therapeutic aspects of age-related neurodegenerative illnesses like AD and may also favor new medication research and development. This review suggests that in order to create successful treatment plans for AD, there is an urgent need for a paradigm shift from a single-target to a multitarget medication approach, as well as the integration of modern and conventional systems.

References

Adler UC, Adler MS (2006) Hahnemann's experiments with 50 millesimal potencies: a further review of his casebooks. Homeopathy 95(3):171–181

Afshari AR, Sadeghnia HR, Mollazadeh H (2016) A review on potential mechanisms of *Terminalia chebula* in Alzheimer's disease. Adv Pharmacol Sci 2016:8964849

Agarwal A, Malini S et al (2002) Effect of *Tinospora cordifolia* on learning and memory in normal and memory deficit rats. Indian J Pharmacol 34:339–349

Akhondzadeh S (2003) *Salvia officinalis* L. extract in the treatment of patients with mild to moderate Alzheimer disease: a double blind, randomized and placebo-controlled trial. J Clin Pharm Ther 28:53–59

Akiyama H, Barger S, Barnum S et al (2000) Inflammation and Alzheimer's disease. Neurobiol Aging 21(3):383–421

Al-Snafi AE (2019) A review on *Lagerstroemia indica*: a potential medicinal plant. IOSR J Pharm 9(6):36–42

Ambegaokar SS, Wu L, Alamshahi K, Lau J, Jazayeri L, Chan S et al (2003) Curcumin inhibits dose-dependently and time-dependently neuroglial proliferation and growth. Neuro Endocrinol Lett 24:469–473

Amraie E, Pouraboli I, Rajaei Z (2020) Neuroprotective effects of *Levisticum officinale* on LPS-induced spatial learning and memory impairments through neurotrophic, anti-inflammatory, and antioxidant properties. Food Funct 11(7):6608–6621

Arpita R (2018) Role of medicinal plants against Alzheimer's disease. Int J Complement Altern Med 11(4):205–208

Bamberger ME, Landreth GE (2002) Inflammation, apoptosis and Alzheimer's disease. Neuroscientist 8:276–283

Begum AN, Jones MR, Lim GP et al (2008) Curcumin structure–function, bioavailability, and efficacy in models of neuroinflammation and Alzheimer's disease. J Pharmacol Exp Ther 326: 196–208

Bhattacharjee SK (2001) Handbook of medicinal plants. Pointer Publishing, Jaipur

Bredesen DE (2014) Reversal of cognitive decline: a novel therapeutic program. Aging (Albany NY) 6(9):707–717

Bredesen DE (2015) Metabolic profiling distinguishes three subtypes of Alzheimer's disease. Aging (Albany NY) 7(8):595–600

Bredesen DE (2016) Inhalational Alzheimer's disease: an unrecognized—and treatable—epidemic. Aging (Albany NY) 8(2):304–313

Bredesen DE, Amos EC, Canick J et al (2016) Reversal of cognitive decline in Alzheimer's disease. Aging (Albany NY) 8(6):1250–1258

Buckle J, Newberg A, Wintering N et al (2008) Measurement of regional cerebral blood flow associated with the M technique-light massage therapy: a case series and longitudinal study using SPECT. J Altern Complement Med 14(8):903–910

Calabrese C, Gregory WL, Leo M, Kraemer D, Bone K, Oken B (2008) Effects of a standardized *Bacopa monnieri* extract on cognitive performance, anxiety, and depression in the elderly: a randomized, double-blind, placebo-controlled trial. J Altern Complement Med 14:707–713

Calderon AI, Simithy-Williams J, Sanchez R, Espinosa A, Valdespino I, Gupta MP (2013) Lycopodiaceae from Panama: a new source of acetylcholinesterase inhibitors. Nat Prod Res 27(4–5):500–505

Cambay Z, Baydas G, Tuzcu M (2011) Pomegranate (*Punica granatum* L.) flowers improves learning and memory performances impaired by diabetes mellitus in rats. Acta Physiol Hung 98: 409–420

Castro-Silva ES, Bello M, Rosales-Hernández MC, Correa-Basurto J, Hernández-Rodríguez M, Villalobos-Acosta D et al (2020) Fucosterol from *Sargassum horridum* as an amyloid-beta ($A\beta_{1-42}$) aggregation inhibitor: in vitro and in silico studies. J Biomol Struct Dyn 39(4): 1271–1283. https://doi.org/10.1080/07391102.2020.1729863

Chainani N (2020) Safety and anti-inflammatory activity of curcumin component of turmeric (*Curcuma longa*). J Altern Complement Med 9:161–168

Chaudhari KS, Tiwari NR, Tiwari RR, Sharma RS (2017) Neurocognitive effect of nootropic drug Brahmi (*Bacopa monnieri*) in Alzheimer's disease. Ann Neurosci 24(2):111–122

Chethana KR, Senol FS, Orhan IE, Anilakumar KR, Keri RS (2017) *Cassia tora* Linn.: a boon to Alzheimer's disease for its anti-amyloidogenic and cholinergic activities. Phytomedicine 33: 43–52

Chethana KR, Sasidhar BS, Naika M, Keri S (2018) Phytochemical composition of *Caesalpinia crista* extract as potential source for inhibiting cholinesterase and β-amyloid aggregation: significance to Alzheimer's disease. Asian Pac J Trop Biomed 8(10):500

Choudhury B, Saytode P, Shah V (2014) Neurodegenerative disorders: past, present and future. Int J Appl Pharm Biotechnol 5(2):14–28

Chowdhury S, Kumar S (2020) Alpha-terpinyl acetate: a natural monoterpenoid from *Elettaria cardamomum* as multitarget directed ligand in Alzheimer's disease. J Funct Foods 68:103892

Colcombe S, Kramer AF (2003) Fitness effects on the cognitive function of older adults: a meta-analytic study. Psychol Sci 14(2):125–130

Das A, Shanker G, Nath C, Pal R, Singh S, Singh HK (2002) A comparative study in rodents of standardized extracts of *Bacopa monnieri* and *Ginkgo biloba*: anticholinesterase and cognitive enhancing activities. Pharmacol Biochem Behav 73:893–900

Dhanasekaran M, Holcomb LA, Hitt AR, Tharakan B, Porter JW, Young KA, Manyam BV (2009) *Centella asiatica* extract selectively decreases amyloid beta levels in hippocampus of Alzheimer's disease animal model. Phytother Res 23:14–19

Dhingra D, Sharma A (2006) Antidepressant-like activity of *Glycyrrhiza glabra* L. in mouse models of immobility tests. Prog Neuro-Psychopharmacol Biol Psychiatry 30(3):449–454

Dhouafli Z, Ben Jannet H, Mahjoub B, Leri M, Guillard J, Saidani Tounsi M et al (2019) 1,2,4-Trihydroxynaphthalene-2*O*-β-D-glucopyranoside: a new powerful antioxidant and inhibitor of Aβ42 aggregation isolated from the leaves of *Lawsonia inermis*. Nat Prod Res 33(10): 1406–1414

Di Patre PL, Read SL, Cummings JL, Tomiyasu U, Vartavarian LM, Secor DL et al (1999) Progression of clinical deterioration and pathological changes in patients with Alzheimer's disease evaluated at biopsy and autopsy. Arch Neurol 56:1254–1261

Downey LA, Kean J, Nemeh F et al (2013) An acute, double-blind, placebo-controlled crossover study of 320 mg and 640 mg doses of a special extract of *Bacopa monnieri* (CDRI 08) on sustained cognitive performance. Phytother Res 27:1407–1413

Dubey T, Chinnathambi S (2019) Brahmi (*Bacopa monnieri*): an ayurvedic herb against the Alzheimer's disease. Arch Biochem Biophys 676:108153

Elufioye TO, Chinaka CG, Oyedeji AO (2019) Antioxidant and anticholinesterase activities of *Macrosphyra longistyla* (DC) Hiern relevant in the management of Alzheimer's disease. Antioxidants 8(9):400

Eugenie N, Kabra A, Rao M, Prajapati P (2019) Traditional ayurvedic and herbal remedies for Alzheimer's disease: from bench to bedside. Expert Rev Neurother 19(5):359–374. https://doi.org/10.1080/14737175.2019.1596803

Eyre HA, Acevedo B, Yang H et al (2016) Changes in neural connectivity and memory following a yoga intervention for older adults: a pilot study. J Alzheimers Dis 52(2):673–684

Eze FN, Ingkaninan K, Prapunpoj P (2019) Transthyretin anti-amyloidogenic and fibril disrupting activities of *Bacopa monnieri* (L.) Wettst (Brahmi) extract. Biomol Ther 9(12):845

Falode JA, Akinmoladun AC, Olaleye MT, Akindahunsi AA (2017) Sausage tree (*Kigelia africana*) flavonoid extract is neuroprotective in $AlCl_3$-induced experimental Alzheimer's disease. Pathophysiology 24(4):251–259

Ferreira L, Tanaka K, Santos-Galduroz RF, Galduroz JC (2015) Respiratory training as strategy to prevent cognitive decline in aging: a randomized controlled trial. Clin Interv Aging 10:593–603

Fiala M, Liu PT, Espinosa-Jeffrey A, Rosenthal MJ, Bernard G, Ringman JM et al (2007) Innate immunity and transcription of MGAT-III and Toll-like receptors in Alzheimer's disease patients are improved by bisdemethoxycurcumin. Proc Natl Acad Sci U S A 104:12849–12854

Finley JW, Gao S (2017) A perspective on *Crocus sativus* L. (saffron) constituent crocin: a potent water-soluble antioxidant and potential therapy for Alzheimer's disease. J Agric Food Chem 65(5):1005–1020

Garcia-Alloza M, Borrelli LA, Rozkalne A, Hyman BT, Bacskai BJ (2007) Curcumin labels amyloid pathology in vivo, disrupts existing plaques and partially restores distorted neurites in an Alzheimer mouse model. J Neurochem 102:1095–1104

Gharibi A, Khalili M, Kiasalari Z et al (2013) The effect of *Zingiber officinalis* L. on learning and memory in rats. J Basic Clin Pathophysiol 2:2013–2014

Gokani T (2014) Ayurveda—the science of healing. Headache 54(6):1103–1106

Goswami S, Saoji A, Kumar N, Thawani V, Tiwari M, Thawani M (2011) Effect of *Bacopa monnieri* on cognitive functions in Alzheimer's disease patients. Int J Collab Res Int Med Public Health 3:285–293

Gujran S, Singh SP, Pal R et al (2007) An extract of *Commiphora wightii* with lipid lowering properties, has protective effects against streptozotocin induced memory deficits in mice. Pharmacol Biochem Behav 8:797–805

Hanif K, Kumar M, Singh N, Shukla R (2015) Effect of homeopathic Lycopodium clavatum on memory functions and cerebral blood flow in memory-impaired rats. Homeopathy 104 (01):24–28

Hashim SA, VanItallie TB (2014) Ketone body therapy: from the ketogenic diet to the oral administration of ketone ester. J Lipid Res 55(9):1818–1826

Hatziagapiou K, Kakouri E, Lambrou GI, Bethanis K, Tarantilis PA (2019) Antioxidant properties of *Crocus sativus* L. and its constituents and relevance to neurodegenerative diseases; focus on Alzheimer's and Parkinson's disease. Curr Neuropharmacol 17(4):377–402

Hritcu L, Noumedem JA, Cioanca O et al (2014) Methanolic extract of *Piper nigrum* fruits improves memory impairment by decreasing brain oxidative stress in amyloid beta(1–42) rat model of Alzheimer's disease. Cell Mol Neurobiol 34(3):437–444

Illum L (2003) Nasal drug delivery—possibilities, problems and solutions. J Control Release 87(1–3):187–198

Ingale SP, Kasture SB (2017) Protective effect of standardized extract of *Passiflora incarnata* flower in Parkinson's and Alzheimer's disease. Ancient Sci Life 36(4):200

Jiang J, Wang W, Sun YJ, Hu M, Li F, Zhu DY (2007) Neuroprotective effect of curcumin on focal cerebral ischemic rats by preventing blood–brain barrier damage. Eur J Pharmacol 30:54–62

Jones NS, Quraishi S, Mason JD (1997) The nasal delivery of systemic drugs. Int J Clin Pract 51(5): 308–311

Joo HB, Kim JM, Park SY (2016) Anti-amyloidogenic constituents from roots of *Dryopteris crassirhizoma* in Alzheimer's disease cellular model. Planta Med 82(S 01):P472

Joshi H, Parle M (2006) Antiamnesic effects of Desmodium gangeticum in mice. Yakugaku Zasshi 126:795–804

Joy J (2015) The Ayurvedic perspective on the etiopathogenesis of neurodegeneration. J Health Educ Res Dev 3:137

Kausalya J, Vidhyalakshmi S, Vadivel V, Sriram S, Rajalakshmi P, Brindha P (2017) Phytochemical levels and antioxidant activity of traditionally processed Indian herbal mixture (Acorus calamus, Curcuma aromatica and Zingiber officinale). Trop J Nat Prod Res 1(6):262–269

Kim DJ, Kim MS, Kim S, Hwang KW, Park SY (2017) Anti-amyloidogenic effects of *Perilla frutescens* var. acuta on beta-amyloid aggregation and disaggregation. J Food Biochem 41(5): e12393

Kleinrichert K, Alappat B (2019) Comparative analysis of antioxidant and anti-amyloidogenic properties of various polyphenol rich phytoceutical extracts. Antioxidants 8(1):13

Kramer AF, Hahn S, Cohen NJ et al (1999) Ageing, fitness and neurocognitive function. Nature 400(6743):418–419

Kulkarni SK, Dhir A (2008) *Withania somnifera*: an Indian ginseng. Prog Neuro-Psychopharmacol Biol Psychiatry 32:1093105

Kumar N, Abichandani L, Thawani V, Gharpure K, Naidu M, Venkat Ramana G (2016) Efficacy of standardized extract of *Bacopa monnieri* (Bacognize®) on cognitive functions of medical students: a 6-week, randomized placebo-controlled trial. Evid Based Complement Alternat Med 2016:4103423

Lad V (2006) Textbook of Ayurveda—a complete guide to clinical assessment—Nidana Panchakam; the five clinical barometers. The Ayurvedic Press, Albuquerque, pp 1–21

Lavretsky H, Epel ES, Siddarth P et al (2013) A pilot study of yogic meditation for family dementia caregivers with depressive symptoms: effects on mental health, cognition, and telomerase activity. Int J Geriatr Psychiatry 28(1):57–65

Lee JE, Kim N, Yeo JY, Seo DG, Kim S, Lee JS et al (2019) Anti-amyloidogenic effects of asarone derivatives from *Perilla frutescens* leaves against beta-amyloid aggregation and nitric oxide production. Molecules 24(23):4297

Liddle M, Hull C, Liu C, Powell D (2006) Contact urticaria from curcumin. Dermatitis 17:196–197

Linde K, Clausius N, Ramirez G et al (1997) Are the clinical effects of homoeopathy placebo effects? A meta-analysis of placebocontrolled trials. Lancet 350:834–843

Lindgren L, Westling G, Brulin C et al (2012) Pleasant human touch is represented in pregenual anterior cingulate cortex. NeuroImage 59(4):3427–3432

Liou KT, Shen YC, Chen CF, Tsao CM, Tsai SK (2003) The anti-inflammatory effect of honokiol on neutrophils: mechanisms in the inhibition of reactive oxygen species production. Eur J Pharmacol 475(1–3):19–27

Liu QS, Jiang HL, Wang Y, Wang LL, Zhang JX, He CH et al (2018) Total flavonoid extract from *Dracocephalum moldavica* L. attenuates β-amyloid-induced toxicity through anti-amyloidogenesic and neurotrophic pathways. Life Sci 193:214–225

Longo FM, Massa SM (2004) Neuroprotective strategies in Alzheimer's disease. NeuroRx 1(1): 117–127

Lopa SS, Hasan MK, Ahammed MS, Islam KM, Alam AK, Rahman MAA et al (2019) *Typhonium trilobatum* demonstrates both antioxidant and acetylcholinesterase inhibitory activities in vitro. Bangl Pharm J 22(1):92–98

Luders E, Clark K, Narr KL, Toga AW (2011) Enhanced brain connectivity in long-term meditation practitioners. NeuroImage 57(4):1308–1316

Mahboubi M (2019) *Melissa officinalis* and rosmarinic acid in management of memory functions and Alzheimer disease. Asian Pac J Trop Biomed 9(2):47

Manyam BV, Kumar A (2013) Ayurvedic constitution (prakruti) identifies risk factor of developing Parkinson's disease. J Altern Complement Med 19(7):644–649

Mazouz W, Haouli NEH, Gali L, Vezza T, Bensouici C, Mebrek S et al (2020) Antioxidant, anti-Alzheimer, anti-diabetic, and anti-inflammatory activities of the endemic halophyte *Limonium spathulatum* (Desf.) kuntze on LPS-stimulated RAW264 macrophages. S Afr J Bot 135:101–108

McCarty MF, DiNicolantonio JJ, O'Keefe JH (2015) Ketosis may promote brain macroautophagy by activating Sirt1 and hypoxia-inducible factor-1. Med Hypotheses 85(5):631–639

Mishra S, Palanivelu K (2008) The effect of curcumin (turmeric) on Alzheimer's disease: an overview. Ann Indian Acad Neurol 11(1):13–19

Mishra L, Singh BB, Dagenais S (2001) Healthcare and disease management in ayurveda. Altern Ther Health Med 7(2):44–50

Mishra S, Mishra M, Seth P, Sharma SK (2011) Tetrahydrocurcumin confers protection against amyloid beta-induced toxicity. Neuroreport 22:23–27

Mohaddese M, Mahdizadeh E, Heidary TR (2019) Chemical composition and antimicrobial activity of F. excelsior L. seeds essential oil. Jundishapur J Nat Pharma Prod 14(1):e61105

Morgan A, Stevens J (2010) Does *Bacopa monnieri* improve memory performance in older persons? Results of a randomized, placebo-controlled, double-blind trial. J Altern Complement Med 16:753–759

Murphy T, Dias GP, Thuret S (2014) Effects of diet on brain plasticity in animal and human studies: mind the gap. Neural Plast 2014:563160

Murugesa Mudaliar GS (1998) Gunapadam mooligai vaguppu-Siddha material medica, 5th edn. Department of Indian Medicine and Homeopathy Publishers, Chennai, p 625

Narlawar R, Pickhardt M, Leuchtenberger S, Baumann K, Krause S, Dyrks T et al (2008) Curcumin-derived pyrazoles and isoxazoles: Swiss army knives or blunt tools for Alzheimer's disease? Chem Med Chem 3:165–172

Newberg AB, Wintering N, Khalsa DS, Roggenkamp H, Waldman MR (2012) Meditation effects on cognitive function and cerebral blood flow in subjects with memory loss: a preliminary study. Ann Neurosci 19(2):81

Ng TP, Chiam PC, Lee T, Chua HC, Lim L, Kua EH (2006) Curry consumption and cognitive function in the elderly. Am J Epidemiol 164:898–906

Ngandu T, Lehtisalo J, Solomon A et al (2015) A 2 year multidomain intervention of diet, exercise, cognitive training, and vascular risk monitoring versus control to prevent cognitive decline in at-risk elderly people (FINGER): a randomised controlled trial. Lancet 385(9984):2255–2263

Ninivaggi FJ (2010) A comprehensive guide to traditional Indian medicine for the west. Rowman & Littlefield Publishers, Lanham, pp 201–220

Oboh G, Adedayo BC, Adetola MB, Oyeleye IS, Ogunsuyi OB (2022) Characterization and neuroprotective properties of alkaloid extract of *Vernonia amygdalina* Delile in experimental models of Alzheimer's disease. Drug Chem Toxicol 45:731–740

Ono K, Hasegawa K, Naiki H, Yamada MJ (2004) Curcumin has potent anti-amyloidogenic effects for Alzheimer's beta fibrils in vitro. Neurosci Res 75:742–750

Ouchi Y, Kanno T, Okada H et al (2006) Changes in cerebral blood flow under the prone condition with and without massage. Neurosci Lett 407(2):131–135

Pandav R, Belle SH, DeKosky ST (2000) Apolipoprotein E polymorphism and Alzheimer's disease: the Indo-US cross-national dementia study. Arch Neurol 57:824–830

Park SY, Kim DS (2002) Discovery of natural products from *Curcuma longa* that protect cells from beta-amyloid insult: a drug discovery effort against Alzheimer's disease. J Nat Prod 65:1227–1231

Patil DN, Yadav SR, Patil S, Bapat VA, Jadhav JP (2020) Multidimensional studies of *Pancratium parvum* Dalzell against acetylcholinesterase: a potential enzyme for Alzheimer's management. J Am Coll Nutr 39(7):601–618

Pendurthi UR, Rao LV (2000) Suppression of transcription factor Egr-1 by curcumin. Thromb Res 97:179–189

Penumala M, Zinka RB, Shaik JB, Mallepalli SK, Vadde R, Amooru DG (2018) Phytochemical profiling and in vitro screening for anticholinesterase, antioxidant, antiglucosidase and neuroprotective effect of three traditional medicinal plants for Alzheimer's disease and diabetes mellitus dual therapy. BMC Complement Altern Med 18(1):77

Perry NSL, Houghton PJ, Theobald A et al (2000) In-vitro inhibition of erythrocyte acetylcholinesterase by *Salvia lavandulaefolia* essential oil and constituent terpenes. J Pharm Pharmacol 52(7):1347–1356

Pires A, Fortuna A, Alves G, Falcao A (2009) Intranasal drug delivery: how, why and what for? J Pharm Pharm Sci 12(3):288–311

Pramodinee D, Mahesh M, Niranjan D et al (2011) Memory enhancing activity of *Cissampelos pareira* in mice. Int J Pharm Sci 3:206–211

Purvya MC, Meena MS (2011) A review on role of prakriti in aging. Ayu 32(1):20–24

Rai KS, Murthy KD, Karanth KS et al (2002) *Clitoria ternatea* root extract enhances acetylcholine content in rat hippocampus. Fitoterapia 73:685–689

Rao RV (2012) Ayurvedic approach to Alzheimer's disease. Ayurveda J Health 11(Fall-1):3–13

Rao RV, Descamps O, John V, Bredesen DE (2012) Ayurvedic medicinal plants for Alzheimer's disease: a review. Alzheimers Res Ther 4(3):22

Ravi SK, Ramesh BN, Mundugaru R, Vincent B (2018) Multiple pharmacological activities of *Caesalpinia crista* against aluminium-induced neurodegeneration in rats: relevance for Alzheimer's disease. Environ Toxicol Pharmacol 58:202–211

Roodenrys S, Booth D, Bulzomi S, Phipps A, Micallef C, Smoker J (2002) Chronic effects of Brahmi (*Bacopa monnieri*) on human memory. Neuropsychopharmacology 27:279–281

Shahinozzaman M, Taira N, Ishii T, Halim MA, Hossain MA, Tawata S (2018) Anti-inflammatory, anti-diabetic, and anti-Alzheimer's effects of prenylated flavonoids from *Okinawa propolis*: an investigation by experimental and computational studies. Molecules 23(10):2479

Shantala P, Nagarjun V (2013) Alzheimer's disease and ayurveda: first descriptions in medical literature. Ayurveda J Health 11(Spring-3):29–35

Sharma R, Amin H (2015) Rasayana therapy: ayurvedic contribution to improve quality of life. World J Pharmacol Res Technol 4:23–33

Sharma R, Chaturvedi C, Tewari P (1987) Efficacy of *Bacopa monniera* in revitalizing intellectual functions in children. J Res Educ Ind Med 1:12

Sharma R, Kuca K, Nepovimova E, Kabra A, Rao MM, Prajapati PK (2019) Traditional ayurvedic and herbal remedies for Alzheimer's disease: from bench to bedside. Expert Rev Neurother 19(5):359–374

Sheeja Malar D, Beema Shafreen R, Karutha Pandian S, Pandima Devi K (2017) Cholinesterase inhibitory, antiamyloidogenic and neuroprotective effect of the medicinal plant *Grewia tiliaefolia*—an in vitro and in silico study. Pharm Biol 55(1):381–393

Siddha Noi Nadal (2023) Part-1 Book 7

Sierpina VS, Wollschlaeger B, Blumenthal M (2003) *Gingko biloba*. Am Fam Physician, 68 923–926

Singh M, Ramassamy C (2017) In vitro screening of neuroprotective activity of Indian medicinal plant *Withania somnifera*. J Nutr Sci 6:e54

Singh SK, Srivastav S, Castellani RJ, Plascencia-Villa G, Perry G (2019) Neuroprotective and antioxidant effect of *Ginkgo biloba* extract against AD and other neurological disorders. Neurotherapeutics 16(3):666–674

Sireeratawong S, Jaijoy K, Khonsung P, Lertprasertsuk N et al (2016) Acute and chronic toxicities of *Bacopa monnieri* extract in Sprague-Dawley rats. BMC Complement Altern Med 16:249

Soumyanath A (2012) *Centella asiatica* extracts improves behavioural deficits in a mouse model of Alzheimer's disease. Int J Alzheimers Dis 2012:9

Sowndhararajan K, Deepa P, Kim M, Park SJ, Kim S (2018) An overview of neuroprotective and cognitive enhancement properties of lignans from *Schisandra chinensis*. Biomed Pharmacother 97:958–968

Sudarsanam SR, Moonandi M (2014) Potency of Kara Sooda Sathu Parpam, a herbo mineral Siddha drug in the management of Kalladaippu noi (urolithiasis): a drug review. Int J Res Ayurveda Pharm 5(3):372–378
Tan MA, Lagamayo MWD, Alejandro GJD (2019) An SSA anti-amyloidogenic and cyclooxygenase inhibitory activity of *Guettarda speciosa*. Molecules 24(22):4112
Tang YY, Holzel BK, Posner MI (2015) The neuroscience of mindfulness meditation. Nat Rev Neurosci 16(4):213–225
Taslimi P, Köksal E, Gören AC, Bursal E, Aras A, Kılıç Ö et al (2020) Anti-Alzheimer, antidiabetic and antioxidant potential of *Satureja cuneifolia* and analysis of its phenolic contents by LC-MS/MS. Arab J Chem 13(3):4528–4537
Tepe AS, Ozaslan M (2020) Anti-Alzheimer, anti-diabetic, skin-whitening, and antioxidant activities of the essential oil of *Cinnamomum zeylanicum*. Ind Crops Prod 145:112069
Thakkar VT, Deshmukh A, Hingorani L et al (2017) Development and optimization of dispersible tablet of *Bacopa monnieri* with improved functionality for memory enhancement. J Pharm Bioallied Sci 9:208–215
Thakur M, Virk R, Sangha PS, Saxena V (2019) The effects of turmeric on Alzheimer's patients. J Food Sci Nutr Res 2:347–353
Theendakara V, Peters-Libeu CA, Spilman P, Poksay KS, Bredesen DE, Rao RV (2016) Direct transcriptional effects of Apolipoprotein E. J Neurosci 36(3):685–700
Tolppanen AM, Solomon A, Kulmala J et al (2015) Leisure-time physical activity from mid- to late life, body mass index, and risk of dementia. Alzheimers Dement 11(4):434–443
Thomas NP (2001) Homeopathy in the USA. Br Homeopath J 90(02):99–103
Uddin MJ, Abdullah-Al-Mamun M, Biswas K, Asaduzzaman M, Rahman MM (2016) Assessment of anticholinesterase activities and antioxidant potentials of *Anisomeles indica* relevant to the treatment of Alzheimer's disease. Orient Pharm Exp Med 16(2):113–121
Vohora SB, Shah SA, Dandiya PC (1990) Central nervous system studies on an ethanol extract of *Acorus calamus* rhizomes. J Ethnopharmacol 28:53–62
Wang DJ, Rao H, Korczykowski M et al (2011) Cerebral blood flow changes associated with different meditation practices and perceived depth of meditation. Psychiatry Res 191(1):60–67
Wang J, Tan L, Yu JT (2016) Prevention trials in Alzheimer's disease: current status and future perspectives. J Alzheimers Dis 50(4):927–945
Wangchuk P, Sastraruji T, Taweechotipatr M, Keller PA, Pyne SG (2016) Anti-inflammatory, anti-bacterial and anti-acetylcholinesterase activities of two isoquinoline alkaloids—scoulerine and cheilanthifoline. Nat Prod Commun 11(12):1934578X1601101207
Yang F, Lim GP, Begum AN, Ubeda OJ, Simmons MR, Ambegaokar SS et al (2005) Curcumin inhibits formation of amyloid beta oligomers and fibrils, binds plaques, and reduces amyloid in vivo. J Biol Chem 280:5892–5901
Zhang L, Fiala M, Cashman J, Sayre J, Espinosa A, Mahanian M et al (2006) Curcuminoids enhance amyloid-beta uptake by macrophages of Alzheimer's disease patients. J Alzheimers Dis 10:1–7

Chapter 13
Effect of Diet Patterns in the Prevention of Alzheimer's Disease

Daniela S. Cosio, Suset Rodriguez, Arumugam R. Jayakumar, and Michael J. Paidas

Abstract There are over 55 million people living with dementia worldwide. Alzheimer's disease (AD) is the most common type of dementia and likely comprises up to 60–70% of dementia cases. With a worldwide increase in the age and longevity of the population, a dramatic rise in AD prevalence is expected in the coming decades. A lack of mechanistic understanding of AD, as well as a lack of effective treatment strategies, particularly in the early stages of disease progression, further aggravates the incidence. In the past several years, there has been great interest in the usefulness of dietary interventions to prevent or diminish the progression of AD. In this review, we summarize some of the main mechanisms by which the Western Pattern Diet (WPD) can promote the development and progression of AD and review the evidence for the use of the Mediterranean diet (MD), the Dietary Approaches to Stop Hypertension (DASH) diet, the ketogenic diet (KD), and the Mediterranean-DASH Intervention for Neurodegenerative Delay (MIND) diet in the prevention and treatment of AD. While the existing findings are promising, the overall lack of randomized case-control studies, varying study end points, and differing diet definitions and consumption measurement methods make it difficult to determine a cause-and-effect relationship. Thus, we recommend that further studies, especially randomized control trials, involving all four of these diets be conducted to clarify the effects of these diet patterns on the development and pathogenesis of AD, as well as on clinically relevant cognitive measures. Additional studies are also needed to determine the optimal diet duration, as well as the rate of retention, adherence, and safety of these diets. Lastly, longer term studies conducted at different points in the human life span (early, mid-, and late life) would elucidate when best to implement these diet patterns.

D. S. Cosio · S. Rodriguez · A. R. Jayakumar (✉) · M. J. Paidas (✉)
Department of Obstetrics, Gynecology and Reproductive Sciences, University of Miami Miller School of Medicine, Miami, FL, USA
e-mail: ajayakumar@med.miami.edu; mxp1440@med.miami.edu

A. Justin Thenmozhi, T. Manivasagam (eds.), *Nutraceuticals for Alzheimer's Disease: A Promising Therapeutic Approach*, Nutritional Neurosciences,
https://doi.org/10.1007/978-981-99-0677-2_13

Keywords Alzheimer's disease · Dementia · Diet patterns · Western diet · Mediterranean diet · DASH diet · Ketogenic diet · MIND diet · Advanced glycation end products · Inflammation · Gut microbiome

Abbreviations

AD	Alzheimer's disease
AGEs	Advanced glycation end products
BBB	Blood–brain barrier
CRP	c-Reactive protein
DASH	Dietary Approaches to Stop Hypertension
FDA	Food and Drug Administration
IFN-γ	Interferon gamma
IL-1β	Interleukin one beta
IL-6	Interleukin six
KD	Ketogenic diet
MD	Mediterranean diet
MIND	Mediterranean-DASH Intervention for Neurodegenerative Delay
NF-κB	Nuclear factor kappa beta
SCFAs	Short-chain fatty acids
TNF-α	Tumor necrosis factor alpha
WPD	Western Pattern Diet

13.1 Introduction

13.1.1 The Burden of Alzheimer's Disease

There are more than 55 million people living with dementia worldwide, and about ten million new cases of dementia are diagnosed each year (World Health Organization 2021, https://www.who.int/news-room/fact-sheets/detail/dementia). Alzheimer's disease (AD) is the most common cause of dementia and likely makes up 60–70% of dementia cases. According to the Alzheimer's Association (2021), more than six million Americans are reported to be living with AD. In fact, one in nine people aged 65 and older (11.3%) has AD (Alzheimer's Association 2021, https://www.alz.org/media/documents/alzheimers-facts-and-figures.pdf) Because AD is most prevalent in those aged 65 and older, as the life expectancy in the USA continues to increase, the number of people who develop AD, and those whose AD progresses to later, more severe stages, will also increase.

Between 2000 and 2019, the number of deaths from Alzheimer's disease in the USA, as recorded on death certificates, has increased 145%, while the number of deaths from heart disease (the number one cause of death in the USA) decreased

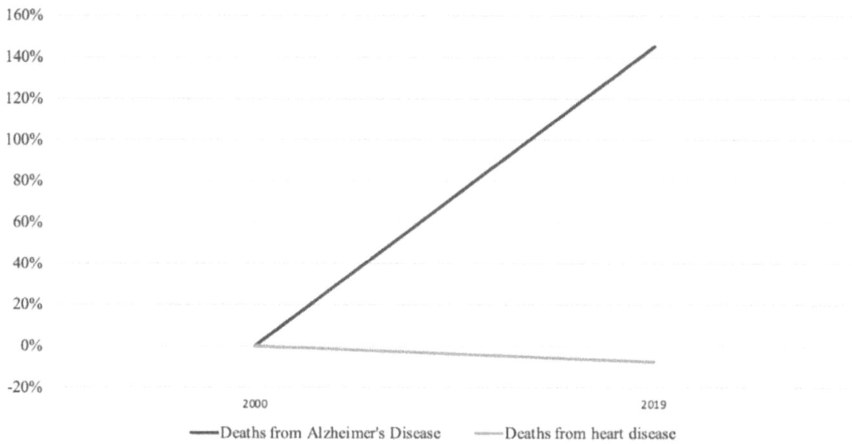

Fig. 13.1 Percent change from 2000 to 2019 in the number of deaths attributable to Alzheimer's disease vs. heart disease

7.3% (Fig. 13.1) (Alzheimer's Association 2021). Among people aged 70, twice as many (61%) of those with AD are expected to die within 10 years compared to those without AD (30%) (Alzheimer's Association 2021). These reports suggest that AD is not only a debilitating condition, but also a deadly one, and that while medical advancements have been made in the prevention and treatment of other fatal conditions, such as heart disease, little advancement seems to have been made in the prevention and curative treatment of AD.

13.1.2 The Alzheimer's Disease Continuum

While the underlying pathophysiology of AD is not fully understood, much research has focused on describing the prominent features of the disease. These include the accumulation of amyloid plaques (for review, see Hardy and Allsop 1991; Hardy and Higgins 1992; O'Brien and Wong 2011; Brito-Moreira et al. 2017; Gallardo and Holtzman 2019), presence of tau neurofibrillary tangles (for review see Hardy and Allsop 1991; Small and Duff 2008; Takashima 2009; Meraz-Ríos et al. 2010; Gallardo and Holtzman 2019; Busche and Hyman 2020), impaired glucose metabolism and brain insulin resistance (for review see Abolhassani et al. 2017; Arnold et al. 2018; Butterfield and Halliwell 2019; Kellar and Craft 2020), and neuroinflammation and neuronal cell death (for review see Herrup 2012; Heneka et al. 2015; Fricker et al. 2018; Calvo-Rodriguez and Bacskai 2021).

These changes in the brain start much earlier than the development of symptoms. In fact, the brain changes associated with AD are thought to begin, on average, about 20 years before the development of symptoms such as memory loss or cognitive dysfunction (Alzheimer's Association 2021, and references therein). The

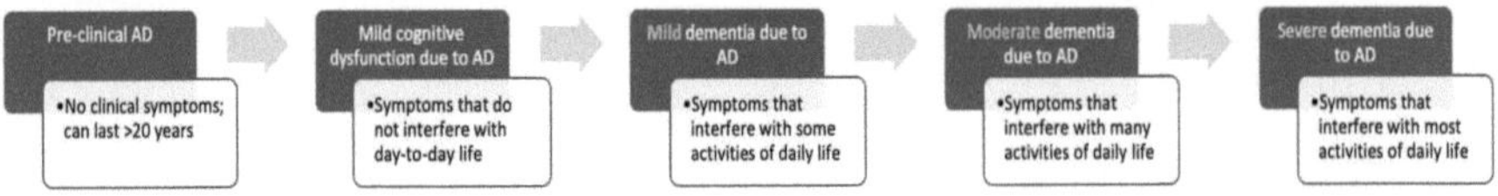

Fig. 13.2 The Alzheimer's disease continuum

progression of the disease from preclinical neurologic changes to the most significant degree of dementia has been termed the "Alzheimer's disease continuum" (Aisen et al. 2017; Alzheimer's Association 2021) (Fig. 13.2). The progression of the disease through the phases in the continuum is variable, and it is not known how long it will take each patient to progress to the next phase.

13.1.3 *Current Medical Treatment Options*

As of March 2021, the U.S. Food and Drug Administration (FDA) has approved seven drugs for the treatment of AD and/or its associated symptoms. The majority of these drugs only temporarily alleviate symptoms of AD, such as cognitive decline (donepezil, rivastigmine, galantine, memantine, and donepezil + memantine) and insomnia (suvorexant), but do not effectively prevent nor alter the course of the disease. This is because the aforementioned drugs can only slow, but do not stop, the effects caused by progressive damage to neurons in AD. They do this by either increasing the amount of neurotransmitters in the brain or, in the case of memantine, by blocking NMDA receptors in the brain that cause further, excitation-induced, neuronal damage (Table 13.1).

Only one of the FDA-approved drugs, aducanumab, a monoclonal antibody that removes amyloid plaques, is directed at a suspected component of the underlying pathophysiology of AD (Table 13.1). However, this drug showed conflicting results in the two phase III trials conducted by Biogen (Budd Haeberlein et al. 2020). While the level of amyloid plaques in the brain was successfully and consistently reduced in the participants of both trials, the primary end point, reduction in clinical decline, was met in only one of the two trials. The FDA ultimately decided to approve aducanumab under its Accelerated Approval pathway, "a pathway intended to provide earlier access to potentially valuable therapies for patients with serious diseases where there is an unmet need, and where there is an expectation of clinical benefit despite some residual uncertainty regarding that benefit" (Cavazzoni 2021). Under this pathway, approval is given based on a surrogate end point, in this case the reduction in brain amyloid plaque, that is believed to predict clinical benefit. As part of the Accelerated Approval pathway, the FDA requires that companies confirm the anticipated clinical benefit of the drug within 9 years. This means evidence is still pending on the efficacy of this drug to consistently reduce clinical decline.

Because the majority of the FDA-approved drugs treat only the clinical symptoms of AD, they are only useful later in the disease course, once symptoms have

Table 13.1 Mechanisms of action of the FDA-approved drugs for the treatment of AD and its clinical symptoms

General category of drug	Drug name	Mechanism of action	Effect
Cholinesterase inhibitors	Donepezil Rivastigmine Galantamine	Prevent the breakdown of the neurotransmitter, acetylcholine	Inhibition of acetylcholine's breakdown allows for a longer duration of action of the neurotransmitter, a chemical messenger in the brain that is important for memory and learning
Glutamate regulator	Memantine	Blockade of the glutamate NMDA receptor	Blockage of the glutamate NMDA receptor decreases excitation-induced neuronal damage (excitotoxicity)
Cholinesterase inhibitor + glutamate regulator	Donepezil + memantine	Both prevent the breakdown of acetylcholine and block the NMDA receptor	Same effect as its component drugs, but with the added advantage of fewer pills, an extended-release capsule, and an easy-to-open capsule, allowing patients or caregivers to mix the powder into food
Orexin receptor antagonist	Suvorexant	Blocks the binding of orexin A and B to their receptors	Orexin is a neurotransmitter that promotes wakefulness. Blocking its binding to its receptor dampens the wakefulness signal
Monoclonal antibody	Aducanumab	Antibody against amyloid in the brain	This drug promotes the removal of amyloid from the brain, one of the hypothesized contributors to neuronal death in AD

developed (Fig. 13.3). There is a notable gap in prevention and treatment options earlier in the disease course.

13.1.4 The Push to Address Modifiable Risk Factors as a Means of Prevention and Treatment of AD

As the prevalence of AD continues to increase, so too does the need for effective treatment and prevention strategies, especially at the earlier stages of the disease. With drug intervention lacking in this regard, prevention and treatment efforts have begun to focus on addressing the modifiable risk factors for the development of AD, such as diet, exercise, and sleep. This is because, while biological markers such as amyloid and tau are useful indicators of risk of progression to AD, many people with these biomarkers never develop dementia, indicating that environmental factors

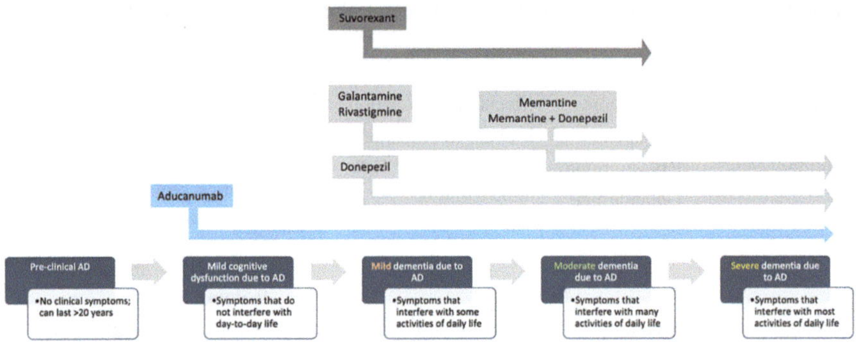

Fig. 13.3 FDA-approved medications for the treatment of AD and where in the continuum of AD they are approved for use. Blue, intended to delay clinical decline. Light grey, used for the symptomatic treatment of cognitive decline caused by AD. Dark grey, used for the symptomatic treatment of insomnia caused by AD

likely also play a role in the development of the disease (Livingston et al. 2020). In fact, it has been suggested that addressing these modifiable risk factors could prevent or delay up to 40% of dementia cases (Livingston et al. 2020).

One of the strongest, evidence-based modifiable risk factors for the prevention of AD is diet (for review, see Baumgart et al. 2015; Xu et al. 2015; World Health Organization 2019; Więckowska-Gacek et al. 2021). Here, we will discuss the mechanisms of diet in the development and progression of AD and will review the evidence for the use of the Mediterranean diet (MD), the Dietary Approaches to Stop Hypertension (DASH) diet, the ketogenic diet (KD), and the Mediterranean-DASH Intervention for Neurodegenerative Delay (MIND) diet in the prevention and treatment of AD.

13.2 The Western Pattern Diet and Alzheimer's Disease

The Western Pattern Diet (WPD), the predominant dietary pattern in the USA and a growing diet globally, is characterized by a high intake of processed grains, prepackaged and processed foods, red meat, processed meat, candy/sweets, high-sugar drinks, fried foods, butter, high-fat dairy products, eggs, potatoes, corn, and high-fructose corn syrup, along with low intake of fruits, vegetables, whole grains, fish, nuts, and seeds (Table 13.2). In animal studies, the WPD is mimicked using high fats (30–65%), saturated fatty acids, simple sugars, and cholesterol (Więckowska-Gacek et al. 2021). Evidence from both rodent (Kanoski and Davidson 2011; Francis and Stevenson 2013; Thériault et al. 2016; Sah et al. 2017; Ke et al. 2020) and human studies (Francis and Stevenson 2011; Gibson et al. 2013; Attuquayefio et al. 2017; Stevenson and Francis 2017) has shown that the WPD can contribute to pathological features seen in the brain in AD (Barnard et al. 2014; Jacka

Table 13.2 Defining the Western Pattern Diet (WPD), Mediterranean diet, Dietary Approaches to Stop Hypertension (DASH) diet, ketogenic diet, and Mediterranean-DASH Intervention for Neurodegenerative Delay (MIND) diet

	Western Pattern Diet (WPD)	Mediterranean Diet (MD)	Dietary Approaches to Stop Hypertension (DASH) Diet	Ketogenic Diet (KD)	Mediterranean-DASH Intervention for Neurodegenerative Delay (MIND) Diet
High / Moderate Intake Foods	Processed grains Pre-packaged foods Processed foods Red meat Processed meat Candy/sweets High-sugar drinks Fried foods Butter High-fat dairy Eggs Potatoes Corn High-fructose corn syrup	High intake: Fruits Vegetables Whole grains Nuts Legumes Cereals Seeds Moderate intake: Fish Poultry Red wine	High intake: Fruits Vegetables Whole grains Nuts Moderate intake: Fish	Naturally high-fat foods	Leafy green / other vegetables Nuts Berries Beans Whole grains Fish Poultry Olive oil Wine
Low Intake Foods / Foods to Avoid	Fruits Vegetables Whole grains Fish Nuts Seeds	Red and processed meats	High-fat dairy Dietary sodium Alcohol	Low carbohydrate foods	Red meats Butter and stick margarines Cheese Pastries/sweets Fried or fast food
Special Notes	The name of this diet is a generalization and does not represent the diet of all Western cultures.	Use of olive oil as the main source of fat.	Was first introduced, and is now accepted, as a non-pharmacological treatment for hypertension (Appel et al., 1997).	Was originally used to treat refractory epilepsy (Huttenlocher, 1976).	A hybrid of the Mediterranean and DASH diets.

Yellow: foods that appear in the high/moderate intake row for the intervention diets, but in the low intake row of the Western diet. Green: foods that appear in the low/avoid row for the intervention diets but in the high intake row for the Western diet. Red: significant difference between the MD/MIND diet and the DASH diet

et al. 2015). Additionally, this diet can contribute to the development of medical conditions that are considered risk factors for the development of AD, such as cerebrovascular diseases, diabetes, hypertension, obesity, and dyslipidemia (Silva et al. 2019). Further, certain components of the WPD, such as high saturated fats, trans fats, and low levels of antioxidants, are associated with a higher risk of developing AD (Scarmeas et al. 2009; de Wilde et al. 2017; Liyanage et al. 2019; Silva et al. 2019; Baranowski et al. 2020). Some of the proposed mechanisms by which consistent consumption of the WPD can contribute to AD include the production of advanced glycation end products (AGEs), disruption of the gut microbiome, and development of a chronic inflammatory state.

13.2.1 Advanced Glycation End Products (AGEs)

Part of the problem with the WPD is the way in which food is processed. In order to enhance flavor and texture, food is often preprocessed using high-temperature techniques. Food prepared in this way undergoes the Maillard reaction in which the carbonyl group of a reducing sugar (glucose, galactose, and fructose) binds with the amino group of a protein, producing an intermediate product, which then undergoes a conformational change to ultimately become an advanced glycation end product (AGE) (Gill et al. 2019). AGEs can then tightly bind surrounding proteins. In low quantities, the binding of surrounding proteins is temporary and reversible, as AGEs are cleared by a natural detoxification process. But in higher quantities, or in combination with disease states, the detoxification process is overwhelmed and serum levels of AGEs become elevated, a risk factor for a variety of disorders, including AD (Bucala et al. 1994; Vitek et al. 1994; Smith 2017; Rhee and Kim 2018; Merhi et al. 2019). While AGEs are also produced endogenously, it is the exogenously produced ones that contribute most to the total pool of AGEs (Delgado-Andrade and Fogliano 2018), and it is the pre-prepared and processed foods, a large component of the WPD, that have the highest amount of exogenous AGEs compared to freshly prepared and unprocessed food (Kent and Uribarri 2014).

In the brain, AGEs bind to β-amyloid and stimulate its aggregation and accumulation (Fig. 13.4), part of the pathophysiology associated with AD (Vitek et al. 1994; Hsu et al. 2019). The binding of AGEs to their receptors in the brain also activates a pro-inflammatory pathway by stimulating nuclear factor kappa beta (NF-κB), tumor necrosis factor alpha (TNF-α), interleukin six (IL-6), and C-reactive protein (CRP) that further leads to neuronal damage (Fig. 13.4) (Abate et al. 2017; Uribarri 2018). In order to reduce AGEs by up to 50%, it is recommended that food be fresh/non-processed and that cooking should primarily involve stewing or boiling, as opposed to grilling, roasting, or deep-frying (Uribarri et al. 2010). This concept of consuming

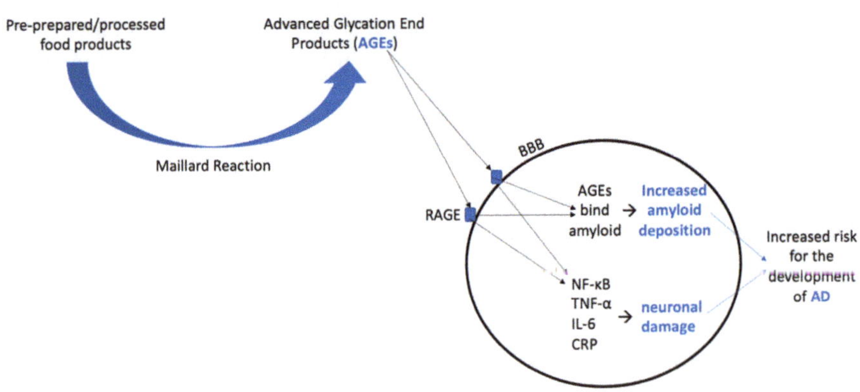

Fig. 13.4 Production of advanced glycation end products (AGEs) and the effect in the brain. *BBB* blood–brain barrier, *CRP* C-reactive protein, *IL-6* interleukin six, *NF-κB* nuclear factor kappa beta, *RAGE* receptor for advanced glycation end products, *TNF-α* tumor necrosis factor alpha

mainly non-processed, fresh food is a common component of most of the diet patterns proposed to reduce the risk of AD.

13.2.2 The Gut Microbiome

The human gastrointestinal (GI) tract contains thousands of species of microorganisms, culminating in about 10^{14} microorganisms, the most found anywhere in the human body (Jiang et al. 2017). These organisms are referred to collectively as the gut microbiome and alterations in their composition/environment (dysbiosis) have been shown to alter immune and brain function, augment neuroinflammation, and contribute to neurodegeneration (Cryan and Dinan 2012; Tillisch 2014; Sharon et al. 2016; Dinan and Cryan 2017; Duffy et al. 2019; Spencer et al. 2019). Diet, and more specifically the WPD, has been found to cause dysbiosis in the gut microbiome (Bruce-Keller et al. 2015; Noble et al. 2017; Christ et al. 2019; Lustig 2020; Martínez Leo and Segura Campos 2020; Leblhuber et al. 2021; Więckowska-Gacek et al. 2021), which has been implicated in the pathophysiology of AD (Fig. 13.5) (Harach et al. 2017; Jiang et al. 2017; Vogt et al. 2017; Liu et al. 2019; Sochocka et al. 2019; Al Bander et al. 2020; Chu et al. 2022).

The gut microbiome produces short-chain fatty acids (SCFAs) by fermenting nondigestible carbohydrates from the diet. These SCFAs (mainly acetate, propionate, and butyrate) act as signaling molecules in many organ systems, including the central nervous system, inhibiting neuroinflammation by regulating the synthesis of neurotransmitters and their receptor expression (i.e., dopamine and γ-aminobutyric acid) (Jena et al. 2018; Martínez Leo and Segura Campos 2020). Butyrate has been

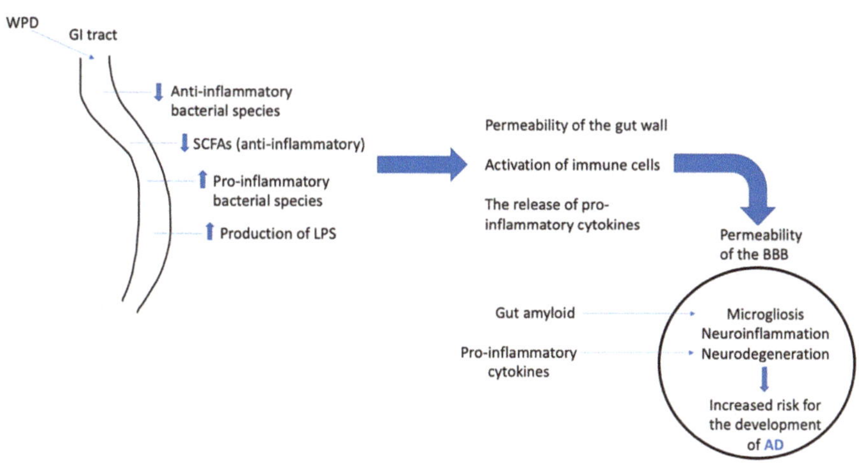

Fig. 13.5 The effect of the Western Pattern Diet (WPD) on the gut microbiome. *GI* gastrointestinal, *SCFAs* short-chain fatty acids, *LPS* lipopolysaccharides, *BBB* blood–brain barrier, *AD* Alzheimer's disease

shown to augment gut barrier function by upregulation of claudin-1 (a tight junction protein) transcription (Wang et al. 2012) and by increasing the activity of adenosine monophosphate-activated protein kinase, which increases the rate of tight junction assembly (Peng et al. 2009).

Diet-induced gut microbiome dysbiosis can reduce the bacterial species that produce SCFAs, such as *Faecalibacterium*, *Roseburia*, and *Eubacterium* (main producers of butyrate), leading to increased gut, systemic, and central nervous system inflammation (Fig. 13.5) (Sanz et al. 2010; Bruce-Keller et al. 2015; Chakraborti 2015; Di Lorenzo et al. 2019). Additionally, the reduction in anti-inflammatory gut microbial species allows for the overgrowth of pro-inflammatory species, which produce large amounts of lipopolysaccharides (LPS), a bacterial endotoxin that can lead to increased permeability of the gut wall, activation of immune cells (macrophages, neutrophils, and dendritic cells), and release of pro-inflammatory cytokines (interleukin one alpha, IL-1β, TNF-α, and IL-6) (Fig. 13.5) (Raetz and Whitfield 2002; Vighi et al. 2008; Chakraborti 2015; Di Lorenzo et al. 2019; Marizzoni et al. 2020). Studies in humans have shown that the gut microbiome of patients with AD, as compared to healthy controls, had decreased numbers of anti-inflammatory bacteria (*Firmicutes* and *Actinobacteria*) and increased numbers of pro-inflammatory bacteria (*Proteobacteria* and *Bacteroidetes*) and that there was a correlation between increasing amounts of the pro-inflammatory bacteria and the clinical severity of the AD (Vogt et al. 2017; Liu et al. 2019).

The chronic production and systemic spread of pro-inflammatory cytokines can impair the integrity of the BBB, which can result in increased influx and decreased clearance of β-amyloid, via changes in transport proteins such as P-glycoprotein, low-density lipoprotein receptor-related protein 1, and receptor for advanced glycation end products (Fig. 13.5) (Jaeger et al. 2009; Erickson et al. 2012; Weintraub et al. 2014; Varatharaj and Galea 2017). The accumulation of β-amyloid in the brain then promotes neuroinflammation via the increased generation of pro-inflammatory cytokines, such as interferon gamma (IFN-γ), interleukin one beta (IL-1β), and TNF-α, leading to microgliosis, neuroinflammation, and ultimately neurodegeneration, all factors increasing the risk for the development of AD (Fig. 13.5) (Sastre et al. 2006; Lee et al. 2010; Tagliabue and Elli 2013; Tan et al. 2013; Heneka et al. 2015; Heppner et al. 2015; Sampson and Mazmanian 2015; Zhao et al. 2015; Heras-Sandoval et al. 2016; Noble et al. 2017; Westfall et al. 2017; De-Paula et al. 2018; Gąsiorowski et al. 2018; Komanduri et al. 2019; Kowalski and Mulak 2019; Liu et al. 2019; Sochocka et al. 2019; He et al. 2020; Khan et al. 2020; Martínez Leo and Segura Campos 2020; Zhang et al. 2020; Leblhuber et al. 2021).

13.2.3 Inflammation

The inflammatory response produced in large part by one's diet (as explained above), in its acute form, is useful as a defense mechanism to eliminate harmful factors. But if the inflammatory response becomes cyclic and chronic, for example through the consistent consumption of the WPD, it can lead to many of the brain findings seen in AD such as decreased brain volume (Gu et al. 2015), cerebral vascular events (Gu et al. 2018; Swardfager et al. 2017), aggregation of amyloid β (Szczechowiak et al. 2019), neuronal damage/neurodegeneration (Holmes et al. 2009; Heneka et al. 2015), and neuronal death (Akiyama et al. 2000; Block et al. 2007).

As nutrition/diet has been shown to modulate the immune system and inflammatory processes, it has become a target of efforts to prevent the neuroinflammation implicated in the pathogenesis of AD (Heneka et al. 2015; Heppner et al. 2015; Calder et al. 2017; McGrattan et al. 2019).

13.3 Evidence for the Use of Specific Diet Patterns in the Prevention and Treatment of AD

Studies investigating the effect of nutrition on cognitive function have found that a lower ratio of omega-6 to omega-3 fatty acid intake can improve neuroinflammation, spatial memory, and overall cognitive status and that polyphenols, unsaturated fats, and antioxidant vitamins can prevent neuroinflammation and oxidative stress (Devassy et al. 2016; Andruchow et al. 2017; Monacelli et al. 2017; Miquel et al. 2018). Fish oil, fruits, vegetables, and folate have also been found to be neuroprotective (Smith and Blumenthal 2010; Silva et al. 2019). While these studies on individual dietary components have found promising results, more recently the idea of using entire diet systems to prevent/treat cognitive disorders, including AD, has gained traction due to the potential for a synergistic effect when neuroprotective foods are consumed together (Pistollato et al. 2018; Szczechowiak et al. 2019). In fact, a recent systematic review and meta-analysis on the effect of different diets on the reduction of AD biomarkers found a significant effect (Hill et al. 2019), thus supporting the continued investigation of diet/nutrition as a form of prevention and treatment for AD.

13.3.1 The Mediterranean and Dietary Approaches to Stop Hypertension (DASH) Diets

The Mediterranean diet (MD) describes a dietary pattern traditionally consumed in Mediterranean countries, albeit with regional differences. The diet consists of a high intake of fruits, vegetables, whole grains, nuts, legumes, cereals, and seeds; moderate intake of fish, poultry, and red wine; low intake of red and processed meats; and use of olive oil as the main source of fat (Table 13.2) (for review, see Davis et al. 2015; Solfrizzi et al. 2017). The Dietary Approaches to Stop Hypertension (DASH) diet was first introduced (Appel et al. 1997), and is now accepted, as a non-pharmacological treatment for hypertension. Like the MD, the DASH diet recommends a high intake of fruits, vegetables, nuts, and whole grains and a moderate consumption of fish (Table 13.2). However, unlike the MD, the DASH diet places greater emphasis on low-fat dairy foods, low dietary sodium, and limited alcohol (Table 13.2).

Both the MD and the DASH diet have been investigated as potential preventive and therapeutic strategies for dementia because of their proven anti-inflammatory and immunomodulatory effects (Wengreen et al. 2013; Minihane et al. 2015; Casas et al. 2016, 2017; Bailey and Holscher 2018; Soltani et al. 2018). While the exact mechanisms are not fully understood, these effects may be due to the synergistic action of various high-intake components of these two diets, such as fiber from whole grains, antioxidants from fruits and vegetables, and plant-derived flavonoids. Both diets are high in natural sources of fiber, which are thought to positively influence the gut microbiome by increasing SCFA-producing bacteria and decreasing the amount of pro-inflammatory bacteria, the opposite effect of the WPD (Fig. 13.5) (Weickert and Pfeiffer 2008; Bailey and Holscher 2018; Garcia-Mantrana et al. 2018). In fact, adherence to the MD has been found to be positively correlated with SCFA concentrations (De Filippis et al. 2016) and negatively correlated with LPS concentrations (Pastori et al. 2015). Antioxidants from fruits and vegetables, plant-derived flavonoids, and long-chain omega-3 fatty acids all have been found to dampen neuroinflammatory processes by inhibiting cytokine production in the brain, another step in the development of AD (Lau et al. 2007; Bazinet and Laye 2014; Rendeiro et al. 2015; Devassy et al. 2016; Mohammadzadeh Honarvar et al. 2017; Monacelli et al. 2017; Flanagan et al. 2018).

In neuroimaging studies, the MD has been shown to be protective of brain structures whose defect is linked to the development of AD. These studies have shown an association between the consumption of an MD and decreased loss of cortical thickness (Mosconi et al. 2014; Staubo et al. 2017), decreased loss in brain volume and a slower rate of hippocampal atrophy (Gu et al. 2015, 2018), improved structural connectivity (Pelletier et al. 2015), and less beta amyloid accumulation (Berti et al. 2018; Rainey-Smith et al. 2018).

In observational studies, adherence to an MD by older adults has been associated with increased cognitive performance (Katsiardanis et al. 2013; Kesse-Guyot et al. 2013; McEvoy et al. 2017), and both the MD and the DASH diet have been associated with slower rates of cognitive decline (Tangney et al. 2014; Galbete et al. 2015; Qin et al. 2015; Anastasiou et al. 2018) and a reduced risk of the development of AD (Scarmeas et al. 2009; Lourida et al. 2013; Singh et al. 2014; Wu and Sun 2017; McGrattan et al. 2019).

While many of these studies are promising, results have not necessarily been consistent (Samieri et al. 2013; Olsson et al. 2015; Haring et al. 2016), likely due to differences in study populations and in the varying methods used to assess diet and cognition (McGrattan et al. 2019). For example, one study showed improved cognitive function in response to a modified, calorie-restricted DASH diet among overweight adults with hypertension (Smith and Blumenthal 2010), while another study demonstrated an improvement in cognition only when the DASH diet was combined with aerobic exercise (Blumenthal et al. 2019).

Additionally, evidence shows that the effect of these diets may depend on the length of time the diet is implemented and how closely the diet is followed. While a modest beneficial effect of increased MD adherence over 4–6 years on cognitive function was demonstrated (Casas et al. 2016), the adoption of an MD over a lesser period of time—6 months (Knight et al. 2016) and 1 year (Marseglia et al. 2018)—had no effect on cognitive function. Additionally, participants with the greatest MD adherence demonstrated improved global cognition and episodic memory compared to those with low adherence (Marseglia et al. 2018), factors that decline in AD, showing that not only the length of time but also the level of adherence to the diet may affect results.

The number of studies involving direct diet intervention on the development of AD (as opposed to a secondary end point like cognitive function or memory) has been limited, making it difficult to prove a causal relationship. In a diet and AD prospective study of 923 participants, ages 58–98 years, followed on average 4.5 years, only the highest of the three levels of adherence to the MD and the DASH diet were associated with lower AD risk (MD: HR = 0.46, 95% CI 0.26, 0.79; DASH diet: HR = 0.61, 95% CI 0.38, 0.97) (Morris et al. 2015a). A recent review of the existing literature regarding the effect of different diets on cognitive decline and the risk of developing AD found a similar trend: higher adherence to the MD was associated with better cognitive scores in 9 of 12 cross-sectional studies, 17 of 25 longitudinal studies, and 1 of 3 trials (van den Brink et al. 2019). The same review found that higher adherence to the MD was associated with a lower risk of AD in 1 case-control study and 6 of 8 longitudinal studies and that higher adherence to the DASH diet was associated with better cognitive function in 1 cross-sectional study, 2 of 5 longitudinal studies, and 1 trial and was associated with a lower AD risk in 1 longitudinal study (van den Brink et al. 2019).

13.3.2 The Ketogenic Diet

The ketogenic diet (KD) is a high-fat, low-carbohydrate diet, which brings the body into a fasting-like state of ketosis (the production of ketone bodies for energy) (Table 13.2). Under normal conditions, the brain primarily utilizes glucose to generate ATP. However, when glucose is not readily available, the brain can and does use ketone bodies as a source of energy. The KD was originally used to treat refractory epilepsy (Huttenlocher 1976; Pinto et al. 2018), but more recent studies have shown that it may also be useful in preventing/treating neurodegenerative diseases, including AD (Reger et al. 2004; Van der Auwera et al. 2005). Preclinical studies have found that the consumption of a KD can reduce inflammation (Milder et al. 2010; Dupuis et al. 2015), provide an alternative source of energy (Kim et al. 2012), improve mitochondrial dysfunction (Sullivan et al. 2004; Van der Auwera et al. 2005), reduce oxidative stress (Sullivan et al. 2004; Van der Auwera et al.

2005), increase cerebral ATP concentrations (Sullivan et al. 2004), decrease amyloid deposition (Van der Auwera et al. 2005; Kashiwaya et al. 2013; Zhang et al. 2013; Pawlosky et al. 2017), and ameliorate tau pathology (Kashiwaya et al. 2013), all components believed to play a role in the development of AD (Markesbery 1997; Hensley 2010; Mosconi et al. 2009). It is thought that the KD likely prevents the development of AD via its effects on these brain functions. Preclinical studies have also shown that the KD reduces pro-inflammatory gut bacteria in young healthy mice (Ma et al. 2018), which is considered another mechanism by which the KD may reduce the risk of AD (Fig. 13.5). However, not all preclinical studies on the effects of the KD in AD were consistent. There have been mouse models of Alzheimer's disease that have found the implementation of a ketogenic diet to have no improvement on cognition and no effect on β-amyloid levels (Beckett et al. 2013; Brownlow et al. 2013; Ohnuma et al. 2016). Therefore, more studies which use consistent methods, tools of measurement, and similar end points need to be conducted.

As has been seen with the effects of the MD and the DASH diet on AD, evidence shows that the effect of the KD on AD likely depends on the length of time the diet is implemented. In a study on the effect of a medium-chain triglyceride (MCT)-based ketogenic formula on cognitive function in patients with mild-to-moderate stages of AD, although the patients' plasma levels of ketone bodies were successfully increased 120 min after a single intake of ketogenic formula, there was no significant difference in any cognitive test results between the group that received the ketogenic formula and those that received placebo formula (Ota et al. 2019). However, in the subsequent chronic intake trial of the ketogenic formula (12-week regimen) in patients with mild-to-moderate stages of AD, the patients showed significant improvement in memory tests compared to their baseline scores (Ota et al. 2019). Additionally, Krikorian et al. (2012) showed that ketone body levels positively correlated with memory performance, indicating that the level of ketone bodies present, in addition to the length of time the diet is followed, could be a mitigating factor in the effectiveness of the KD on brain pathology/function.

The few clinical studies on the effects of the KD on cognition showed improvement in memory and cognitive function (Reger et al. 2004; Henderson et al. 2009; Rebello et al. 2015; Taylor et al. 2017; Ota et al. 2019). However, these studies were not all conducted in patients with AD or using the primary end point of AD. Additionally, some of the studies had a small number of participants (i.e., safety trials) and some used a ketogenic supplement instead of a true KD intervention. Therefore, while these findings show promise in the KD's potential as a preventive/therapeutic option for AD, further studies on the effect of the KD on the prevention and treatment of AD itself must be conducted.

Lastly, the KD has been associated with more side effects than other diet patterns. Up to 50% of people who primarily consume a KD experience nausea, vomiting, constipation, decreased appetite, and even increased risk of intestinal disease (Tuck and Staudacher 2019). Until further safety studies are performed, especially in the elderly, a KD is only recommended for limited periods of time.

13.3.3 Mediterranean-DASH Intervention for Neurodegenerative Delay (MIND) Diet

The Mediterranean-DASH Intervention for Neurodegenerative Delay (MIND) diet is a hybrid (combination) of the Mediterranean and DASH diets. The MIND diet was developed around the incorporation of ten brain-healthy foods (leafy green vegetables, other vegetables, nuts, berries, beans, whole grains, fish, poultry, olive oil, and wine) and the avoidance of five unhealthy foods (red meats, butter and stick margarines, cheese, pastries and sweets, and fried or fast food) (Table 13.2), all of which have been studied and shown to play a role in the prevention of dementia and cognitive decline (Koch and Jensen 2016; Solfrizzi et al. 2017; Cremonini et al. 2019; van den Brink et al. 2019). For example, epidemiologic studies have found a significant association with increased consumption of green leafy vegetables and slower cognitive decline (Kang et al. 2005; Morris et al. 2006), while cell models of neurotoxicity have shown that berries have anti-inflammatory and antioxidant activity (Tavares et al. 2013; Cásedas et al. 2017), an effect attributed to their high content of polyphenols (Pandareesh et al. 2015; Kelly et al. 2017), such as flavonoids, which have also been linked to slower rates of cognitive decline in humans (Devore et al. 2012). Because the MIND diet is the newest of the diets we have discussed, it has been investigated to a more limited extent in the context of AD. Still, because it combines two diet patterns that have shown great potential in the prevention and treatment of AD (the MD and the DASH diet), and because it has been shown to have many of the same anti-inflammatory and immunomodulatory effects, it should be studied further.

Like the other diets discussed, the effect of the MIND diet on cognition seems to be correlated with the level of adherence to the diet. Greater adherence to the MIND diet has been associated with better cognitive function and decreased risk of cognitive decline (McEvoy et al. 2017; Shakersain et al. 2018; Hosking et al. 2019), as well as a slower decline in global cognition (Morris et al. 2015a) in all but one study (Berendsen et al. 2018). More specifically, a prospective study found that a moderate-to-high adherence to the MIND diet was associated with a statistically significant reduction in AD rate compared to low adherence (Morris et al. 2015b). This is particularly important because this same study found that only the highest level of adherence to the DASH diet and MD was significantly associated with a reduced rate of developing AD, indicating that the MIND diet may provide a less stringent, yet just as effective, preventive strategy than other diets.

No randomized controlled trials to date have been published investigating the effects of the MIND diet on the prevention/treatment of AD. However, there is one ongoing phase III randomized control trial evaluating the effects of a 3-year intervention of the MIND diet on cognitive decline and brain neurodegeneration, conclusions from which have not yet been published (http://www.clinicaltrials.gov/NCT02817074). The results from this trial are expected to add an understanding of the role of a dietary pattern intervention, and more specifically the MIND diet, as a preventive strategy for AD (Liu et al. 2021).

13.4 Conclusion

As the prevalence, cost, and death toll of AD continue to increase, so too does the need for effective prevention and treatment strategies, especially at the earlier stages of the disease. With drug intervention lacking in this regard, efforts have turned to addressing the modifiable risk factors for the development of AD, especially diet. Above, we discussed some of the main mechanisms by which the WPD can promote the development and progression of AD and reviewed the evidence for the use of the Mediterranean diet (MD), the Dietary Approaches to Stop Hypertension (DASH) diet, the ketogenic diet (KD), and the Mediterranean-DASH Intervention for Neurodegenerative Delay (MIND) diet in the prevention and treatment of AD.

Current scientific evidence suggests that diet likely plays both a preventive and symptom-modulating role in AD, likely through decreased production of AGEs, stabilization of the gut microbiome, decreased neuroinflammation, and altering of brain energy utilization. Further, this effect seems to be contingent on the duration and level of adherence to the abovementioned diets, where the strongest associations have been observed for the MIND diet, although the KD was not included in these studies. The KD has been shown to have the most side effects of any of the proposed diet patterns.

The studies that have been conducted so far are promising, but the overall lack of randomized case-control studies, varying study end points, and differing diet definitions and consumption measurement methods make it difficult to come to a consensus on the cause-and-effect relationship of these diet patterns and AD. Further studies, especially randomized control trials, involving all four of these diets now need to be conducted to clarify the effects of these diets on the development and pathogenesis of AD, as well as on clinically relevant cognitive measures. Additional studies are also needed to determine the duration of diet needed for optimal effect, as well as the rate of retention, adherence, and safety of these diets. Longer term studies conducted at different points in the human life span (early, mid-, and late life) are needed to better understand when best to implement these diet patterns, given the AD continuum that spans decades.

Acknowledgements We apologize for having limited references to review articles and more recently published articles, as we were limited in space.

References

Abate G, Marziano M, Rungratanawanich W, Memo M, Uberti D (2017) Nutrition and AGE-ing: focusing on Alzheimer's disease. Oxid Med Cell Longev 2017:7039816

Abolhassani N, Leon J, Sheng Z, Oka S, Hamasaki H, Iwaki T, Nakabeppu Y (2017) Molecular pathophysiology of impaired glucose metabolism, mitochondrial dysfunction, and oxidative DNA damage in Alzheimer's disease brain. Mech Ageing Dev 161(Pt A):95–104

Aisen PS, Cummings J, Jack CR Jr, Morris JC, Sperling R, Frölich L et al (2017) On the path to 2025: understanding the Alzheimer's disease continuum. Alzheimers Res Ther 9(1):6

Akiyama H, Barger S, Barnum S, Bradt B, Bauer J, Cole GM et al (2000) Inflammation and Alzheimer's disease. Neurobiol Aging 21(3):383–421

Al Bander Z, Nitert MD, Mousa A, Naderpoor N (2020) The gut microbiota and inflammation: an overview. Int J Environ Res Public Health 17:7618

Alzheimer's Association (2021) 2021 Alzheimer's disease facts and figures. Alzheimers Dement 17(3):327–406

Anastasiou CA, Yannakoulia M, Kontogianni MD, Kosmidis MH, Mamalaki E, Dardiotis E (2018) Mediterranean lifestyle in relation to cognitive health: results from the HELIAD study. Nutrients 10(10):1557

Andruchow ND, Konishi K, Shatenstein B, Bohbot VD (2017) A lower ratio of omega-6 to omega-3 fatty acids predicts better hippocampus-dependent spatial memory and cognitive status in older adults. Neuropsychology 31(7):724–734

Appel LJ, Moore TJ, Obarzanek E, Vollmer WM, Svetkey LP, Sacks FM et al (1997) A clinical trial of the effects of dietary patterns on blood pressure. DASH Collaborative Research Group. N Engl J Med 336(16):1117–1124

Arnold SE, Arvanitakis Z, Macauley-Rambach SL, Koenig AM, Wang HY, Ahima RS et al (2018) Brain insulin resistance in type 2 diabetes and Alzheimer disease: concepts and conundrums. Nat Rev Neurol 14(3):168–181

Attuquayefio T, Stevenson RJ, Oaten MJ, Francis HM (2017) A 4-day Western-style dietary intervention causes reductions in hippocampal-dependent learning and memory and interoceptive sensitivity. PLoS One 12(2):e0172645

Bailey MA, Holscher HD (2018) Microbiome-mediated effects of the Mediterranean diet on inflammation. Adv Nutr 9(3):193–206

Baranowski BJ, Marko DM, Fenech RK, Yang AJT, MacPherson REK (2020) Healthy brain, healthy life: a review of diet and exercise interventions to promote brain health and reduce Alzheimer's disease risk. Appl Physiol Nutr Metab 45(10):1055–1065

Barnard ND, Bush AI, Ceccarelli A, Cooper J, de Jager CA, Erickson KI et al (2014) Dietary and lifestyle guidelines for the prevention of Alzheimer's disease. Neurobiol Aging 35(Suppl 2): S74–S78

Baumgart M, Snyder HM, Carrillo MC, Fazio S, Kim H, Johns H (2015) Summary of the evidence on modifiable risk factors for cognitive decline and dementia: a population-based perspective. Alzheimers Dement 11(6):718–726

Bazinet RP, Laye S (2014) Polyunsaturated fatty acids and their metabolites in brain function and disease. Nat Rev Neurosci 15(12):771–785

Beckett TL, Studzinski CM, Keller JN, Paul Murphy M, Niedowicz DM (2013) A ketogenic diet improves motor performance but does not affect β-amyloid levels in a mouse model of Alzheimer's disease. Brain Res 1505:61–67

Berendsen AM, Kang JH, Feskens EJM, de Groot CPGM, Grodstein F, van de Rest O (2018) Association of long-term adherence to the MIND diet with cognitive function and cognitive decline in American women. J Nutr Health Aging 22(2):222–229

Berti V, Walters M, Sterling J, Quinn CG, Logue M, Andrews R et al (2018) Mediterranean diet and 3-year Alzheimer brain biomarker changes in middle-aged adults. Neurology 90(20):e1789–e1798

Block ML, Zecca L, Hong JS (2007) Microglia-mediated neurotoxicity: uncovering the molecular mechanisms. Nat Rev Neurosci 8(1):57–69

Blumenthal JA, Smith PJ, Mabe S, Hinderliter A, Lin PH, Liao L et al (2019) Lifestyle and neurocognition in older adults with cognitive impairments: a randomized trial. Neurology 92(3):e212–e223

Brito-Moreira J, Lourenco MV, Oliveira MM, Ribeiro FC, Ledo JH, Diniz LP et al (2017) Interaction of amyloid-β (Aβ) oligomers with neurexin 2α and neuroligin 1 mediates synapse damage and memory loss in mice. J Biol Chem 292(18):7327–7337

Brownlow ML, Benner L, D'Agostino D, Gordon MN, Morgan D (2013) Ketogenic diet improves motor performance but not cognition in two mouse models of Alzheimer's pathology. PLoS One 8(9):e75713

Bruce-Keller AJ, Salbaum JM, Luo M, Blanchard E 4th, Taylor CM, Welsh DA et al (2015) Obese-type gut microbiota induce neurobehavioral changes in the absence of obesity. Biol Psychiatry 77(7):607–615

Bucala R, Makita Z, Vega G, Grundy S, Koschinsky T, Cerami A et al (1994) Modification of low density lipoprotein by advanced glycation end products contributes to the dyslipidemia of diabetes and renal insufficiency. Proc Natl Acad Sci U S A 91:9441–9445

Budd Haeberlein S, von Hehn C, Tian Y, Chalkias S, Kandadi Muralidharan K, Chen T et al (2020) EMERGE and ENGAGE topline results: two phase 3 studies to evaluate aducanumab in patients with early Alzheimer's disease. Biogen 16(s9):e047259

Busche MA, Hyman BT (2020) Synergy between amyloid-β and tau in Alzheimer's disease. Nat Neurosci 23(10):1183–1193

Butterfield DA, Halliwell B (2019) Oxidative stress, dysfunctional glucose metabolism and Alzheimer disease. Nat Rev Neurosci 20(3):148–160

Calder PC, Bosco N, Bourdet-Sicard R, Capuron L, Delzenne N, Doré J, Franceschi C, Lehtinen MJ, Recker T, Salvioli S, Visioli F (2017) Health relevance of the modification of low grade inflammation in ageing (inflammageing) and the role of nutrition. Ageing Res Rev 40:95–119

Calvo-Rodriguez M, Bacskai BJ (2021) Mitochondria and calcium in Alzheimer's disease: from cell signaling to neuronal cell death. Trends Neurosci 44(2):136–151

Casas R, Sacanella E, Urpí-Sardà M, Corella D, Castañer O, Lamuela-Raventos RM et al (2016) Long-term immunomodulatory effects of a Mediterranean diet in adults at high risk of cardiovascular disease in the PREvención con DIeta MEDiterránea (PREDIMED) randomized controlled trial. J Nutr 146(9):1684–1693

Casas R, Urpi-Sardà M, Sacanella E, Arranz S, Corella D, Castañer O et al (2017) Anti-inflammatory effects of the Mediterranean diet in the early and late stages of atheroma plaque development. Mediat Inflamm 2017:3674390

Cásedas G, Les F, Gómez-Serranillos MP, Smith C, López V (2017) Anthocyanin profile, antioxidant activity and enzyme inhibiting properties of blueberry and cranberry juices: a comparative study. Food Funct 8(11):4187–4193

Cavazzoni, P (2021) FDA's decision to approve new treatment for Alzheimer's disease. Silver Spring. Food and Drug Administration

Chakraborti CK (2015) New-found link between microbiota and obesity. World J Gastrointest Pathophysiol 6(4):110–119

Christ A, Lauterbach M, Latz E (2019) Western diet and the immune system: an inflammatory connection. Immunity 51(5):794–811

Chu CQ, Yu LL, Qi GY, Mi YS, Wu WQ, Lee YK et al (2022) Can dietary patterns prevent cognitive impairment and reduce Alzheimer's disease risk: exploring the underlying mechanisms of effects. Neurosci Biobehav Rev 135:104556

Cremonini AL, Caffa I, Cea M, Nencioni A, Odetti P, Monacelli F (2019) Nutrients in the prevention of Alzheimer's disease. Oxidative Med Cell Longev 2019:9874159

Cryan JF, Dinan TG (2012) Mind-altering microorganisms: the impact of the gut microbiota on brain and behaviour. Nat Rev Neurosci 13(10):701–712

Davis C, Bryan J, Hodgson J, Murphy K (2015) Definition of the Mediterranean diet; a literature review. Nutrients 7(11):9139–9153

De Filippis F, Pellegrini N, Vannini L, Jeffery IB, La Storia A, Laghi L et al (2016) High-level adherence to a Mediterranean diet beneficially impacts the gut microbiota and associated metabolome. Gut 65(11):1812–1821

de Wilde MC, Vellas B, Girault E, Yavuz AC, Sijben JW (2017) Lower brain and blood nutrient status in Alzheimer's disease: results from meta-analyses. Alzheimers Dement 3:416–431

Delgado-Andrade C, Fogliano V (2018) Dietary advanced glycosylation end-products (dAGEs) and melanoidins formed through the Maillard reaction: physiological consequences of their intake. Annu Rev Food Sci Technol 9:271–291

De-Paula VDJR, Forlenza AS, Forlenza OV (2018) Relevance of gut microbiota in cognition, behaviour and Alzheimer's disease. Pharmacol Res 136:29–34

Devassy JG, Leng S, Gabbs M, Monirujjaman M, Aukema HM (2016) Omega-3 polyunsaturated fatty acids and oxylipins in neuroinflammation and management of Alzheimer disease. Adv Nutr 7(5):905–916

Devore EE, Kang JH, Breteler MM, Grodstein F (2012) Dietary intakes of berries and flavonoids in relation to cognitive decline. Ann Neurol 72(1):135–143

Di Lorenzo F, De Castro C, Silipo A, Molinaro A (2019) Lipopolysaccharide structures of Gram-negative populations in the gut microbiota and effects on host interactions. FEMS Microbiol Rev 43(3):257–272

Dinan TG, Cryan JF (2017) Gut instincts: microbiota as a key regulator of brain development, ageing and neurodegeneration. J Physiol 595(2):489–503

Duffy CM, Hofmeister JJ, Nixon JP, Butterick TA (2019) High fat diet increases cognitive decline and neuroinflammation in a model of orexin loss. Neurobiol Learn Mem 157:41–47

Dupuis N, Curatolo N, Benoist JF, Auvin S (2015) Ketogenic diet exhibits anti-inflammatory properties. Epilepsia 56(7):e95–e98

Erickson MA, Hartvigson PE, Morofuji Y, Owen JB, Butterfield DA, Banks WA (2012) Lipopolysaccharide impairs amyloid β efflux from brain: altered vascular sequestration, cerebrospinal fluid reabsorption, peripheral clearance and transporter function at the blood–brain barrier. J Neuroinflamm 9:150

Flanagan E, Müller M, Hornberger M, Vauzour D (2018) Impact of flavonoids on cellular and molecular mechanisms underlying age-related cognitive decline and neurodegeneration. Curr Nutr Rep 7(2):49–57

Francis H, Stevenson R (2013) The longer-term impacts of Western diet on human cognition and the brain. Appetite 63:119–128

Francis HM, Stevenson RJ (2011) Higher reported saturated fat and refined sugar intake is associated with reduced hippocampal-dependent memory and sensitivity to interoceptive signals. Behav Neurosci 125(6):943–955

Fricker M, Tolkovsky AM, Borutaite V, Coleman M, Brown GC (2018) Neuronal cell death. Physiol Rev 98(2):813–880

Galbete C, Toledo E, Toledo JB, Bes-Rastrollo M, Buil-Cosiales P, Marti A et al (2015) Mediterranean diet and cognitive function: the sun project. J Nutr Health Aging 19(3):305–312

Gallardo G, Holtzman DM (2019) Amyloid-β and tau at the crossroads of Alzheimer's disease. Adv Exp Med Biol 1184:187–203

Garcia-Mantrana I, Selma-Royo M, Alcantara C, Collado MC (2018) Shifts on gut microbiota associated to Mediterranean diet adherence and specific dietary intakes on general adult population. Front Microbiol 9:890

Gąsiorowski K, Brokos B, Echeverria V, Barreto GE, Leszek J (2018) RAGE-TLR crosstalk sustains chronic inflammation in neurodegeneration. Mol Neurobiol 55(2):1463–1476

Gibson EL, Barr S, Jeanes YM (2013) Habitual fat intake predicts memory function in younger women. Front Hum Neurosci 7:838

Gill V, Kumar V, Singh K, Kumar A, Kim JJ (2019) Advanced glycation end products (AGEs) may be a striking link between modern diet and health. Biomol Ther 9(12):888

Gu Y, Brickman AM, Stern Y, Habeck CG, Razlighi QR, Luchsinger JA et al (2015) Mediterranean diet and brain structure in a multiethnic elderly cohort. Neurology 85(20):1744–1751

Gu Y, Gutierrez J, Meier IB, Guzman VA, Manly JJ, Schupf N et al (2018) Circulating inflammatory biomarkers are related to cerebrovascular disease in older adults. Neurol Neuroimmunol Neuroinflamm 6(1):e521

Harach T, Marungruang N, Duthilleul N, Cheatham V, Mc Coy KD, Frisoni G, Neher JJ, Fåk F, Jucker M, Lasser T, Bolmont T (2017) Reduction of Abeta amyloid pathology in APPPS1 transgenic mice in the absence of gut microbiota. Sci Rep 7:41802

Hardy J, Allsop D (1991) Amyloid deposition as the central event in the aetiology of Alzheimer's disease. Trends Pharmacol Sci 12(10):383–388

Hardy JA, Higgins GA (1992) Alzheimer's disease: the amyloid cascade hypothesis. Science 256(5054):184–185

Haring B, Wu C, Mossavar-Rahmani Y, Snetselaar L, Brunner R, Wallace RB et al (2016) No association between dietary patterns and risk for cognitive decline in older women with 9-year follow-up: data from the Women's Health Initiative Memory Study. J Acad Nutr Diet 116(6): 921–930.e1

He Y, Li B, Sun D, Chen S (2020) Gut microbiota: implications in Alzheimer's disease. J Clin Med 9(7):2042

Henderson ST, Vogel JL, Barr LJ, Garvin F, Jones JJ, Costantini LC (2009) Study of the ketogenic agent AC-1202 in mild to moderate Alzheimer's disease: a randomized, double-blind, placebo-controlled, multicenter trial. Nutr Metab (Lond) 6:31

Heneka MT, Carson MJ, El Khoury J, Landreth GE, Brosseron F, Feinstein DL et al (2015) Neuroinflammation in Alzheimer's disease. Lancet Neurol 14(4):388–405

Hensley K (2010) Neuroinflammation in Alzheimer's disease: mechanisms, pathologic consequences, and potential for therapeutic manipulation. J Alzheimers Dis 21(1):1–14

Heppner FL, Ransohoff RM, Becher B (2015) Immune attack: the role of inflammation in Alzheimer disease. Nat Rev Neurosci 16(6):358–372

Heras-Sandoval D, Pedraza-Chaverri J, Pérez-Rojas JM (2016) Role of docosahexaenoic acid in the modulation of glial cells in Alzheimer's disease. J Neuroinflamm 13:61

Herrup K (2012) The contributions of unscheduled neuronal cell cycle events to the death of neurons in Alzheimer's disease. Front Biosci (Elite Ed) 4:2101–2109

Hill E, Goodwill AM, Gorelik A, Szoeke C (2019) Diet and biomarkers of Alzheimer's disease: a systematic review and meta-analysis. Neurobiol Aging 76:45–52

Holmes C, Cunningham C, Zotova E, Woolford J, Dean C, Kerr S et al (2009) Systemic inflammation and disease progression in Alzheimer disease. Neurology 15(4):581–589

Hosking DE, Eramudugolla R, Cherbuin N, Anstey KJ (2019) MIND not Mediterranean diet related to 12-year incidence of cognitive impairment in an Australian longitudinal cohort study. Alzheimers Dement 15(4):581–589

Hsu YH, Chen YW, Wu MH, Tu LH (2019) Protein glycation by glyoxal promotes amyloid formation by islets amyloid polypeptide. Biophys J 116:2304–2313

Huttenlocher PR (1976) Ketonemia and seizures: metabolic and anticonvulsant effects of two ketogenic diets in childhood epilepsy. Pediatr Res 10(5):536–540

Jacka FN, Cherbuin N, Anstey KJ, Sachdev P, Butterworth P (2015) Western diet is associated with a smaller hippocampus: a longitudinal investigation. BMC Med 13:215

Jaeger LB, Dohgu S, Sultana R, Lynch JL, Owen JB, Erickson MA et al (2009) Lipopolysaccharide alters the blood–brain barrier transport of amyloid beta protein: a mechanism for inflammation in the progression of Alzheimer's disease. Brain Behav Immun 23(4):507–517

Jena PK, Sheng L, Di Lucente J, Jin LW, Maezawa I, Wan YY (2018) Dysregulated bile acid synthesis and dysbiosis are implicated in Western diet-induced systemic inflammation, microglial activation, and reduced neuroplasticity. FASEB J 32(5):2866–2877

Jiang C, Li G, Huang P, Liu Z, Zhao B (2017) The gut microbiota and Alzheimer's disease. J Alzheimers Dis 58(1):1–15

Kang JH, Ascherio A, Grodstein F (2005) Fruit and vegetable consumption and cognitive decline in aging women. Ann Neurol 57(5):713–720

Kanoski SE, Davidson TL (2011) Western diet consumption and cognitive impairment: links to hippocampal dysfunction and obesity. Physiol Behav 103(1):59–68

Kashiwaya Y, Bergman C, Lee JH, Wan R, King MT, Mughal MR et al (2013) A ketone ester diet exhibits anxiolytic and cognition-sparing properties, and lessens amyloid and tau pathologies in a mouse model of Alzheimer's disease. Neurobiol Aging 34(6):1530–1539

Katsiardanis K, Diamantaras AA, Dessypris N, Michelakos T, Anastasiou A, Katsiardani KP, Kanavidis P, Papadopoulos FC, Stefanadis C, Panagiotakos DB, Petridou ET (2013) Cognitive impairment and dietary habits among elders: the Velestino study. J Med Food 16(4):343–350

Ke X, Fu Q, Sterrett J, Hillard CJ, Lane RH, Majnik A (2020) Adverse maternal environment and western diet impairs cognitive function and alters hippocampal glucocorticoid receptor promoter methylation in male mice. Physiol Rep 8(8):e14407

Kellar D, Craft S (2020) Brain insulin resistance in Alzheimer's disease and related disorders: mechanisms and therapeutic approaches. Lancet Neurol 19(9):758–766

Kelly E, Vyas P, Weber JT (2017) Biochemical properties and neuroprotective effects of compounds in various species of berries. Molecules 23(1):26

Kent R, Uribarri J (2014) Processing contaminants: advanced glycation end products (AGEs). In: Encyclopedia of food and safety. Elsevier, Amsterdam, pp 371–375

Kesse-Guyot E, Andreeva VA, Lassale C, Ferry M, Jeandel C, Hercberg S et al (2013) Mediterranean diet and cognitive function: a French study. Am J Clin Nutr 97(2):369–376

Khan MS, Ikram M, Park JS, Park TJ, Kim MO (2020) Gut microbiota, its role in induction of Alzheimer's disease pathology, and possible therapeutic interventions: special focus on anthocyanins. Cell 9(4):853

Kim DY, Hao J, Liu R, Turner G, Shi FD, Rho JM (2012) Inflammation-mediated memory dysfunction and effects of a ketogenic diet in a murine model of multiple sclerosis. PLoS One 7(5):e35476

Knight A, Bryan J, Wilson C, Hodgson JM, Davis CR, Murphy KJ (2016) The Mediterranean diet and cognitive function among healthy older adults in a 6-month randomized controlled trial: the MedLey study. Nutrients 8(9):579

Koch M, Jensen MK (2016) Association of the MIND diet with cognition and risk of Alzheimer's disease. Curr Opin Lipidol 27(3):303–304

Komanduri M, Gondalia S, Scholey A, Stough C (2019) The microbiome and cognitive aging: a review of mechanisms. Psychopharmacology 236(5):1559–1571

Kowalski K, Mulak A (2019) Brain–gut–microbiota axis in Alzheimer's disease. J Neurogastroenterol Motil 25(1):48–60

Krikorian R, Shidler MD, Dangelo K, Couch SC, Benoit SC, Clegg DJ (2012) Dietary ketosis enhances memory in mild cognitive impairment. Neurobiol Aging 33(2):425.e19–425.e27

Lau FC, Shukitt-Hale B, Joseph JA (2007) Nutritional intervention in brain aging: reducing the effects of inflammation and oxidative stress. Subcell Biochem 42:299–318

Leblhuber F, Ehrlich D, Steiner K, Geisler S, Fuchs D, Lanser L, Kurz K et al (2021) The immunopathogenesis of Alzheimer's disease is related to the composition of gut microbiota. Nutrients 13(2):361

Lee YJ, Han SB, Nam SY, Oh KW, Hong JT (2010) Inflammation and Alzheimer's disease. Arch Pharm Res 33(10):1539–1556

Liu P, Wu L, Peng G, Han Y, Tang R, Ge J et al (2019) Altered microbiomes distinguish Alzheimer's disease from amnestic mild cognitive impairment and health in a Chinese cohort. Brain Behav Immun 80:633–643

Liu X, Morris MC, Dhana K, Ventrelle J, Johnson K, Bishop L et al (2021) Mediterranean-DASH intervention for neurodegenerative delay (MIND) study: rationale, design and baseline characteristics of a randomized control trial of the MIND diet on cognitive decline. Contemp Clin Trials 102:106270

Livingston G, Huntley J, Sommerlad A, Ames D, Ballard C, Banerjee S et al (2020) Dementia prevention, intervention, and care: 2020 report of the Lancet Commission. Lancet 396(10248): 413–446

Liyanage SI, Vilekar P, Weaver DF (2019) Nutrients in Alzheimer's disease: the interaction of diet, drugs and disease. Can J Neurol Sci 46(1):23–34

Lourida I, Soni M, Thompson-Coon J, Purandare N, Lang IA, Ukoumunne OC, Llewellyn DJ (2013) Mediterranean diet, cognitive function, and dementia: a systematic review. Epidemiology 24(4):479–489
Lustig RH (2020) Ultraprocessed food: addictive, toxic, and ready for regulation. Nutrients 12(11): 3401
Ma D, Wang AC, Parikh I, Green SJ, Hoffman JD, Chlipala G et al (2018) Ketogenic diet enhances neurovascular function with altered gut microbiome in young healthy mice. Sci Rep 8(1):6670
Marizzoni M, Cattaneo A, Mirabelli P, Festari C, Lopizzo N, Nicolosi V et al (2020) Short-chain fatty acids and lipopolysaccharide as mediators between gut dysbiosis and amyloid pathology in Alzheimer's disease. J Alzheimers Dis 78(2):683–697
Markesbery WR (1997) Oxidative stress hypothesis in Alzheimer's disease. Free Radic Biol Med 23(1):134–147
Marseglia A, Xu W, Fratiglioni L, Fabbri C, Berendsen AAM, Bialecka-Debek A et al (2018) Effect of the NU-AGE diet on cognitive functioning in older adults: a randomized controlled trial. Front Physiol 9:349
Martínez Leo EE, Segura Campos MR (2020) Effect of ultra-processed diet on gut microbiota and thus its role in neurodegenerative diseases. Nutrition 71:110609
McEvoy CT, Guyer H, Langa KM, Yaffe K (2017) Neuroprotective diets are associated with better cognitive function: the health and retirement study. J Am Geriatr Soc 65(8):1857–1862
McGrattan AM, McGuinness B, McKinley MC, Kee F, Passmore P, Woodside J et al (2019) Diet and inflammation in cognitive ageing and Alzheimer's disease. Curr Nutr Rep 8(2):53–65
Meraz-Ríos MA, Lira-De León KI, Campos-Peña V, De Anda-Hernández MA, Mena-López R (2010) Tau oligomers and aggregation in Alzheimer's disease. J Neurochem 112(6):1353–1367
Merhi Z, Kandaraki EA, Diamanti-Kandarakis E (2019) Implications and future perspectives of AGEs in PCOS pathophysiology. Trends Endocrinol Metab 30:150–162
Milder JB, Liang LP, Patel M (2010) Acute oxidative stress and systemic Nrf2 activation by the ketogenic diet. Neurobiol Dis 40(1):238–244
Minihane AM, Vinoy S, Russell WR, Baka A, Roche HM, Tuohy KM, Teeling JL, Blaak EE, Fenech M, Vauzour D, McArdle HJ, Kremer BHA, Sterkman L, Vafeiadou K, Benedetti MM, Williams CM, Calder PC et al (2015) Low-grade inflammation, diet composition and health: current research evidence and its translation. Br J Nutr 114(7):999–1012
Miquel S, Champ C, Day J, Aarts E, Bahr BA, Bakker M (2018) Poor cognitive ageing: vulnerabilities, mechanisms and the impact of nutritional interventions. Ageing Res Rev 42:40–55
Mohammadzadeh Honarvar N, Saedisomeolia A, Abdolahi M, Shayeganrad A, Taheri Sangsari G, Hassanzadeh Rad B, Muench G et al (2017) Molecular anti-inflammatory mechanisms of retinoids and carotenoids in Alzheimer's disease: a review of current evidence. J Mol Neurosci 61(3):289–304
Monacelli F, Acquarone E, Giannotti C, Borghi R, Nencioni A (2017) Vitamin C, aging and Alzheimer's disease. Nutrients 9(7):670
Morris MC, Evans DA, Tangney CC, Bienias JL, Wilson RS (2006) Associations of vegetable and fruit consumption with age-related cognitive change. Neurology 67(8):1370–1376
Morris MC, Tangney CC, Wang Y, Sacks FM, Barnes LL, Bennett DA et al (2015a) MIND diet slows cognitive decline with aging. Alzheimers Dement 11(9):1015–1022
Morris MC, Tangney CC, Wang Y, Sacks FM, Bennett DA, Aggarwal NT (2015b) MIND diet associated with reduced incidence of Alzheimer's disease. Alzheimers Dement 11(9): 1007–1014
Mosconi L, Mistur R, Switalski R, Tsui WH, Glodzik L, Li Y et al (2009) FDG-PET changes in brain glucose metabolism from normal cognition to pathologically verified Alzheimer's disease. Eur J Nucl Med Mol Imaging 36(5):811–822
Mosconi L, Murray J, Tsui WH, Li Y, Davies M, Williams S et al (2014) Mediterranean diet and magnetic resonance imaging-assessed brain atrophy in cognitively normal individuals at risk for Alzheimer's disease. J Prev Alzheimers Dis 1(1):23–32

Noble EE, Hsu TM, Kanoski SE (2017) Gut to brain dysbiosis: mechanisms linking Western diet consumption, the microbiome, and cognitive impairment. Front Behav Neurosci 11:9

O'Brien RJ, Wong PC (2011) Amyloid precursor protein processing and Alzheimer's disease. Annu Rev Neurosci 34:185–204

Ohnuma T, Toda A, Kimoto A, Takebayashi Y, Higashiyama R, Tagata Y et al (2016) Benefits of use, and tolerance of, medium-chain triglyceride medical food in the management of Japanese patients with Alzheimer's disease: a prospective, open-label pilot study. Clin Interv Aging 11: 29–36

Olsson E, Karlström B, Kilander L, Byberg L, Cederholm T, Sjögren P (2015) Dietary patterns and cognitive dysfunction in a 12-year follow-up study of 70 year old men. J Alzheimers Dis 43(1): 109–119

Ota M, Matsuo J, Ishida I, Takano H, Yokoi Y, Hori H et al (2019) Effects of a medium-chain triglyceride-based ketogenic formula on cognitive function in patients with mild-to-moderate Alzheimer's disease. Neurosci Lett 690:232–236

Pandareesh MD, Mythri RB, Srinivas Bharath MM (2015) Bioavailability of dietary polyphenols: factors contributing to their clinical application in CNS diseases. Neurochem Int 89:198–208

Pastori D, Carnevale R, Nocella C, Novo M, Santulli M, Cammisotto V et al (2015) Gut-derived serum lipopolysaccharide is associated with enhanced risk of major adverse cardiovascular events in atrial fibrillation: effect of adherence to Mediterranean diet. J Am Heart Assoc 6(6): e005784

Pawlosky RJ, Kemper MF, Kashiwaya Y, King MT, Mattson MP, Veech RL (2017) Effects of a dietary ketone ester on hippocampal glycolytic and tricarboxylic acid cycle intermediates and amino acids in a 3xTgAD mouse model of Alzheimer's disease. J Neurochem 141(2):195–207

Pelletier A, Barul C, Feart C, Helmer C, Bernard C, Periot O et al (2015) Mediterranean diet and preserved brain structural connectivity in older subjects. Alzheimers Dement 11(9):1023–1031

Peng L, Li Z-R, Green RS, Holzman IR, Lin J (2009) Butyrate enhances the intestinal barrier by facilitating tight junction assembly via activation of AMP-activated protein kinase in Caco-2 cell monolayers. J Nutr 139(9):1619–1625

Pinto A, Bonucci A, Maggi E, Corsi M, Businaro R (2018) Anti-oxidant and anti-inflammatory activity of ketogenic diet: new perspectives for neuroprotection in Alzheimer's disease. Antioxidants (Basel) 7(5):63

Pistollato F, Iglesias RC, Ruiz R, Aparicio S, Crespo J, Lopez LD (2018) Nutritional patterns associated with the maintenance of neurocognitive functions and the risk of dementia and Alzheimer's disease: a focus on human studies. Pharmacol Res 131:32–43

Qin B, Adair LS, Plassman BL, Batis C, Edwards LJ, Popkin BM et al (2015) Dietary patterns and cognitive decline among Chinese older adults. Epidemiology 26(5):758–768

Raetz CRH, Whitfield C (2002) Lipopolysaccharide endotoxins. Annu Rev Biochem 71:635–700

Rainey-Smith SR, Gu Y, Gardener SL, Doecke JD, Villemagne VL, Brown BM et al (2018) Mediterranean diet adherence and rate of cerebral Aβ-amyloid accumulation: data from the Australian imaging, biomarkers and lifestyle study of ageing. Transl Psychiatry 8(1):238

Rebello CJ, Keller JN, Liu AG, Johnson WD, Greenway FL (2015) Pilot feasibility and safety study examining the effect of medium chain triglyceride supplementation in subjects with mild cognitive impairment: a randomized controlled trial. BBA Clin 3:123–125

Reger MA, Henderson ST, Hale C, Cholerton B, Baker LD, Watson GS, Hyde K, Chapman D, Craft S et al (2004) Effects of beta-hydroxybutyrate on cognition in memory-impaired adults. Neurobiol Aging 25(3):311–314

Rendeiro C, Rhodes JS, Spencer JP (2015) The mechanisms of action of flavonoids in the brain: direct versus indirect effects. Neurochem Int 89:126–139

Rhee SY, Kim YS (2018) The role of advanced glycation end products in diabetic vascular complications. Diabetes Metab J 42:188–195

Sah SK, Lee C, Jang JH, Park GH (2017) Effect of high-fat diet on cognitive impairment in triple-transgenic mice model of Alzheimer's disease. Biochem Biophys Res Commun 493(1): 731–736

Samieri C, Grodstein F, Rosner BA, Kang JH, Cook NR, Manson JE et al (2013) Mediterranean diet and cognitive function in older age. Epidemiology 24(4):490–499
Sampson TR, Mazmanian SK (2015) Control of brain development, function, and behavior by the microbiome. Cell Host Microbe 17:565–576
Sanz Y, Santacruz A, Gauffin P (2010) Gut microbiota in obesity and metabolic disorders. Proc Nutr Soc 69(3):434–441
Sastre M, Klockgether T, Heneka MT (2006) Contribution of inflammatory processes to Alzheimer's disease: molecular mechanisms. Int J Dev Neurosci 24(2–3):167–176
Scarmeas N, Stern Y, Mayeux R, Manly JJ, Schupf N, Luchsinger JA (2009) Mediterranean diet and mild cognitive impairment. Arch Neurol 66(2):216–225
Shakersain B, Rizzuto D, Larsson SC, Faxén-Irving G, Fratiglioni L, Xu WL (2018) The Nordic Prudent Diet reduces risk of cognitive decline in the Swedish older adults: a population-based cohort study. Nutrients 10(2):229
Sharon G, Sampson TR, Geschwind DH, Mazmanian SK (2016) The central nervous system and the gut microbiome. Cell 167(4):915–932
Silva MVF, Loures CMG, Alves LCV, de Souza LC, Borges KBG, Carvalho MDG (2019) Alzheimer's disease: risk factors and potentially protective measures. J Biomed Sci 26(1):33
Singh B, Parsaik AK, Mielke MM, Erwin PJ, Knopman DS, Petersen RC et al (2014) Association of Mediterranean diet with mild cognitive impairment and Alzheimer's disease: a systematic review and meta-analysis. J Alzheimers Dis 39(2):271–282
Small SA, Duff K (2008) Linking Abeta and tau in late-onset Alzheimer's disease: a dual pathway hypothesis. Neuron 60(4):534–542
Smith PJ, Blumenthal JA (2010) Diet and neurocognition: review of evidence and methodological considerations. Curr Aging Sci 3(1):57–66
Smith PK (2017) Do advanced glycation end-products cause food allergy? Curr Opin Allergy Clin Immunol 17:325–331
Sochocka M, Donskow-Łysoniewska K, Diniz BS, Kurpas D, Brzozowska E, Leszek J (2019) The gut microbiome alterations and inflammation-driven pathogenesis of Alzheimer's disease—a critical review. Mol Neurobiol 56(3):1841–1851
Solfrizzi V, Custodero C, Lozupone M, Imbimbo BP, Valiani V, Agosti P et al (2017) Relationships of dietary patterns, foods, and micro- and macronutrients with Alzheimer's disease and late-life cognitive disorders: a systematic review. J Alzheimers Dis 59(3):815–849
Soltani S, Chitsazi MJ, Salehi-Abargouei A (2018) The effect of dietary approaches to stop hypertension (DASH) on serum inflammatory markers: a systematic review and meta-analysis of randomized trials. Clin Nutr 37(2):542–550
Spencer SJ, Basri B, Sominsky L, Soch A, Ayala MT, Reineck P et al (2019) High-fat diet worsens the impact of aging on microglial function and morphology in a region-specific manner. Neurobiol Aging 74:121–134
Staubo SC, Aakre JA, Vemuri P, Syrjanen JA, Mielke MM (2017) Mediterranean diet, micronutrients and macronutrients, and MRI measures of cortical thickness. Alzheimers Dement 13(2):168–177
Stevenson RJ, Francis HM (2017) The hippocampus and the regulation of human food intake. Psychol Bull 143(10):1011–1032
Sullivan PG, Rippy NA, Dorenbos K, Concepcion RC, Agarwal AK, Rho JM (2004) The ketogenic diet increases mitochondrial uncoupling protein levels and activity. Ann Neurol 55(4):576–580
Swardfager W, Yu D, Ramirez J, Cogo-Moreira H, Szilagyi G, Holmes MF et al (2017) Peripheral inflammatory markers indicate microstructural damage within periventricular white matter hyperintensities in Alzheimer's disease: a preliminary report. Alzheimers Dement (Amst) 7: 56–60
Szczechowiak K, Diniz BS, Leszek J (2019) Diet and Alzheimer's dementia—nutritional approach to modulate inflammation. Pharmacol Biochem Behav 184:172743
Tagliabue A, Elli M (2013) The role of gut microbiota in human obesity: recent findings and future perspectives. Nutr Metab Cardiovasc Dis 23(3):160–168

Takashima A (2009) Amyloid-beta, tau, and dementia. J Alzheimers Dis 17(4):729–736

Tan MS, Yu JT, Jiang T, Zhu XC, Tan L (2013) The NLRP3 inflammasome in Alzheimer's disease. Mol Neurobiol 48(3):875–882

Tangney CC, Li H, Wang Y, Barnes L, Schneider JA, Bennett DA et al (2014) Relation of DASH- and Mediterranean-like dietary patterns to cognitive decline in older persons. Neurology 83(16): 1410–1416

Tavares L, Figueira I, McDougall GJ, Vieira HL, Stewart D, Alves PM (2013) Neuroprotective effects of digested polyphenols from wild blackberry species. Eur J Nutr 52(1):225–236

Taylor MK, Sullivan DK, Mahnken JD, Burns JM, Swerdlow RH (2017) Feasibility and efficacy data from a ketogenic diet intervention in Alzheimer's disease. Alzheimers Dement (N Y) 4:28–36

Thériault P, ElAli A, Rivest S (2016) High fat diet exacerbates Alzheimer's disease-related pathology in APPswe/PS1 mice. Oncotarget 7(42):67808–67827

Tillisch K (2014) The effects of gut microbiota on CNS function in humans. Gut Microbes 5(3): 404–410

Tuck CJ, Staudacher HM (2019) The keto diet and the gut: cause for concern? Lancet Gastroenterol Hepatol 4(12):908–909

Uribarri J (2018) Dietary AGE and their role in health and disease. CRC Press, Boca Raton

Uribarri J, Woodruff S, Goodman S, Cai W, Chen X, Pyzik R et al (2010) Advanced glycation end products in foods and a practical guide to their reduction in the diet. J Am Diet Assoc 110(6): 911–16.e12

van den Brink AC, Brouwer-Brolsma EM, Berendsen AAM, van de Rest O (2019) The Mediterranean, dietary approaches to stop hypertension (DASH), and Mediterranean-DASH intervention for neurodegenerative delay (MIND) diets are associated with less cognitive decline and a lower risk of Alzheimer's disease—a review. Adv Nutr 10(6):1040–1065

Van der Auwera I, Wera S, Van Leuven F, Henderson ST (2005) A ketogenic diet reduces amyloid beta 40 and 42 in a mouse model of Alzheimer's disease. Nutr Metab (Lond) 2:28

Varatharaj A, Galea I (2017) The blood–brain barrier in systemic inflammation. Brain Behav Immun 60:1–12

Vighi G, Marcucci F, Sensi L, Di Cara G, Frati F (2008) Allergy and the gastrointestinal system. Clin Exp Immunol 153(Suppl 1):3–6

Vitek MP, Bhattacharya K, Glendening JM, Stopa E, Vlassara H, Bucala R et al (1994) Advanced glycation end products contribute to amyloidosis in Alzheimer disease. Proc Natl Acad Sci U S A 91:4766–4770

Vogt NM, Kerby RL, Dill-McFarland KA, Harding SJ, Merluzzi AP, Johnson SC, Carlsson CM, Asthana S, Zetterberg H, Blennow K, Bendlin BB, Rey FE (2017) Gut microbiome alterations in Alzheimer's disease. Sci Rep 7(1):13537

Wang HB, Wang PY, Wang X, Wan YL, Liu YC (2012) Butyrate enhances intestinal epithelial barrier function via up-regulation of tight junction protein Claudin-1 transcription. Dig Dis Sci 57(12):3126–3135

Weickert MO, Pfeiffer AF (2008) Metabolic effects of dietary fiber consumption and prevention of diabetes. J Nutr 138(3):439–442

Weintraub MK, Kranjac D, Eimerbrink MJ, Pearson SJ, Vinson BT, Patel J et al (2014) Peripheral administration of poly I:C leads to increased hippocampal amyloid-beta and cognitive deficits in a non-transgenic mouse. Behav Brain Res 266:183–187

Wengreen H, Munger RG, Cutler A, Quach A, Bowles A, Corcoran C et al (2013) Prospective study of dietary approaches to stop hypertension- and Mediterranean-style dietary patterns and age-related cognitive change: the Cache County study on memory, health and aging. Am J Clin Nutr 98(5):1263–1271

Westfall S, Lomis N, Kahouli I, Dia SY, Singh SP, Prakash S (2017) Microbiome, probiotics and neurodegenerative diseases: deciphering the gut brain axis. Cell Mol Life Sci 74:3769–3787

Więckowska-Gacek A, Mietelska-Porowska A, Wydrych M, Wojda U (2021) Western diet as a trigger of Alzheimer's disease: from metabolic syndrome and systemic inflammation to neuroinflammation and neurodegeneration. Ageing Res Rev 70:101397

World Health Organization (2019) Risk reduction of cognitive decline and dementia: WHO guidelines. World Health Organization, Geneva

World Health Organization (2021) Dementia. World Health Organization, Geneva

Wu L, Sun D (2017) Adherence to Mediterranean diet and risk of developing cognitive disorders: an updated systematic review and meta-analysis of prospective cohort studies. Sci Rep 7:41317

Xu W, Tan L, Wang HF, Jiang T, Tan MS, Tan L et al (2015) Meta-analysis of modifiable risk factors for Alzheimer's disease. J Neurol Neurosurg Psychiatry 86(12):1299–1306

Zhang J, Cao Q, Li S, Lu X, Zhao Y, Guan JS, Chen JC, Wu Q, Chen GQ et al (2013) 3-Hydroxybutyrate methyl ester as a potential drug against Alzheimer's disease via mitochondria protection mechanism. Biomaterials 34(30):7552–7562

Zhang M, Zhao D, Zhou G, Li C (2020) Dietary pattern, gut microbiota, and Alzheimer's disease. J Agric Food Chem 68(46):12800–12809

Zhao Y, Dua P, Lukiw WJ (2015) Microbial sources of amyloid and relevance to amyloidogenesis and Alzheimer's disease (AD). J Alzheimers Dis Parkinsonism 5(1):177

Chapter 14
Probiotic and Dietary Interventions in Alzheimer's Disease

John Jeyaraj Kiruba, Thimma Gopalram Vardhani, Amuthavel Aarthy, Sundaramoorthy Iswaryalakshmi, Ponnaiah Brundha, Muthukannan Aishwarya Lakshmi, Uthandakalaipandian Ramesh, Arokiasamy Justin Thenmozhi, and Mariakuttikan Jayalakshmi

Abstract Alzheimer's disease (AD), a sickness that results in the degradation of brain cells, leads to various complications including dementia, which is characterised by a decline in thinking and independence in daily tasks. Additionally, it comes with a number of risk factors. Numerous researchers have reported links between dietary habits and prevalence of AD, proving that nutrition is a significant modifiable risk factor. Dietary variations can either hasten or delay the neurodegeneration seen in AD. It has also been noted that reduced AD symptoms are associated with better gut microbiota. Use of probiotics is suggested as a prophylactic measure to preserve intestinal microbial balance and improve well-being. Studies done to evaluate probiotic supplementation in AD have shown promising results. Consuming a diet rich in grains, legumes, fruits, and vegetables that is plant based appears to be good for human health because it encourages the growth of diverse and stable microbial communities, consequently helping in the decline of AD progression.

J. J. Kiruba · T. G. Vardhani · A. Aarthy · S. Iswaryalakshmi · P. Brundha · M. A. Lakshmi · M. Jayalakshmi (✉)
Department of Immunology, School of Biological Sciences, Madurai Kamaraj University, Madurai, Tamil Nadu, India
e-mail: jayalakshmi@mkuniversity.ac.in

U. Ramesh
Department of Molecular Biology, School of Biological Sciences, Madurai Kamaraj University, Madurai, Tamil Nadu, India

A. Justin Thenmozhi
Department of Biochemistry, School of Biological Sciences, Madurai Kamaraj University, Madurai, Tamil Nadu, India

Department of Biochemistry and Biotechnology, Faculty of Science, Annamalai University, Chidambaram, Tamil Nadu, India

A. Justin Thenmozhi, T. Manivasagam (eds.), *Nutraceuticals for Alzheimer's Disease: A Promising Therapeutic Approach*, Nutritional Neurosciences,
https://doi.org/10.1007/978-981-99-0677-2_14

Keywords Alzheimer's disease · Gut microbiota · Probiotics · Mediterranean diet · Constipation · Dietary fibre

14.1 Introduction

Alzheimer's disease (AD) is a neurological disorder leading to degradation of brain cells, and this plays a vital role in precipitating dementia. It is characterised by a decline in thinking and increase in dependence towards daily tasks. Several risk factors (Fig. 14.1) including the increasing age, gender, genetic factors, head injuries, infections, cardiovascular diseases, environmental factors, lifestyle, obesity, diabetics, etc. play a major role in the disease (Breijyeh and Karaman 2020). According to studies, there is a link between diet intake and AD. Certain foods especially probiotics promoting a healthy gut microbiome have been thought to alleviate the progression of AD and reduce its complications. This chapter consists of three main components: foods increasing gut microbiota, foods reducing risk factors for AD, and foods reducing constipation, which is one of the complications of AD.

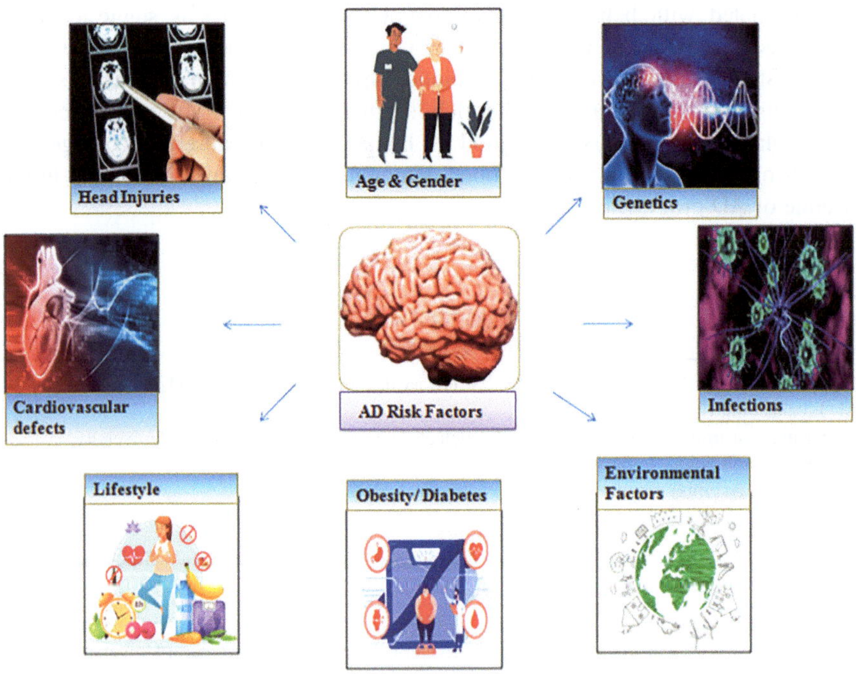

Fig. 14.1 Major risk factors for Alzheimer's disease

14.2 Foods Increasing Gut Microbiota

14.2.1 Gut Microbiota

The entire population of microorganisms that colonises gut is referred to as gut microbiota (Jandhyala et al. 2015). Human gastrointestinal tract contains approximately 100 trillion microorganisms. Microbiome encodes over three million genes that produce thousands of metabolites, replacing many of the functions of the host, which has an impact on the host's fitness, phenotype, and health (Valdes et al. 2018). *Escherichia coli* and *Streptococcus* spp. are two examples of facultative anaerobic strains of bacteria that typically colonise the large gut. Any oxygen present in the gut is broken down by these first colonisers, resulting in a strong anaerobic environment.

Microorganisms start to colonise the gut of humans right from the infant stage. The infant's feeding pattern strongly influences the subsequent colonisation of microorganisms. In addition to being wholesome, complete nourishment for infants, human milk also causes noticeable changes in the probiotic levels of the infant gut. The quantity and frequency at which different species colonise the infant intestine are directly influenced by factors such as the microbiota of the female genital tract, hygienic circumstances, obstetric procedures, vaginal or caesarean mode of delivery, and type of feeding (Wallace et al. 2011).

14.2.2 Gut Microbiota and Health

In a healthy person, the gut microbiota maintains a symbiotic interaction with the gut mucosa and contribute significantly to their metabolism, immunity, and gut defence. Figure 14.2 depicts the role of gut microbiota in health. The gut mucosal immune system must be tolerant of the helpful commensals while preventing the proliferation of the resident pathogens since a healthy gut microbiota is necessary for normal homeostasis (Jandhyala et al. 2015). Dietary fibres and endogenous intestinal mucus are majorly the non-digestible substrates that can be fermented by the microbiota of the gut (Valdes et al. 2018). Thus, gut microbiota plays a major role in our healthy being.

14.2.3 Altered Gut Microbiota and Diseases

The microbiota has a direct impact on many facets of typical host physiology, including nutritional status, behaviour, and stress response. They may also play a major role in the development of numerous diseases that affect both nearby and distant organ systems. Imbalance of gut microorganisms leads to chaos, thereby

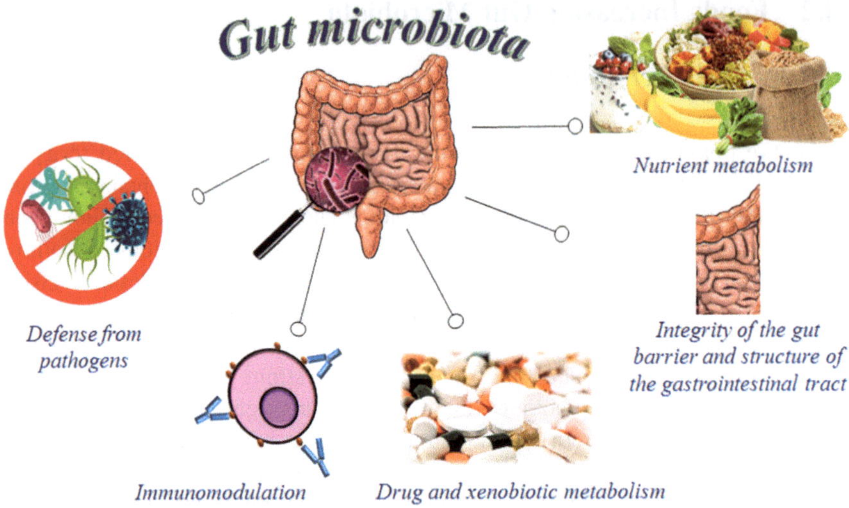

Fig. 14.2 Role of gut microbiota in health

leading to the overpopulation of pathogenic intestinal tract intruders (Sekirov et al. 2010).

Numerous human disorders, including metabolic diseases like obesity and diabetes, luminal diseases like inflammatory bowel diseases and irritable bowel syndrome, allergy diseases, and neurodegenerative illnesses have all been linked to dysregulated gut microbiota (Castelli et al. 2021). According to a clinical investigation, the gut microbiota of AD patients exhibits altered bacterial abundance, including a decline in *Firmicutes* and *Bifidobacterium*, an increase in *Bacteroidetes*, as well as reduced microbial diversity (Naomi et al. 2021).

The knowledge horizon of gut microbiota has been expanded significantly after the advent of metagenomics. The gut microbiota is influenced by a variety of circumstances. It has been hypothesised that environmental factors can modify the human gut microbiota. Each human individual has a distinct gut microbiome because the gut microbial population is so complicated. Thus, the gut microbiota forms a dynamic ecosystem that is continually threatened by a variety of factors, including noise, altitude, temperature, toxicants, pollution, and physical activity (Gubert et al. 2020).

14.2.4 Diet and Gut Microbiota

Food is a significant source for the synthesis of metabolites. In reality, diet modifies the gut microbiota because the nutrients from dietary intake travel to the gut microbiota, which affects the ecosystem and microbial metabolic profile (Vernocchi et al. 2020). Gut microflora uses dietary carbohydrates as a major source of nutrients.

Colonic organisms like *Bacteroides*, *Roseburia*, *Bifidobacterium*, *Faecalibacterium*, and *Enterobacteria* produce short-chain fatty acids (SCFA) like butyrate, propionate, and acetate. They are the rich sources of energy for the host, through the fermentation of carbohydrates that escaped proximal digestion and indigestible oligosaccharides (Jandhyala et al. 2015).

A diet high in grains, legumes, fruits, and vegetables that is plant based appears to be good for human health because it encourages the growth of diverse and stable microbial communities. In order to reduce *Clostridia*, fructose and lactose (Seo et al. 2020) have been emphasised. Non-digestible carbohydrates cause a decrease in *Clostridium* and *Enterococcus* species while increasing *Ruminococcus*, *Eubacterium rectale*, lactic acid bacteria (LAB), and *Roseburia* (Vernocchi et al. 2020). Gut microbiota's composition is strongly influenced by macronutrients, fibre, polyphenols, and prebiotics. The gut microbiota is regarded as a crucial regulator of centrally mediated events, such as immune response, neurological disorders, and metabolic balance (Wong et al. 2018).

Modifying one's diet has been advocated as a treatment to alleviate several disease symptoms. An underlying idea of such diet-based interventions is that nutrition and metabolic substrates might improve dysfunctional metabolic balance while also having positive impacts on neuro-inflammation and neuronal function (Gubert et al. 2020).

Adopting modern eating practices has raised more and more health concerns in recent years due to their high correlation with obesity and related metabolic illnesses, promotion of inflammation, and structural and behavioural changes in the gut flora (Rinninella et al. 2019).

14.2.5 Role of Probiotics in Gut Microbiota

Probiotics are live non-pathogenic microorganisms that promote health benefits when consumed in adequate quantity. Consuming probiotics can be advised as a prophylactic measure to preserve the balance of gut microorganisms and so improve "well-being" (Gibson 2004). They consist of *Saccharomyces* sp. yeast (ingredient in fermented foods and beverages) or lactic acid bacteria (ingredient in dairy products), such as *Lactobacillus* and *Bifidobacterium* species, and are regulated as dietary supplements and foods (Williams 2010). Probiotics exert their beneficial effects via improvement of intestinal health through microbiota regulation, immune system stimulation and development, protection for intestinal lining integrity, function as antibiotics, and enhancement of brain-derived neurotrophic factor (BDNF) (Naomi et al. 2021) increasing and synthesising nutrient bioavailability, minimising lactose intolerance symptoms, and lowering the risk of some additional disorders (Nagpal et al. 2012) by a number of methods, such as lowering intestinal pH, reducing harmful microbial colonisation and invasion, and altering the host immunological response (Williams 2010). Examples of probiotics include *Lactococcus*,

Streptococcus, *Enterococcus*, *Bacillus clausii*, and *Enterococcus faecium* SF68 (Naomi et al. 2021).

14.2.6 Probiotic Mechanisms of Action in Preventing AD

It has been suggested that probiotics can help with age-associated changes of immune features. It has been shown that probiotic therapies can improve immune responses by altering cytokine production and boosting the distribution and performance of T cells, macrophages, and natural killer cells, as well as mucosal and systemic antibody responses (Wong et al. 2018).

Through neurological communication (vagal nerve), endocrine signalling [hypothalamus–pituitary–adrenal (HPA) axis], and immune system (cytokines), the gut can interact with the brain, regulating brain function, behaviour, and most interestingly cognition. Some factors, particularly specialised diets like the ketogenic and Mediterranean diets and omega-3 supplements, have been shown to positively affect these systems and their bidirectional communication (Gubert et al. 2020).

Recent clinical trials and numerous in vivo studies have demonstrated the usefulness of some bacterial strains in slowing AD progression. Through the pathways of neuro-inflammation, Aβ abnormality, tau phosphorylation, neurotransmitter dysregulation, and oxidative stress, the gut microbiota interacts with the development of AD. Following a change in the microbiota's composition, these pathways are dysregulated and linked to an increase in blood–brain barrier (BBB) permeability, which encourages neuro-inflammation, neuronal cell death, and ultimately AD (Naomi et al. 2021).

Stress and worry have a strong correlation with the risk of dementia. Stress from the outside or the environment can cause psychological distress, which can be made worse by oxidative damage and inflammation. Neuro-inflammation plays a significant part in the pathogenetic process of AD, according to an increasing number of studies. The aberrant accumulation of inflammatogenic molecules that causes inflammation in the brain is brought on by the activation of innate and acquired immunity. The relationship between the gut microbiota and the expression of central immune cells has been documented in a number of experiments. Antibiotic-treated germ-free animals show impaired microglial development and immunological responsiveness to bacterial stimulation. Aβ oligomer can go from the intestine to the brain as a result of changes in the gut microbiota activating proinflammatory cytokines and increasing intestinal permeability. Probiotics have been shown to alter the gut microbiota, reduce oxidative stress, and regulate the inflammatory process (Naomi et al. 2021).

The BDNF boosted by probiotics is composed of a particular kind of protein that helps neurons survive and differentiate. As a result, it is essential for neurological development. If these components of the brain are absent, problems such as learning deficits and memory impairments frequently result. Probiotics have positive effects on the central nervous system (CNS) by changing the composition of gut microbiota, by increasing the diversity of the good bacterial composition, and by consequently improving CNS functions. Apart from brain neurotrophic factor, probiotics have a favourable prognosis for treating memory deficiencies and mental illnesses by directly altering brain biochemical elements including serotonin, γ-aminobutyric acid (GABA), and dopamine. As a result, probiotics have been shown to improve gut microbial balance, which slows the course of AD (Naomi et al. 2021).

Based on the results of both the animal and clinical experiments, ingestion of probiotics has a beneficial effect on AD. It is evidently adequate to consume probiotics for 4 weeks in humans and 2 weeks in animals to elicit noticeable effects. The most often utilised single- or multi-strain AD preparations for animal models include *Bifidobacteria infantis*, *Bifidobacteria longum*, *Lactobacillus acidophilus*, *Lactobacillus plantarum*, and *Lactobacillus casei*. These probiotics are all known to contain "good" bacteria that can boost immunity and stop the growth of dangerous bacteria (Naomi et al. 2021).

14.2.7 What Makes Probiotics Different from Other Microbes?

The two genera of probiotic bacteria that are most frequently utilised are *Lactobacillus* and *Bifidobacterium.* Since they do not contain lipopolysaccharides (LPS), eating them does not result in any kind of inflammation. The unique probiotic *Escherichia coli* Nissle 1917 strain produces semi-rough LPS without P and S fimbrial adhesions. It is unique because it lacks these characteristics, which render it non-pathogenic and a perfect microbe for probiotics (Naomi et al. 2021).

14.2.8 Prebiotics in Enhancing Probiotics

Prebiotics are non-viable dietary components that the good bacteria in intestine preferentially break down. To increase the amount and/or activity of probiotics, notably *Lactobacillus* and *Bifidobacterium* species, dietary manipulation of the gut flora by prebiotics is intended to promote health. Perhaps, prebiotics can be ingested more organically through diet. Prebiotic oligosaccharides like fructooligosaccharides are present in a variety of fruits and vegetables like onion, garlic, banana, asparagus, leek, Jerusalem artichoke, and chicory (Manning and Gibson 2004).

At least three requirements must be met for a dietary substrate to be classified as a prebiotic: The substrate must meet the following requirements: (1) it cannot be hydrolysed or absorbed in the stomach or small intestine; (2) it must be selective for good commensal bacteria in the colon, such as *Bifidobacteria*; and

(3) fermentation of the substrate must result in favourable luminal/systemic effects in the host (Manning and Gibson 2004).

14.2.9 Experimental Validation of Probiotics

Uncertainty surrounds the exact pathophysiology of AD. However, growing evidence suggests that the gut microbiota plays a role in the neuropathology of AD (Naomi et al. 2021). Sixty AD patients participated in a randomised, double-blind, and controlled clinical trial to examine the effects of probiotic supplementation on cognitive function and metabolic condition. The study showed that probiotic supplementation for 12 weeks improved various metabolic conditions and cognitive function in AD patients (Akbari et al. 2016).

It is widely established that giving multispecies probiotics to Alzheimer's patients can modify both the microbial dysbiosis and the serum tryptophan metabolism (Leblhuber et al. 2018). Ji and Shen (2021) have discussed 35 papers in total, including 9 human trials with AD and probiotics and 26 animal model trials. In the 26 animal model studies, mice were employed in 24 of them, whereas AD models from *Caenorhabditis elegans* and *Drosophila melanogaster* were used in 2 of them, respectively. Probiotic therapy has neuroprotective effects, might lessen cognitive impairment, and could control gut microbiota dysbiosis, which may be connected to oxidative and inflammatory pathways. Probiotics therefore appear to be a promising strategy to combat AD.

14.3 Foods Reducing Risk Factors for AD

AD is the leading cause of dementia, and the increasing global prevalence of AD is a major public health concern (Hu et al. 2013). In addition to genetic predisposition to AD, the impact of other lifestyle-related risk factors on the development of AD has also been highlighted.

According to epidemiological research, certain food components are crucial for preventing AD. It is well recognised that people with neurodegenerative illnesses, especially AD, benefit by eating enough food and maintaining a healthy, balanced diet. Therefore, one of the main goals is adequate nutrition, and a properly balanced diet can prevent AD and support pharmacological treatment in the elderly. Mainly quitting smoking; maintaining healthy blood pressure, cholesterol, glucose, and homocysteine levels; preventing overweight and obesity; reducing stress; engaging in mental gymnastics; and engaging in physical exercise are changes in daily routines that can protect against dementia, including AD (Fig. 14.3).

Another component of prevention against neurodegenerative diseases is the introduction of specific nutrients into the diet or diet according to prescribed nutritional models. It can help prevent neurodegenerative diseases and support

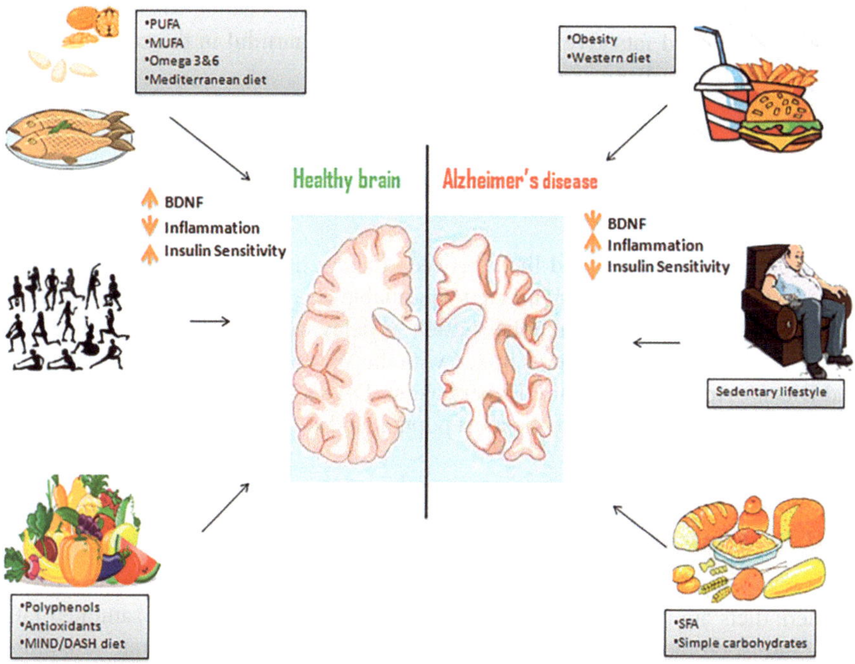

Fig. 14.3 Influences of nutrition and exercise on AD risk and brain health. Increased BDNF, decreased inflammation, and improved insulin sensitivity are all results of exercise and Mediterranean diet components on the brain. Sedentary lifestyles and elements of the Western diet raise the risk of AD through decreases in BDNF, elevated inflammation, and diminished insulin sensitivity. *DASH* Dietary Approaches to Stop Hypertension, *MUFA* monounsaturated fatty acids, *PUFA* polyunsaturated fatty acids, *SFA* saturated fatty acids

basic management of AD patients. Incorporating omega-3 fatty acids, antioxidant vitamins, B vitamins, folic acid, plant polyphenols, fish, and vegetables, together with a moderate intake of fried meat, is thought to lower the risk of AD (Kępka et al. 2022).

Few studies have addressed the potential role of diet in the progression of AD. There is evidence that certain dietary habits, such as the Mediterranean diet and vitamin supplementation, prevent the development of neurodegenerative diseases. More precisely, amyloidogenesis, oxidative stress, and inflammation are some of the fundamental characteristics of AD development that have been linked to nutrition. It has been extensively researched how certain dietary elements may contribute to AD development or control. For example, elevated cholesterol levels lead to increased activity of the APP-cleaving enzymes g-secretase and BACE1, leading to increased production of amyloid-beta, and a conformational change from a helical-rich Aβ structure to an aggregation-prone β-pleated sheet is promoted. Statin therapy-assisted therapeutic lowering of cholesterol has also been linked to a reduction in amyloid beta build-up and the onset of AD. The complexity of AD pathogenesis is demonstrated by dementias, which often show a progressive drop in serum cholesterol. Studies on fatty acids have produced conflicting results, but in general, omega-3 fatty acids have shown some potential benefits, whereas omega-6, trans, and saturated fats give little benefit or may be harmful in the context of AD (Thelen and Brown-Borg 2020).

14.4 Diet as a Modifiable Risk Factor

Numerous studies have found links between dietary habits and prevalence of AD, proving that nutrition is a significant modifiable risk factor. According to these studies, dietary variations can either hasten or delay the neurodegeneration seen in AD. The primary dietary patterns linked with the Western, Mediterranean, Japanese, and Dietary Approaches to Stop Hypertension (DASH)-style diets will be examined in this review because they are the most popular and widely studied diets (Fig. 14.4).

14.4.1 Western Diet

Western diets are typically high in calories, carbohydrates, trans and saturated fats, salt, and food additives. Following a Western diet can accelerate cognitive

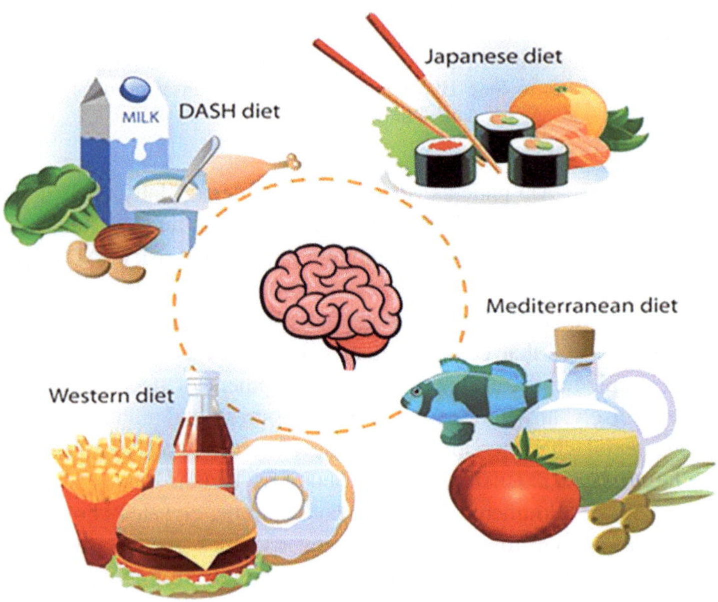

Fig. 14.4 Eating habits that affect the risk of AD. The risk of AD may be lowered by a healthy diet, the DASH diet, the Mediterranean diet, and the Japanese diet. The risk of AD may be increased by a Western diet

impairment and raise the risk of developing AD. Fatty acids can be divided into saturated fatty acids (SFAs) and unsaturated fatty acids and further subdivided into mono- (MUFA) and polyunsaturated (PUFA) fatty acids. High SFA intake has been linked to an increased risk of AD as well as metabolic diseases like type 2 diabetes mellitus (T2DM). Some human research shows a correlation between SFA consumption and cognitive deterioration; nevertheless, animal models must be used to evaluate whether SFA can actually change AD-related indicators. There is proof that feeding C57BL6/J mice high SFA levels causes a rise in BACE1 content and activity, APP content, and Aβ peptides. There is little epidemiological study on the relationship between long-term use of simple carbohydrates and cognitive function in human population. However, evidence suggests that simple sugars have a negative impact on cognition and increase the risk of AD.

Numerous physiological changes brought on by consuming a Western diet may elevate protein markers linked to AD pathogenesis either directly or indirectly. Neuro-inflammation, insulin resistance, and a decrease in neurotrophin content are just a few of the changes that take place in the brain with continued adherence to a Western diet and speed up the onset of AD.

14.4.1.1 Neuro-Inflammation

Inflammation is strongly associated with cognitive impairment and pathogenesis of AD. This is supported by the large number of activated microglial cells and proinflammatory cytokines in post-mortem brain tissue of AD patients. Interleukin (IL-1, IL-6) and tumour necrosis factor (TNF-α) are examples of inflammatory cytokines that are known to impair hippocampal synaptic plasticity and promote amyloidosis, neuronal death, cortical thinning, and brain volume loss. The neuro-inflammation seen in AD is likely influenced by dietary variables, which can affect inflammatory responses.

The Western diet most likely causes neuro-inflammation via two mechanisms: (1) direct brain effects and (2) inflammatory effects on peripheral tissues. The impact of simple carbohydrate and SFA diet on neuro-inflammation is clearly demonstrated in animal models. In rats, simple carbs cause inflammation in the hippocampal region, whereas C57BL6 mice on a high-SFA diet ad libitum exhibit signs of neuro-inflammation, including increased levels of TNF-α, IL-6, and chemokine monocyte chemoattractant protein-1 in the cortical brain region. Furthermore, long-term Western food consumption causes gliosis and worsens plaque load in the mouse hippocampus. c-Jun N-terminal kinase (JNK) activation has been found to be higher in post-mortem brains from AD patients and mice fed high-fat, high-sucrose diets, which is a sign of increased inflammation. These findings point to a possible mechanism whereby a Western diet can cause neuro-inflammation without any other factors being involved. Dysbiosis compromises the integrity of the gut barrier and encourages the release of toxic microbial products into the bloodstream, which causes systemic inflammation. A Western diet can also result in obesity and persistent whole-body inflammation due to the growing adipose tissue's release of proinflammatory cytokines. These peripherally circulating inflammatory cytokines have the ability to cross the blood–brain barrier, get inside the brain, and activate microglia.

14.4.1.2 Insulin Resistance

Through a number of downstream effectors, intact insulin signalling suppresses the development of Aβ plaques and tau hyperphosphorylation. Thus, the development of Aβ plaques and neurofibrillary tangles can be attributed to an insulin-resistant Western diet. Studies on human AD patients, post-mortem AD brains, and animal models of AD all support the disturbance of insulin signalling in the brain and its link to AD pathogenesis.

Numerous clinical and epidemiological research since the first Rotterdam study that suggested an elevated risk of AD in people with T2DM have further established the link between insulin resistance and AD. Remarkably, it has been noted that diminishing insulin sensitivity is linked to poor cognitive function and that >80% of AD patients have T2DM or abnormal blood glucose levels. Patients with AD exhibit higher fasting plasma insulin levels, lower cerebrospinal fluid insulin levels, and higher Aβ concentrations. Reduced insulin and insulin receptor content have been seen in post-mortem AD brains. Animal studies offer more conclusive proof of how a Western diet affects brain insulin signalling and AD markers. In the prefrontal brain of miniature swine fed a high-fat, high-fructose corn syrup, and high-cholesterol diet for 10 weeks, there was less insulin-stimulated Akt signalling. Furthermore, when given an SFA and sucrose diet or sweetened water (60% fructose or sucrose), hamsters and wild-type mice exhibit decreased hippocampus and cortical insulin signalling. In addition to worsened memory impairments and amyloidosis in animal models of AD, sweetened water reduces insulin signalling. Tauopathy can potentially be caused by disruptions in the insulin signalling pathway. For instance, neurofibrillary tangle accumulation was found in the hippocampus of a mouse model of T2DM, and tau phosphorylation was elevated in the brains of insulin receptor knockout mice that only affected neurons.

14.4.1.3 Brain-Derived Neurotrophic Factor (BDNF)

A neurotrophin, BDNF is important for neuronal survival and function as well as synaptic plasticity. The BDNF content and synaptic plasticity in the parietal cortex and hippocampus are both decreased in AD. Additionally, rapid cognitive impairment in AD patients has been linked to decreased circulating BDNF. Since post-mortem brain biopsies are the only way to directly analyse BDNF in human brain tissue, rodent models have aided in the current understanding of BDNF in AD. For instance, diets rich in SFA and simple carbohydrates lower BDNF levels, which lowers synaptic plasticity and neurogenesis in the mouse hippocampus. The amount of BDNF in the brain was also decreased in C57BL/6J mice fed with a 7-week SFA diet. It is significant that these declines in BDNF content are also accompanied by declines in cognition (Baranowski et al. 2020).

14.4.2 *Mediterranean Diets*

The Mediterranean diet, a typical diet found in the Mediterranean area, is defined by a high intake of fruits, vegetables, cereals, bread, potatoes, chicken, beans, nuts, and fish as well as a moderate intake of alcohol and a low intake of red meat and dairy products. Following a Mediterranean diet may impact not only one's risk of developing AD but also one's mortality from AD. Overall mortality and neurodegenerative illnesses were much lower in people who ate a Mediterranean-style diet. Numerous studies backed up the positive relationship between AD risk and Mediterranean diet adherence, since fruits, vegetables, seafood, and moderate alcohol use lower the risk of AD (Hu et al. 2013).

14.4.3 *Japanese Diet*

Increased consumption of fish and plant foods (soybean products, seaweed, vegetables, and fruits) and a reduction in the consumption of refined carbs and animal fats (meat) define the traditional Japanese diet. In a population-based study, a diet with a high consumption of vegetables, algae, milk, and dairy products and a low intake of rice was linked to a lower risk of AD (Hu et al. 2013).

14.4.4 *DASH-Style Diets*

The DASH diet emphasises consuming a large amount of plant-based foods, nuts, fruits, vegetables, fish, poultry, whole grains, and low-fat dairy products while limiting the intake of red meat, sodium, sweets, and beverages with added sugar. People on the DASH diet showed larger neurocognitive improvements when compared to normal subjects in a randomised clinical trial for high blood pressure. It is scientifically conceivable that DASH could lower the risk of AD because hypertension is linked to an elevated risk for the disease.

14.5 The Effects of Foods and Beverages on the Risk of AD

Fish Consuming fish may lower the incidence of dementia and AD, particularly in APOE epsilon 4 non-carriers, according to epidemiological studies.

Eicosapentaenoic acid and docosahexaenoic acid, which are marine long-chain omega-3 fatty acids, are thought to be connected to the favourable relationship. There is a growing consensus that eating fish in general is healthy. A prospective study with 815 participants, aged 65–94, found that eating fish more frequently than once per week was associated with a 60% lower risk of AD than eating fish infrequently or never.

Fruits and Vegetables Regular fruit and vegetable diet may reduce the risk of dementia and AD. A lower incidence of AD and dementia was linked to eating a diet that contained a medium to high proportion of fruits and vegetables as opposed to none or little. If the association between eating fruits and vegetables and risk of AD is confirmed, one possible explanation for this association is that fruits and vegetables are low in saturated fats and rich sources of antioxidants and bioactive compounds (such as vitamin E, vitamin C, carotenoids, and flavonoids).

Dairy Poor cognitive performance has been linked to a lesser intake of milk or dairy products. By reducing circulatory changes and structural changes to the brain that accompany cognitive decline, dairy products that are high in vitamin D, phosphorus, and magnesium may lower the risk of cognitive impairment. However, older adults' cognitive deterioration may be linked to their diet of whole-fat dairy products. At midlife, moderate consumption of unsaturated fats from milk products and spreads reduced the risk of AD, whereas midlife consumption of saturated fats from milk products and spreads was linked to an elevated risk of AD.

Coffee Drinking coffee may lower the risk of developing AD. Due to a component in coffee that interacts with caffeine to specifically boost plasma cytokines, coffee may be the greatest source of caffeine to prevent AD. According to research drinking coffee reduces the risk of developing AD compared to not drinking it.

Tea According to observational studies, consuming tea was linked to lower chances of cognitive decline and impairment, and the protective effect was not exclusive to any one type of tea. Compared to a placebo, black tea was found to considerably improve auditory and visual attention. Green tea epigallocatechin-3-gallate (EGCG) has been demonstrated to lower amyloid formation and sarkosyl-soluble phosphorylated tau isoforms in AD mice models. Green tea polyphenols may decrease cognitive decline by reducing oxidative stress. The catechins, L-theanine, polyphenols, and other substances found in tea leaves may be responsible for the neuroprotective effects of tea drinking (Hu et al. 2013).

14.6 Implementation of Healthy Food Practice in AD

The healthy lifestyle and nutrition chart are a manner of outlining the ideas that, when put into practice, can lead to correct growth, active minds and bodies, and a long life of good health. Following the adult nutritional guidelines can assist in maintaining good health and lowering the risk of developing dementia, but poor eating habits can have long-term detrimental consequences on one's health (Table 14.1). The volume and frequency of food groups ingested should also be taken into consideration. Physical exercise also has a great effect on health. A healthy diet includes eating a variety of foods from various food groups on a regular basis (Kępka et al. 2022).

14.7 Foods Reducing Constipation

Food is the chief energy reservoir for living beings. Consumed food will undergo metabolism and provides energy; a part of the energy is used and another part is stored in the body. The stored energy should be automatically used by the system during fasting or several other conditions. The undigested food will excrete as waste to maintain the stability of the system. Any miss in these steps, ingestion, digestion, absorption, storage, and excretion, will ultimately result in illness. The changes in diet or routine food intake and inadequate intake of fibre may lead to difficulty in excretion, which ultimately leads to constipation.

14.7.1 Bowel Movement and Constipation

To lead a healthy life, bowel movement is crucial. The bowel is an important part of the digestive or gastrointestinal system and helps the body to absorb nutrients and fluids from food intake. Emptying intestines (bowel) comfortably is an important part of healthy gut and can help maintain the muscles that control the active and strong intestines. The less frequent bowel movement and difficulty in stool passing may lead to constipation. It has been reported by Kim et al. (2019) that constipation has been observed in 4.3–17.2% of Alzheimer's patients.

In order to control the bowel, one has to be aware of the sensation to empty the bowel. This consciousness occurs when the stools move around the rectum, leading it to expand and send messages to the brain through the sensory pathways that the intestine needs to be emptied (Fig. 14.5). At this stage, the finely tuned nerve endings are able to differentiate whether the stool is solid or liquid so that one can react accordingly. The ageing process, lifestyle, and treatment for other medical conditions may predispose some people with dementia unable to sense the signal, thereby leading to constipation. Sometimes, the effect of dementia itself may also lead a person to become constipated (Alzheimer's Scotland 2011).

Table 14.1 Implementation of healthy food practice in AD (Kępka et al. 2022)

S. no.	Food group/ activities	Recommended	Avoid	Comments
1	 Physical exercise	30–45 min minimum daily	Reduce the time spent sitting down	Cycling, swimming, walking, gardening, and walking up the stairs and housework
2	 Fluids	Tea, coffee, freshly squeezed fruit, vegetable juice, and water (still, medium, or strongly mineralised)	Limit the intake of sweetened beverages and flavoured fluids, as well as boiled or sparkling water	Drink 1.5–2 L water daily. After waking up, one should immediately drink a glass of water. Between meals, sip on water (1 glass at least 15 min before meals and 15 min after meals). The primary hydration source should be pure mineral water
3	 Vegetables and fruits	Either raw or barely cooked, preferably	Limit the intake of sugar and other sweets (replace them with fruit and nuts, pumpkin seeds, sunflower)	Divided into 5 parts of no less than 400 g of fruits and vegetables (one portion equals one cup). They ought to make up at least 50% of the daily allotment of food (one portion may be a glass of freshly squeezed juice). Keep in mind the proper ratios: around three-quarters of the plate should be vegetables, and one-quarter should be fruits. Significant variety
4	 Grain products	Items made from whole grains (whole-grain brown rice, whole-wheat noodles, and cereal groats including buckwheat and barley)	Consume minimally processed foods	All cereals consumed should contain at least 50% whole grains. Dietary fibre controls how the digestive system works, makes it easier to

(continued)

Table 14.1 (continued)

S. no.	Food group/ activities	Recommended	Avoid	Comments
				maintain a healthy weight, avoids constipation and development of colon cancer, and lowers blood cholesterol levels
5	 Milk and milk products	Yoghurt, kefir, buttermilk, and partially cottage cheese, together with milk (with up to 2% fat)	Avoid dairy products that are already flavoured and contain additional flavourings (sugar, aromas, and dyes)	Minimum of two glasses per day, at least one of which should contain cheese, such as one cup (200 mL) of kefir or yoghurt, 280–400 g of semi-skimmed cheese, or one slice (30 g) of yellow cheese. Less frequently should the rennet cheeses be consumed (due to their higher fat and higher energy content)
6	 Meat and meat products	Fish (salmon, tuna, herring, mackerel, cod), poultry, lean meat (ham, sirloin, fillet, pork loin)	Limit the weekly intake of meat to 0.5 kg, especially red and processed meats. Avoid consuming meat preparations since they have low meat content and a lot of salt, phosphates, nitrites, water, colours, smells, flavour enhancers, sugar, starch, and other additives	Eggs and legumes (beans, lentils, peas, and soybeans) are meat substitutes that are high in protein; they should be consumed 1–2 times per week. The meat should undergo as little processing as possible, preferably cooking it in foil or an ovenproof dish while stewing it rather than frying it
7	 Vegetable oils	Oils: olive, canola, soybean, sunflower, peanut, and other vegetable oils and margarines without trans-fatty acids	Animal fats	Vegetable oils, nuts, and seeds can be used in place of animal fats. Consume as an ingredient in salads or other dishes in moderation and preferably in raw

(continued)

Table 14.1 (continued)

S. no.	Food group/ activities	Recommended	Avoid	Comments
				form. Use rapeseed or olive oil for quick frying. Saturated and monounsaturated fats from deep-frying (lard, clarified butter, coconut oil)
8	 Herbs	Fried and dried	Prepared spice mixtures	Use spices and herbs on a daily basis, such as cinnamon, turmeric, oregano, thyme, basil, garlic, ginger, and rosemary. The flavours and beneficial chemicals found in herbs and other natural spices, such as antioxidants, are also present
9	 Salt (NaCl)	Salt replacements made from potassium or magnesium	Use only as much salt as is necessary for cooking and preparation	Consumption of salt should not exceed 5 g per day, including bread, sausages, cheese, salty snacks, and salting-out (approximately a flat teaspoon). Use iodised salt and rock salt. Limit the intake of foods high in sodium, such as meats, canned fish, and meat; blue cheese, rennet, silage, smoked goods, marinated vegetables, soups, and powdered sauces; spice blends; broth cubes; and salty snacks (chips, sticks, pretzels, crackers, peanuts)

(continued)

Table 14.1 (continued)

S. no.	Food group/ activities	Recommended	Avoid	Comments
10	 Sugar	Can be replaced by fruit and nuts, brown sugar (unrefined), natural sweeteners, i.e. stevia, xylitol, maple and date syrup, honey	Limit the intake of sweets, artificial sweeteners, and white sugar	Limit to 10% of total energy: 200 kcal, or 10 teaspoons of sugar, is less than 10% of 2000 kcal (50 g). Consume sugar, sweeteners, added sugars, and naturally occurring sugars in honey and fruit juices sparingly

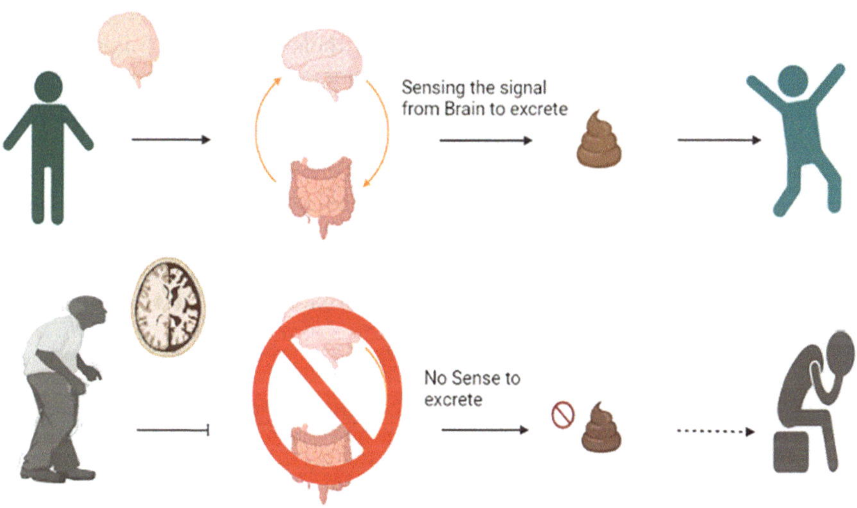

Fig. 14.5 Sensation of bowel movement in normal and AD patients

14.7.2 Association Between Constipation and Neurological Diseases

During 1987, in the USA, the frequency distribution of the International Classification of Diseases (ICD) codes of 8.8 million Medicare patients was compared for those with and without constipation. Most of the strongest associations were found between constipation and neurological diseases, the majority of which were attributable to damage to the CNS (Johanson et al. 1992).

Depression and anxiety were the primary health problems associated with AD, at least 9 years prior to the clinical diagnosis. Constipation and abnormal weight loss were also commonly noticed (7 years before the index date). These findings facilitate differentiation between risk factors associated with early stages of illness and probable co-morbid conditions occurring within a few years of diagnosis (Nedelec et al. 2022). Probiotics have been experimentally used to alleviate the problems of constipation (Martínez-Martínez et al. 2017). Using the experimental mice model Tg2576, Kim et al. (2019) demonstrated that constipation could be considered as a symptom for the AD progressors.

14.7.3 Role of Dietary Fibres in AD Patients Enduring with Constipation

Fibre-rich diets that contain phytosterols have been shown to lower low-density lipoprotein and increase high-density lipoprotein cholesterol, which is implicated in membrane cholesterol and amyloid beta (Aβ) homeostasis. Plasma cholesterol is closely related and regulated by phytosterols. Fibre-rich diets also contain a variety of fatty acids including SCFA. Therefore, understanding the synergistic effects of SCFA and phytosterols in glucose regulation and cholesterol homeostasis is important for the maintenance of a healthy lifestyle. Diet, heredity, and medications may play a vital role in delaying the onset of AD. Nutritional therapy has become a new norm for controlling appetite, as early nutritional therapy can help the genes in delaying liver and brain diseases (Martins and Fernando 2014).

14.8 Conclusion

AD has a complex aetiology, and thus many factors are involved in the development and progression of the disease. However, there is no proper treatment option available for this disease. According to research, alteration in gut microbial diversity and defects in gut brain axis are linked to AD. Probiotics are recognised to be one of the best preventative measures against cognitive decline in AD. They have advantageous effects on the regulation of microbiota in the gut. Improving the quality of life by having a socialisable healthy lifestyle and food habits may have a positive impact on this neurodegenerative disease. Observational studies have also stressed the importance of the Mediterranean diet, dietary antioxidants, homocysteine-related vitamins, higher intake of polyunsaturated fatty acids, fish and its related fatty acids, and lower caloric and fat intake in lowering the risk of AD.

References

Akbari E, Asemi Z, Daneshvar Kakhaki R, Bahmani F, Kouchaki E, Tamtaji OR, Salami M (2016) Effect of probiotic supplementation on cognitive function and metabolic status in Alzheimer's disease: a randomized, double-blind and controlled trial. Front Aging Neurosci 8:256

Alzheimer's Scotland Action on dementia, constipation and faecal impaction–information sheet IS41, 2011

Baranowski BJ, Marko DM, Fenech RK, Yang AJ, MacPherson RE (2020) Healthy brain, healthy life: a review of diet and exercise interventions to promote brain health and reduce Alzheimer's disease risk. Appl Physiol Nutr Metab 45(10):1055–1065

Breijyeh Z, Karaman R (2020) Comprehensive review on Alzheimer's disease: causes and treatment. Molecules 25(24):5789

Castelli V, D'Angelo M, Quintiliani M, Benedetti E, Cifone MG, Cimini A (2021) The emerging role of probiotics in neurodegenerative diseases: new hope for Parkinson's disease? Neural Regen Res 16(4):628

Gibson GR (2004) From probiotics to prebiotics and a healthy digestive system. J Food Sci 69(5): M141–M143

Gubert C, Kong G, Renoir T, Hannan AJ (2020) Exercise, diet and stress as modulators of gut microbiota: implications for neurodegenerative diseases. Neurobiol Dis 134:104621

Hu N, Yu JT, Tan L, Wan YL, Sun L, Tan L (2013) Nutrition and the risk of Alzheimer's disease. Biomed Res Int 2013:524820

Jandhyala SM, Talukdar R, Subramanyam C, Vuyyuru H, Sasikala M, Reddy DN (2015) Role of the normal gut microbiota. World J Gastroenterol 21(29):8787–8803

Ji HF, Shen L (2021) Probiotics as potential therapeutic options for Alzheimer's disease. Appl Microbiol Biotechnol 105(20):7721–7730

Johanson JF, Sonnenberg A, Koch TR, Mccarty DJ (1992) Association of constipation with neurologic diseases. Dig Dis Sci 37(2):179–186

Kępka A, Ochocińska A, Borzym-Kluczyk M, Chojnowska S, Skorupa E, Przychodzeń M, Waszkiewicz N (2022) Healthy food pyramid as well as physical and mental activity in the prevention of Alzheimer's disease. Nutrients 14(8):1534

Kim JE, Park JJ, Lee MR, Choi JY, Song BR, Park JW, Hwang DY (2019) Constipation in Tg2576 mice model for Alzheimer's disease associated with dysregulation of mechanism involving the mAChR signaling pathway and ER stress response. PLoS One 14(4):e0215205

Leblhuber F, Steiner K, Schuetz B, Fuchs D, Gostner JM (2018) Probiotic supplementation in patients with Alzheimer's dementia—an explorative intervention study. Curr Alzheimer Res 15(12):1106–1113

Manning TS, Gibson GR (2004) Prebiotics. Best Pract Res Clin Gastroenterol 18(2):287–298

Martínez-Martínez MI, Calabuig-Tolsá R, Cauli O (2017) The effect of probiotics as a treatment for constipation in elderly people: a systematic review. Arch Gerontol Geriatr 71:142–149

Martins IJ, Fernando WM (2014) High fibre diets and Alzheimer's disease. Food Nutr Sci 5(4): 410–424

Nagpal R, Kumar A, Kumar M, Behare PV, Jain S, Yadav H (2012) Probiotics, their health benefits and applications for developing healthier foods: a review. FEMS Microbiol Lett 334(1):1–15

Naomi R, Embong H, Othman F, Ghazi HF, Maruthey N, Bahari H (2021) Probiotics for Alzheimer's disease: a systematic review. Nutrients 14(1):20

Nedelec T, Couvy-Duchesne B, Monnet F, Daly T, Ansart M, Gantzer L, Durrleman S (2022) Identifying health conditions associated with Alzheimer's disease up to 15 years before diagnosis: an agnostic study of French and British health records. Lancet Digit Health 4(3):e169–e178

Rinninella E, Cintoni M, Raoul P, Lopetuso LR, Scaldaferri F, Pulcini G, Mele MC (2019) Food components and dietary habits: keys for a healthy gut microbiota composition. Nutrients 11(10): 2393

Sekirov I, Russell SL, Antunes LCM, Finlay BB (2010) Gut microbiota in health and disease. Physiol Rev 90(3):859–904

Seo YS, Lee HB, Kim Y, Park HY (2020) Dietary carbohydrate constituents related to gut dysbiosis and health. Microorganisms 8(3):427

Thelen M, Brown-Borg HM (2020) Does diet have a role in the treatment of Alzheimer's disease? Front Aging Neurosci 12:617071

Valdes AM, Walter J, Segal E, Spector TD (2018) Role of the gut microbiota in nutrition and health. BMJ 361:k2179

Vernocchi P, Del Chierico F, Putignani L (2020) Gut microbiota metabolism and interaction with food components. Int J Mol Sci 21(10):3688

Wallace TC, Guarner F, Madsen K, Cabana MD, Gibson G, Hentges E, Sanders ME (2011) Human gut microbiota and its relationship to health and disease. Nutr Rev 69(7):392–403

Williams NT (2010) Probiotics. Am J Health Syst Pharm 67(6):449–458

Wong CB, Kobayashi Y, Xiao JZ (2018) Probiotics for preventing cognitive impairment in Alzheimer's disease. In: Gut microbiota-brain axis. IntechOpen, London, pp 85–104

Chapter 15
Advantages and Disadvantages of Nutraceuticals

Ramamoorthy Rajalakshmi, Miranda A. Melians, Fay F. Pon, Daniela S. Cosio, Venugopal Buvarahamurthy, Arumugam R. Jayakumar, and Michael J. Paidas

Abstract Nutraceuticals are bioactive compounds used to enhance health-supporting effects and to treat and prevent various diseases. They are largely used to control and prevent chronic diseases like diabetes, cardiovascular diseases, respiratory diseases, cancer, and gastrointestinal and neurological disorders. Thus, the nutraceutical market is increasing at a fast rate. Nutraceuticals can be prescribed by a qulified health professional based on a patient's health issues or can be used without a prescription. While nutraceuticals are good for health in general and have various advantages including antioxidant, anti-inflammatory, anticancer, etc., the significant disadvantages are underreported as available literature expresses concerns over the authentic source of their raw materials, purity of the compound, presence of other active compounds, quality, lack of experimental evidence, false advertising, contamination with heavy metals, and interactions between supplements and drugs. One of the major limitations of nutraceuticals is questionable drug bioavailability, which is the accessibility of the nutrients by the body after digestion, absorption, and transportation. Poor bioavailability may result in little or no effect in lessening the advantage of nutraceuticals. People with underlying diseases requiring multidrug regimens often simultaneously self-prescribe nutraceuticals. Meanwhile, they are unaware of drug–drug interactions that not only diminish the efficacy of nutraceuticals but also lead to potentially toxic side effects. Side effects of nutraceuticals due to misuse or overuse may vary from mild to moderate and severe. Therefore, it is important to understand the bioavailability, toxicity, and side effect

R. Rajalakshmi · F. F. Pon · D. S. Cosio · A. R. Jayakumar (✉) · M. J. Paidas (✉)
Departments of Obstetrics, Gynecology and Reproductive Sciences, University of Miami Miller School of Medicine, Miami, FL, USA
e-mail: ajayakumar@med.miami.edu; mxp1440@med.miami.edu

M. A. Melians
School of Nursing and Health Studies, University of Miami, Coral Gables, Miami, FL, USA

V. Buvarahamurthy
Dr. A.L.M. Post-Graduate Institute of Basic Medical Sciences, University of Madras, Chennai, Tamil Nadu, India

A. Justin Thenmozhi, T. Manivasagam (eds.), *Nutraceuticals for Alzheimer's Disease: A Promising Therapeutic Approach*, Nutritional Neurosciences,
https://doi.org/10.1007/978-981-99-0677-2_15

profile of nutraceuticals before use. This review focuses on highlighting the advantages and disadvantages of some commonly used nutraceuticals in various conditions related to heart, liver, kidney, gastrointestinal, and pulmonary disorders. We also explore the potential advantages and disadvantages of nutraceuticals in various neurological conditions, and the mechanisms of their action, particularly their role in the blood–brain barrier in transporting these nutraceuticals into the brain to regulate brain homeostasis and thereby other organs.

Keywords Absorption · Bioavailability · Blood–brain barrier · Dosage · Nutraceuticals · Side effects · Toxicity

Abbreviations

AD	Alzheimer's disease
ALD	Alcoholic liver disease
BBB	Blood–brain barrier
BHA	Butylated hydroxyanisole
BHT	Butylated hydroxytoluene
CKD	Chronic kidney disease
CNS	Central nervous system
COPD	Chronic obstructive pulmonary disease
CVD	Cardiovascular disease
DHA	Docosahexaenoic acid
DILI	Drug-induced liver disease
EPA	Eicosapentaenoic acid
GERD	Gastroesophageal reflux disease
GIT	Gastrointestinal tract
HIE	Hypoxic-ischemic encephalopathy
IBD	Inflammatory bowel disease
IBS	Irritable bowel syndrome
LDL	Low-density lipoprotein
NAFLD	Nonalcoholic fatty liver disease
PD	Parkinson's disease
PG	Propyl gallate
ROS	Reactive oxygen species

15.1 Introduction

Nutraceuticals are a group of food products that provide medical/health benefits. The term "nutraceuticals" is derived from the words "nutrition" and "pharmaceuticals" and was first coined by Dr. Stephen De Felice, Founder and Chairman of the

Foundation for Innovation in Medicine (FIM) in 1989 (Maddi et al. 2007; Zeisel 1999; Brower 1998; Mandala et al. 2010). The active contents of nutraceuticals were not only involved in the improvement of overall health (Mueller 1999; Zeisel 1999) but also used to prevent and treat many chronic diseases like cardiovascular disease, obesity, diabetes, cancer, inflammation, and neurodegenerative disorders (Kalra 2003; Pandey et al. 2010; Nasri et al. 2014). Although nutraceuticals have historically been used in many cultural practices for centuries, recent nutritional and medical evidence has garnered the attention and footing of the nutraceutical sales industry (Dillard and German 2000). Nutraceuticals have quickly spread all over the world as an alternative to pharmaceuticals (Da Costa 2017). These nutraceuticals are sold in the market in the form of pills, powders, and other medicinal forms (Bull 2000). According to the nutrition business journal report, the United States is the largest market in the world followed by Japan (Nwosu and Ubaoji 2020).

The costs and severe side effects associated with modern pharmaceutical drugs have led to the search for an alternative approach to the prevention and treatment of diseases that is more affordable and accessible (Adelaja and Schilling 1999). Nutraceuticals are the foremost alternatives to pharmaceutical drugs and are considered safe without any side effects (Foster 2016). Nutraceuticals for medicinal purposes are available to the public for purchase, as well as through prescription. While nutraceuticals are considered safe, those who have suffered from nutraceutical toxicity are often dealing with the consequences of inadvertent complications due to misuse or overuse. In addition, consumers are unaware of food legislation and regulations surrounding nutraceuticals (Shaw et al. 1997). Importantly, nutraceuticals are not patent protected and are sold without any FDA approval (Chauhan et al. 2013). Thus, it is essential to understand the advantages and disadvantages of some of the most used nutraceuticals before consumption. This chapter also discusses nutraceuticals and their advantages.

Thus, it is essential to understand the advantages and disadvantages of nutraceuticals before consumption. This chapter also discusses the benefits and disadvantages of nutraceuticals in various disease conditions. Additionally, special attention was given to the use of nutraceuticals in brain disorders such as Parkinson's disease, Alzheimer's disease, epilepsy, and hypoxic-ischemic encephalopathy since nutraceuticals have been widely used in these disorders. Further, we will also discuss the potential role of the blood–brain barrier in transporting nutraceuticals. Moreover, it is crucial to understand the factors which affect the efficacy (i.e., bioavailability and dosage). If the dosage is not correct, then consumption will result in side effects like bleeding, diarrhea, weakness, antibiotic resistance, or toxicity. Overall, this review is a comprehensive analysis of the beneficial effect of nutraceuticals on health and disease and their bioavailability, dosage, side effects, and related toxicity, as compared to pharmaceuticals.

15.2 Classification of Nutraceuticals

Nutraceuticals are generally classified into two types: conventional and nonconventional (AlAli et al. 2021). Conventional nutraceuticals are simply taken from traditional food with no changes, and these include a variety of fibers, probiotics, prebiotics, polyunsaturated fatty acids, vitamins, herbals, and phytochemicals found in natural fruits, vegetables, and animal products. These conventional nutraceuticals are good for antioxidant, anti-inflammatory, and immune-enhancing properties (Weaver et al. 2012). Conventional nutraceuticals have been further classified based on the source, mechanism of action, and chemical nature (Kokate et al. 2002; Kalia 2005). The main advantages of traditional nutraceuticals are cost-effectiveness, tolerability, safety, and availability (Kim and Kim 2010). Nonconventional nutraceuticals have been artificially prepared through the help of biotechnology. Further, recent advances in biotechnology and plant breeding have made their mass production possible. The nonconventional nutraceuticals are further divided into fortified and recombinant products (Gupta et al. 2010; Anjali Garg et al. 2018).

Fortified foods are food products fortified with vitamins and/or minerals to maximize their efficiency. For example, omega-3 and omega-6 fatty acids are added to infant formula for the growth and neural development benefits (Carlson 1999; Ruchi 2017). Another example of this is the milk fortified with vitamin D, which is commonly used in the United States to prevent bone diseases (Datta and Vitolins 2016). Similarly, recombinant nutraceuticals are produced through genetic modification or recombination (Singh and Sinha 2012). A common example of recombinant nutraceutical is the fermented product produced from genetically modified organisms that are then extracted and used as nutraceuticals such as resveratrol (Ruchi 2017). Another best example of this is the golden rice produced with vitamin A (https://solutionpharmacy.in/classification-of-nutraceuticals). The different classes of nutraceuticals are shown in Fig. 15.1.

15.3 Advantages and Disadvantages of Nutraceuticals

Nutraceuticals have various advantages on multiple aspects of our health when consumed in appropriate quantities. Commonly known advantages include improved immune function from vitamin C, prevention of cardiovascular disorders from fatty acids, anti-inflammatory properties from carotenoids, risk reduction of neurological disorders from phytochemicals, and many more. As studies showing the advantages of nutraceuticals have increased, as well as their advertisement from the nutrition and health industries, the intake of nutraceuticals through a regular diet or in supplement form has become very popular. Often, people self-prescribe various nutraceuticals in the hopes of gaining these health benefits alone or in conjunction with other prescribed medications (Shrestha et al. 2021). As such, people may

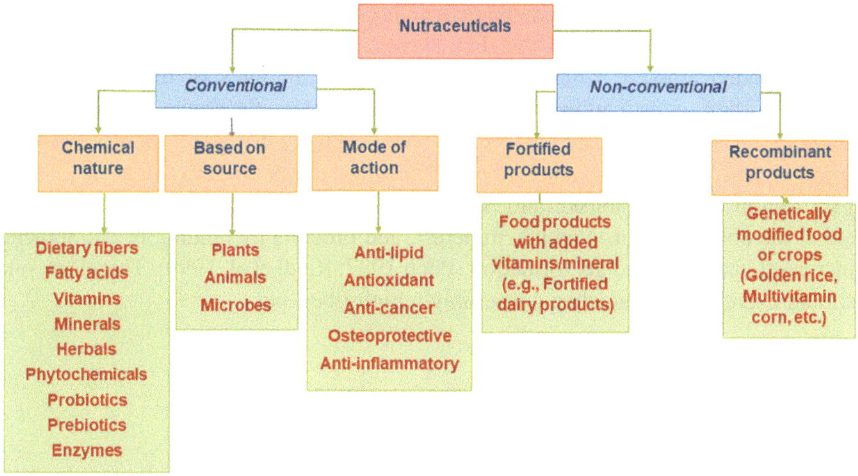

Fig. 15.1 Classification of nutraceuticals. Nutraceuticals are generally divided into conventional and nonconventional types. Conventional nutraceuticals exist naturally in foods without modifications. They can be described by their chemical nature, source, and mode of action. The nonconventional nutraceuticals are nutraceuticals that have modified/added nutrients, viz. fortified products and recombinant products. Fortified products are foods that have added nutraceuticals (e.g., orange juice with added calcium). Recombinant products are the result of biotechnologies and engineering that enhance foods with nutraceuticals

unintentionally misuse or overuse nutraceuticals, which can have harmful side effects. The same components that have health benefits are also toxic when consumed in excess. Due to the general population shifting towards nutraceuticals as alternative health options, the advantages and disadvantages of commonly used nutraceuticals need to be thoroughly and comprehensively analyzed. Here, we also discuss the properties, dosages, and side effects of several types of nutraceuticals.

15.3.1 Dietary Fibers

Dietary fibers are non-starchy polysaccharides, found mainly in fruits, vegetables, wheat, etc. (Leclere et al. 1994). Dietary fibers are digested only by gut microbiota and thus impact the bacterial population and its metabolic activities in the gastrointestinal tract. Our gut microbiota contains a multitude of healthy and protective microbes, which helps in maintaining a homeostatic relationship with our intestinal immune system and other organ systems (Schroeder and Bäckhed 2016). However, any disruption to the microbiota is harmful, potentially leading to the development of chronic diseases. Consumption of dietary fibers is associated with richer gut microbiota and therefore improves the prevention of chronic diseases such as IBS, diabetes, heart disease, food allergies, obesity, gastrointestinal disorders, and certain cancers (Liu et al. 2015; Makki et al. 2018). Some of the soluble fibers were also found to enhance the immune function in cells (Anderson et al. 2009). While dietary fibers should be included in the diet, there are disadvantages to exceeding the recommended amount. Consumption of excessive amounts of dietary fiber can cause diarrhea, constipation, bloating, abdominal pain, flatulence, reduced blood sugar levels, and weight gain likely due to their capacity in lowering energy density, although few studies have shown that a high-fiber diet is protective against weight gain (Saibil 1989; Grabitske and Slavin 2009). Dietary fiber can also negatively affect the absorption of vitamins, minerals, and proteins by binding or entrapment, increased viscosity, and gel formation (Pilch 1987; Riedl et al. 1999). Therefore, one should exercise caution when consuming a high-fiber diet.

15.3.2 Prebiotics and Probiotics

Prebiotics are nutrients that are selectively utilized by host probiotic bacteria in the gut to confer health benefits (Valdes et al. 2018). The most used prebiotics are indigestible oligosaccharides such as galactose-containing oligosaccharides, xylose-containing oligosaccharides, and fructo-oligosaccharides (Kerry et al. 2018). They serve as an energy source for probiotics/gut microbiota and maintain gut health (Liu et al. 2017). The recommended dose of oligofructose is between 10 and 20 g. However, overconsumption may cause diarrhea, stomach flatus, and abdominal complaints (Mundi et al. 2017).

Probiotics are live or heat-killed microbes or microbial components used mainly for balance and restoration of the gut microbiota and are also important for protection against pathogens, immunomodulation, and maintenance of intestinal barrier integrity (Piqué et al. 2019). In recent years, these probiotics are added as an ingredient in dietary supplements, food products, and infant formulations (Pamer 2016). The most used bacteria are *Lactobacillus* and *Bifidobacterium*, which are the most common bacteria of gut microbiota (Fuller 1991; Linares et al. 2017). The other bacteria used as probiotics are *Streptococcus*, *Bacillus*, and *Enterococcus* (Mills et al. 2018). Probiotics are considered a potential therapeutic agent for gastrointestinal disorders and extraintestinal disorders like hepatic encephalopathy (Liu et al. 2018). They are also helpful in preventing infection by producing antimicrobial compounds against pathogens (Van Zyl et al. 2020). Dosage of probiotics depends on the specificity of strain, duration of administration, and its survival drives through the gastrointestinal tract (GIT) (Bezkorovainy 2001). According to a Cochrane review, five billion colony-forming units per day were found significantly effective (Wilkins and Sequoia 2017). However, translocation of probiotics to tissues and blood causes bacteremia. In addition, the development of antibiotic resistance can develop and lead to excessive stimulation of the immune system in susceptible individuals (Sanders et al. 2010). It is important to monitor the occurrence of any unintended consequences for use in vulnerable populations including newborns before consumption.

In recent days, synbiotics, a combination of probiotics and prebiotics, have been used to enhance immunity and gut health. Further, these products help compete with invading pathogens by producing antitoxins, antibiotics, and enzymes (Singh et al. 2017). The prebiotics enhance the adherence of probiotic bacteria in the gastrointestinal tract. At the same time, the probiotics use prebiotics as a nutrient source and help them to stay longer in the gut. In addition, probiotics are important for maintaining the resident gut microbiota, intestinal epithelial cells, and host immune system (Nguyen et al. 2022). Hence, these synbiotics are considered more beneficial for people with gastrointestinal disorders.

15.3.3 Polyunsaturated Fatty Acids (PUFAs)

Polyunsaturated fatty acids or essential fatty acids are important for body function, which include omega-3 and omega-6 fatty acids, and the only source in humans is through consumption. In a regular diet, omega-3s are found in fish oil and fish with high-fat content while omega-6s come from most vegetable oils (Das et al. 2012). PUFAs are one of the commonly used nutraceuticals for the prevention and treatment of cardiovascular disease, asthma, diabetes, and multiple sclerosis (Simopoulos 2002). The balanced intake of omega-6-fatty acids/omega-3-fatty acids is important to decrease the risk of heart diseases (Simopoulos 2008). The FDA-recommended maximum intake of eicosapentaenoic acid (EPA) and docosahexaenoic acid (DHA) is 3 g/day and omega-3 fatty acids is 2 g/day as a dietary supplement. However, bleeding is reported as one of the drawbacks of omega-3 fatty acids due to its interaction with warfarin, an anticoagulant medication taken by people with stroke and heart attack, which results in neurosurgical intervention (Gross et al. 2017). Although the PUFAs are good for people with cardio diseases, patients taking anticoagulants must limit the use of PUFAs.

15.3.4 Vitamins

Vitamins are important micronutrients sold in the market either in the form of single vitamins or as multivitamin complexes or combinations of vitamin and mineral supplements (Woo 2007). It is one of the most used supplements by adults in the United States (Knapik et al. 2014). Vitamins are mainly important for the primary and secondary prevention of many chronic diseases (Guallar et al. 2013). For example, vitamin C is most popularly consumed to improve the immune system. It has antioxidant properties and plays a significant role in phagocytosis and chemotaxis (Carr and Maggini 2017). Similarly, vitamin E is used to prevent oxidative stress in carcinogenesis due to its highest antioxidant activity and is one of the extensively studied vitamins for cancer prevention (Yang et al. 2008). The recommended dose of vitamin E is 800–1200 mg/day or less. However, more than

1200 mg/day results in bleeding due to antiplatelet action and leads to diarrhea, weakness, blurred vision, and gonadal dysfunction (Ziegler and Filer Jr 1996). Vitamin D is considered an immune stimulator and is used as an antiviral agent. Recently, vitamin D efficiency in mitigating morbidity from Covid-19 infection was reported (the functional medical approach). Vitamin A is considered an anti-inflammatory vitamin and is considered important for the prevention of measles-associated pneumonia and lower respiratory tract infections. Vitamin B is a group of vitamins important for respiratory function, immunity, and endothelial integrity. Recent reports show the effect of vitamin B complex in reducing the risk and organ failure associated with Covid-19 (Shakoor et al. 2020). However, the report says that the toxicity associated with overconsumption of fat-soluble vitamins is more prevalent than that of water-soluble vitamins. Photosensitivity and neurotoxicity are common side effects associated with the overconsumption of vitamin B6 although it is a very important coenzyme involved in various metabolic pathways and reduces advanced glycation end (AGE) products (Ziegler and Filer Jr 1996). As vitamins are sold as supplements in many convenience and grocery stores, people are at a greater risk of misusing these nutraceuticals when self-prescribing supplements as they may be unaware of the recommended dose and the potentially toxic nature (Petroczi et al. 2011). Vitamin deficiency is rare in developed countries. However, vitamins can only be taken after prescription for people with deficiency to avoid hypervitaminosis and related toxicity.

15.3.5 Animal Products

Collagen is one of the animal proteins from bovine connective tissues which is used as a nutraceutical, since it has antioxidant, antiaging, anti-tumor, anti-inflammatory, and anti-obesity properties in humans (Song and Li 2017). The role of collagen and collagen peptide (CP) in improving skin elasticity (Kim et al. 2018; Bolke et al. 2019) was reported. Marine collagen peptide (MCP), a collagen peptide derived from marine fish, is used for its antioxidant, anti-aging, and antihypertensive activities (Liang et al. 2010). Collagen peptides are considered safe for humans at doses of 10 g/day for up to 5 months (Khatri et al. 2021). Recently, Kim et al. reported that the collagen peptide and collagen tripeptide modulate the age-associated sarcopenia in mice model by decreasing myostatin expression and increasing insulin-like growth factor 1 (IGF-1) (Kim et al. 2022). In addition, the collagen peptide was found to increase the bone mineral density in postmenopausal women (Konig et al. 2018). Hence, collagen and collagen-derived peptides were considered good for skin and bone. Many collagen products are available for the skin. However, further studies are needed for bone applications. Although there are no major side effects, some people experience gastrointestinal symptoms, skin rashes, kidney stones, and rarely liver abnormalities through oral collagen (Martini 2019; Wang 2021).

15.3.6 Herbals and Phytochemicals

Herbal nutraceuticals not only are considered effective in promoting health and longevity but also enhance the overall quality of life. Most importantly, they are considered a powerful tool against malnutrition-induced acute and chronic diseases such as cancers, heart disease, and diabetes (Arts and Hollman 2005; Chauhan et al. 2013). Phytochemicals are a group of chemicals used as nutraceuticals in either isolated or purified form from plants as an alternative to the direct consumption of plants, fruits, and herbals. For example, carotenoid is a group of pigments found abundantly in plants, fruits, vegetables, and algae (Lee et al. 2019; Shardell et al. 2011), which are commonly used to improve vision and cognitive and immune functions. They are also used to prevent cancer and heart problems through anticancer, anti-inflammatory, and antioxidant properties (Chew et al. 2014). Among various carotenoids, β-carotene and astaxanthin were found to inhibit *Helicobacter pylori*-mediated inflammation (Kang and Kim 2017). In another study, quercetin was found to be effective in reducing blood vessel damage from oxidative stress in diabetic patients (Kruger et al. 2002). Although there is no proof of toxicity, there is no/less clinical evidence to prove the benefits of these plant-based nutraceuticals. However, overconsumption or misuse of some of the herbal compounds is found to be toxic. For example, both ingestion and injection of genistein can affect the development of the reproductive system, decrease thymic weight, lead to delayed-type hypersensitivity response, modulate the immune response, or reduce thyroid peroxidase (Guo et al. 2005). Some phytochemicals induce estrogenicity and thereby increase the risk of estrogen-sensitive cancers. Hence, users and providers should consult before starting a new nutraceutical therapy. Table 15.1 shows some of the phytochemicals used as nutraceuticals and their properties.

15.4 Nutraceuticals in the Treatment of Organ-Specific Diseases

In the following work, the role of nutraceuticals in the prevention and treatment of various organs and their diseases is discussed.

15.4.1 Brain

The brain uses a large amount of energy without affecting homeostasis. Nutrient deficiency leads to disturbances in the central and peripheral nervous system, which results in neurological disorders (Williams et al. 2015). The intake of several dietary components has an influence on cognitive function. In recent years, many reports have been published about the role of different nutraceuticals and supplements and

Table 15.1 Phytochemicals and its property, medicinal benefits with references

Phytochemicals	Properties	Medical benefits	References
Quercetin	Antioxidant	Controls diabetes, heart disease	Kruger et al. (2002)
Hesperidin	Anti-inflammatory	Venous insufficiency, hemorrhoids	Garg et al. (2001)
Phytosterols	Induce bile acid synthesis	Lower blood cholesterol, hypertension	Li and Zhang (2001)
Beta-carotene	Antioxidant	Cancer	Stahl and Sies (2005)
Epigallocatechin gallate	Antioxidant	Cancer	Thomasset et al. (2007)
Saponins	Anti-tumor	Cancer	Li et al. (2018)
Tannins	Antioxidant Anticarcinogen	Cancer	Li et al. (2018)
Sulfur compounds	Anti-tumor	Cancer	Cerella et al. (2011)
Curcumin	Antioxidant	Cancer	Rahmani et al. (2018)
Isoflavones	Antioxidant	Cancer	Vitale et al. (2013)
Lutein and zeaxanthin	Antioxidant	Age-related macular degeneration (AMD)	Brookmeyer et al. (2007)
Resveratrol	Sirtuin-like activity	Inflammatory diseases	Jang et al. (1997)
Gentianine	Anti-inflammatory	Inflammatory diseases, arthritis	Kwak et al. (2005)
Bromelain	Anti-inflammatory	Cancer, diabetes	Pavan et al. (2012)
Capsaicin	Anti-obese	Obesity	Rubin and Levin (1994)
Eugenol	Antioxidant, anticancer	Cancer	Ghosh et al. (2005)
Catechin	Antiangiogenic	Cancer	Juneja et al. (2013)

their influence on brain function. Some preliminary reports show the promising effect of nutraceuticals on neurodegenerative diseases and other diseases of the brain (Rubin and Levin 1994). Nutraceuticals tend to modify the misfolded proteins important for brain function and thereby prevent neurodegeneration. Recently, it was identified that the active molecules present in green tea, coffee, cauliflower, and wine influence symptoms of neurodegenerative disorders such as Alzheimer's disease and Parkinson's disease as well as psychosis (Makkar et al. 2020). Figure 15.2 shows different neuro disorders and nutraceuticals used in the suppression of neurodegenerative disorders. Although there are many diseases related to neurodegeneration and neuropsychiatry conditions, only some of the diseases and the role of nutraceuticals are discussed further. The reason for this is the limitation of pharmacological therapies for certain age-related disorders and cognitive diseases.

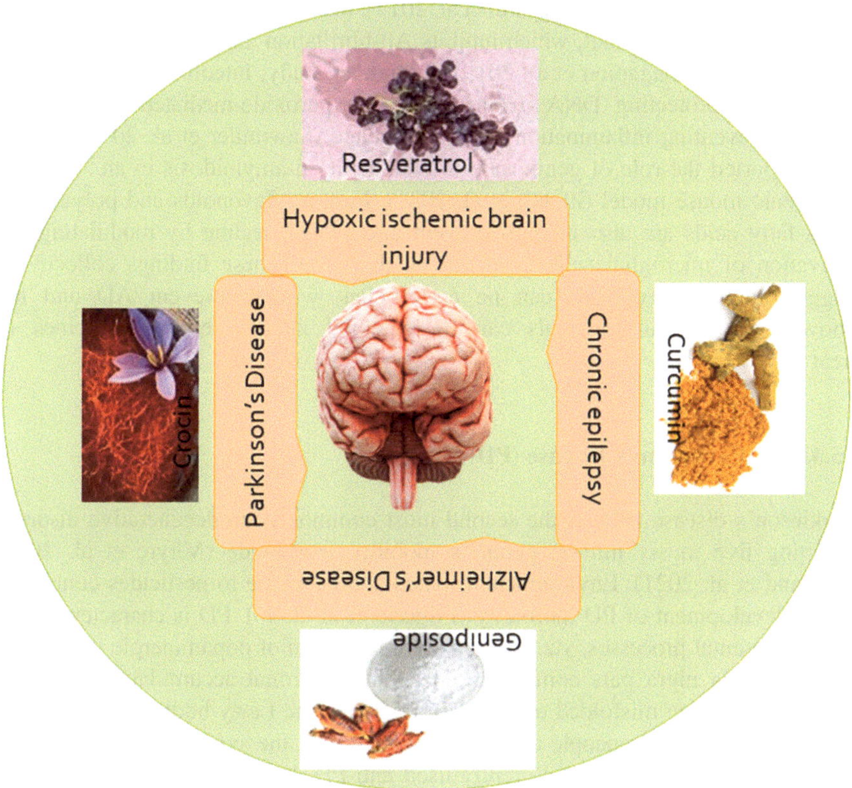

Fig. 15.2 Examples of neurodegenerative diseases and the corresponding nutraceuticals known to be an effective treatment: Parkinson's disease—crocin; hypoxic-ischemic brain injury—resveratrol; chronic epilepsy–curcumin; Alzheimer's disease—geniposide

15.4.1.1 Alzheimer's Disease (AD)

Alzheimer's disease (AD) is one of the forms of dementia among the aged population worldwide (Zhang et al. 2020). Nutritional deficiency is largely responsible for cognitive decline in the elderly population (Dominguez and Barbagallo 2018; Zhao et al. 2018). However, healthy lifestyle behavior and change in nutrition play an important role in the enhancement of brain function (Mecocci et al. 2014). Recent studies have shown the potential role of nutraceuticals like prebiotics and probiotics in preventing AD in humans (Pluta et al. 2020). Withanine, an alkaloid from the plant *Withania somnifera*, is used mainly for enhancing memory (Bhattarai et al. 2013). Specifically, it inhibits the enzyme acetylcholine esterase, which is the most reliable method to prevent AD today. Therefore, *Withania somnifera* is one of the highly recommended compounds in the management of AD (Chauhan and Mehla 2015). Crocin was also found to improve mild-to-moderate AD by altering the levels of oxidative markers and enzymes like catalase, glutathione, glutathione peroxidase, and superoxide dismutase (Boccardi et al. 2017). In addition, the anti-amyloidogenic property is very important, which inhibits Aβ-fibrillation and prevents Tau protein aggregation (Hatziagapiou et al. 2019). In another study, luteolin was found to be effective in protecting DNA against hydrogen peroxide-mediated toxicity and thereby preventing inflammation and cell damage (Sawmiller et al. 2014). Zhang et al. reported the role of geniposide in attenuating β-amyloidosis in an APP/PS1 transgenic mouse model (Zhang et al. 2021). Further, flavonoids and polyunsaturated fatty acids are also used to prevent age-related decline by modulating the activation of microglial cells (Vauzour et al. 2017). These findings collectively suggest that nutraceuticals can be an efficient way to prevent AD and the abovementioned nutraceuticals can be potential alternatives to medicines to treat AD.

15.4.1.2 Parkinson's Disease (PD)

Parkinson's disease (PD) is the second most common neurodegenerative disorder affecting five to six million people's mobility worldwide (Mhyre et al. 2012; Aarsland et al. 2021). Environmental toxins and exposure to pesticides contribute to the development of PD morbidity (Frigerio et al. 2006). PD is characterized by two detrimental processes, viz. a progressive reduction of dopaminergic neurons in the substantia nigra pars compacta and an intraneuronal accumulation of Lewy bodies, containing misfolded α-synuclein (α-Syn). The Lewy bodies are present in neurons of vulnerable people and are associated with the symptoms of PD (Braak and Del Tredici 2008). The currently used anti-PD drugs are only to prevent the disease progression. Thus, alternative medicine is needed to completely prevent the disease. The EGCG, a flavonoid from green tea, was found to cross the BBB and help prevent Parkinson's disease (Dutta and Mohanakumar 2015). Similarly, crocin, a pigment from *Crocus sativus*, was found to attenuate PD due to its anti-inflammatory activity (Zhang et al. 2021). Crocin protected the dopaminergic neurons in the ventral tegmental area. Several vitamins such as vitamins B3, B6, B9, B12, D, E, and C were also found to be effective against PD (Zhao et al. 2019).

15.4.1.3 Epilepsy

Epilepsy is a common neurological disorder of both genders at all age. Germline and somatic mutations and autoimmune-mediated dysregulation of inflammation can elicit seizures, which result in prolonged loss of consciousness, uncontrolled muscle contractions, and spasms (Scheffer et al. 2017). Epilepsy is the reason for most premature death (Beghi et al. 2015). Although 70% of epilepsy patients manage seizures with conventional medication, the remaining 30% will suffer from intractable epilepsy. Hence, it is important to search for alternative therapy to manage epilepsy. Curcumin, the most common compound of turmeric, was found to show beneficial effects in the treatment of chronic epilepsy (Kaur et al. 2015). The antiepileptogenic activity of curcumin reduces spontaneous recurrent seizures (Dhir 2018). Similarly, vitamins like B6, C, D3, and E and omega-3 PUFA were also found to show potential benefits in managing seizures by improving cognitive functions (Gaby 2007; Kim and Cho 2019). However, at higher doses, pyridoxine causes a neurotoxic effect. Hence, optimal dosage is important for functional benefits (Kim and Cho 2019).

15.4.1.4 Ischemia

Ischemia is brain damage that occurs during late pregnancy and childbirth and is one of the major causes of neonatal mortality worldwide. Newborns who survive with ischemic brain damage may have severe visual and hearing impairment, seizures, epilepsy, mental retardation, cerebral palsy, or other communication problems (Shaw and Yager 2019). The generation of free radicals and oxidative stress is fatal to the neonatal brain (Bracci 2001). In addition, increased calcium triggers nitric oxide production by the nitric oxide synthase, also responsible for the brain damage (Kalogeris et al. 2014).

We have been actively investigating many drugs to prevent HIE. Polyphenols were found to play an important role in the prevention of several neurodegenerative disorders. Resveratrol was found to involve in the microglial activation and thereby helps in the control of ischemic stroke and traumatic brain injury in a mouse model (Lopez et al. 2015). Similarly, piceatannol, a resveratrol derivative, was also found to be effective in the regulation of brain metabolism, reduction in neuronal death, and brain damage through maternal lactation (Dumont et al. 2019). The in vitro and in vivo studies of quercetin show numerous neuroprotective effects on focal ischemia and oxygen–glucose deprivation injury. However, there is a question about its bioavailability due to rapid metabolism and excretion (Dajas 2012). Still, further studies are needed to prove the effect of nutraceuticals in the prevention and treatment of brain disorders.

15.4.1.5 Blood–Brain Barrier

The blood–brain barrier (BBB) is a complex, dynamic interface between the blood and central nervous system (CNS) important for the delivery of drugs to the central nervous system (Neuwelt et al. 2008). The BBB is primarily involved in the regulation of brain microenvironment for neuronal functions. Further, it acts as an interface between the brain and periphery to transport nutrients to the brain and protects the CNS from pathogens and toxic chemicals (Abbott et al. 2010). The BBB is mainly composed of brain microvascular endothelial cells, which are surrounded by neurons, microglia, pericytes, and astrocytes. These cells interact and form a functional unit called neurovascular unit (Neuwelt et al. 2008). The endothelial cells of BBB are connected via tight junctions, which are responsible for the resistance of the materials that pass through BBB to the brain (Boonstra et al. 2015). Generally, low molecular weight and positively charged lipid-soluble molecules can cross the BBB (Tosi et al. 2013). All the other nutrients and metabolites must cross the BBB to reach the brain through the specialized transport systems. There are six major pathways that are involved in the transport of nutrients to the brain, viz. lipid-mediated diffusion, paracellular diffusion, carrier-mediated diffusion, receptor-mediated transcytogenesis, absorptive mediated transcytogenesis, and active efflux treatment. Among the various pathways, carrier-mediated diffusion was found to play an important role in the transport of glucose, amino acids, fatty acids, metal ions, and vitamins. The small lipophilic molecules are transported through lipid-mediated diffusion, and larger molecules such as proteins and peptides are transported via receptor-mediated transport. The damage and dysfunction of BBB result in various neurological and CNS disorders (Campos-Bedolla et al. 2014). Hence, BBB is the focus of research for CNS diseases. Several models have been developed to examine and understand the BBB permeability to nutrients and nutraceuticals. Among the various models, in silico analyses, in vitro trials, and in vivo animal models are used to estimate the molecules that transferred across the BBB (Shityakov et al. 2013). Several new transport systems and carriers are also discovered through novel technologies.

Recent report says that the administration of natural neuroprotective compounds minimizes neuronal damage (Ayuso and Montaner 2015). Specifically, the neuroprotective diet includes bioactive compounds with antioxidant and anti-inflammatory activities. However, the nutraceuticals must be transferred through the BBB to use this as a potential therapy for brain disorders. The BBB plays a central role in the entry of nutraceuticals to the CNS by active influx and efflux transport systems. Many neuroprotective molecules maintain and protect the function and integrity of BBB (Campos-Bedolla et al. 2014). Atorvastatin and pravastatin were found to prevent BBB disturbances (Pallebage-Gamarallage et al. 2012). Takechi et al. also studied the role of anti-inflammatory nutraceuticals in the inhibition of fat-induced BBB disturbances (Takechi et al. 2013). Chevalier et al. identified the BBB crossing ability of cholecalciferol, niacin, resveratrol, *N*-acetyl cysteine, and sulforaphane using log BB in silico prediction method (Chevalier 2018). Figure 15.3 shows the common nutraceuticals that cross the BBB and the mechanism of transport of nutraceuticals. Some of the nutraceuticals or their metabolites can directly cross the BBB as its structure is similar to the compound already present in the brain. For example, omega-3 fatty acids like DHA, EPA, and arachidonic acid freely diffuse across the membranes of BBB because of their smaller size, few hydrogen bonds, and lipophilicity, which are considered key factors that help penetrate the BBB (Leclerc et al. 2021). But some of the natural products are present outside of the brain in the circulation and exert their action via regulating brain metabolic hormones or peripheral metabolism (Leclerc et al. 2021). However, the exact mechanisms of transport of phenolic metabolites through BBB need to be investigated. In recent years, nanosystems are used to help drug molecules to cross BBB. Again, the nanocarriers must be nontoxic and able to interact with BBB receptors (Andrade et al. 2018).

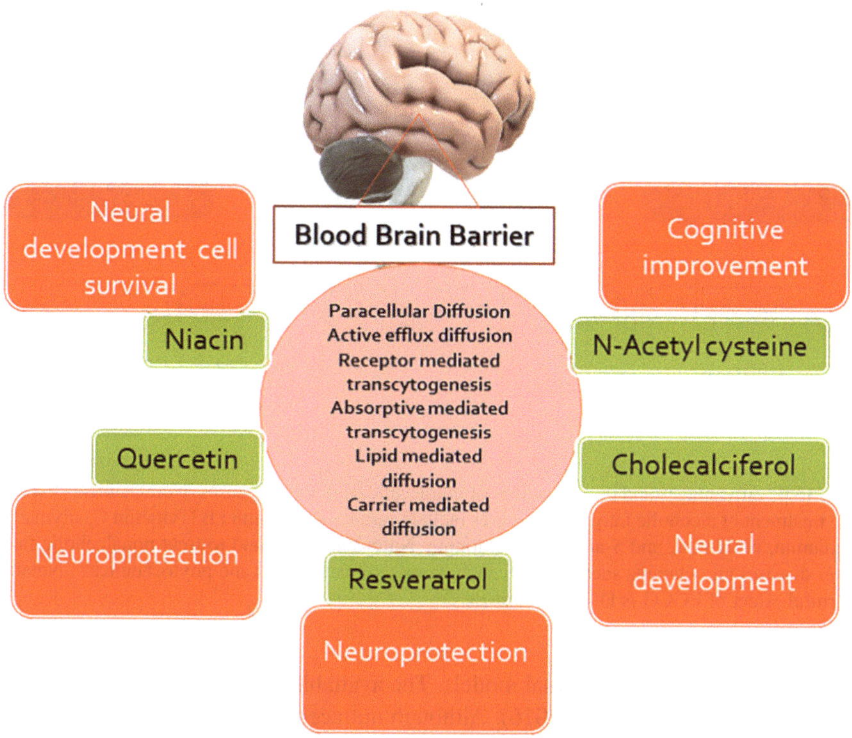

Fig. 15.3 The nutraceuticals that cross the blood–brain barrier and their associated prevention or treatment of neurological conditions. The transport mechanisms of these nutraceuticals across the BBB are also listed. Nutraceuticals quercetin and resveratrol provide neuroprotection in hypoxic-ischemic brain injury. Niacin and cholecalciferol enhance neural development. *N*-Acetyl cysteine improves cognitive function

15.4.2 Liver

The liver is the largest gland in the human body, which is important for the metabolism of carbohydrates and lipids. It is also important for the synthesis of proteins essential for osmolarity and coagulation (Leung 2014). Liver pathologies increase due to sedentary lifestyle and nutritional imbalances.

15.4.2.1 Alcoholic Liver Disease (ALD)

Alcoholic liver disease (ALD) is one of the second major causes of liver disease (Osna et al. 2017). There is no FDA-approved drug or nutritional therapy for ALD. Cessation of alcohol intake and liver transplantation are recommended for end-stage ALD. However, there is evidence that compounds like carotenoids, vitamin B3, vitamin C, silymarin, curcumin, probiotics, and *S*-adenosylmethionine are useful in

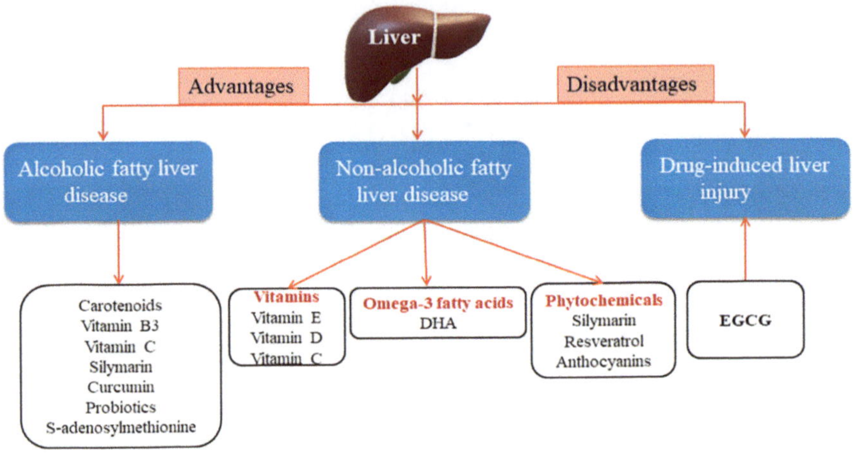

Fig. 15.4 Advantages and disadvantages of nutraceuticals in liver diseases. Nutraceuticals used in the treatment of alcoholic fatty liver diseases include carotenoids, vitamin B3, vitamin C, silymarin, curcumin, probiotics, and *S*-adenosylmethionine. Nutraceuticals found to treat nonalcoholic fatty liver diseases are vitamins such as E, D, and C; omega-3 fatty acids; and phytochemicals. Notable harmful effect of EGCG is DILI at overdose

ameliorating the ALD in animal models. The available number of clinical studies is also limited (Ghorbani et al. 2016). Although nutraceuticals are advantageous in the treatment of liver diseases, overuse or long-term use of nutraceuticals (e.g., EGCG and catechin) results in enhanced oxidative stress and liver injury (Mazzanti et al. 2009). Figure 15.4 shows the advantages and disadvantages of nutraceuticals used for the treatment of liver diseases.

15.4.2.2 Nonalcoholic Fatty Liver Disease (NAFLD)

Worldwide, 25% of the population is affected by chronic nonalcoholic fatty liver disease (NAFLD), which is one of the growing medical issues devoid of pharmaceutical solutions (Vernon et al. 2011; Younossi et al. 2016). It starts with a simple accumulation of lipid in hepatocytes (steatosis) and progresses to severe nonalcoholic steatohepatitis (NASH). Finally, this NAFL raises the risk of developing hepatocellular carcinoma through liver fibrosis and cirrhosis (Loomba and Sanyal 2013; Noureddin and Rinella 2015). Currently, there is no approved treatment for NAFLD except lifestyle modifications, weight loss, and combinations of the drug pioglitazone and vitamin E with limited efficacy (Hardy et al. 2016; Hung and Bodenheimer Jr. 2018). Hence, nutraceuticals are the drug of choice to treat NAFLD. Polyphenols like resveratrol, silymarin, EGCG, and curcumin have been found to show promising effects in ameliorating liver diseases by counteracting oxidative stress by its antioxidant effect (Charytoniuk et al. 2017; Gillessen and

Schmidt 2020; Tang et al. 2021; Mokgalaboni et al. 2021). Vitamin E at normal than higher doses was found to reduce inflammation and liver fibrosis (Li et al. 2016).

15.4.2.3 Drug-Induced Liver Injury (DILI)

Drug-induced liver injury (DILI) is one of the problems associated with certain prescription and nonprescription medications, which leads to acute liver failure. Specifically, the overuse of acetaminophen and carbon tetrachloride leads to depletion of antioxidative mechanisms, and it is converted into toxic products by cytochrome P450 2E1 (CYP2E1). Oleuropein from olive oil and epicatechin from cocoa were found helpful in the treatment of DILI and ALD (Jemai et al. 2020; Huang et al. 1994). Luteolin protects one from DILI by restoring antioxidant compounds and reducing the inflammatory markers (Tai et al. 2015). Polyphenols like resveratrol, EGCG, and curcumin are the common nutraceuticals for liver diseases. Sometimes, long-term use of nutraceuticals like curcumin and EGCG results in DILI (Imam et al. 2019).

15.4.3 Lungs

Respiratory diseases (RD) are one of the major causes of the health burden worldwide. The different types of respiratory diseases start from airway inflammation to severe asthma, COPD, and pulmonary fibrosis. The RD is mainly triggered by indoor and outdoor environmental pollutants, allergens, smoking, and microbes, which results in both reversible and irreversible airway inflammation and obstruction (Chellappan et al. 2020). Treatment options to prevent these pulmonary diseases are limited. Considering the highly favorable and nontoxic nature, nutraceuticals can be useful to prevent or at least ameliorate various pulmonary disorders. Below, we discuss one or more potential nutraceuticals that are used to prevent pulmonary conditions in experimental models, as well as in humans.

15.4.3.1 Airway Inflammation

Chronic inflammation due to failure in cellular immune systems results in the progression of various respiratory diseases (Le Rouzic et al. 2017). A wide variety of nutraceuticals are used for respiratory health including dietary fiber, fatty acids, quercetin, curcumin, glycyrrhizin, vitamins E and D, and EGCG. Nutraceuticals are found to be effective in maintaining lung health and preventing inflammation. Among different nutraceuticals, fish oil with omega-3 fatty acids, DHA, and EPA shows good results in the treatment of age-related respiratory diseases and bleomycin-induced pulmonary fibrosis (Hwang and Ho 2018). Intake of high dietary fibers reduces asthmatic airway inflammation, contraction, coughing, and wheezing.

This is mainly due to the anti-inflammatory activity of dietary fiber via the production of short-chain fatty acids (SCFAs), which can improve tolerance to immune response and regulate immune pathways that help to decrease the inflammation caused by asthma allergen (Williams et al. 2019).

15.4.3.2 Asthma

Quercetin was found to be effective in the treatment of ovalbumin (OVA)-induced asthmatic mouse model due to its anti-allergic effects (Park et al. 2009). Another phytochemical, glycyrrhizin from *Glycyrrhiza glabra*, effectively alleviates OVA-induced asthma in mice model by reducing inflammation and eosinophils (Ram et al. 2006). Sulforaphane (SFN), a phytoconstituent, was found effective in the treatment of various respiratory diseases through antioxidant and anti-inflammatory activities (Harvey et al. 2011; Guerrero-Beltrán et al. 2012; Al-Harbi et al. 2019). However, overnutrition leads to immunometabolism-mediated adverse effects. The mechanism and relationship between immunometabolism mediators and respiratory diseases must be studied for further understanding of therapeutic actions (Berthon and Wood 2015). Figure 15.5 shows the various nutraceuticals used in the treatment of respiratory diseases and their potential disadvantages.

15.4.3.3 Chronic Obstructive Pulmonary Diseases (COPD)

According to the World Health Organization (WHO), COPD is considered one of the deadliest forms of respiratory diseases responsible for 6% of global deaths every year (Fuller-Thomson and Lacombe-Duncan 2016). Any probiotics with

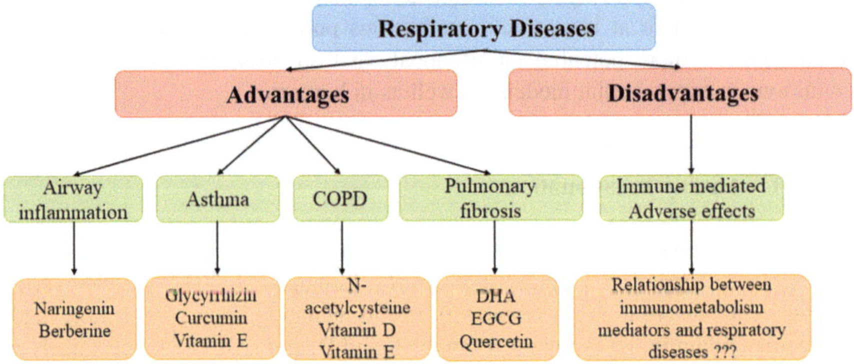

Fig. 15.5 Advantages and disadvantages of nutraceuticals in the treatment of respiratory diseases. Nutraceuticals such as naringenin and berberine for airway inflammation; glycyrrhizin, curcumin, and vitamin E for asthma; *N*-acetylcysteine and vitamins D and E for COPD; and DHA, EGCG, and quercetin for pulmonary fibrosis are effectively used to treat respiratory disorders. Disadvantages stem from overnutrition, which leads to immunometabolism-mediated adverse effects

anti-inflammatory effects can be effectively used in the management of COPD (Mortaz et al. 2015). *Bifidobacterium breve* (Mortaz et al. 2015) and *Lactobacillus rhamnosus* (Carvalho et al. 2020) were found to be effectively involved in ameliorating the inflammation. Vitamins such as vitamin C were found to be effective in treating COPD by restoring antioxidant defense mechanisms (Silva et al. 2018). In another study, vitamin D was found to reduce the expression of pro-inflammatory cytokines and help suppress COPD.

15.4.3.4 Pulmonary Fibrosis (PF)

Pulmonary fibrosis develops after exposure to certain drugs, toxicants, and asbestos. These agents recruit inflammatory mediators and profibrotic responses. The continuous production of cytokines and growth factors leads to the development of fibrosis (Wilson and Wynn 2009; Luzina et al. 2015). Quercetin was found to be effective in preventing PF by inhibiting mainly the MAPKs and NF-kβ inflammatory pathways (Gausauge 1997). DHA, EPA, and EGCG are the most important nutraceuticals that have been tested for the prevention of PF (Hwang and Ho 2018).

15.4.4 Kidney

Kidneys play vital roles in the health of an individual by maintaining homeostasis. Various inflammatory and infectious conditions and metabolic derangements affect the renal function, which results in acute renal injury, which progresses to chronic renal disease (Gwaltney-Brant 2021). However, sometimes, overuse of vitamin C and certain herbals is nephrotoxic, resulting in renal failure (Asif 2012). Figure 15.6

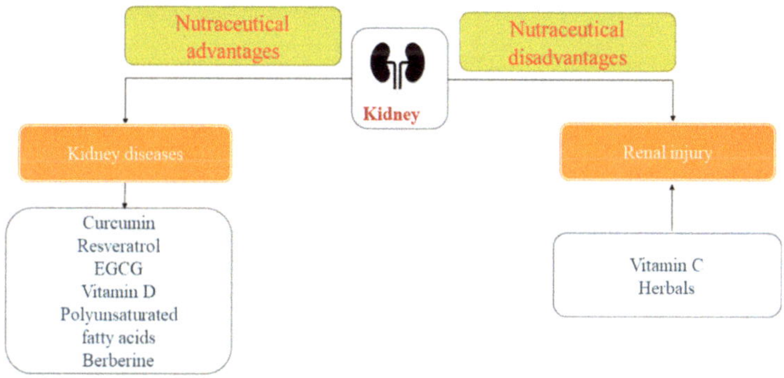

Fig. 15.6 Several nutraceuticals have proven advantages for the treatment and prevention of kidney diseases. These are curcumin, resveratrol, EGCG, vitamin D, polyunsaturated fatty acids, and berberine and advantages and disadvantages of nutraceuticals used for kidney diseases. At the same time, high dose of vitamin C and some of the herbals induces renal injury

shows the positive and negative effects of nutraceuticals on the kidney. Hence, strict controls must be maintained to ensure safety while consuming herbal nutraceuticals.

15.4.4.1 Chronic Kidney Disease (CKD)

Worldwide, 8–16% of the population suffers from chronic kidney disease (CKD) (Keith et al. 2004; Coresh et al. 2007). CKD leads to progressive loss of kidney function and consequent accumulation of toxins (Moradi et al. 2013), and the precursor to these toxins is in the GI tract (Vanholder et al. 2008). Various factors including oxidative stress and pro-inflammatory and inflammatory mediators are responsible for the progression of kidney diseases (Stenvinkel et al. 2005). Further, diabetes mellitus and intestinal microbiota also contribute to the development of kidney diseases (Le Chatelier et al. 2013; Evenepoel et al. 2009). Therefore, nutraceuticals that prevent inflammation and promote rich microbiota are useful for maintaining healthy kidneys. Studies on a healthy and well-balanced diet along with some nutraceutical agents like fatty acids, dietary fibers, and some phytocompounds, such as curcumin, steviol glycosides, and resveratrol, were found to reduce the risk of kidney diseases (Sabatino et al. 2017). Wang et al. found the renal protective effect of EGCG through the regulation of NF-kβ and Nrf2 signaling pathways in a unilateral ureteral obstruction (UUO) mice model (Wang et al. 2015).

15.4.5 Heart

Cardiovascular disease is the primary heart-related disease due to modifiable risk factors like diabetes, dyslipidemia, and hypertension. Figure 15.7 shows different types of heart-related problems and associated nutraceuticals.

15.4.5.1 Cardiovascular Disease (CVD)

Cardiovascular disease (CVD) is the leading cause of death and is a growing health concern worldwide. Age, gender, family history, metabolic diseases, lifestyle modifications (sedentary lifestyle), and dietary factors are predominantly involved in CVD (McGill Jr 1979). Intake of adequate quantity of vitamins, PUFAs, and probiotics was found to improve cardiac output and prevent heart failure. Coenzyme Q_{10} (CoQ10) or ubiquinone is one of the most used nutraceutical antioxidants to modify the plasma lipid profile and reduce the risk of CVD and hypertension. Andersson et al. found the effect of plant sterol in reducing the blood lipid concentrations, thereby reducing cardiovascular effects (Andersson et al. 2004). In another study, vitamin E at a dose of 400 or 800 IU/day was found to effectively decrease cardiac events (Stephens et al. 1996). Vitamin D is also involved in the regulation of vascular health and blood pressure (Vimaleswaran et al. 2014).

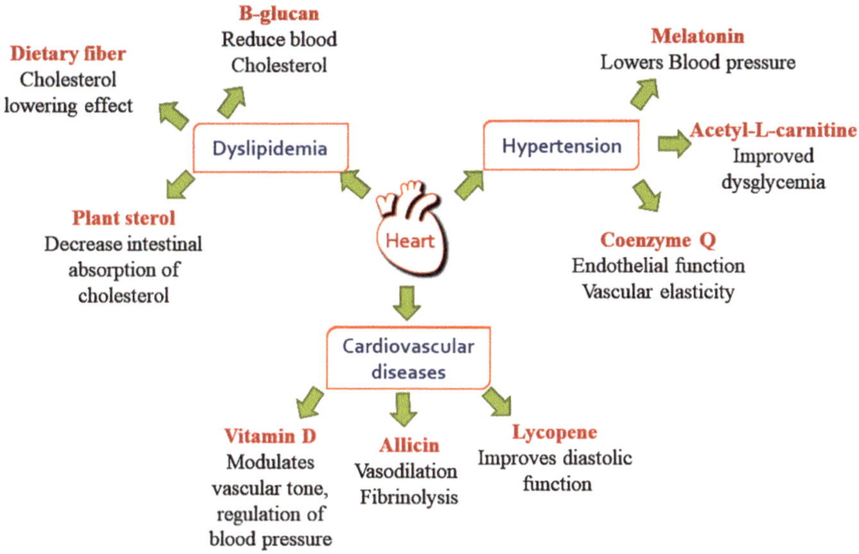

Fig. 15.7 The nutraceuticals in the treatment of heart diseases. Dyslipidemia can be improved or prevented by plant sterols, dietary fiber, and β-glucan by decreasing cholesterol levels. Hypertension can be treated with the use of melatonin, acetyl-L-carnitine, and coenzyme Q. Other vascular diseases can benefit from vitamin D, allicin, and lycopene

15.4.5.2 Ischemic-Reperfusion Injury

Ischemic-reperfusion injury is due to depletion of oxygen to the heart muscle, which results in increased ROS. CoQ10 is found to preserve ischemic-reperfusion injury by increasing antioxidant capacity and limiting oxidative damage. Further, it also increases ATP concentration and reduces apoptosis (Madmani et al. 2014).

15.4.5.3 Dyslipidemia

Dyslipidemia or lipoprotein abnormalities are one of the major factors responsible for the development of CVD. Plant sterols, dietary fibers, beta-glucan, and psyllium are some of the nutraceuticals used to lower the absorption of cholesterol (Cicero et al. 2021).

15.4.6 Gastrointestinal (GI) Disorders

Nutraceuticals are most generally used for the prevention and treatment of gastrointestinal (GI) disorders and overall health of GI (Gao et al. 2020). Disturbances in the structural and functional integrity of GI result in severe impairment of life quality

(Schonberg et al. 2015). The prebiotics which we have considered are the primary candidates selectively utilized by gut microbiota and that confer health benefits to GI system (Gibson et al. 2017). Similarly, probiotics play an important role in the metabolism and absorption of nutrients by interacting both with the host and the gut microbiota (Suez et al. 2018). Modulation of gut microbiota with probiotics itself is considered as the primary therapeutic means of prevention and treatment of many gut-related diseases (Spisni et al. 2022). In addition, synbiotics are used to fight against multidrug-resistant organisms and maintain the intestinal pH, and to restore the intestinal mucosal barrier (Newman and Arshad 2020; Li et al. 2020). Figure 15.8 shows the nutraceuticals involved in the treatment of gastrointestinal disorders.

15.4.6.1 Colon Cancer

Colon cancer is one of the most dangerous forms of cancer with the ability to spread to other parts of the body such as liver, lung, ovaries, and other organs (Donaldson 2004). Consumption of foods with secondary metabolites like flavonoids, alkaloids, saponins, phenolics, vitamins, and minerals helps control colon cancer generation. Luteolin, a flavonoid, was found to be effective in decreasing the multiplicity of tumor (Lim et al. 2007). Curcumin from turmeric was clinically proven for its

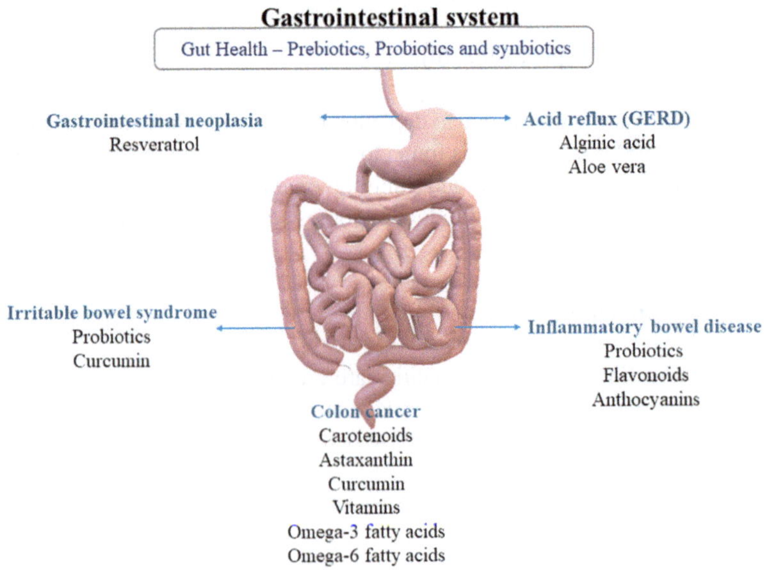

Fig. 15.8 Nutraceuticals are used in the treatment of various diseases related to the gastrointestinal system. Overall gut health is maintained through prebiotics, probiotics, and synbiotics. Gastrointestinal neoplasia can be treated with resveratrol. Acid reflux can be treated with alginic acid and aloe vera. Irritable bowel syndrome can be improved with probiotics and curcumin. Inflammatory bowel disease can benefit from probiotics, flavonoids, and anthocyanins. Colon cancer can be improved with carotenoids, astaxanthin, curcumin, vitamins, and omega-3 and omega-6 fatty acids

anticarcinogenic effect, which was found to modulate protein and immune molecule expression (Huang et al. 1994). Eugenol, omega-3 fatty acids, and vitamins like vitamins B6 and B12 can also reduce the risk of colon cancer (Seeram et al. 2006; Larsson et al. 2004; Lee et al. 2003).

15.4.6.2 Irritable Bowel Syndrome (IBS)

Irritable bowel syndrome is a highly prevalent functional GI disorder with predominant constipation, diarrhea, or mixed bowel habits (Drossman 2016). Generally, the IBS is associated with the psychological conditions such as anxiety and depression. Probiotic *Saccharomyces cerevisiae* is found to be effective in the improvement of abdominal pain and bloating symptoms in patients with IBS (Spiller and Major 2016).

15.4.6.3 Gastroesophageal Reflux Disease (GERD)

Acid reflux or GERD is due to backflow of stomach acid into esophagus that causes heartburn. This is associated with the problems in digestion of certain irritating foods or drinks (Richter and Castell 1982). Panahi et al. found the effect of aloe vera gel in reducing the effect of GERD symptoms in a randomized controlled study (Panahi et al. 2015).

15.4.6.4 Inflammatory Bowel Disease (IBD)

Inflammatory bowel diseases are characterized by inflammation and mucosal damage of intestine and are one of the chronic GI disorders associated with nausea, diarrhea, fatigue, abdominal pain, and rectal bleeding. Ulcerative colitis (UC) and Crohn's disease are the two forms of IBD. Currently, there is no effective therapy to completely cure IBD (Pithadia and Jain 2011). Probiotics, vitamins, fatty acids, and phytochemicals are used to inhibit the inflammation effect against IBD (Al Mijan and Lim 2018).

15.5 Nutraceuticals in Pregnancy

15.5.1 Clinical Context: Safety and Efficacy of Nutraceutical Use During Pregnancy

For the obstetrician, an important part of prenatal counseling at each prenatal visit is assessing patient compliance with daily prenatal vitamins. The American College of

Obstetricians and Gynecologists (ACOG), the governing body for clinical practice recommendations, in fact, supports starting prenatal vitamins, particularly folate, at least 3 months before pregnancy if the pregnancy is planned. The prepregnancy and first trimester periods are critical time points in which the fetus is vulnerable to maternal malnutrition and consumption of toxins. It becomes imperative, therefore, that nutraceuticals marketed for prenatal consumption are free of contaminants and are being used at correct doses and frequencies (ACOG bulletin).

15.5.2 Contamination and Toxicity of Poorly Regulated Prenatal Vitamins

Due to the lack of patent protection and regulation, heavy metal contamination in prenatal vitamins poses a major concern for maternal-fetal health. One study sampled 51 brands of prenatal vitamins and found that all vitamins contained some arsenic, lead, and cadmium in concentrations that varied from acceptable for consumption to levels exceeding safety guidelines (Schwalfenberg et al. 2018). Testing of multiple same-brand batches showed a lack of internal consistency (Genuis et al. 2012). Heavy metals are heavily linked to the cognitive slowing in the newborn (Schwalfenberg et al. 2018). A process for more stringent regulation to ensure the safety and purity of prenatal vitamins is needed.

15.5.3 Dosage and Timing of Prenatal Vitamins on Mother and Fetus

Patient education is needed on the accurate timing and dosage of prenatal vitamins. While it is common knowledge that folic acid supplementation can reduce the incidence of neural tube defects, many women are unclear about the proper trimester for consuming folate supplements (Al Arifi et al. 2022). Similarly, while the literature supports maternal dietary micronutrients and omega-3 fatty acids playing major roles in the proper development of the fetal and neonatal immune system, little is known about when to begin supplementation: prior to pregnancy or during the first trimester (Rees et al. 2022). Not only are nutraceuticals heavily used in preventing birth defects, but supplements can also be used as an intervention after the fetus is exposed to certain teratogens. Specifically, evidence supports the use of prenatal and postnatal choline supplements to offset the neonatal neurocognitive impairments associated with fetal alcohol exposure (Ernst et al. 2022). Figure 15.9 shows the effect of contaminated and non-contaminated nutraceuticals.

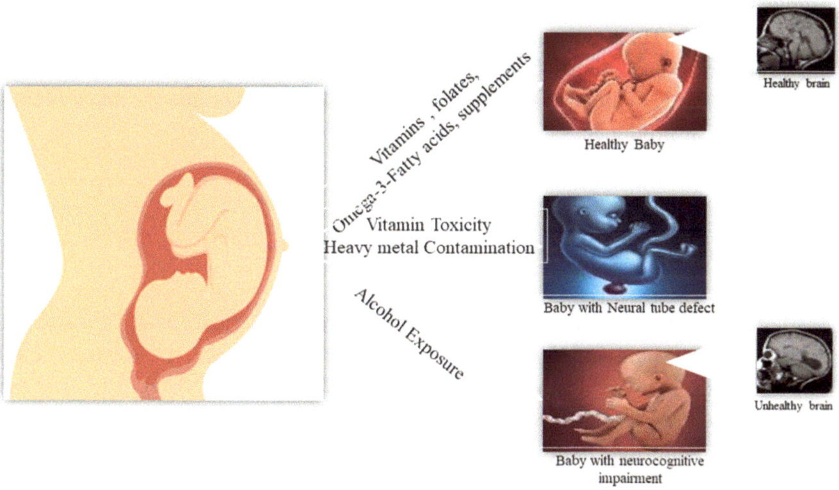

Fig. 15.9 Prior to pregnancy as well as the first trimester are the most important periods to prevent maternal malnutrition and consumption of toxins. Advantages and disadvantages of nutraceuticals for prenatal care include folates for neural tube development, and omega-3 fatty acids aid in the healthy development of the fetus immune system. However, some vitamins and heavy metal contamination can have harmful outcomes such as neural tube defects. Alcohol exposure consequences to the fetus can be improved by using prenatal and postnatal choline

15.6 Factors Affecting the Efficacy of Nutraceuticals

There are many factors that affect the efficacy of nutraceuticals. The primary factors which affect the efficacy of nutraceuticals are quality, impurities, and heavy metal contaminants (Foster 2016). For example, a common herb "ginseng" has several varieties like Brazilian, California, and Malaysian ginseng. Hence, it is very difficult to identify the real ginseng (Foster 2016). Similarly, some of the varieties of *Illicium verum* (star anise) were found to contain neurotoxins (Mathon et al. 2013). Moreover, herbal products derived from plants are a complex mixture that contains various bioactive compounds that are difficult to characterize due to variation in composition (Schmitt and Ferro 2013). Hence, good manufacturing practices (GMP) must be followed to prepare nutraceuticals without any contamination. Further, poor permeability, ineffective targeting, low solubility, fast metabolism, and short half-life after intake are considered as other factors that negatively affect efficacy (Zaki 2014). Before being marketed, the nutraceuticals are assessed for their efficacy. The assessment of efficacy depends on in vitro and in vivo studies. Consequently, many studies are required to prove the efficacy of nutraceuticals. Further, clearly evidenced high-level clinical trials are important to investigate the long-term effect and safety of nutraceuticals (AlAli et al. 2021).

Similarly, there is less evidence to prove the toxicity and to evaluate the therapeutic potential of herbal nutraceuticals because of the complexity of chemicals. The complexity further increases due to environmental conditions and fertilizers and pesticides used. These are also considered as factors that affect the efficacy of nutraceuticals. Thus, there should be proper guidelines for pharmacokinetic and pharmacodynamic evaluations of nutraceuticals to prevent variations and adulterants in nutraceutical products.

15.6.1 *Bioavailability*

Most of the nutraceuticals are taken through the oral route. The bio-efficiency depends on the bioavailability of the nutraceuticals. Hence, it is important to understand the bioavailability of nutraceuticals (Dima et al. 2020). The term bioavailability is used to describe the nutrients that are digested, absorbed, and metabolized through normal pathways (Shangraw 1990). According to pharmacology, bioavailability is the rate and extent to which a compound or drug is absorbed and becomes available at the site of action (Center for Drug Evaluation and Research (CDER) 2003). Therefore, bioavailability is "the proportion of a nutrient capable of being absorbed and available for use or stored" (Fairweather-Tait et al. 1987). Bioavailability is a complex process involving several different stages like liberation, absorption, distribution, metabolism, and elimination (Rein et al. 2013).

Bioaccessibility is the first step of bioavailability, which involves the release of compounds from the food matrix and made available for intestinal absorption (Saura-Calixto et al. 2007). The bioaccessibility starts with mastication in the mouth, followed by digestion with digestive fluids in the gastrointestinal lumen (Gropper et al. 2009). Next, the compounds are absorbed and transported to various organs. The absorption of a compound depends on solubility, interaction with other dietary ingredients, molecular transformations, cellular transporters, and its interaction with gut microbiota (Neilson and Ferruzzi 2011). The absorption of nutrients also varies from person to person depending on genetic predisposition and diet. Further, the absorption mechanism differs for both hydrophilic and lipophilic compounds (Richelle et al. 2006). The absorption mechanism of lipid-related compounds is complex due to the intestinal barrier (Fernández-García et al. 2012) because the larger lipids must first be converted into micelles using bile acids and amphiphilic nutrients through the emulsification process (Singh et al. 2009). Then, the lipids are transported by enterocytes via passive and facilitated diffusion (Cansell et al. 2003). In the enterocytes, the lipids are re-esterified by chylomicrons before being secreted into the lymphatic system (Niot et al. 2009). Nevertheless, the absorption of hydrophilic compounds is simple by enzymatic hydrolysis in the intestine followed by glucosidase-mediated hydrolysis by enterocytes (Del Rio et al. 2010). Figure 15.10 illustrates the different steps involved in the bioavailability of nutraceuticals.

However, the bioavailability of nutraceuticals was found to be affected by structural complexity and isomeric configuration of compounds, mechanism of transport of bioactive molecules, metabolic enzymes, and finally food–drug interactions (Williamson and Manach 2005; Brand et al. 2006; Scholz and Williamson

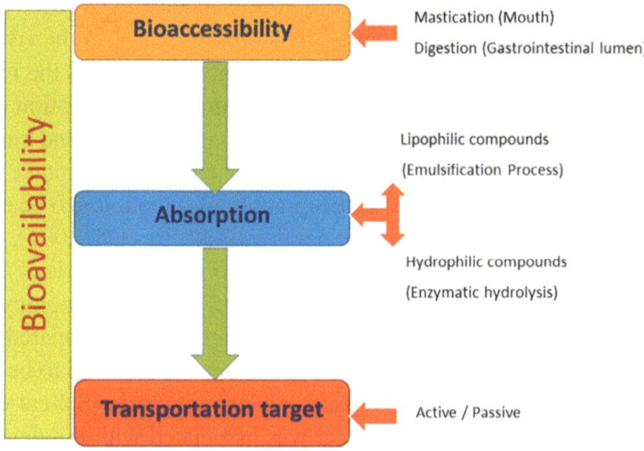

Fig. 15.10 The bioavailability process for nutraceuticals. Bioavailability is used to describe the nutrients that are digested and become available at the site of action. It entails the accessibility of these nutrients through digestion, absorption, and transportation to target areas

2007; Ottaviani et al. 2011). Among various factors, molecular weight and chemical structures are considered two important parameters that affect the bioavailability of nutraceuticals. Solubility and permeability characteristics differ based on molecular and physicochemical properties, and these also affect the bioavailability of nutraceuticals (Dahan et al. 2009). Based on the solubility, nutraceuticals are further divided into four groups: type I, type II, type III, and type IV (Porter et al. 2008). In recent years, several developments have been made to improve the solubility, bioavailability, and efficacy of nutraceuticals through chemical modification using colloidal and nanosystems (Yang et al. 2008).

15.6.2 Dosage, Side Effects, and Toxicity Potential of Nutraceuticals

People consume thousands of plant-based nutraceuticals to meet their nutritional needs. Nutraceuticals like herbal products, phytochemicals, and spices are considered safe and beneficial without any toxic effects when taken in the right quantities and doses (Riccioni et al. 2018). The safety of nutraceuticals is determined largely based on the type of compound, time, and quantity taken. The presence of contaminants such as heavy metals, toxic pesticides, fertilizers, and mycotoxins will result in some adverse health issues (Gul et al. 2016). Misuse or overuse also results in toxicity (Pirillo and Catapano 2015). In addition, nutraceuticals can interact with concurrent medications that patients with chronic illnesses are taking. The drug–nutraceutical interactions may result in the inhibition of drug efficacy or lead to a higher incidence of toxicity and severe side effects (Diamond and Bailey 2013).

From a toxicology point of view, it is important to identify the dose-response of a person. For that, it is important to conduct various in vitro and in vivo studies to identify optimal dosage response. However, a recent report shows the interaction of herbal products with pharmaceutical drugs. For example, consumption of very high doses of garlic or garlic oil can be toxic to the liver, kidney, heart, and lung (Banerjee et al. 2003; Ali et al. 2000). Similarly, the recommended daily intake of phenolic compounds may be 1 g/day. However, the quantity of flavonoids to be taken should be no more than a few tens of milligrams per day (Pokorny et al. 2001). Aloe vera is considered safe and is used for its wide range of properties for various applications. But recent evidence showed carcinogenic effects of unknown etiology (Boudreau et al. 2013). The aloe leaf extract was found to be highly toxic compared to aloe vera gel (Pandiri et al. 2011). Danthron, an aloe constituent, was found to be toxic and is responsible for DNA damage and caspase-induced apoptosis (Sehgal et al. 2013). Similarly, the long-term use of goldenseal was associated with the development of hepatocarcinoma/adenoma in F344/N rats (Maeda et al. 2014). Recent evidence also proves that the purified form of isoflavones leads to uterine hypertrophy or reproductive tract malformations, inhibits androgen production, reduces fertility, and stimulates estrogen-dependent tumor growth (Allred et al. 2001; Ronis et al. 2016). Isoflavones also lead to endometriosis and estrogen-sensitive cancers in women taking these products (Ronis et al. 2016).

Synthetic nutraceuticals were found to be more toxic than natural ones. The synthetic antioxidants like butylated hydroxyanisole (BHA), *tertiary*-butyl hydroquinone (TBHQ), and propyl gallate (PG) are more toxic because they can cause damage to the double-helical structure of DNA due to their interaction with nucleic acids, which results in the formation of molecular complexes (Dolatabadi and Kashanian 2010). However, the acceptability of synthetic products among consumers is diminished because of the accumulation of these compounds in cells, tissues, organs, and the body (Kulawik et al. 2013; Anraku et al. 2018). BHA and butylated hydroxytoluene (BHT) have been restricted by legislative regulations due to toxic and carcinogenic effects (Wichi 1988). In addition, PG has been shown to exhibit liver toxicity and enhance carcinogenesis (Shahidi and Ambigaipalan 2015).

15.7 Perspectives

Nutraceuticals are considered part of the diet because of the limitless benefits in the enhancement of health and in disease treatment. The benefits of nutraceuticals are explored every day for their role in biological processes like cell proliferation, inducing gene expression and antioxidant defenses and thereby helping in the treatment of various life-threatening diseases. Nutraceuticals are used for physiological illnesses to psychological ailments. It was observed that herbal nutraceuticals are the most common form for the treatment of various diseases. Since oxidative stress and related inflammation constitute a major component in various chronic disease conditions, nutraceuticals with high antioxidant and anti-inflammatory properties are considered a favorable option for the treatment of these disorders. The predominant reasons why people are focusing on nutraceuticals are safety, cost-effectiveness, availability, and ease to obtain (i.e., no prescription needed), as compared to acquiring pharmaceutical drugs which have many impediments including mild-to-moderate toxicity issues and severe side effects, such as nausea, heartburn, fatigue, dizziness, drowsiness, chest pain, loss of appetite, leg pain, constipation, bloating, stomach upset, tinnitus, behavioral changes such as paranoia, aggressiveness, hallucinations, addiction, impaired judgment, impulsiveness, and loss of self-control.

Nutraceuticals are not regulated by the FDA like pharmaceuticals. While pharmaceutical drugs have a wide range of negative impacts, unfortunately, nutraceuticals also show different drawbacks and many limitations. These include ineffective targeting, poor solubility, low permeability, fast metabolism, and other limitations. Further, taking more than needed will cost more and might raise the risk of side effects (e.g., too much vitamin A can cause headaches/liver damage, reduce bone strength, and cause birth defects, and excess iron can induce nausea and vomiting and may damage the liver and other organs). Nutraceuticals are also toxic and lead to the death of individuals even at minimally increased concentrations than the recommended dosage. Therefore, it is important to limit the use of nutraceuticals only at the time of illness.

Moreover, there is no standard for the preparation of nutraceuticals (e.g., the collagen isolated from animal and marine origin will show distinct properties, and the phytochemicals prepared are different in different countries because of the origin, species, and environmental conditions). Further, the quality of nutraceuticals can also be changed due to their contamination by pesticides, metals, and minerals. Accordingly, one should be aware of taking the nutraceuticals before consumption due to the abovementioned disadvantages. Specifically, people with allergic reactions and heart problems and those already on medications must consult a physician before including the nutraceuticals in their daily diet.

Acknowledgements We thank Jayakumar Nithura, Medical Academy of Science and Technology, Homestead FL, 33030, for assistance with manuscript preparation. We apologize for having limited references to review articles and more recently published articles, as we were limited in space.

References

Aarsland D, Batzu L, Halliday GM, Geurtsen GJ, Ballard C, Ray Chaudhuri K, Weintraub D (2021) Parkinson disease-associated cognitive impairment [Erratum in: Nat Rev Dis Primers 2021 Jul 13;7(1):53]. Nat Rev Dis Prim 7(1):47. https://doi.org/10.1038/s41572-021-00280-3

Abbott NJ, Patabendige AA, Dolman DE, Yusof SR, Begley DJ (2010) Structure and function of the blood–brain barrier. Neurobiol Dis 7(1):13–25

Adelaja AO, Schilling BJ (1999) Nutraceuticals: blurring the line between food and drugs in the twenty-first century. Choices 14(4):35–39

Al Arifi MN, Alqahtani AM, Alotaibi AN, Babelghaith SD, Alwhaibi A, Alsanea S, Alghadeer SM, Al-Arifi NM (2022) Knowledge and practice of childbearing women in Saudi Arabia towards folic acid supplement-evidence from a cross-sectional study. Nutrients 14(2):357. https://doi.org/10.3390/nu14020357

Al Mijan M, Lim BO (2018) Diets, functional foods, and nutraceuticals as alternative therapies for inflammatory bowel disease: present status and future trends. World J Gastroenterol 24(25): 2673

AlAli M, Alqubaisy M, Aljaafari MN, AlAli AO, Baqais L, Molouki A, Abushelaibi A, Lai KS, Lim SH (2021) Nutraceuticals: transformation of conventional foods into health promoters/ disease preventers and safety considerations. Molecules 26(9):2540

Al-Harbi NO, Nadeem A, Ahmad SF, AlThagfan SS, Alqinyah M, Alqahtani F, Ibrahim KE, Al-Harbi MM (2019) Sulforaphane treatment reverses corticosteroid resistance in a mixed granulocytic mouse model of asthma by upregulation of antioxidants and attenuation of Th17 immune responses in the airways. Eur J Pharmacol 855:276–284

Ali M, Thomson M, Afzal M (2000) Garlic and onions: their effect on eicosanoid metabolism and its clinical relevance. Prostaglandins Leukot Essent Fatty Acids 62(2):55–73

Allred CD, Allred KF, Ju YH, Virant SM, Helferich WG (2001) Soy diets containing varying amounts of genistein stimulate growth of estrogen-dependent (MCF-7) tumors in a dose-dependent manner. Cancer Res 61(13):5045–5050

Anderson JW, Baird P, Davis RH, Ferreri S, Knudtson M, Koraym A, Waters V, Williams CL (2009) Health benefits of dietary fiber. Nutr Rev 67(4):188–205

Andersson SW, Skinner J, Ellegård L, Welch AA, Bingham S, Mulligan A, Andersson H, Shaw KT (2004) Intake of dietary plant sterols is inversely related to serum cholesterol concentration in men and women in the EPIC Norfolk population: a cross-sectional study. Eur J Clin Nutr 58(10):1378–1385

Andrade S, Ramalho MJ, Pereira MDC, Loureiro JA (2018) Resveratrol brain delivery for neurological disorders prevention and treatment. Front Pharmacol 20(9):1261

Anjali Garg V, Dhiman A, Dutt R, Ranga S (2018) Health benefits of nutraceuticals. Pharm Innov J 7:178–181

Anraku M, Gebicki JM, Iohara D, Tomida H, Uekama K, Maruyama T, Hirayama F, Otagiri M (2018) Antioxidant activities of chitosans and its derivatives in in vitro and in vivo studies. Carbohydr Polym 199:141–149

Arts IC, Hollman PC (2005) Polyphenols and disease risk in epidemiologic studies. Am J Clin Nutr 81(1 Suppl):317S–325S. https://doi.org/10.1093/ajcn/81.1.317S

Asif M (2012) A brief study of toxic effects of some medicinal herbs on kidney. Adv Biomed Res 1: 44. https://doi.org/10.4103/2277-9175.100144

Ayuso MI, Montaner J (2015) Advanced neuroprotection for brain ischemia: an alternative approach to minimize stroke damage. Expert Opin Investig Drugs 24(9):1137–1142

Banerjee SK, Mukherjee PK, Maulik SK (2003) Garlic as an antioxidant: the good, the bad and the ugly. Phytother Res 7(2):97–106

Beghi E, Giussani G, Sander JW (2015) The natural history and prognosis of epilepsy. Epilept Disord 17(3):243–253. https://doi.org/10.1684/epd.2015.0751

Berthon BS, Wood LG (2015) Nutrition and respiratory health—feature review. Nutrients 7(3): 1618–1643. https://doi.org/10.3390/nu7031618

Bezkorovainy A (2001) Probiotics: determinants of survival and growth in the gut. Am J Clin Nutr 73(2):399s–405s

Bhattarai JP, Park SJ, Han SK (2013) Potentiation of NMDA receptors by *Withania somnifera* on hippocampal CA1 pyramidal neurons. Am J Chin Med 41(03):503–513

Boccardi V, Tinarelli C, Mecocci P (2017) Nutraceuticals and cognitive dysfunction: focus on Alzheimer's disease. In: Neuroprotective effects of phytochemicals in neurological disorders, vol 28. Wiley, New York, pp 561–579

Bolke L, Schlippe G, Gerß J, Voss W (2019) A collagen supplement improves skin hydration, elasticity, roughness, and density: results of a randomized, placebo-controlled. Blind Study Nutr 11(10):2494

Boonstra E, de Kleijn R, Colzato LS, Alkemade A, Forstmann BU, Nieuwenhuis S (2015) Neurotransmitters as food supplements: the effects of GABA on brain and behavior. Front Psychol 6(6):1520

Boudreau MD, Beland FA, Nichols JA, Pogribna M (2013) Toxicology and carcinogenesis studies of a nondecolorized [corrected] whole leaf extract of *Aloe barbadensis* Miller (Aloe vera) in F344/N rats and B6C3F1 mice (drinking water study). Natl Toxicol Prog Tech Rep Ser 577:1–266

Braak H, Del Tredici K (2008) Invited article: nervous system pathology in sporadic Parkinson disease. Neurology 70(20):1916–1925

Bracci GB (2001) Free radicals and brain damage in the newborn. Biol Neonate 79:180–186

Brand W, Schutte ME, van Williamson G, Zanden JJ, Cnubben NHP, Groten JP, Van Bladeren PJ, Rietjens IMCM (2006) Flavonoid-mediated inhibition of intestinal ABC transporters may affect the oral bioavailability of drugs, food-borne toxic compounds and bioactive ingredients. Biomed Pharmacother 60:508–519

Brookmeyer R, Johnson E, Ziegler-Graham K, Arrighi HM (2007) Forecasting the global burden of Alzheimer's disease. Alzheimers Dement 3(3):186–191

Brower V (1998) Nutraceuticals: poised for a healthy slice of the healthcare market? Nat Biotechnol 16:728–731. https://doi.org/10.1038/nbt0898-728

Bull E (2000) What is nutraceutical? Pharm J 265:57–58

Campos-Bedolla P, Walter FR, Veszelka S, Deli MA (2014) Role of the blood–brain barrier in the nutrition of the central nervous system. Arch Med Res 45(8):610–638

Cansell M, Nacja F, Combe N (2003) Marine lipid-based liposomes increase in vivo FA bioavailability. Lipids 38:551–559

Carlson SE (1999) Long-chain polyunsaturated fatty acids and development of human infants. Acta Paediatr 88:72–77

Carr AC, Maggini S (2017) Vitamin C and immune function. Nutrients 9(11):1211

Carvalho JL, Miranda M, Fialho AK, Castro-Faria-Neto H, Anatriello E, Keller AC, Aimbire F (2020) Oral feeding with probiotic *Lactobacillus rhamnosus* attenuates cigarette smoke-induced COPD in C57Bl/6 mice: relevance to inflammatory markers in human bronchial epithelial cells. PLoS One 15(4):e0225560

Center for Drug Evaluation and Research (CDER) (2003) Guidance for industry: bioavailability and bioequivalence studies for orally administrated drug products general considerations. Food and Drug Administration (FDA), Rockville

Cerella C, Dicato M, Jacob C, Diederich M (2011) Chemical properties and mechanisms determining the anti-cancer action of garlic-derived organic sulfur compounds. Anti Cancer Agents Med Chem 11(3):267–271

Charytoniuk T, Drygalski K, Konstantynowicz-Nowicka K, Berk K, Chabowski A (2017) Alternative treatment methods attenuate the development of NAFLD: a review of resveratrol molecular mechanisms and clinical trials. Nutrition 34:108–117. https://doi.org/10.1016/j.nut.2016.09.001

Chauhan NB, Mehla J (2015) Ameliorative effects of nutraceuticals in neurological disorders. In: Bioactive nutraceuticals and dietary supplements in neurological and brain disease, vol 1. Elsevier, Amsterdam, pp 245–260

Chauhan B, Kumar G, Kalam N, Ansari SH (2013) Current concepts and prospects of herbal nutraceutical: a review. J Adv Pharm Technol Res 4:4–8

Chellappan DK, Yee LW, Xuan KY, Kunalan K, Rou LC, Jean LS, Ying LY, Wie LX, Chellian J, Mehta M, Satija S (2020) Targeting neutrophils using novel drug delivery systems in chronic respiratory diseases. Drug Dev Res 81(4):419–436

Chevalier R (2018) Penetration assessment of dietary supplements and drugs through the blood–brain barrier for potential treatment of Parkinson's disease. bioRxiv 2018:362541

Chew EY, Clemons TE, Sangiovanni JP, Danis RP, Ferris FL 3rd, Elman MJ, Antoszyk AN, Ruby AJ, Orth D, Bressler SB, Fish GE, Hubbard GB, Klein ML, Chandra SR, Blodi BA, Domalpally A, Friberg T, Wong WT, Rosenfeld PJ, Agrón E, Toth CA, Bernstein PS, Sperduto

RD, Age-Related Eye Disease Study 2 (AREDS2) Research Group (2014) Secondary analyses of the effects of lutein/zeaxanthin on age-related macular degeneration progression: AREDS2 report No. 3. JAMA Ophthalmol 132(2):142–149. https://doi.org/10.1001/jamaophthalmol.2013.7376

Cicero AFG, Fogacci F, Stoian AP, Vrablik M, Al Rasadi K, Banach M, Toth PP, Rizzo M (2021) Nutraceuticals in the management of Dyslipidemia: which, when, and for whom? Could nutraceuticals help low-risk individuals with non-optimal lipid levels? Curr Atheroscler Rep 23(10):57. https://doi.org/10.1007/s11883-021-00955-y

Coresh J, Selvin E, Stevens LA, Manzi J, Kusek JW, Eggers P, Van Lente F, Levey AS (2007) Prevalence of chronic kidney disease in the United States. JAMA 298(17):2038–2047

Da Costa JP (2017) A current look at nutraceuticals—key concepts and prospects. Trends Food Sci Technol 62:68–78

Dahan A, Miller JM, Amidon GL (2009) Prediction of solubility and permeability class membership: provisional BCS classification of the world's top oral drugs. AAPS J 11(4):740–746

Dajas F (2012) Life or death: neuroprotective and anticancer effects of quercetin. J Ethnopharmacol 143(2):383–396

Das L, Bhaumik E, Raychaudhuri U, Chakraborty R (2012) Role of nutraceuticals in human health. J Food Sci Technol 49(2):173–183. https://doi.org/10.1007/s13197-011-0269-4

Datta M, Vitolins MZ (2016) Food fortification and supplement use: are there health implications? Crit Rev Food Sci Nutr 56(13):2149–2159

Del Rio D, Borges G, Crozier A (2010) Berry flavonoids and phenolics: bioavailability and evidence of protective effects. Br J Nutr 104:S67–S90

Dhir A (2018) Curcumin in epilepsy disorders. Phytother Res 32(10):1865–1875. https://doi.org/10.1002/ptr.6125

Diamond BJ, Bailey MR (2013) Ginkgo biloba: indications, mechanisms, and safety. Psychiatr Clin 36(1):73–83

Dillard CJ, German JB (2000) Phytochemicals: nutraceuticals and human health. J Sci Food Agric 80:1744–1756

Dima C, Assadpour E, Dima S, Jafari SM (2020) Bioavailability of nutraceuticals: role of the food matrix, processing conditions, the gastrointestinal tract, and nanodelivery systems. Compr Rev Food Sci Food Saf 19(3):954–994. https://doi.org/10.1111/1541-4337.12547

Dolatabadi JE, Kashanian S (2010) A review on DNA interaction with synthetic phenolic food additives. Food Res Int 43(5):1223–1230

Dominguez LJ, Barbagallo M (2018) Nutritional prevention of cognitive decline and dementia. Acta Biomed 89(2):276–290. https://doi.org/10.23750/abm.v89i2.7401

Donaldson MS (2004) Nutrition and cancer: a review of the evidence for an anti-cancer diet. Nutr J 3(1):1–21

Drossman DA (2016) Functional gastrointestinal disorders: history, pathophysiology, clinical features, and Rome IV. Gastroenterology 150(6):1262–1279

Dumont U, Sanchez S, Olivier B, Chateil JF, Pellerin L, Beauvieux MC, Bouzier-Sore AK, Roumes H (2019) Maternal consumption of piceatannol: a nutritional neuroprotective strategy against hypoxia-ischemia in rat neonates. Brain Res 1717:86–94

Dutta D, Mohanakumar KP (2015) Tea and Parkinson's disease: constituents of tea synergize with antiparkinsonian drugs to provide better therapeutic benefits. Neurochem Int 89:181–190

Ernst AM, Gimbel BA, de Water E, Eckerle JK, Radke JP, Georgieff MK, Wozniak JR (2022) Prenatal and postnatal choline supplementation in fetal alcohol spectrum disorder. Nutrients 14(3):688. https://doi.org/10.3390/nu14030688

Evenepoel P, Meijers BK, Bammens BR, Verbeke K (2009) Uremic toxins originating from colonic microbial metabolism. Kidney Int Suppl 114:S12–S19. https://doi.org/10.1038/ki.2009.402

Fairweather-Tait SJ, Balmer SE, Scott PH, Minski MJ (1987) Lactoferrin and iron absorption in newborn infants. Pediatr Res 22(6):651–654. https://doi.org/10.1203/00006450-198712000-00007

Fernández-García E, Carvajal-Lérida I, Jarén-Galán M, Garrido-Fernández J, Pérez-Gálvez A, Hornero-Méndez D (2012) Carotenoids bioavailability from foods: from plant pigments to efficient biological activities. Food Res Int 46:438–450

Foster S (2016) Towards an understanding of ginseng adulteration: the tangled web of names, history, trade, and perception. HerbalGram 3:36–57

Frigerio R, Sanft KR, Grossardt BR, Peterson BJ, Elbaz A, Bower JH, Ahlskog JE, De Andrade M, Maraganore DM, Rocca WA (2006) Chemical exposures and Parkinson's disease: a population-based case-control study. Mov Disord 21(10):1688–1692

Fuller R (1991) Probiotics in human medicine. Gut 32(4):439

Fuller-Thomson E, Lacombe-Duncan A (2016) Understanding the association between chronic obstructive pulmonary disease and current anxiety: a population-based study. COPD J Chron Obstruct Pulmon Dis 13(5):622–631

Gaby AR (2007) Natural approaches to epilepsy. Altern Med Rev 12(1):9–24

Gao X, Liu J, Li L, Liu W, Sun M (2020) A brief review of nutraceutical ingredients in gastrointestinal disorders: evidence and suggestions. Int J Mol Sci 21(5):1822

Garg A, Garg S, Zaneveld LJ, Singla AK (2001) Chemistry and pharmacology of the citrus bioflavonoid hesperidin. Phytother Res 15(8):655–669

Gausauge S (1997) Interleukin 1β-converting enzyme (Caspase-1) is overexpressed in adenocarcinoma of the pancreas. Cancer Res 58:2703–2708

Genuis SJ, Schwalfenberg G, Siy AK, Rodushkin I (2012) Toxic element contamination of natural health products and pharmaceutical preparations. PLoS One 7(11):e49676

Ghorbani Z, Hajizadeh M, Hekmatdoost A (2016) A dietary supplementation in patients with alcoholic liver disease: a review on current evidence. Hepatobiliary Pancreat Dis Int 15(4): 348–360

Ghosh R, Nadiminty N, Fitzpatrick JE, Alworth WL, Slaga TJ, Kumar AP (2005) Eugenol causes melanoma growth suppression through inhibition of E2F1 transcriptional activity. J Biol Chem 280(7):5812–5819

Gibson GR, Hutkins R, Sanders ME, Prescott SL, Reimer RA, Salminen SJ, Scott K, Stanton C, Swanson KS, Cani PD, Verbeke K (2017) Expert consensus document: the International Scientific Association for Probiotics and Prebiotics (ISAPP) consensus statement on the definition and scope of prebiotics. Nat Rev Gastroenterol Hepatol 14(8):491–502

Gillessen A, Schmidt HH (2020) Silymarin as supportive treatment in liver diseases: a narrative review. Adv Ther 37(4):1279–1301. https://doi.org/10.1007/s12325-020-01251-y

Grabitske HA, Slavin JL (2009) Gastrointestinal effects of low-digestible carbohydrates. Crit Rev Food Sci Nutr 49(4):327–360

Gropper SS, Smith JL, Groff JL (2009) The digestive system: mechanism for nourishing the body. Advanced nutrition and human metabolism, 5th edn. Cengage Learning, Belmont, pp 33–62

Gross BW, Gillio M, Rinehart CD, Lynch CA, Rogers FB (2017) Omega-3 fatty acid supplementation and warfarin: a lethal combination in traumatic brain injury. J Trauma Nurs 24(1):15–18

Guallar E, Stranges S, Mulrow C, Appel LJ, Miller ER III (2013) Enough is enough: stop wasting money on vitamin and mineral supplements. Ann Intern Med 159(12):850–851

Guerrero-Beltrán CE, Calderón-Oliver M, Pedraza-Chaverri J, Chirino YI (2012) Protective effect of sulforaphane against oxidative stress: recent advances. Exp Toxicol Pathol 64(5):503–508

Gul K, Singh AK, Jabeen R (2016) Nutraceuticals and functional foods: the foods for the future world. Crit Rev Food Sci Nutr 56(16):2617–2627

Guo TL, Chi RP, Germolec DR, White KL Jr (2005) Stimulation of the immune response in B6C3F1 mice by genistein is affected by exposure duration, gender, and litter order. J Nutr 135(10):2449–2456

Gupta S, Chauhan D, Mehla K, Sood P, Nair A (2010) An overview of nutraceuticals: current scenario. J Basic Clin Pharm 1(2):55

Gwaltney-Brant SM (2021) Nutraceuticals in renal diseases. In: Nutraceuticals. Academic Press, New York, pp 131–139

Hardy T, Oakley F, Anstee QM, Day CP (2016) Nonalcoholic fatty liver disease: pathogenesis and disease spectrum. Annu Rev Pathol 11:451–496. https://doi.org/10.1146/annurev-pathol-012615-044224

Harvey CJ, Thimmulappa RK, Sethi S, Kong X, Yarmus L, Brown RH, Feller-Kopman D, Wise R, Biswal S (2011) Targeting Nrf2 signaling improves bacterial clearance by alveolar macrophages in patients with COPD and in a mouse model. Sci Transl Med 3(78):78ra32

Hatziagapiou K, Kakouri E, Lambrou GI, Bethanis K, Tarantilis PA (2019) Antioxidant properties of *Crocus sativus* L. and its constituents and relevance to neurodegenerative diseases; focus on Alzheimer's and Parkinson's disease. Curr Neuropharmacol 17(4):377–402. https://doi.org/10.2174/1570159X16666180321095705

Huang MT, Lou YR, Ma W, Newmark HL, Reuhl KR, Conney AH (1994) Inhibitory effects of dietary curcumin on forestomach, duodenal, and colon carcinogenesis in mice. Cancer Res 54(22):5841–5847

Hung CK, Bodenheimer HC Jr (2018) Current treatment of nonalcoholic fatty liver disease/nonalcoholic steatohepatitis. Clin Liver Dis 22(1):175–187. https://doi.org/10.1016/j.cld.2017.08.012

Hwang YY, Ho YS (2018) Nutraceutical support for respiratory diseases. Food Sci Hum Wellness 7(3):205–208

Imam Z, Khasawneh M, Jomaa D, Iftikhar H, Sayedahmad Z (2019) Drug induced liver injury attributed to a curcumin supplement. Case Rep Gastrointest Med 2019:6029403. https://doi.org/10.1155/2019/6029403

Jang M, Cai L, Udeani GO, Slowing KV, Thomas CF, Beecher CW, Fong HH, Farnsworth NR, Kinghorn AD, Mehta RG, Moon RC (1997) Cancer chemopreventive activity of resveratrol, a natural product derived from grapes. Science 275(5297):218–220

Jemai H, Mahmoudi A, Feryeni A, Fki I, Bouallagui Z, Choura S, Chamkha M, Sayadi S (2020) Hepatoprotective effect of oleuropein-rich extract from olive leaves against cadmium-induced toxicity in mice. Biomed Res Int 2020:4398924. https://doi.org/10.1155/2020/4398924

Juneja LR, Kapoor MP, Okubo T, Rao T (2013) Green tea polyphenols: nutraceuticals of modern life. CRC Press, Boca Raton

Kalia AN (2005) Textbook of industrial pharmacognosy. CBS Publisher, New Delhi, pp 204–208

Kalogeris T, Bao Y, Korthuis RJ (2014) Mitochondrial reactive oxygen species: a double-edged sword in ischemia/reperfusion vs preconditioning. Redox Biol 2:702–714

Kalra EK (2003) Nutraceutical—definition and introduction. AAPS Pharm Sci 5:E25

Kang H, Kim H (2017) Astaxanthin and β-carotene in *Helicobacter pylori*-induced gastric inflammation: a mini-review on action mechanisms. J Cancer Prevent 22(2):57

Kaur H, Patro I, Tikoo K, Sandhir R (2015) Curcumin attenuates inflammatory response and cognitive deficits in experimental model of chronic epilepsy. Neurochem Int 89:40–50

Keith DS, Nichols GA, Gullion CM, Brown JB, Smith DH (2004) Longitudinal follow-up and outcomes among a population with chronic kidney disease in a large managed care organization. Arch Intern Med 164(6):659–663

Kerry RG, Patra JK, Gouda S, Park Y, Shin HS, Das G (2018) Benefaction of probiotics for human health: a review. J Food Drug Anal 26(3):927–939

Khatri M, Naughton RJ, Clifford T, Harper LD, Corr L (2021) The effects of collagen peptide supplementation on body composition, collagen synthesis, and recovery from joint injury and exercise: a systematic review. Amino Acids 53(10):1493–1506

Kim JE, Cho KO (2019) Functional nutrients for epilepsy. Nutrients 11(6):1309. https://doi.org/10.3390/nu11061309

Kim M, Kim Y (2010) Hypocholesterolemic effects of curcumin via up-regulation of cholesterol 7a-hydroxylase in rats fed a high fat diet. Nutr Res Pract 4:191–195

Kim DU, Chung HC, Choi J, Sakai Y, Lee BY (2018) Oral intake of low-molecular-weight collagen peptide improves hydration, elasticity, and wrinkling in human skin: a randomized, double-blind, placebo-controlled study. Nutrients 10(7):826

Kim JE, Kwon EY, Han Y (2022) A collagen hydrolysate containing tripeptides ameliorates sarcopenia in middle-aged mice. Molecules 27(9):2718. https://doi.org/10.3390/molecules27092718

Knapik JJ, Steelman RA, Hoedebecke SS, Farina EK, Austin KG, Lieberman HR (2014) A systematic review and meta-analysis on the prevalence of dietary supplement use by military personnel. BMC Complement Altern Med 14(1):1–9

Kokate CK, Purohit AP, Gokhale SB (2002) Nutraceutical and cosmeceutical. In: Pharmacognosy, 21st edn. Nirali Prakashan, Pune, pp 542–549

Konig D, Oesser S, Scharla S, Zdzieblik D, Gollhofer A (2018) Specific collagen peptides improve bone mineral density and bone markers in postmenopausal women—a randomized controlled study. Nutrients 10(1):97. https://doi.org/10.3390/nu10010097

Kruger CL, Murphy M, DeFreitas Z, Pfannkuch F, Heimbach J (2002) An innovative approach to the determination of safety for a dietary ingredient derived from a new source: case study using a crystalline lutein product. Food Chem Toxicol 40(11):1535–1549

Kulawik P, Ozogul F, Glew R, Ozogul Y (2013) Significance of antioxidants for seafood safety and human health. J Agric Food Chem 61(3):475–491

Kwak WJ, Kim JH, Ryu KH, Cho YB, Jeon SD, Moon CK (2005) Effects of gentianine on the production of pro-inflammatory cytokines in male Sprague-Dawley rats treated with lipopolysaccharide (LPS). Biol Pharm Bull 28(4):750–753

Larsson SC, Kumlin M, Ingelman-Sundberg M, Wolk A (2004) Dietary long-chain n-3 fatty acids for the prevention of cancer: a review of potential mechanisms. Am J Clin Nutr 79(6):935–945

Le Chatelier E, Nielsen T, Qin J, Prifti E, Hildebrand F, Falony G, Almeida M, Arumugam M, Batto JM, Kennedy S, Leonard P (2013) Richness of human gut microbiome correlates with metabolic markers. Nature 500(7464):541–546

Le Rouzic O, Pichavant M, Frealle E, Guillon A, Si-Tahar M, Gosset P (2017) Th17 cytokines: novel potential therapeutic targets for COPD pathogenesis and exacerbations. Eur Respir J 50(4):1602434

Leclerc M, Dudonné S, Calon F (2021) Can natural products exert neuroprotection without crossing the blood–brain barrier? Int J Mol Sci 22(7):3356

Leclere CL, Champ M, Boillot J, Guille G, Lecannu G, Molis C, Bornet F, Krempff M, Delort-Laval J (1994) Role of viscous guar gum in lowering the glycemic response after a solid meal. Am J Nutr 59:914–921

Lee KW, Lee HJ, Surh YJ, Lee CY (2003) Vitamin C and cancer chemoprevention: reappraisal. Am J Clin Nutr 78(6):1074–1078

Lee Y, Hu S, Park YK, Lee JY (2019) Health benefits of carotenoids: a role of carotenoids in the prevention of non-alcoholic fatty liver disease. Prev Nutr Food Sci 24:103–113

Leung PS (ed) (2014) The gastrointestinal system: gastrointestinal, nutritional and hepatobiliary physiology. Springer Science & Business, New York, p 29

Li SQ, Zhang QH (2001) Advances in the development of functional foods from buckwheat. Crit Rev Food Sci Nutr 41(6):451–464

Li J, Cordero P, Nguyen V, Oben JA (2016) The role of vitamins in the pathogenesis of non-alcoholic fatty liver disease. Integr Med Insights 11:19–25

Li X, Li X, Huang N, Liu R, Sun R (2018) A comprehensive review and perspectives on pharmacology and toxicology of saikosaponins. Phytomedicine 50:73–87. https://doi.org/10.1016/j.phymed.2018.09.174. Epub 2018 Sep 17. PMID: 30466994; PMCID: PMC7126585

Li C, Niu Z, Zou M, Liu S, Wang M, Gu X, Lu H, Tian H, Jha R (2020) Probiotics, prebiotics, and synbiotics regulate the intestinal microbiota differentially and restore the relative abundance of specific gut microorganisms. J Dairy Sci 103(7):5816–5829

Liang J, Pei XR, Wang N, Zhang ZF, Wang JB, Li Y (2010) Marine collagen peptides prepared from chum salmon (*Oncorhynchus keta*) skin extend the life span and inhibit spontaneous tumor incidence in Sprague-Dawley rats. J Med Food 13(4):757–770

Lim DY, Jeong Y, Tyner AL, Park JH (2007) Induction of cell cycle arrest and apoptosis in HT-29 human colon cancer cells by the dietary compound luteolin. Am J Physiol Gastrointest Liver Physiol 292(1):G66–G75

Linares DM, Gómez C, Renes E, Fresno JM, Tornadijo ME, Ross RP, Stanton C (2017) Lactic acid bacteria and bifidobacteria with potential to design natural biofunctional health-promoting dairy foods. Front Microbiol 8:846. https://doi.org/10.3389/fmicb.2017.00846

Liu L, Wang S, Liu J (2015) Fiber consumption and all-cause, cardiovascular, and cancer mortalities: a systematic review and meta-analysis of cohort studies. Mol Nutr Food Res 59(1): 139–146

Liu F, Li P, Chen M, Luo Y, Prabhakar M, Zheng H, He Y, Qi Q, Long H, Zhang Y, Sheng H (2017) Fructooligosaccharide (FOS) and galactooligosaccharide (GOS) increase Bifidobacterium but reduce butyrate producing bacteria with adverse glycemic metabolism in healthy young population. Sci Rep 7(1):11789

Liu Y, Tran DQ, Rhoads JM (2018) Probiotics in disease prevention and treatment. J Clin Pharmacol 58:S164–S179

Loomba R, Sanyal AJ (2013) The global NAFLD epidemic. Nat Rev Gastroenterol Hepatol 10(11): 686–690

Lopez MS, Dempsey RJ, Vemuganti R (2015) Resveratrol neuroprotection in stroke and traumatic CNS injury. Neurochem Int 89:75–82. https://doi.org/10.1016/j.neuint.2015.08.009. Epub 2015 Aug 12. PMID: 26277384; PMCID: PMC4587342

Luzina IG, Todd NW, Sundararajan S, Atamas SP (2015) The cytokines of pulmonary fibrosis: much learned, much more to learn. Cytokine 74(1):88–100

Maddi VS, Aragade PD, Digge VG, Nitaliker MN (2007) Importance of nutraceuticals in health management. Pharmacogn Rev 1:377–379

Madmani ME, Solaiman AY, Agha KT, Madmani Y, Shahrour Y, Essali A, Kadro W (2014) Coenzyme Q10 for heart failure. Cochrane Database Syst Rev 6:CD008684

Maeda J, Kijima A, Inoue K, Ishii Y, Ichimura R, Takasu S, Kuroda K, Matsushita K, Kodama Y, Saito N et al (2014) In vivo genotoxicity of *Ginkgo biloba* extract in gpt delta mice and constitutive androstane receptor knockout mice. Toxicol Sci 140:298–306

Makkar R, Behl T, Bungau S, Zengin G, Mehta V, Kumar A, Uddin MS, Ashraf GM, Abdel-Daim MM, Arora S, Oancea R (2020) Nutraceuticals in neurological disorders. Int J Mol Sci 21(12): 4424

Makki K, Deehan EC, Walter J, Bäckhed F (2018) The impact of dietary fiber on gut microbiota in host health and disease. Cell Host Microbe 23(6):705–715

Mandala F, Guida R, Guida GF (2010) Nutraceuticals: what they are and how they work. Eur Soc Cardiol Pract 9:11

Martini N (2019) Collagen supplements. J Prim Health Care 11(4):385–386

Mathon C, Bongard B, Duret M, Ortelli D, Christen P, Bieri S (2013) Analysis of the neurotoxin anisatin in star anise by LC-MS/MS. Food Addit Contam Part A 30(9):1598–1605

Mazzanti G, Menniti-Ippolito F, Moro PA, Cassetti F, Raschetti R, Santuccio C, Mastrangelo S (2009) Hepatotoxicity from green tea: a review of the literature and two unpublished cases. Eur J Clin Pharmacol 65(4):331–341

McGill HC Jr (1979) Relationship of dietary cholesterol to serum cholesterol concentration and to atherosclerosis in man. Am J Clin Nutr 32(12 Suppl):2664–2702

Mecocci P, Tinarelli C, Schulz RJ, Polidori MC (2014) Nutraceuticals in cognitive impairment and Alzheimer's disease. Front Pharmacol 5:147. https://doi.org/10.3389/fphar.2014.00147

Mhyre TR, Boyd JT, Hamill RW, Maguire-Zeiss KA (2012) Parkinson's disease. Subcell Biochem 65:389–455. https://doi.org/10.1007/978-94-007-5416-4_16

Mills JP, Rao K, Young VB (2018) Probiotics for prevention of *Clostridium difficile* infection. Curr Opin Gastroenterol 34:3–10

Mokgalaboni K, Ntamo Y, Ziqubu K, Nyambuya TM, Nkambule BB, Mazibuko-Mbeje SE, Gabuza KB, Chellan N, Tiano L, Dludla PV (2021) Curcumin supplementation improves biomarkers of oxidative stress and inflammation in conditions of obesity, type 2 diabetes and

NAFLD: updating the status of clinical evidence. Food Funct 12(24):12235–12249. https://doi.org/10.1039/d1fo02696h

Moradi H, Sica DA, Kalantar-Zadeh K (2013) Cardiovascular burden associated with uremic toxins in patients with chronic kidney disease. Am J Nephrol 38(2):136–148

Mortaz E, Adcock IM, Ricciardolo FL, Varahram M, Jamaati H, Velayati AA, Folkerts G, Garssen J (2015) Anti-inflammatory effects of *Lactobacillus rhamnosus* and *Bifidobacterium breve* on cigarette smoke activated human macrophages. PLoS One 10(8):e0136455. https://doi.org/10.1371/journal.pone.0136455

Mueller C (1999) The regulatory status of medical foods and dietary supplements in the United States. Nutrition 15:249–251

Mundi M, Mikal KM, Ahmed OH, Sarbini SR (2017) A review on the effects of prebiotics on cell toxicity and integrity. Int J Food Prop 20(Suppl 1):S1045–S1052

Nasri H, Baradaran A, Shirzad H, Rafieian-Kopaei M (2014) New concepts in nutraceuticals as alternative for pharmaceuticals. Int J Prev Med 5(12):1487

Neilson AP, Ferruzzi MG (2011) Influence of formulation and processing on absorption and metabolism of flavan-3-ols from tea and cocoa. Annu Rev Food Sci Technol 2:125–151

Neuwelt E, Abbott NJ, Abrey L, Banks WA, Blakley B, Davis T, Engelhardt B, Grammas P, Nedergaard M, Nutt J, Pardridge W, Rosenberg GA, Smith Q, Drewes LR (2008) Strategies to advance translational research into brain barriers. Lancet Neurol 7:84–96. https://doi.org/10.1016/S1474-4422(07)70326-5

Newman AM, Arshad M (2020) The role of probiotics, prebiotics and synbiotics in combating multidrug-resistant organisms. Clin Ther 42(9):1637–1648

Nguyen TT, Nguyen PT, Pham MN et al (2022) Synbiotics: a new route of self-production and applications to human and animal health. Probiot Antimicro Proteins 14(5):980–993. https://doi.org/10.1007/s12602-022-09960-2

Niot I, Poirier H, Tran TTT, Besnard P (2009) Intestinal absorption of long-chain fatty acids: evidence and uncertainties. Prog Lipid Res 48:101–115

Noureddin M, Rinella ME (2015) Nonalcoholic fatty liver disease, diabetes, obesity, and hepatocellular carcinoma. Clin Liver Dis 19(2):361–379

Nwosu OK, Ubaoji KI (2020) Nutraceuticals: history, classification and market demand. In: Functional foods and nutraceuticals. Springer, Cham, pp 13–22

Osna NA, Donohue TM Jr, Kharbanda KK (2017) Alcoholic liver disease: pathogenesis and current management. Alcohol Res 38(2):147

Ottaviani JI, Momma TY, Heiss C, Kwik-Uribe C, Schroeter H, Keen CL (2011) The stereochemical configuration of flavanols influences the level and metabolism of flavanols in humans and their biological activity in vivo. Free Radic Biol Med 50:237–244

Pallebage-Gamarallage M, Lam V, Takechi R, Galloway S, Clark K, Mamo J (2012) Restoration of dietary-fat induced blood-brain barrier dysfunction by anti-inflammatory lipid-modulating agents. Lipids Health Dis 7(11):117

Pamer EG (2016) Resurrecting the intestinal microbiota to combat antibiotic-resistant pathogens. Science 352(6285):535–538

Panahi Y, Khedmat H, Valizadegan G, Mohtashami R, Sahebkar A (2015) Efficacy and safety of aloe vera syrup for the treatment of gastroesophageal reflux disease: a pilot randomized positive-controlled trial. J Tradit Chin Med 35(6):632–636

Pandey M, Verma RK, Saraf SA (2010) Nutraceuticals: new era of medicine and health. Asian J Pharm Clin Rep 3:11–15

Pandiri AR, Sills RC, Hoenerhoff MJ, Peddada SD, Ton TVT, Hong HHL, Flake GP, Malarkey DE, Olson GR, Pogribny IP et al (2011) Aloe vera non-decolorized whole leaf extract-induced large intestinal tumors in F344 rats share similar molecular pathways with human sporadic colorectal tumors. Toxicol Pathol 39:1065–1074

Park HJ, Lee CM, Jung ID, Lee JS, Jeong YI, Chang JH, Chun SH, Kim MJ, Choi IW, Ahn SC, Shin YK, Yeom SR, Park YM (2009) Quercetin regulates Th1/Th2 balance in a murine model of asthma. Int Immunopharmacol 9(3):261–267. https://doi.org/10.1016/j.intimp.2008.10.021. Epub 2008 Dec 4. PMID: 19061976

Pavan R, Jain S, Kumar A (2012) Properties and therapeutic application of bromelain: a review. Biotechnol Res Int 2012:976203

Petroczi A, Taylor G, Naughton DP (2011) Mission impossible? Regulatory and enforcement issues to ensure safety of dietary supplements. Food Chem Toxicol 49(2):393–402

Pilch SM (1987) Physiological effects and health consequences of dietary fiber. Agris, Washington, DC

Piqué N, Berlanga M, Miñana-Galbis D (2019) Health benefits of heat-killed (Tyndallized) probiotics: an overview. Int J Mol Sci 20(10):2534

Pirillo A, Catapano AL (2015) Berberine, a plant alkaloid with lipid- and glucose-lowering properties: from in vitro evidence to clinical studies. Atherosclerosis 243:449–461

Pithadia AB, Jain S (2011) Treatment of inflammatory bowel disease (IBD). Pharmacol Rep 63(3): 629–642

Pluta R, Ułamek-Kozioł M, Januszewski S, Czuczwar SJ (2020) Gut microbiota and pro/prebiotics in Alzheimer's disease. Aging (Albany NY) 12(6):5539–5550. https://doi.org/10.18632/aging.102930

Pokorny J, Trojakova L, Takacsova M (2001) The use of natural antioxidants in food products of plant origin. In: Antioxidants in food: practical applications. Woodhead Publishing Ltd, Cambridge, pp 355–372

Porter CJ, Wasan KM, Constantinides P (2008) Lipid-based systems for the enhanced delivery of poorly water-soluble drugs. Adv Drug Deliv Rev 6(60):615–616

Rahmani AH, Alsahli MA, Aly SM, Khan MA, Aldebasi YH (2018) Role of curcumin in disease prevention and treatment. Adv Biomed Res 7:38. https://doi.org/10.4103/abr.abr_147_16

Ram A, Mabalirajan U, Das M, Bhattacharya I, Dinda AK, Gangal SV, Ghosh B (2006) Glycyrrhizin alleviates experimental allergic asthma in mice. Int Immunopharmacol 6(9): 1468–1477

Rees G, Brough L, Orsatti GM, Lodge A, Walker S (2022) Do micronutrient and omega-3 fatty acid supplements affect human maternal immunity during pregnancy? A scoping review. Nutrients 14(2):367. https://doi.org/10.3390/nu14020367

Rein MJ, Renouf M, Cruz-Hernandez C, Actis-Goretta L, Thakkar SK, da Silva PM (2013) Bioavailability of bioactive food compounds: a challenging journey to bioefficacy. Br J Clin Pharmacol 75(3):588–602. https://doi.org/10.1111/j.1365-2125.2012.04425.x

Riccioni G, Gammone MA, Currenti W, D'Orazio N (2018) Effectiveness and safety of dietetic supplementation of a new nutraceutical on lipid profile and serum inflammation biomarkers in hypercholesterolemic patients. Molecules 23:1168

Richelle M, Sabatier M, Steiling H, Williamson G (2006) Skin bioavailability of dietary vitamin E, carotenoids, polyphenols, vitamin C, zinc and selenium. Br J Nutr 96:227–238

Richter JE, Castell DO (1982) Gastroesophageal reflux. Pathogenesis, diagnosis, and therapy. Ann Intern Med 97(1):93–103. https://doi.org/10.7326/0003-4819-97-1-93

Riedl J, Linseisen J, Hoffmann J, Wolfram G (1999) Some dietary fibers reduce the absorption of carotenoids in women. J Nutr 129(12):2170–2176

Ronis MJ, Gomez-Acevedo H, Blackburn ML, Cleves MA, Singhal R, Badger TM (2016) Uterine responses to feeding soy protein isolate and treatment with 17β-estradiol differ in ovariectomized female rats. Toxicol Appl Pharmacol 297:68–80

Rubin SA, Levin ER (1994) The endocrinology of vasoactive peptides: synthesis to function. J Clin Endocrinol Metab 78(1):6–10

Ruchi S (2017) Role of nutraceuticals in health care: a review. Int J Green Pharm 11(3):S386

Sabatino A, Regolisti G, Cosola C, Gesualdo L, Fiaccadori E (2017) Intestinal microbiota in type 2 diabetes and chronic kidney disease. Curr Diabetes Rep 17(3):1–9

Saibil F (1989) Diarrhea due to fiber overload. N Engl J Med 320(9):599

Sanders ME, Akkermans LM, Haller D, Hammerman C, Heimbach J, Hörmannsperger G, Huys G, Levy DD, Lutgendorff F, Mack D, Phothirath P, Solano-Aguilar G, Vaughan E (2010) Safety assessment of probiotics for human use. Gut Microbes 1(3):164–185. https://doi.org/10.4161/gmic.1.3.12127. Epub 2010 Mar 4. PMID: 21327023; PMCID: PMC3023597

Saura-Calixto F, Serrano J, Goñi I (2007) Intake and bioaccessibility of total polyphenols in a whole diet. Food Chem 101:492–501

Sawmiller D, Li S, Shahaduzzaman M, Smith AJ, Obregon D, Giunta B, Borlongan CV, Sanberg PR, Tan J (2014) Luteolin reduces Alzheimer's disease pathologies induced by traumatic brain injury. Int J Mol Sci 15(1):895–904

Scheffer IE, Berkovic S, Capovilla G, Connolly MB, French J, Guilhoto L, Hirsch E, Jain S, Mathern GW, Moshé SL, Nordli DR (2017) ILAE classification of the epilepsies: position paper of the ILAE Commission for Classification and Terminology. Epilepsia 58(4):512–521

Schmitt J, Ferro A (2013) Nutraceuticals: is there good science behind the hype? Br J Clin Pharmacol 75(3):585

Scholz S, Williamson G (2007) Interactions affecting the bioavailability of dietary polyphenols in vivo. Int J Vitam Nutr Res 77:224–235

Schonberg MA, Breslau ES, Hamel MB, Bellizzi KM, McCarthy EP (2015) Colon cancer screening in U.S. adults aged 65 and older according to life expectancy and age. J Am Geriatr Soc 63:750–756

Schroeder B, Bäckhed F (2016) Signals from the gut microbiota to distant organs in physiology and disease. Nat Med 22:1079–1089. https://doi.org/10.1038/nm.4185

Schwalfenberg G, Rodushkin I, Genuis SJ (2018) Heavy metal contamination of prenatal vitamins. Toxicol Rep 5:390–395. https://doi.org/10.1016/j.toxrep.2018.02.015

Seeram NP, Adams LS, Zhang Y, Lee R, Sand D, Scheuller HS, Heber D (2006) Blackberry, black raspberry, blueberry, cranberry, red raspberry, and strawberry extracts inhibit growth and stimulate apoptosis of human cancer cells in vitro. J Agric Food Chem 54(25):9329–9339

Sehgal I, Winters WD, Scott M, Kousoulas K (2013) An in vitro and in vivo toxicologic evaluation of a stabilized aloe vera gel supplement drink in mice. Food Chem Toxicol 55:363–370

Shahidi F, Ambigaipalan P (2015) Phenolics and polyphenolics in foods, beverages and spices: antioxidant activity and health effects—a review. J Funct Foods 18:820–897

Shakoor H, Feehan J, Mikkelsen K, Al Dhaheri AS, Ali HI, Platat C, Ismail LC, Stojanovska L, Apostolopoulos V (2021) Be well: a potential role for vitamin B in COVID-19. Maturitas 144:108–111. https://doi.org/10.1016/j.maturitas.2020.08.007. Epub 2020 Aug 15. PMID: 32829981; PMCID: PMC7428453

Shangraw RF (1990) Standards for vitamins and nutritional supplements-who and when. Pharmacopeial Forum 16(4):751–758

Shardell MD, Alley DE, Hicks GE, El-Kamary SS, Miller RR, Semba RD, Ferrucci L (2011) Low-serum carotenoid concentrations and carotenoid interactions predict mortality in US adults: the Third National Health and Nutrition Examination Survey. Nutr Res 31(3):178–189

Shaw OE, Yager JY (2019) Preventing childhood and lifelong disability: maternal dietary supplementation for perinatal brain injury. Pharmacol Res 139:228–242

Shaw D, Leon C, Kolev S, Murray V (1997) Traditional remedies and food supplements. A 5-year toxicological study. Drug Saf 17:342–356

Shityakov S, Neuhaus W, Dandekar T, Förster C (2013) Analysing molecular polar surface descriptors to predict blood–brain barrier permeation. Int J Comput Biol Drug Des 6(1–2): 146–156

Shrestha R, Shrestha S, Badri KC, Shrestha S (2021) Evaluation of nutritional supplements prescribed, its associated cost and patients knowledge, attitude and practice towards nutraceuticals: a hospital based cross-sectional study in Kavrepalanchok, Nepal. PLoS One 16(6):e0252538

Silva BS, Lira FS, Ramos D, Uzeloto JS, Rossi FE, Freire AP, Silva RN, Trevisan IB, Gobbo LA, Ramos EM (2018) Severity of COPD and its relationship with IL-10. Cytokine 106:95–100

Simopoulos AP (2002) Omega-3 fatty acids in inflammation an autoimmune diseases. J Am Coll Nutr 21:495–505

Simopoulos AP (2008) The importance of the omega-6/omega-3 fatty acid ratio in cardiovascular disease and other chronic diseases. Exp Biol Med 233(6):674–688

Singh J, Sinha S (2012) Classification, regulatory acts and applications of nutraceuticals for health. Int J Pharm Bio Sci 2:177–187

Singh H, Ye A, Horne D (2009) Structuring food emulsions in the gastrointestinal tract to modify lipid digestion. Prog Lipid Res 48:92–100

Singh J, Ancheria RK, Khinchi MP, Nama N, Singh SP (2017) A review on food supplement-nutraceuticals. Asian J Pharm Res Dev 1:1–7

Song H, Li B (2017) Beneficial effects of collagen hydrolysate: a review on recent developments. Biomed J Sci Tech Res 1:458–461

Spiller R, Major G (2016) IBS and IBD—separate entities or on a spectrum? Nat Rev Gastroenterol Hepatol 10:613–621

Spisni E, Turroni S, Alvisi P, Spigarelli R, Azzinnari D, Ayala D, Imbesi V, Valerii MC (2022) Nutraceuticals in the modulation of the intestinal microbiota: current status and future directions. Front Pharmacol 2022:780

Stahl W, Sies H (2005) Bioactivity and protective effects of natural carotenoids. Biochim Biophys Acta (BBA) Mol Basis Dis 1740(2):101–107

Stenvinkel P, Ketteler M, Johnson RJ, Lindholm B, Pecoits-Filho R, Riella M, Heimbürger O, Cederholm T, Girndt M (2005) IL-10, IL-6, and TNF-α: central factors in the altered cytokine network of uremia—the good, the bad, and the ugly. Kidney Int 67(4):1216–1233

Stephens NG, Parsons A, Brown MJ, Schofield PM, Kelly F, Cheeseman K, Mitchinson MJ (1996) Randomised controlled trial of vitamin E in patients with coronary disease: Cambridge Heart Antioxidant Study (CHAOS). Lancet 347(9004):781–786

Suez J, Zmora N, Zilberman-Schapira G, Mor U, Dori-Bachash M, Bashiardes S, Zur M, Regev-Lehavi D, Brik RB, Federici S, Horn M (2018) Post-antibiotic gut mucosal microbiome reconstitution is impaired by probiotics and improved by autologous FMT. Cell 174(6): 1406–1423

Tai M, Zhang J, Song S, Miao R, Liu S, Pang Q, Wu Q, Liu C (2015) Protective effects of luteolin against acetaminophen-induced acute liver failure in mouse. Int Immunopharmacol 27(1): 164–170. https://doi.org/10.1016/j.intimp.2015.05.009

Takechi R, Pallebage-Gamarallage MM, Lam V, Giles C, Mamo JC (2013) Nutraceutical agents with anti-inflammatory properties prevent dietary saturated-fat induced disturbances in blood–brain barrier function in wild-type mice. J Neuroinflamm 19(10):73

Tang G, Xu Y, Zhang C, Wang N, Li H, Feng Y (2021) Green tea and epigallocatechin gallate (EGCG) for the management of nonalcoholic fatty liver diseases (NAFLD): insights into the role of oxidative stress and antioxidant mechanism. Antioxidants (Basel) 10(7):1076

Thomasset SC, Berry DP, Garcea G, Marczylo T, Steward WP, Gescher AJ (2007) Dietary polyphenolic phytochemicals—promising cancer chemopreventive agents in humans? A review of their clinical properties. Int J Cancer 120(3):451–458

Tosi G, Bortot B, Ruozi B, Dolcetta D, Vandelli MA, Forni F, Severini GM (2013) Potential use of polymeric nanoparticles for drug delivery across the blood–brain barrier. Curr Med Chem 20(17):2212–2225

Valdes AM, Walter J, Segal E, Spector TD (2018) Role of the gut microbiota in nutrition and health. BMJ 13:361

Van Zyl WF, Deane SM, Dicks LM (2020) Molecular insights into probiotic mechanisms of action employed against intestinal pathogenic bacteria. Gut Microbes 12(1):1831339

Vanholder R, Van Laecke S, Glorieux G (2008) What is new in uremic toxicity? Pediatr Nephrol 23(8):1211–1221. https://doi.org/10.1007/s00467-008-0762-9

Vauzour D, Camprubi-Robles M, Miquel-Kergoat S, Andres-Lacueva C, Bánáti D, Barberger-Gateau P, Bowman GL, Caberlotto L, Clarke R, Hogervorst E, Kiliaan AJ (2017) Nutrition for the ageing brain: towards evidence for an optimal diet. Ageing Res Rev 35:222–240

Vernon G, Baranova A, Younossi ZM (2011) Systematic review: the epidemiology and natural history of non-alcoholic fatty liver disease and non-alcoholic steatohepatitis in adults. Aliment Pharmacol Ther 34(3):274–285

Vimaleswaran KS, Cavadino A, Berry DJ, LifeLines Cohort Study Investigators, Jorde R, Dieffenbach AK, Lu C, Alves AC, Heerspink HJ, Tikkanen E, Eriksson J, Wong A, Mangino M, Jablonski KA, Nolte IM, Houston DK, Ahluwalia TS, van der Most PJ, Pasko D, Zgaga L, Thiering E, Vitart V, Fraser RM, Huffman JE, de Boer RA, Schöttker B, Saum KU, McCarthy MI, Dupuis J, Herzig KH, Sebert S, Pouta A, Laitinen J, Kleber ME, Navis G, Lorentzon M, Jameson K, Arden N, Cooper JA, Acharya J, Hardy R, Raitakari O, Ripatti S, Billings LK, Lahti J, Osmond C, Penninx BW, Rejnmark L, Lohman KK, Paternoster L, Stolk RP, Hernandez DG, Byberg L, Hagström E, Melhus H, Ingelsson E, Mellström D, Ljunggren O, Tzoulaki I, McLachlan S, Theodoratou E, Tiesler CM, Jula A, Navarro P, Wright AF, Polasek O, International Consortium for Blood Pressure (ICBP); Cohorts for Heart and Aging Research in Genomic Epidemiology (CHARGE) Consortium; Global Blood Pressure Genetics (Global BPGen) Consortium, Hayward C, Wilson JF, Rudan I, Salomaa V, Heinrich J, Campbell H, Price JF, Karlsson M, Lind L, Michaëlsson K, Bandinelli S, Frayling TM, Hartman CA, Sørensen TI, Kritchevsky SB, Langdahl BL, Eriksson JG, Florez JC, Spector TD, Lehtimäki T, Kuh D, Humphries SE, Cooper C, Ohlsson C, März W, de Borst MH, Kumari M, Kivimaki M, Wang TJ, Power C, Brenner H, Grimnes G, van der Harst P, Snieder H, Hingorani AD, Pilz S, Whittaker JC, Järvelin MR, Hyppönen E (2014) Association of vitamin D status with arterial blood pressure and hypertension risk: a Mendelian randomisation study. Lancet Diabetes Endocrinol 2(9):719–729. https://doi.org/10.1016/S2213-8587(14)70113-5

Vitale DC, Piazza C, Melilli B, Drago F, Salomone S (2013) Isoflavones: estrogenic activity, biological effect and bioavailability. Eur J Drug Metab Pharmacokinet 38(1):15–25

Wang H (2021) A review of the effects of collagen treatment in clinical studies. Polymers (Basel) 13(22):3868. https://doi.org/10.3390/polym13223868

Wang Y, Wang B, Du F, Su X, Sun G, Zhou G, Bian X, Liu N (2015) Epigallocatechin-3-gallate attenuates oxidative stress and inflammation in obstructive nephropathy via NF-κB and Nrf2/HO-1 signalling pathway regulation. Basic Clin Pharmacol Toxicol 117(3):164–172

Weaver CM, Alekel DL, Ward WE, Ronis MJ (2012) Flavonoid intake and bone health. J Nutr Gerontol Geriatr 31(3):239–253

Wichi HP (1988) Enhanced tumor development of butylated hydroxyanisole (BHA) from the prospective of effect on fore stomach and oesophageal squamous epithelium. Food Chem Toxicol 26:717–723

Williams RJ, Mohanakumar KP, Beart PM (2015) Neuro-nutraceuticals: the path to brain health via nourishment is not so distant. Neurochem Int 89:1–6. https://doi.org/10.1016/j.neuint.2015.08.012

Wilkins T, Sequoia J (2017) Probiotics for gastrointestinal conditions: a summary of the evidence. Am Fam Physician 96(3):170–178. PMID: 28762696

Williams LM, Scott HA, Wood LG (2019) Soluble fibre as a treatment for inflammation in asthma. J Nutr Intermed Metab 18:100108. https://doi.org/10.1016/j.jnim.2019.100108

Williamson G, Manach C (2005) Bioavailability and bioefficacy of polyphenols in humans. II. Review of 93 intervention studies. Am J Clin Nutr 81(1):243S–255S

Wilson MS, Wynn TA (2009) Pulmonary fibrosis: pathogenesis, etiology and regulation. Mucosal Immunol 2(2):103–121

Woo JJ (2007) Adverse event monitoring and multivitamin–multimineral dietary supplements. Am J Clin Nutr 85(1):323S–324S

Yang CS, Sang S, Lambert JD, Lee MJ (2008) Bioavailability issues in studying the health effects of plant polyphenolic compounds. Mol Nutr Food Res 52:S139–S151

Younossi ZM, Koenig AB, Abdelatif D, Fazel Y, Henry L, Wymer M (2016) Global epidemiology of nonalcoholic fatty liver disease—meta-analytic assessment of prevalence, incidence, and outcomes. Hepatology 64(1):73–84

Zaki NM (2014) Progress and problems in nutraceuticals delivery. J Bioequiv Bioavailab 6(3):75
Zeisel SH (1999) Regulation of "nutraceuticals". Science 285:185–186
Zhang H, Hardie L, Bawajeeh AO, Cade J (2020) Meat consumption, cognitive function and disorders: a systematic review with narrative synthesis and meta-analysis. Nutrients 12(5): 1528. https://doi.org/10.3390/nu12051528
Zhang W, Zhang F, Hu Q, Xiao X, Ou L, Chen Y, Luo S, Cheng Y, Jiang Y, Ma X, Zhao Y (2021) The emerging possibility of the use of geniposide in the treatment of cerebral diseases: a review. Chin Med 16(1):1–6
Zhao C, Noble JM, Marder K, Hartman JS, Gu Y, Scarmeas N (2018) Dietary patterns, physical activity, sleep, and risk for dementia and cognitive decline. Curr Nutr Rep 7(4):335–345. https://doi.org/10.1007/s13668-018-0247-9
Zhao X, Zhang M, Li C, Jiang X, Su Y, Zhang Y (2019) Benefits of vitamins in the treatment of Parkinson's disease. Oxid Med Cell Longev 2019:9426867
Ziegler EE, Filer LJ Jr (1996) Present knowledge in nutrition, 7th edn. International Life Sciences Institute-Nutrition Foundation, Washington, DC

Chapter 16
Clinical Application and Trials with Nutraceuticals

Jonathan Newell, Jayakumar Nithura, Joven Isaac, Miranda A. Melians, Arokiasamy Justin Thenmozhi, Natarajan Sampath, and Hussain Hussain

Abstract Pharmaceutical drugs are chemical compounds that are designed to prevent, diagnose, and treat a specific disorder. Nutraceuticals are products that come from food sources which help in aiding physiological benefits and preventing various diseases. Nutraceuticals have major benefits over pharmaceuticals. These include no adverse effects, being economically affordable, ease of access to the public, multiple therapeutic effects, and increasing the health value by improving the medical condition. Many foods and drugs that classify as nutraceuticals offer health and medical benefits to the consumer. These include dietary supplements, probiotics, vitamins, and medicinal foods. While offering multiple health benefits, many of these nutraceuticals also offer some protection against chronic diseases. These include vitamin C, which has been studied to increase the manageability of various

J. Newell
NSU University School, Davie, FL, USA

J. Nithura
Medical Academy for Science and Technology, Homestead, FL, USA

J. Isaac
The University of South Florida, Tampa, FL, USA

M. A. Melians
School of Nursing and Health Studies, University of Miami, Miami, FL, USA

A. J. Thenmozhi
Department of Biochemistry, School of Biological Sciences, Madurai Kamaraj University, Madurai, Tamil Nadu, India

Department of Biochemistry and Biotechnology, Faculty of Science, Annamalai University, Chidambaram, Tamil Nadu, India

N. Sampath
School of Chemical and Biotechnology, SASTRA Deemed University, Thanjavur, Tamil Nadu, India

H. Hussain (✉)
Department of Internal Medicine and Infectious Disease, Larkin Community Hospital, Miami, FL, USA

A. Justin Thenmozhi, T. Manivasagam (eds.), *Nutraceuticals for Alzheimer's Disease: A Promising Therapeutic Approach*, Nutritional Neurosciences,
https://doi.org/10.1007/978-981-99-0677-2_16

conditions including the more recent pandemic of SARS-CoV-2 infection and reduce the risk of Alzheimer's disease. These nutraceuticals have also been proven to aid in the treatment of numerous cancers by strengthening the immune system while undergoing chemotherapy. While evidence shows a beneficial effect of these nutraceuticals on human health and disease, dietary supplements are not reviewed and approved by FDA instead FDA issues regulations about ingredients, labeling, registration, etc., and it is the industry's responsibility to comply with these regulations. Further, there are many clinical trials for the consumption of nutraceuticals. Examples are the use of resveratrol, omega-3, alpha-lipoic acid, curcumin, coenzyme Q10, and L-arginine. This review summarizes the current treatment strategies and clinical trials of nutraceuticals in various conditions with focus on CNS disorders emphasizing its therapeutic value in AD. We believe this information will assist industries and R & D sectors interested in developing nutraceuticals for the treatment of various diseases, as well as patients who suffer from chronic diseases.

Keywords Clinical trial · Dietary supplements · Herbal products · Nutraceuticals · Omega-3 · Probiotics · Treatment · FDA regulations

16.1 Introduction

Unlike pharmaceuticals, nutraceuticals are more commonly used in everyday life, such as nutritional choices and dietary supplements. Many nutraceuticals are naturally occurring, or used in many common food items, allowing them to be more convenient for the consumers. The utilization of nutraceutical is increasing every day due to their benefits and minimal associated adverse effects. There has been an increase in demand and supply for nutraceuticals due to the COVID-19 pandemic, however, even before the pandemic, the nutraceutical industry was growing at a rate of about 6–7% year after year. Some of the major nutraceuticals in use include ginseng, an antioxidant that strengthens the immune system, and omega 3, which provides many cardiovascular benefits such as lowering blood pressure and reducing the chances of myocardial infarction or stroke. Another major nutraceutical used is cod liver oil, which provides a wide variety of health benefits including bone health improvement, eye health, reduced risk of heart disease, and improved symptoms of depression and anxiety.

Nutraceuticals are used to fight major health issues such as obesity, cardiovascular diseases, arthritis, cancer, diabetes, high blood pressure, and cholesterol. Nutraceuticals, functional foods, and micronutrients prevent potential cancer cell growth and boost a person's immune system after undergoing cancer treatment such as chemotherapy. However, the FDA classifies nutraceuticals as foods and not drugs, therefore, they are not reviewed for approval by the FDA, instead, FDA issues regulations about ingredients, labeling, registration, etc. On the other hand, numerous nutraceuticals are currently undergoing clinical trials. In a recent analysis of clinical trials, a report stated that GTCs in nutraceuticals help in reducing body weight by increasing energy expenditure, reducing nutrient absorption, and by stimulating fatty acid oxidation. For lung disorders, nutraceuticals containing vitamins such as vitamins A, B, C, and D as well as the intake of carotenoids, flavonoids, curcumins, magnesium, and resveratrol as well as omega-3 fatty acids, show effects on protecting the lungs from disease and by improving lung function. In the prevention of kidney diseases, nutraceutical resveratrol is used to improve vascular functions in patients that suffer from chronic kidney diseases. However, based on human clinical trials, evidence is too scattered to draw any sort of conclusion. Nutraceuticals are both more sustainable and less costly than many pharmaceuticals, therefore after numerous clinical trials confirm their effectiveness in treating diseases and conditions, nutraceuticals will be utilized more as treatment in the future. This review summarizes the current treatment strategies and clinical trials regarding nutraceuticals.

16.2 Current Treatment Strategies with Nutraceuticals

16.2.1 Nutraceuticals in Neurological Disorder

The connection between diet and neurological conditions has been studied for years, and so has the relationship between nutraceuticals and dietary supplements with various neurological conditions. Certain nutraceuticals are essential components for neuronal integrity and function.

16.2.1.1 Omega-3 on Insomnia, Mood, Autism, and Dementia

While one study found that an omega-3 PUFA deficiency can result in disturbed nocturnal sleep and found a positive relation between omega-3 PUFA and overall sleep wellness, another study raised opposite findings stating high-EPA fish oil supplements are likely associated with a disturbance of sleeping after successful treatment of depression; the symptoms disappeared after cessation of supplementation, although such negative reports are rare (Zhao et al. 2020). With this information, researchers found that the nutraceutical, omega-3 PUFA, especially docosahexaenoic and eicosatetraenoic acids, are necessary for the normal brain function and visual development and regulation of behavior and mood in children diagnosed with autism spectrum disorder (Sivamaruthi et al. 2020). It was also found that compared to normal children, those with autism spectrum disorder have a low level of omega-3 (Sivamaruthi et al. 2020).

Although the results of this study found that implementing the nutraceutical omega-3 into children with autism spectrum disorder's diets improves their symptoms, the researchers involved believe that further study on the topic is needed to establish strong evidence for the beneficial effect of omega-3 on children diagnosed with autism spectrum disorder (Sivamaruthi et al. 2020). In the study regarding the usage of nutraceuticals in the treatment of dementia, researchers found omega-3 to be beneficial in the treatment of dementia symptoms (Wells et al. 2017).

16.2.1.2 Curcumin on Alzheimer's Disease and Depression

In a study performed, isolating the nutraceutical curcumin and its potential as a treatment method for many conditions, researchers found that the common nutraceutical, vitamin E can aid in the treatment of Alzheimer's disease (Kunnumakkara et al. 2017). Researchers then found that curcumin treatment resulted in elevated levels of vitamin E without causing any adverse reactions through the antioxidant effects of curcumin (Kunnumakkara et al. 2017). Along with Alzheimer's disease, curcumin was also found to improve symptoms of depression (Sanmukhani et al. 2014) in many patients. The study confirmed curcumin to be effective and safe to treat the major depressive disorder without concurrent suicidal ideation or other psychotic disorders in those patients (Kunnumakkara et al. 2017).

16.2.1.3 Different Nutraceuticals on Alzheimer's Disease

Alzheimer's disease is one of the most common progressive neurodegenerative diseases affecting human beings. Currently, more than 47 million patients are suffering from Alzheimer's and the incidence will increase to 130 million by 2050 (Tiwari et al. 2019). The exact pathophysiology of Alzheimer's disease is still complex (Tiwari et al. 2019). The accumulation of beta-amyloid, tau protein, and hyperphosphorylated transactive response DNA-binding protein 43 (TDP-43) leads to degenerative changes in the neurons (Tiwari et al. 2019).

A randomized clinical trial was conducted on 79 patients with Alzheimer's disease to receive selenium combined with a probiotic (*Bifidobacterium bifidum*, *Lactobacillus acidophilus*, and *Bifidobacterium longum*) for 12 weeks (Tamtaji et al. 2019). The study found a combination of probiotics with selenium showed significant improvement in cognitive function (Tamtaji et al. 2019).

A total of 340 patients with Alzheimer's disease were conducted in a randomized clinical trial on a high dose of vitamin B for 1.5 years (Aisen et al. 2008). The cognitive function was assessed through Alzheimer's Disease Assessment Scale (ADAS-cog). The authors exhibited no significant improvement in cognitive function with the use of vitamin B (Aisen et al. 2008).

Phase I multicentric, randomized, double-blind clinical trial was conducted 82 patients with mild-moderate Alzheimer's disease to receive a nutraceutical BrainUp-10 for 12 weeks (Guzman-Martinez et al. 2021). This component was manufactured to block the pathway of tau accumulation in Alzheimer's disease. The data of the study presented significant improvement in cognitive function (Guzman-Martinez et al. 2021).

Ginkgo Biloba is commonly used for memory and cognition. A randomized clinical trial was conducted on 3069 normal individuals aged above 75 years to investigate the efficacy of Ginkgo Biloba in prevention of Alzheimer's disease (DeKosky et al. 2008). The scientists concluded no significant prevention of Alzheimer's disease with the use of Ginkgo Biloba (DeKosky et al. 2008).

A randomized clinical trial was conducted on 210 patients with Alzheimer's disease. The patients received 12-month 800 IU daily of vitamin D (Jia et al. 2019). The study found an improvement in cognitive function and decreased the beta-amyloid biomarkers in those patients (Jia et al. 2019).

A study conducted patients with Alzheimer's disease to assess the efficacy of nutraceutical polyamine spermidine (Wirth et al. 2018). The participants received this nutraceutical for three consecutive months (Wirth et al. 2018). The data showed beneficial effect of polyamine spermidine on cognitive function (Wirth et al. 2018).

Phase II, cohort, randomized clinical trial was conducted on 106 patients with Alzheimer's disease to a nutraceutical formulation such as acetyl L-carnitine, folate, vitamin B12, methionine, alpha-tocopherol, and *N*-acetyl cysteine for a total of 12 months (Remington et al. 2015). In phase I, this nutraceutical showed significant improvement in cognitive function. But in phase II, the study found no beneficial effect with utilization of these components on cognitive function in Alzheimer's disease (Remington et al. 2015).

Souvenaid has been linked with Alzheimer's disease in the last couple decades, by improving the conductivity of the neurons (Scheltens et al. 2012). A randomized clinical trial was conducted patients with mild Alzheimer's disease to assess the efficacy of Souvenaid on cognitive function (Scheltens et al. 2012). The participants received Souvenaid for 24 weeks. The authors demonstrated significant beneficial effect of this component on cognitive function (Scheltens et al. 2012).

16.2.1.4 Tryptophan on Insomnia

Tryptophan is the precursor for serotonin, which plays an important role in sleep regulation and mood. It has been found to improve insomnia and other sleep disturbance (Zhao et al. 2020).

16.2.1.5 Vitamin B on Intellectual Abnormality in Down Syndrome

A study by Mazurek and Wyka (2015) demonstrates that patients with Down Syndrome (DS) present with vitamin B deficiency and abnormal blood homocysteine levels exhibit a decrease in the rate of intellectual development. The authors further report those early dietary interventions by parents of DS children afford an opportunity for decreasing the risk or delaying some of the DS associated-behavioral changes (Mazurek and Wyka 2015).

16.2.1.6 MIG-99 on Migraine

MIG-99 has shown beneficial effects in the treatment of migraine. Some side effects were noticed including arthralgias and oral ulcers (Wells et al. 2017).

16.2.2 Nutraceuticals in Cancer

Nutraceuticals have the potential to reduce cancer growth by inducing cancer cell apoptosis and inhibiting proliferation. Resveratrol, genistein, capsaicin, curcumin, flavopiridol, and caffeic acid are some of the common nutraceuticals that have anticancer properties. Nutraceuticals are currently used for the treatment of various cancers, as well as in clinical trials, given along with chemotherapy to strengthen the immune system (Harvie 2014). Below we summarize the current treatment strategies and clinical trials on nutraceuticals for cancer.

16.2.2.1 Omega-3 and Vitamin E

In an analysis performed by Critical Reviews in Oncology/Hematology, the use of multiple nutraceuticals including omega-3 fatty acids and vitamins was examined in clinical trials and treatment due to their anticancer effects (Vernieri et al. 2018). The authors conclude that specific dietary supplementation provides potential cancer prevention and reduction in tumor recurrences as well as improves mortality and morbidity, while the authors also noted that patients supplemented with nutraceuticals in advanced malignancies are uncertain, as well as discouraged to use due to lack of sufficient safety and efficacy data. In another study, a group of 60 patients with generalized tumors including GI, breast, lung liver, and pancreas were investigated for the effect of a combination of omega-3 fatty acids and vitamin E on overall health (Gogos et al. 1998; Mochamat et al. 2017). However, there was a significant increase in the survival rate of patients who had been taking these two dietary supplements together (Gogos et al. 1998; Mochamat et al. 2017). Further, these authors failed to differentiate between the specific impact of vitamin E supplementation compared with omega-3 fatty acids. In another study, patients experienced side effects such as a metallic taste after magnesium supplementation, diarrhea, and nausea after L-carnitine supplementation, or mild abdominal discomfort and transient diarrhea after a mixture of omega-3 polyunsaturated fatty acids plus vitamin E (Mochamat et al. 2017). Noteworthy, in a clinical trial, vitamin E was found to possibly increase the risk of prostate cancer (Harvie 2014).

16.2.2.2 Vitamin D on Cancer Associated Symptoms

Many cancer patients undergoing treatment have been found to suffer from vitamin D deficiency, therefore vitamin D supplements have been added to many cancer patients' diets to make up for this deficiency (Harvie 2014). There is some evidence for the role of vitamin D in decreasing melanoma risk and reducing progression in some patients and the photoprotective effect of vitamin D as well as improves bone health associated with chemotherapy, although additional high-quality studies are needed to determine appropriate dosing (Thompson and Kim 2021).

16.2.2.3 Anamorelin on Cancer Associated Symptoms

Anamorelin was found to have some positive effects to relieve cancer-related anorexia and improve the quality of life (Zhang et al. 2018).

16.2.2.4 Diet on Cancer Associated Symptoms

When researchers studied which nutraceuticals cancer survivors should take, they focused more directly on breast cancer survivors and found that a dietary pattern high in fruits, vegetables, whole grains, poultry, and fish (Rock et al. 2012) was found to be associated with reduced mortality compared with a dietary pattern characterized by a high intake of refined grains, processed and red meats, desserts, high-fat dairy products, and French fries in the breast cancer survivors studied (Rock et al. 2012). In another study, researchers involved found that multimodality interventions that integrate several factors such as low-saturated fat, plant-based, whole-food diets with exercise, and stress reduction (Zuniga et al. 2020) appear to have the most clinically significant benefit for patients with prostate cancer (Zuniga et al. 2020).

16.2.2.5 Vitamin C on Different Cancers

Vitamin C supplementation was assessed for beneficial effects among 39 patients with stomach (Mochamat et al. 2017), lung (Shorey-Kendrick et al. 2020), liver (Farombi et al. 2005), breast (Suhail et al. 2012), cervix (Fuchs-Tarlovsky et al. 2011), colorectal (Jaffey 1982), and other advance stage malignancies. Vitamin C was substituted intravenously and orally, and patients improved on different subscales of the European Organization for Research and Treatment of Cancer Quality of Life Questionnaire including physical and cognitive function, appetite loss, fatigue, and nausea/vomiting (Mochamat et al. 2017).

16.2.2.6 Glutamine and Arginine on Cancer Associated Symptoms

Arginine and glutamine were also studied in a clinical trial to assess their beneficial effects on cancer symptoms. The authors have noticed an improvement in the body mass index, mood, fatigue, and hematological parameters after 4 weeks (Mochamat et al. 2017).

16.2.2.7 Resveratrol on Different Cancers

Many in vitro and in vivo studies suggest that resveratrol has anticancer properties due to its wide range of mechanisms including antioxidant effects and regulation of pro-apoptotic proteins, and secondary pathways involved in the pathogenesis of malignant cells (Galiniak et al. 2019). Resveratrol is known to reduce the incidence and development of various types of cancer in humans such as cervical (Aggarwal et al. 2004), pancreatic (Djamgoz and Jentzsch 2022), gastric (Aggarwal et al. 2004), breast (Singh et al. 2019), and colorectal (Patel et al. 2010) as well as thyroid cancer (Aggarwal et al. 2004). Research results demonstrated that resveratrol has a protective effect on normal cells while inducing apoptosis in malignant cells (Galiniak et al. 2019).

16.2.2.8 Oleum Fructus bruceas Plus Ganji Decoction on Primary Hepatic Carcinoma

A randomized clinical trial was conducted on 97 patients with primary hepatic carcinoma with oleum fructus bruceas plus ganji decoction, a traditional Chinese herbal medicine (Tian et al. 2010). The patients were subdivided into two groups; case group (received Oleum Fructus bruceas and Ganji Decoction) versus the control group (placebo) (Tian et al. 2010). A significant improvement was observed in those patients post-treatment with these medicines (Tian et al. 2010).

16.2.2.9 Traditional Herbal Medicine's Role in Hepatocellular Carcinoma

A randomized clinical trial was conducted on 364 patients to prevent recurrent hepatocellular carcinoma with the traditional herbal medicine Cinobufacini (Zhai et al. 2018). The patients were divided into two groups: 180 received traditional herbal medicine and 184 patients received trans-arterial chemoembolization (Zhai et al. 2018). The findings revealed that the beneficial effect of traditional herbal medicine was superior to the trans-arterial chemoembolization approach (Zhai et al. 2018).

16.2.3 Nutraceuticals in Cardiovascular Disorder

The most well-known cardiovascular diseases are coronary artery disease, stroke, hypertension, cardiomyopathy, venous thrombosis, arrhythmia, and thromboembolic disease. Nutraceuticals were recently involved in multiple clinical trials to assess their beneficial effects in treating those conditions.

16.2.3.1 Sterols, Polyphenols, and Spirulina

Coronary artery disease arises because of vascular blockage of blood, oxygen, and nutrients to the heart. Compounds found in nutraceuticals such as sterols, polyphenols, and spirulina play an integral role in the treatment of coronary artery disease (Khurana et al. 2013). Sterols work by metabolizing dietary cholesterol which allows cholesterol molecules to be used in circulation in the form of very low-density lipoproteins (Khurana et al. 2013). Polyphenols are involved in signaling alteration that decreases lipid accumulation within the endothelial layer of the arteries, and they also inhibit cytokine and other molecules involved in atherosclerosis (Khurana et al. 2013). Polyphenols are found in a high concentration in pomegranates, a recommendation to eat daily pomegranates in patients with atherosclerosis.

A randomized trial was conducted on 100 patients at risk of suffering from coronary artery disease to compare the efficacy of treatment using low doses of statin and nutraceuticals in high-intensity statin-intolerant patients (Khurana et al. 2013). Many clinical trials have indicated that the use of statins to treat cardiovascular diseases shows significant decreases in the mortality rate and the occurrence of life-threatening cardiac events (Khurana et al. 2013). Although high-intensity statins are usually effective, patients usually discontinue the form of treatment because of the related side effects (Khurana et al. 2013). Therefore, as a solution to the problem, low-intensity statin drugs such as ezetimibe were paired with nutraceuticals containing red yeast rice, policosanol, and berberine. Unfortunately, the trial is yet to prove whether the alternative has any healing effects on high-intensity statin-intolerant patients (Khurana et al. 2013).

16.2.3.2 Selenium on Stroke

Stroke, also referred to as a cerebrovascular accident, occurs when there is an interruption or reduction in the supply of blood to parts of the brain, causing the brain's supply of oxygen and nutrients to be stopped or diminished (Sharifi-Razavi et al. 2022). Nutraceuticals provide preventative measures for reducing the risk of stroke by promoting ischemic tolerance and reducing the effects of consequences experienced after having a stroke, regulating blood pressure, ceasing neurodegeneration, and improving the condition of blood vessels (Sharifi-Razavi et al. 2022). Nutraceuticals explored in clinical trials as therapeutic methods for stroke include selenium. Selenium supplementation in acute ischemic stroke was shown to improve short-term outcomes but cannot influence the long-term outcome (Sharifi-Razavi et al. 2022). This conclusion was drawn after a clinical trial completion of a randomized, parallel, and placebo-controlled study completed on 44 test subjects between the ages of 58 and 78 who suffered ischemic strokes up to 72 h before the administration of selenium (Sharifi-Razavi et al. 2022). The administration period of selenium, given to a group of 22 patients who were chosen randomly, and saline solution, also given to a group of 22 patients of random selection, was 5 days (Sharifi-Razavi et al. 2022).

16.2.3.3 Omega-3, α-Lipoic Acid, Green Tea, Black Tea, Seaweed Wakame, and Arginine on Hypertension

Specific nutraceuticals which can be used to treat hypertension include omega-3, α-lipoic acid, green tea, black tea, seaweed wakame, and arginine (Ghaffari and Roshanravan 2020). An observational clinical trial where 387 participants were studied over 5 years, aimed to provide an analysis of the number of nutraceuticals prescribed, dosages administered, and the conditions which were intended or being treated including hypertension (Ghaffari and Roshanravan 2020). Nutraceuticals administered included berberine (500–2000 mg/day), omega-3 (1–3 g/day), α-lipoic acid (400–800 mg/day), and L-arginine (250 mg/day). No results or outcomes of the trial conducted were provided (Ghaffari and Roshanravan 2020).

16.2.3.4 Niacin on Atherosclerosis

It is known that low levels of high-density lipoprotein (HDL) are a risk in coronary artery diseases, as are increased levels of low-density lipoprotein (LDL) (AIM-HIGH Investigators et al. 2011). Niacin nutraceutical is known to increase the level of HDL, and it is involved in lowering cardiovascular risk as assessed in several clinical studies (AIM-HIGH Investigators et al. 2011). In the Coronary Drug Project of the AIM-HIGH trial, 3414 patients with known coronary artery diseases were randomized to extended-release niacin or placebo (AIM-HIGH Investigators et al. 2011). Patients in the niacin group had more adverse events, including liver function test abnormalities, myopathy, and rhabdomyolysis, compared to placebo (AIM-HIGH Investigators et al. 2011). The Heart Protection Study 2 trial randomized 25,673 patients with known vascular disease to niacin-laropiprant or placebo (AIM-HIGH Investigators et al. 2011). Both groups did not differ significantly in the incidence of MACE (AIM-HIGH Investigators et al. 2011). However, the niacin group had an increased incidence of adverse effects like the AIM-HIGH group (AIM-HIGH Investigators et al. 2011). Patients in the niacin group had more adverse events, including liver function test abnormalities, myopathy, and rhabdomyolysis, compared to placebo (AIM-HIGH Investigators et al. 2011). Both groups did not differ significantly in the incidence of MACE. However, the niacin group had an increase in adverse effects like AIM-HIGH (AIM-HIGH Investigators et al. 2011). A recent clinical study by Jenkins et al. showed that the use of slow-release niacin in patients during therapy does not benefit cardiovascular outcomes but instead does show an increasing trend toward all-cause mortality (AIM-HIGH Investigators et al. 2011).

16.2.4 Nutraceutical in Kidney Disease

The kidney filters about 200 L of fluid per day. It plays a vital function in excreted waste products, toxins, and excess ions (Gounden et al. 1998). In addition, the kidney secretes various hormones, regulates extracellular fluid volume, and maintains serum osmolality and electrolytes concentration. Different endocrine hormones act on the kidney including anti-diuretic hormone, aldosterone, and angiotensin (Gounden et al. 1998). Additionally, the kidney secretes hormones including vitamin D, erythropoietin, dopamine, and prostaglandin (Gounden et al. 1998). Also, there are various diseases affecting the kidney in terms of hypertension, diabetes mellitus, renal failure, nephrotic and nephritic syndromes, as well as cancer and infection (Gounden et al. 1998). Various clinical trials of nutraceuticals in the treatment of renal diseases happened in the last couple decades will be discussed below.

16.2.4.1 Sorghum with Unfermented Probiotic Milk

A clinical trial conducted on patients with chronic renal failure consumed breakfast cereal rich in sorghum (Lopes et al. 2018). The combination of whole sorghum with probiotic milk shown a significant reduction in the inflammatory process associated with chronic kidney injury (Lopes et al. 2018). This meal has a high percentage of carbohydrate (71%), a certain amount of protein (11%), and low fat (0.4%), as well as a large number of fibers, phenolic compounds, and tannin and other antioxidants (Lopes et al. 2018). All of that combination aids in the reduction of C-reactive protein, as well as malondialdehyde serum levels. In addition, the superoxide dismutase has increased largely with this combination (Lopes et al. 2018). Thus, the consumption of sorghum with unfermented milk showed a reduction in inflammation and oxidative stress associated with chronic kidney injury (Lopes et al. 2018).

16.2.4.2 Turmeric and End-Stage Renal Failure

Diabetes mellitus is considered to be one of the most common causes of renal failure. Proteinuria and transforming growth factor beta have been involved in the pathogenesis of end-stage renal failure secondary to diabetes (Khajehdehi et al. 2011). A randomized double-blind clinical trial was conducted on 40 patients diagnosed with diabetic nephropathy (Khajehdehi et al. 2011). The control group received a placebo, while the case group received 500 mg of turmeric for 2 months (Khajehdehi et al. 2011). The proteinuria, interleukin 8, and transforming growth factor-beta were decreased significantly (Khajehdehi et al. 2011).

16.2.4.3 Green Tea Extract in Renal Failure

Green tea extract increases the expression of the receptor for advanced glycation end products (RAGE) (Barocio-Pantoja et al. 2021). A clinical trial was conducted to assess the benefit of increasing RAGE on 39 patients with renal failure (Barocio-Pantoja et al. 2021). After administration of green tea extract for a certain period, an improvement in glomerular filtration rate was noticed in the case group, while no beneficial effect was noticed in the control group (Barocio-Pantoja et al. 2021).

16.2.4.4 Vitamin D and Omega-3 in Chronic Renal Failure

A randomized clinical trial was conducted with 1312 patients who developed chronic renal failure due to diabetes mellitus to assess the benefits of administrating omega-3 and vitamin D among those patients. After 5 years, the study revealed no benefits from using omega 3 and vitamin D in such patients (de Boer et al. 2019). Another randomized clinical trial was conducted in patients with chronic renal failure to investigate the benefit of administration of daily vitamin D for 6 months (Mager et al. 2017). Bone health has improved among those patients, and the serum vitamin D level was back to normal level (Mager et al. 2017).

16.2.4.5 Probiotic Supplementation in Chronic Kidney Injury

A randomized double-blind study was conducted on 16 patients diagnosed with chronic renal failure (Borges et al. 2018). The probiotic was administered for 3 months, and the inflammatory markers and uremic toxins including indoxyl sulfate, *p*-cresyl sulfate, and indole-3-acetic acid were collected (Borges et al. 2018). During the study, the gut profile and inflammatory markers were not changed. In addition, urea levels went up (Borges et al. 2018). Eventually, the probiotic study failed to reveal significant improvement in the condition (Borges et al. 2018).

16.2.4.6 Zinc Supplement in Chronic Renal Failure

A clinical trial was conducted in patients (less than 18 years of age) with chronic kidney disease to assess the nutritional status with zinc (30 mg vs 15 mg) for a specific period of time (Escobedo-Monge et al. 2019). The body mass index, albumin level, and C-reactive protein were assessed for any potential changes during the treatment (Escobedo-Monge et al. 2019). The authors demonstrated significant improvement in the nutritional status of those patients (Escobedo-Monge et al. 2019).

16.2.4.7 Alpha Lipoic Acid Supplement in Autosomal Dominant Polycystic Kidney Disease

A randomized clinical trial on 59 patients with autosomal dominant polycystic kidney disease was conducted to evaluate the efficacy of alpha-lipoic acid on the inflammation associated with this disease (Lai et al. 2020). The authors concluded a significant improvement in the inflammatory markers after alpha-lipoic acid administration (Lai et al. 2020).

16.2.4.8 Vitamin K Supplementation in Chronic Kidney Disease

Chronic kidney injury accelerates atherosclerosis. About 159 patients with chronic kidney disease were conducted in a randomized clinical trial of vitamin K supplement (Witham et al. 2020). A daily supplement of 400 mg of vitamin K for 12 months showed no significant improvement in vascular stiffness among patients with chronic kidney disease (Witham et al. 2020).

16.2.4.9 L-Carnitine Role in Hemodialysis Due to Renal Failure

Japanese patients (80 cases) suffering from chronic renal failure treated with hemodialysis and peritoneal dialysis underwent a randomized clinical trial with L-carnitine. Administration of L-carnitine on daily basis for 12 months showed significant improvement in renal anemia and reduced muscle spasms among those patients (Kuwasawa-Iwasaki et al. 2020).

16.2.4.10 *Curcuma longa* and *Boswellia serrata* Role in Chronic Kidney Disease

A clinical trial was conducted with *Curcuma longa* and *Boswellia serrata* as a nutraceutical supplement for chronic kidney injury in 16 patients (Moreillon et al. 2013). The aim of the study was to reduce inflammation and increase the antioxidants in non-dialysis chronic kidney injury patients (Moreillon et al. 2013). The interleukin-6, tumor necrotic factor alpha, C-reactive protein, and glutathione peroxidase were measured. A significant improvement in the inflammatory profile was identified in those patients posttreatment (Moreillon et al. 2013).

16.2.4.11 Vitamin E Supplementation in Kidney Injury

A randomized double-blind study was conducted on 60 patients diagnosed with diabetic nephropathy (Khatami et al. 2016). Supplementation of vitamin E for

12 weeks revealed significant improvement in the inflammation associated with diabetic nephropathy, reducing the oxidative stress and positive effect on the biomarkers of kidney injury (Khatami et al. 2016). The *P* value was less than 0.001, a large reduction of tumor necrotic factor alpha level, as well as metalloproteinase 2 and 9 levels among those patients (Khatami et al. 2016).

16.2.4.12 Magnesium in Chronic Kidney Injury

One hundred and twenty-eight hypomagnesemic prediabetic patients with high body mass index were involved in a randomized clinical trial for 3 months of daily supplementation of magnesium (Toprak et al. 2017). The study aims to assess the metabolic rate among those patients (Toprak et al. 2017). Insulin resistance, hemoglobin A1c, insulin, uric acid, and waist circumference were significantly decreased post-supplementation (Toprak et al. 2017). Additionally, albumin and magnesium levels were increased when compared with the diseased group.

16.2.4.13 N-3 Fatty Acids and Coenzyme Q10 Roles in Chronic Kidney Injury

A double-blind interventional trial was conducted on 74 patients with chronic kidney injury (Barden et al. 2018). Administration of coenzyme Q10 and N-3 fatty acids revealed positive effects in terms of increasing the level of neutrophil release of leukotriene B5, as well as reduction of myeloperoxidase level (Barden et al. 2018). Both are involved in the chronic inflammation associated with chronic kidney injury (Barden et al. 2018). Reduction of their level by this supplement has a positive impact on such patients in terms of lowering the inflammatory process (Barden et al. 2018).

16.2.4.14 Vitamin C Effect on Kidney Failure

Vitamin C deficiency is common among patients with renal injury. A randomized clinical trial of 3 months was conducted on 99 patients with severe kidney failure (Singer 2011). The study aims to determine whether vitamin C has a positive impact on those patients (Singer 2011). Administration of 250 mg of ascorbic acid three times a week revealed no improvement in the symptoms of vitamin C deficiency (fatigue, gingivitis, cardiovascular instability, and depression) (Singer 2011).

16.2.4.15 Cranberry's Role in Nephrolithiasis

Cranberry has been linked in one study with the development of nephrolithiasis by increasing urinary level of oxalate levels (Redmond et al. 2019). Also, the study has mentioned the association between vitamin C and oxalate levels in urine (Redmond et al. 2019). A randomized clinical trial involved 15 patients subdivided into two groups; some received daily tablets containing vitamin C and cranberry, while the other received only cranberry (Redmond et al. 2019). The level of urinary oxalate increased in both groups, therefore, a recommendation to avoid cranberry in patients at risk of developing urolithiasis (Redmond et al. 2019).

16.2.4.16 Vitamin A Role in Urinary Tract Infection

Acute pyelonephritis leads to tubulointerstitial inflammation and scar formation (Kahbazi et al. 2019). A randomized clinical trial to assess the efficacy of vitamin A shows the prevention of tubulointerstitial inflammation and scar formation. Ninety female patients aged between 2 and 12 years diagnosed with urinary tract infection were involved in this trial (Kahbazi et al. 2019). Supplementation of vitamin A with antibiotics for 10 days revealed a statistically significant P value of less than 0.003 (Kahbazi et al. 2019). At the end of the study, vitamin A showed a significant reduction in the scar formation secondary to tubulointerstitial inflammation, as well as reduced the frequency associated with urinary tract infection (Kahbazi et al. 2019).

16.2.5 Nutraceutical Role in Gastrointestinal Disease

The digestive system involves the gastrointestinal tract and adjunct organs. The function of the gastrointestinal tract is to digest, absorb, and store the waste products to be excreted (Ogobuiro et al. 2022). The adjunct organs including the liver, pancreas, and gall bladder secrete enzymes that aid in the digestive process (Ogobuiro et al. 2022). There are various diseases affecting the gastrointestinal tract including inflammation, infection, malignancy, autoimmune disease, mechanical, and others that needs various treatment and intervention (Ogobuiro et al. 2022). In the last decades, nutraceuticals involved largely in gastrointestinal tract disease as prevention as well as an adjunct treatment or even primary treatment (Ogobuiro et al. 2022).

16.2.5.1 Probiotics Role in the GI of Autism Spectrum Disorders Patients

Frequent gastrointestinal (GI) symptoms associated with autism spectrum disorders linked to gut dysbiosis, favoring the gut–brain axis (Sanctuary et al. 2019). A randomized clinical trial involving children with an autism spectrum disorder undergoes 5 weeks of probiotic–prebiotic supplementation, 2 weeks washout period, and 5 weeks of probiotic only (Sanctuary et al. 2019). At the end of the study, significant improvement in gastrointestinal symptoms was explained by decreased level of interleukin-13 and tumor necrotic factor alpha (Sanctuary et al. 2019).

16.2.5.2 Probiotics in Irritable Bowel Syndrome

A randomized clinical trial was conducted on 74 patients with irritable bowel syndrome to undergo 8 weeks of probiotic treatment with 400 mL fermented milk with *Lactobacillus paracasei* F19, acidophilus La5, and *Bifidobacterium lactis* Bb12 (Simrén et al. 2010). The result suggested no positive effect of the probiotic on gastrointestinal symptoms associated with irritable bowel syndrome in comparison with the control group (Simrén et al. 2010).

16.2.5.3 Simethicone and *Bacillus coagulans* in Irritable Bowel Syndrome

Simethicone is antifoaming functioning to reduce bloating. *Bacillus coagulans* is probiotic (Urgesi et al. 2014). Fifty-two patients with irritable bowel syndrome were involved in a randomized clinical trial for 4 weeks with a combination of simethicone and *Bacillus coagulans* (Urgesi et al. 2014). The study showed significant improvement in the symptoms associated with irritable bowel syndrome post-simethicone and *Bacillus coagulans* (Urgesi et al. 2014).

16.2.5.4 *Bifidobacterium lactis* DN-173010 Role in Constipation Associated with Irritable Bowel Syndrome

Bloating and constipation are commonly associated with irritable bowel syndrome (Agrawal et al. 2009). A randomized clinical trial of 34 patients with irritable bowel syndrome suffering from constipation and bloating involved in a trial to receive daily *Bifidobacterium lactis* DN-173010 (Agrawal et al. 2009). The study uncovered significant improvement in the symptoms with *Bifidobacterium lactis* DN-173010 (Agrawal et al. 2009).

16.2.5.5 Probiotic's Role on Gastrointestinal Symptoms Associated with Systemic Sclerosis

Systemic sclerosis has been linked with the alteration of the gastrointestinal microbiota (Marighela et al. 2019). The probiotics may modulate the microbiota and the immune system (Marighela et al. 2019). Therefore, 73 patients with systemic sclerosis were conducted in a randomized double-blind clinical trial of 8 weeks of probiotics [Bifidobacterium, *Lactobacillus acidophilus*, and rhamnosus, as well as *Lactobacillus paracasei*]. T-cells were monitoring during the course of the study

(Marighela et al. 2019). At the end of the study, T-helper-17 reduced but gastrointestinal symptoms did not improve significantly from the probiotics (Marighela et al. 2019).

16.2.5.6 Probiotics in Small Intestinal Bacterial Overgrowth

Small intestinal bacterial overgrowth can potentially result in various symptoms and complications, particularly malabsorption (Khalighi et al. 2014). A randomized clinical trial involved 30 patients with small intestinal bacterial overgrowth to take probiotics for 5 weeks (Khalighi et al. 2014). Significant prevention of complications associated with such disease occurred at the end of the study (Khalighi et al. 2014).

16.2.5.7 N-3 Fatty Acid Eicosapentaenoic Acid (EPA) Supplementation in Cachectic

N-3 fatty acid from fish oil particularly eicosapentaenoic acid can prevent cachectic complications associated with malignancy (Patursson et al. 2021). Thirty patients were conducted in a randomized clinical trial of administration of EPA and docosahexaenoic acid for 5–7 weeks (Patursson et al. 2021). A statistically significant *P* value of 0.01 was found after 5 weeks of such supplement (Patursson et al. 2021). In contrast, no clinical significance was concluded at the end of the treatment (Patursson et al. 2021).

16.2.5.8 Vitamin D in Ulcerative Colitis

Ulcerative colitis is a chronic autoimmune disease complicated by angiogenesis and inflammation. The study was designed to assess the vitamin D effect on pro-angiogenetic factors (Emami et al. 2020). Therefore, 90 patients with ulcerative colitis were involved in a randomized clinical trial to receive either 300,000 IU of vitamin D as a single dose or normal saline (Emami et al. 2020). In the end, vitamin D aids in reducing the level of proangiogenic factors such as vascular endothelial growth factor (VEGF) among patients with vitamin D deficiency only (Emami et al. 2020).

16.2.5.9 Coenzyme Q10 in Ulcerative Colitis

The main problem with ulcerative colitis is chronic inflammation that may lead to various complications (Farsi et al. 2021). Coenzyme Q10 has a certain role in inflammation, antioxidant, and antimicrobial. A randomized clinical trial was conducted with coenzyme Q10 for 8 weeks in patients (88) with ulcerative colitis. The study results demonstrate a significant reduction in inflammatory signaling posttreatment with coenzyme Q10 (interleukin-17 and NF-kB P65) (Farsi et al. 2021).

16.2.5.10 Vitamin C and E Role in Prevention of Recurrence Colorectal Polyps

Vitamin C and E are involved in the reduction of fecal mutagen levels that are considered to be involved in the pathogenesis of colorectal polyps (McKeown-Eyssen et al. 1988). Patients diagnosed with polyps (137) were conducted in a randomized clinical trial to receive 400 mg of vitamin C and E versus placebo for 2 years. About 41.4% of the patients developed polyps on vitamin supplements, while 50.7% of the patients developed polyps on placebo (McKeown-Eyssen et al. 1988). The rest of the patients lost the follow-up. At the end of the study, the rate of polyp reduction was smaller than the placebo. Therefore, a recommendation is to increase the sample size to ensure more accurate results (McKeown-Eyssen et al. 1988).

16.2.5.11 Fish Oil Role in Sporadic Colonic Adenoma

Omega 3 or fish oil plays an important role in lowering cytokines in rectal mucosa with sporadic colorectal adenoma (Anti et al. 1994). A double-blind study involved 60 patients with sporadic adenoma who received fish oil (2.5, 5.1 and 7.7 g) for 1 month (Anti et al. 1994). Reduction of proliferation was observed in abnormal baseline patterns. Furthermore, scientists have concluded that a low dose of fish oil has short- and long-term benefits by reducing the abnormal proliferation of rectal mucosa (Anti et al. 1994).

16.2.5.12 Bran in Irritable Bowel Syndrome

Bloating is commonly observed in irritable bowel syndrome. The study aims to determine whether bran can improve bloating (Hebden et al. 2002). Twelve patients with irritable bowel syndrome were involved in a randomized clinical trial to receive 15 g of bran per day for 2 weeks (Hebden et al. 2002). The study concluded increased the pain index and bloating, but the small bowel transit was accelerated (Hebden et al. 2002).

16.2.5.13 Rikkunshito on Dyspepsia

Rikkunshito is a traditional Kampo medicine used to treat dyspepsia in Japan. Twenty-three patients with dyspepsia were conducted in a randomized clinical trial to receive Rikkunshito for 4 weeks. By the end of the study, Rikkunshito does not affect gastric motility, yet dyspepsia has improved significantly (Masuy et al. 2020).

16.2.5.14 Ojeok-San Plus Saengmaek-San for Gastroesophageal Reflux Associated with Cough

Gastroesophageal reflux disease is considered one of the most common causes of epigastric pain (Bhang et al. 2020). A herbal component such as Ojeok-san and Saengmaek-san may effectively treat this disease (Bhang et al. 2020). Thirty patients with the gastroesophageal disease were involved in a randomized clinical trial to receive this herbal for 6 weeks (Bhang et al. 2020). Significant improvement was observed in the gastroesophageal symptoms (Bhang et al. 2020).

16.2.6 Nutraceuticals in Liver Disease

Liver is one of the most important vital organs involved in metabolism, immunity, synthesis of substances, detoxification, digestion, and fat-soluble vitamin storage (Kalra et al. 2019). Various pathological diseases affect the liver including inflammation, autoimmune, fibrosis, necrosis, malignancy, and infection (Kalra et al. 2019). The most common chronic liver disease is related to cirrhosis associated with infection particularly hepatitis viruses (Kalra et al. 2019). Furthermore, one of the most common signs related to liver disease is jaundice and fat-soluble vitamin deficiency as well as portal hypertension complications including varices (Kalra et al. 2019). There is various treatment available for hepatic disease which can lead to various adverse events. Also, some of the medications were failed to treat certain diseases (Kalra et al. 2019). Therefore, nutraceutical has involved in the last decades for the treatment of various liver diseases (Kalra et al. 2019).

16.2.6.1 Probiotics on Hepatic Steatosis

Hepatic steatosis is commonly known of nonalcoholic fatty liver disease (NAFLD). Given the pathophysiology of the disease the treatment involved lifestyle modification in terms of low-fat diet, exercise, and weight loss (Mohamad Nor et al. 2021). Without treatment, the condition could progress to liver fibrosis due to chronic inflammation by involvement of T lymphocytes and other inflammatory cells (Mohamad Nor et al. 2021). A randomized clinical trial conducted 39 patients with hepatic steatosis to undergo treatment with probiotics for 6 months, which hypothetically affect T lymphocytes function and reduce the inflammation (Mohamad Nor et al. 2021). At the end of the study, there is no significant improvement of hepatic steatosis with utilization of probiotics (Mohamad Nor et al. 2021).

16.2.6.2 Coffee and Chlorogenic Acid on Hepatic Steatosis

Randomized clinical trial conducted patients with type II diabetes mellitus to assess the benefits of caffeine plus chlorogenic acid on nonalcoholic fatty liver disease (Mansour et al. 2021). The patients were investigated via laboratory such as liver enzyme, C-reactive protein, tumor necrosis factor, and nuclear factor K-B and fibroscan of the liver (Mansour et al. 2021). At the end of the study, neither the caffeine nor the chlorogenic acid was superior to placebo (Mansour et al. 2021). Therefore, no recommendation to use them to treat hepatic steatosis in type II diabetes mellitus (Mansour et al. 2021).

16.2.6.3 Docosahexaenoic Acid, Phosphatidylcholine, Silymarin, Choline, Curcumin, and Tocopherol on Hepatic Steatosis

A mixture of nutraceutical was used for 3 months to assess the benefits of such supplement on hepatic steatosis in a randomized clinical trial (Cerletti et al. 2020). One hundred and thirteen patients were involved in this trial (Cerletti et al. 2020). An improvement of liver function test was confirmed, yet no changes of the metabolic nor the inflammation were observed between the control group and the case group (Cerletti et al. 2020). Therefore, the study failed to demonstrate significance result by using such nutraceutical (Cerletti et al. 2020). But they recommended to increase the sample size in the future this could open a door for future management of hepatic steatosis with nutraceuticals (Cerletti et al. 2020).

16.2.6.4 Vitamin D on Hepatic Cirrhosis with Spontaneous Bacterial Peritonitis

Spontaneous bacterial peritonitis is a very complex disease seen in patients with cirrhosis, the treatment is very complex and requires a longer time than usual peritonitis (Mohamed et al. 2021). Among those patients, liver enzymes were abnormally elevated, and the coagulation profile was also abnormal (Mohamed et al. 2021). A randomized clinical trial was conducted with vitamin D supplements for 6 months in patients with cirrhotic liver disease with spontaneous bacterial peritonitis (Mohamed et al. 2021). A significant clinical improvement in patient's status was found at the end of the study (Mohamed et al. 2021). Therefore, vitamin D use was recommended in patients with cirrhosis and spontaneous bacterial peritonitis (Mohamed et al. 2021).

16.2.6.5 Vitamin D on Alcoholic Liver Cirrhosis

Vitamin D is one of the fat-soluble vitamins that synthesized in the liver. Liver cirrhosis can lead to deficiency in fat-soluble vitamins A, D, E, and K (Savić et al. 2018). Various manifestations of vitamin D can be observed in patients, the most important ones are osteopenia and osteoporosis (Savić et al. 2018). A randomized clinical trial was conducted with daily supplementation of vitamin D for 6 months in 70 patients with alcoholic cirrhosis (Savić et al. 2018). A significant clinical improvement was observed in those patients (Savić et al. 2018). The study, therefore, recommend the use of vitamin D in patient with alcoholic liver cirrhosis (Savić et al. 2018).

16.2.6.6 Vitamin D on Hepatic Steatosis in Children

Vitamin D is deemed to be an anti-inflammatory component. Vitamin D was used in a clinical trial of nonalcoholic fatty liver disease in children (El Amrousy et al. 2022). A randomized clinical trial was conducted in 109 patients with vitamin D to assess the beneficial effect of vitamin D on the inflammation associated with hepatic steatosis (El Amrousy et al. 2022). The authors found that the administration of vitamin D aids in decreasing the inflammation associated with hepatic steatosis (El Amrousy et al. 2022). Vitamin D was also highly recommended in children with hepatic steatosis (El Amrousy et al. 2022).

16.2.6.7 Vitamin D on Chronic Hepatitis C

Patients (75) with chronic hepatitis C underwent a randomized clinical trial for 6 weeks with daily supplementation of vitamin D (Sriphoosanaphan et al. 2021). No significant improvement was identified in fibrogenesis associated with chronic hepatitis C (Sriphoosanaphan et al. 2021). While in general, vitamin D has been proven to be effective in reducing the inflammation and fibrosis in patients with liver dysfunction, the failure of vitamin D in reducing the fibrogenesis associated with chronic hepatitis C strongly suggests the use of a larger sample size and longer treatment period to identify the beneficial effect of vitamin D in chronic hepatitis C patients (Sriphoosanaphan et al. 2021).

16.2.6.8 Biejia Ruangan (Chinese Herbal Medicine) on Hepatic Fibrosis Due to Chronic Hepatitis B

Chronic hepatitis B can potentially lead to hepatic fibrosis and hepatocellular carcinoma (Qu et al. 2014). While various treatment options are available to prevent these complications, none of these options exhibits improvement in patient survival

(Qu et al. 2014). However, a clinical trial that was conducted on 1000 patients with Biejia ruangan, a Chinese herbal medicine, post-chronic hepatitis B (Qu et al. 2014) showed an improvement in hepatic fibrosis (Qu et al. 2014).

16.2.6.9 Sumac (*Rhus coriaria* L) on Hepatic Steatosis

Herbal medicine was involved in the treatment of various diseases. Sumac, a Mediterranean nutraceutical, was selected for a clinical trial that was conducted for 12 weeks on 84 patients with nonalcoholic fatty liver disease (Kazemi et al. 2020). The patients showed a significant reduction in hepatic fibrosis and liver enzymes posttreatment. Therefore, the study strongly suggests the use of Sumac for the treatment of haptic steatosis (Kazemi et al. 2020).

16.2.6.10 Curcumin on Nonalcoholic Fatty Liver

A randomized clinical trial was performed for 8 weeks with curcumin in 55 patients with nonalcoholic fatty liver disease (NAFLD) (hepatic steatosis) (Saberi-Karimian et al. 2020). The result showed improvement in inflammatory markers (interleukins, tumor necrosis factor, interferon-gamma, and vascular endothelial growth factor) associated with this condition (Chashmniam et al. 2019; Mirhafez et al. 2019; Panahi et al. 2016; Saberi-Karimian et al. 2020).

Curcumin was also shown to have a beneficial effect on NAFLD when given along with piperine for a short period or with other nutraceutical mixture [DHA, phosphatidylcholine, silymarin, choline, curcumin, and *d*-alpha-tocopherol (Mirhafez et al. 2021; Panahi et al. 2019)].

16.2.6.11 Zhaoyangwan on Chronic Hepatitis B and Posthepatic Cirrhosis

Patients with hepatitis B and posthepatic cirrhosis (50) were enrolled in a randomized clinical trial for 3 months to receive a nutraceutical Zhaoyangwan (Zhang et al. 2004). The study found that Zhaoyangwan treatment with chronic hepatitis B and posthepatic cirrhosis exhibit a significant clinical improvement (Zhang et al. 2004).

16.2.6.12 Fuzheng Huayu on Posthepatic Cirrhosis

A randomized clinical trial was conducted in patients (180) with posthepatic cirrhosis for 6 months with Fuzheng Huayu (Deng et al. 2013). The study showed significant clinical benefits when using this herbal nutraceutical. This nutraceutical also improves the survival rate among those patients (Deng et al. 2013).

16.2.6.13 L-Carnitine on Hepatic Steatosis

Patients aged between 5 and 15 years old (55) with nonalcoholic fatty liver were involved in a randomized clinical trial to investigate the efficacy of L-carnitine on hepatic steatosis (Saneian et al. 2021). The study showed no significant improvement of L-carnitine on hepatic steatosis (Saneian et al. 2021).

16.3 Summary

Nutraceuticals have received considerable interest in recent years due to their beneficial effect in reducing various diseases including chronic conditions. However, their use has been commonly discouraged for over a period since a lack of sufficient safety and efficacy concerns and they are not patent protected, nor FAD approved. However, the use of omega-3 in malignancy was significant since it reduces mortality. Furthermore, omega-3 was shown to have neuroprotective effects in various neurological conditions including autism spectrum disorder, insomnia, and dementia. Omega-3 was also important in controlling blood pressure, atherosclerosis, and coronary artery diseases. In contrast, omega-3 was not significantly beneficial in chronic renal failure. But in the prevention of colonic adenoma, omega-3 was important in reducing the recurrence rate, the same applies to hepatic steatosis as omega-3 improved the condition. Curcumin benefits neurological disorders including depression, psychotic disorders, and Alzheimer's disease. It was also used in end-stage renal failure, which showed significant benefit in reducing proteinuria. Also, improvement of inflammatory mediators associated with chronic kidney injury was seen in clinical trials. While in hepatic steatosis, curcumin significantly improved the condition.

Vitamin D has been involved in various clinical trials, in cancer, particularly in melanoma. Vitamin D showed a significant reduction in the progression of melanoma. A significant reduction in proangiogenic factors such as VEGF was observed in ulcerative colitis and vitamin D uncovered significant protection against it. In spontaneous bacterial peritonitis, vitamin D improved the outcome of the condition, the same applies to alcoholic liver cirrhosis. For pediatric hepatic steatosis, vitamin D was highly recommended due to its significant benefit in reducing the inflammation associated with this condition. However, vitamin D showed no improvement in chronic kidney injury and the same applies to chronic hepatitis C.

Vitamin E showed an increased risk of prostatic cancer, lower inflammation associated with diabetic nephropathy, and no benefits of using vitamin E in colorectal polyps. Vitamin C improves symptoms associated with cancer. No benefits to using vitamin C for kidney failure and colorectal polyps. The rest of the nutraceutical aids in the improvement of various conditions as listed above. Few clinical trials revealed no significant benefits from utilizing nutraceuticals in patients. Thus, a recommendation to increase the sample size was highly suggested.

Overall, the use of nutraceuticals is strongly recommended for the treatment of various conditions (see Table 16.1), although their safety and toxicity are not well

Table 16.1 Nutraceuticals in the treatment of various conditions

Nutraceutical	Disease and result
Omega-3	Cancer, no benefit
Vitamin E	Cancer, no benefit
Vitamin D	Cancer, benefit in melanoma
Anamorelin	Cancer, benefit in anorexia
Mixture diet	Cancer, showed benefit
Vitamin C	Cancer, showed benefit
Glutamine and arginine	Cancer, showed benefit
Reservatrol	Cancer, showed benefit
Omega-3	Insomnia, mood, autism, and dementia, showed benefit
Curcumin	Alzheimer's disease and depression, showed benefit
Tryptophan	Insomnia, showed benefit
Vitamin B	Intellectual abnormality in down syndrome, showed benefit
MIG-99	Migraine, showed benefit
Sterols, polyphenols, and spirulina	Coronary arterial disease, showed benefit
Selenium	Stroke, showed preventive measures against stroke
Omega-3, α-lipoic acid, green tea, black tea, seaweed wakame, and arginine	Hypertension, no result available due to incompletion
Niacin	Atherosclerosis, showed benefit
Sorghum	Chronic kidney injury, showed benefit
Turmeric	Diabetic nephropathy, showed benefit
Green tea	Renal failure, showed benefit
Vitamin D and Omega-3	Chronic renal failure, no benefit
Probiotic	Chronic renal failure, no benefit
Zinc	Chronic renal failure, showed benefit
Alpha-lipoic acid	Autosomal dominant polycystic kidney disease, showed benefit
Vitamin K	Chronic kidney disease, no benefit
L-Carnitine	Chronic renal failure, showed benefit
Curcumin longa and *Boswellia serrata*	Chronic renal failure, showed benefit
Vitamin E	Diabetic nephropathy, showed benefit
Magnesium	Chronic kidney injury, showed benefit
N-3 fatty acid and coenzyme Q10	Chronic kidney injury, showed benefit
Vitamin C	Severe renal failure, no benefit
Cranberry	Nephrolithiasis, increase risk of nephrolithiasis. Not recommended
Vitamin A	Urinary tract infection, showed benefit
Probiotic	Gastrointestinal in autism spectrum, showed benefit
Probiotic	Irritable bowel syndrome, no benefit
Simethicone and bacillus coagulans	Irritable bowel syndrome, showed benefit

(continued)

Table 16.1 (continued)

Nutraceutical	Disease and result
Bifidobacterium lactis DN-173010	Irritable bowel syndrome, showed benefit
Probiotic	Gastrointestinal in systemic sclerosis, no benefit
Probiotic	Small intestinal bacterial overgrowth, showed benefit
N-3 fatty acid eicosapentaenoic	Cachectic, no benefit
Vitamin D	Ulcerative colitis, showed benefit
Coenzyme Q10	Ulcerative colitis, showed benefit
Vitamin C and E	Colorectal polyps, no benefit
Fish oil	Colonic adenoma, showed benefit
Bran	Bloating in irritable bowel syndrome, no benefit
Rikkunshito	Dyspepsia, no benefit
Ojeok-san and Saengmaek-san	Gastroesophageal reflux disease, showed benefit
Probiotic	Hepatic steatosis, no benefit
Caffeine	Hepatic steatosis, no benefit
Docosahexaenoic acid, phosphatidylcholine, silymarin, choline, curcumin, and tocopherol on hepatic steatosis	Hepatic steatosis, no benefit
Vitamin D	Cirrhotic liver disease with spontaneous bacterial peritonitis, showed benefit
Vitamin D	Liver cirrhosis, shoed benefit
Vitamin D	Nonalcoholic fatty liver disease in children, showed benefit
Vitamin D	Chronic hepatitis C, no benefit
Traditional herbal medicine	Hepatocellular carcinoma, showed benefit
Biejia ruaangan	Hepatic fibrosis due to hepatitis B, showed benefit
Oleum fructus bruceas and Ganji decoction	Primary hepatic carcinoma, showed benefit
Sumac	Hepatic steatosis, showed benefit
Curcumin	Hepatic steatosis, showed benefit
Zhaoyangwan	Chronic hepatitis B and posthepatic cirrhosis, showed benefit
Fuzheng Huayu	Posthepatic cirrhosis, showed benefit
L-Carnitine	Hepatic steatosis, no benefit
Souvenaid	Alzheimer's disease, showed benefit
Acetyl L-carnitine, folate, vitamin B12, methionine, alpha-tocopherol, and *N*-acetyl cysteine	Alzheimer's disease, showed no benefit
Polyamine spermidine	Alzheimer's disease, showed benefit
Ginkgo biloba	Alzheimer's disease, showed no benefit
Vitamin D	Alzheimer's disease, showed benefit
BrainUp-10	Alzheimer's disease, showed benefit
Vitamin B	Alzheimer's disease, showed no benefit
Selenium and probiotic	Alzheimer's disease, showed benefit

defined. Further, only few nutraceuticals are currently in treatment while many are under clinical trials. Therefore, more defined mechanisms, their efficacy, safety, and toxify status should be accelerated in the future and that could bring their usefulness to treat various diseases.

Acknowledgements We thank Dr. Rajalakshmi, PhD, Postdoctoral Fellow, Department of Obstetrics, Gynecology and Reproductive Sciences, University of Miami Miller School of Medicine, Miami, FL 33136 for her for assistance with manuscript preparation.

References

Aggarwal BB, Bhardwaj A, Aggarwal RS, Seeram NP, Shishodia S, Takada Y (2004) Role of resveratrol in prevention and therapy of cancer: preclinical and clinical studies. Anticancer Res 24(5A):2783–2840

Agrawal A, Houghton LA, Morris J, Reilly B, Guyonnet D, Goupil Feuillerat N et al (2009) Clinical trial: the effects of a fermented milk product containing *Bifidobacterium lactis* DN-173 010 on abdominal distension and gastrointestinal transit in irritable bowel syndrome with constipation. Aliment Pharmacol Ther 29(1):104–114

AIM-HIGH Investigators, Boden WE, Probstfield JL, Anderson T, Chaitman BR et al (2011) Niacin in patients with low HDL cholesterol levels receiving intensive statin therapy. N Engl J Med 365(24):2255–2267

Aisen PS, Schneider LS, Sano M, Diaz-Arrastia R, van Dyck CH, Weiner MF, Alzheimer Disease Cooperative Study et al (2008) High-dose B vitamin supplementation and cognitive decline in Alzheimer disease: a randomized controlled trial. JAMA 300(15):1774–1783

Anti M, Armelao F, Marra G, Percesepe A, Bartoli GM, Palozza P et al (1994) Effects of different doses of fish oil on rectal cell proliferation in patients with sporadic colonic adenomas. Gastroenterology 107(6):1709–1718

Barden AE, Shinde S, Burke V, Puddey IB, Beilin LJ, Irish AB et al (2018) The effect of n-3 fatty acids and coenzyme Q10 supplementation on neutrophil leukotrienes, mediators of inflammation resolution and myeloperoxidase in chronic kidney disease. Prostaglandins Other Lipid Mediat 136:1–8

Barocio-Pantoja M, Quezada-Fernández P, Cardona-Müller D, Jiménez-Cázarez MB, Larios-Cárdenas M, González-Radillo OI et al (2021) Green tea extract increases soluble RAGE and improves renal function in patients with diabetic nephropathy. J Med Food 24(12):1264–1270

Bhang YH, Kim KI, Kim J, Ahn J, Jung HS, Yang C et al (2020) Efficacy and safety of Ojeok-san plus Saengmaek-san for gastroesophageal reflux-induced chronic cough: protocol for a pilot, randomized, double-blind, placebo-controlled trial. Trials 21(1):118

Borges NA, Carmo FL, Stockler-Pinto MB, de Brito JS, Dolenga CJ, Ferreira DC et al (2018) Probiotic supplementation in chronic kidney disease: a double-blind, randomized, placebo-controlled trial. J Ren Nutr 28(1):28–36

Cerletti C, Colucci M, Storto M, Semeraro F, Ammollo CT, Incampo F et al (2020) Randomised trial of chronic supplementation with a nutraceutical mixture in subjects with non-alcoholic fatty liver disease. Br J Nutr 123(2):190–197

Chashmniam S, Mirhafez SR, Dehabeh M, Hariri M, Azimi Nezhad M, Nobakht MGBF (2019) A pilot study of the effect of phospholipid curcumin on serum metabolomic profile in patients with non-alcoholic fatty liver disease: a randomized, double-blind, placebo-controlled trial. Eur J Clin Nutr 73(9):1224–1235 de Boer IH, Zelnick LR, Ruzinski J, Friedenberg G, Duszlak J, Bubes VY et al (2019) Effect of vitamin D and omega-3 fatty acid supplementation on kidney function in patients with type 2 diabetes: a randomized clinical trial. JAMA 322(19):1899–1909

DeKosky ST, Williamson JD, Fitzpatrick AL, Kronmal RA, Ives DG, Saxton JA, Lopez OL, Burke G, Carlson MC, Fried LP, Kuller LH, Robbins JA, Tracy RP et al (2008) *Ginkgo biloba* for prevention of dementia: a randomized controlled trial. JAMA 300(19):2253–2262

Deng X, Liang J, Liu ZW, Wu FS, Li X (2013) Treatment of posthepatitic cirrhosis by Fuzheng Huayu Tablet for reinforcing qi and resolving stasis. Chin J Integr Med 19(4):289–296

Djamgoz MBA, Jentzsch V (2022) Integrative Management of Pancreatic Cancer (PDAC): emerging complementary agents and modalities. Nutr Cancer 74(4):1139–1162

El Amrousy D, Abdelhai D, Shawky D (2022) Vitamin D and nonalcoholic fatty liver disease in children: a randomized controlled clinical trial. Eur J Pediatr 181(2):579–586

Emami MR, Sharifi A, Yaseri M, Derakhshanian H, Hosseinzadeh-Attar MJ (2020) Vitamin D suppresses proangiogenic factors in patients with ulcerative colitis: a randomized double blind placebo controlled clinical trial. Complement Ther Clin Pract 39:101086

Escobedo-Monge MF, Ayala-Macedo G, Sakihara G, Peralta S, Almaraz-Gómez A, Barrado E et al (2019) Effects of zinc supplementation on nutritional status in children with chronic kidney disease: a randomized trial. Nutrients 11(11):2671

Farombi EO, Adepoju BF, Ola-Davies OE, Emerole GO (2005) Chemoprevention of aflatoxin B1-induced genotoxicity and hepatic oxidative damage in rats by kolaviron, a natural bioflavonoid of *Garcinia kola* seeds. Eur J Cancer Prev 14(3):207–214

Farsi F, Ebrahimi-Daryani N, Golab F, Akbari A, Janani L, Karimi MY et al (2021) A randomized controlled trial on the coloprotective effect of coenzyme Q10 on immune-inflammatory cytokines, oxidative status, antimicrobial peptides, and microRNA-146a expression in patients with mild-to-moderate ulcerative colitis. Eur J Nutr 60(6):3397–3410

Fuchs-Tarlovsky V, Bejarano-Rosales M, Gutiérrez-Salmeán G, Casillas MA, López-Alvarenga JC, Ceballos-Reyes GM (2011) Efecto de la suplementación con antioxidantes sobre el estrés oxidativo y la calidad de vida durante el tratamiento oncológico en pacientes con cáncer cérvico uterino [Effect of antioxidant supplementation over oxidative stress and quality of life in cervical cancer]. Nutr Hosp 26(4):819–826

Galiniak S, Aebisher D, Bartusik-Aebisher D (2019) Health benefits of resveratrol administration. Acta Biochim Pol 66(1):13–21

Ghaffari S, Roshanravan N (2020) The role of nutraceuticals in prevention and treatment of hypertension: an updated review of the literature. Food Res Int 128:108749

Gogos CA, Ginopoulos P, Salsa B, Apostolidou E, Zoumbos NC, Kalfarentzos F (1998) Dietary omega-3 polyunsaturated fatty acids plus vitamin E restore immunodeficiency and prolong survival for severely ill patients with generalized malignancy: a randomized control trial. Cnacer 82:395–402

Gounden V, Bhatt H, Jialal I (1998) Renal function tests. StatPearls Publishing, Treasure Island (FL)

Guzman-Martinez L, Farías GA, Tapia JP, Sánchez MP, Fuentes P, Gloger S et al (2021) Interventional study to evaluate the clinical effects and safety of the nutraceutical compound BrainUp-10® in a cohort of patients with Alzheimer's disease: a multicenter, randomized, double-blind, and placebo-controlled trial. J Alzheimers Dis 81(3):1231–1241

Harvie M (2014) Nutritional supplements and cancer: potential benefits and proven harms. Am Soc Clin Oncol Educ Book 2014:e478–e486

Hebden JM, Blackshaw E, D'Amato M, Perkins AC, Spiller RC (2002) Abnormalities of GI transit in bloated irritable bowel syndrome: effect of bran on transit and symptoms. Am J Gastroenterol 97(9):2315–2320

Jaffey M (1982) Vitamin C and cancer: examination of the Vale of Leven trial results using broad inductive reasoning. Med Hypotheses 8(1):49–84

Jia J, Hu J, Huo X, Miao R, Zhang Y, Ma F (2019) Effects of vitamin D supplementation on cognitive function and blood Aβ-related biomarkers in older adults with Alzheimer's disease: a randomised, double-blind, placebo-controlled trial. J Neurol Neurosurg Psychiatry 90(12): 1347–1352

Kahbazi M, Sharafkhah M, Yousefichaijan P, Taherahmadi H, Rafiei M, Kaviani P et al (2019) Vitamin A supplementation is effective for improving the clinical symptoms of urinary tract infections and reducing renal scarring in girls with acute pyelonephritis: a randomized, double-blind placebo-controlled, clinical trial study. Complement Ther Med 42:429–437

Kalra A, Yetiskul E, Wehrle CJ (2019) Physiology, liver. StatPearls Publishing, Treasure Island (FL)

Kazemi S, Shidfar F, Ehsani S, Adibi P, Janani L, Eslami O (2020) The effects of sumac (*Rhus coriaria* L.) powder supplementation in patients with non-alcoholic fatty liver disease: a randomized controlled trial. Complement Ther Clin Pract 41:101259

Khajehdehi P, Pakfetrat M, Javidnia K, Azad F, Malekmakan L, Nasab MH et al (2011) Oral supplementation of turmeric attenuates proteinuria, transforming growth factor-β and interleukin-8 levels in patients with overt type 2 diabetic nephropathy: a randomized, double-blind and placebo-controlled study. Scand J Urol Nephrol 45:365–370

Khalighi AR, Khalighi MR, Behdani R, Jamali J, Khosravi A, Kouhestani SH et al (2014) Evaluating the efficacy of probiotic on treatment in patients with small intestinal bacterial overgrowth (SIBO)—a pilot study. Indian J Med Res 140(5):604–608

Khatami PG, Soleimani A, Sharifi N, Aghadavod E, Asemi Z (2016) The effects of high-dose vitamin E supplementation on biomarkers of kidney injury, inflammation, and oxidative stress in patients with diabetic nephropathy: a randomized, double-blind, placebo-controlled trial. J Clin Lipidol 10(4):922–929

Khurana S, Venkataraman K, Hollingsworth A, Piche M, Tai TC (2013) Polyphenols: benefits to the cardiovascular system in health and in aging. Nutrients 5(10):3779–3827

Kunnumakkara AB, Bordoloi D, Padmavathi G, Monisha J, Roy NK, Prasad S et al (2017) Curcumin, the golden nutraceutical: multitargeting for multiple chronic diseases. Br J Pharmacol 174(11):1325–1348

Kuwasawa-Iwasaki M, Io H, Muto M, Ichikawa S, Wakabayashi K, Kanda R et al (2020) Effects of L-carnitine supplementation in patients receiving hemodialysis or peritoneal dialysis. Nutrients 12(11):3371

Lai S, Petramala L, Muscaritoli M, Cianci R, Mazzaferro S, Mitterhofer AP et al (2020) α-Lipoic acid in patients with autosomal dominant polycystic kidney disease. Nutrition 71:110594

Lopes RCSO, de Lima SLS, da Silva BP, Toledo RCL, Moreira MEC, Anunciação PC et al (2018) Evaluation of the health benefits of consumption of extruded tannin sorghum with unfermented probiotic milk in individuals with chronic kidney disease. Food Res Int 107:629–638

Mager DR, Jackson ST, Hoffmann MR, Jindal K, Senior PA (2017) Vitamin D_3 supplementation, bone health and quality of life in adults with diabetes and chronic kidney disease: results of an open label randomized clinical trial. Clin Nutr 36(3):686–696

Mansour A, Mohajeri-Tehrani MR, Samadi M, Qorbani M, Merat S, Adibi H et al (2021) Effects of supplementation with main coffee components including caffeine and/or chlorogenic acid on hepatic, metabolic, and inflammatory indices in patients with non-alcoholic fatty liver disease and type 2 diabetes: a randomized, double-blind, placebo-controlled, clinical trial. Nutr J 20(1): 35

Marighela TF, Arismendi MI, Marvulle V, Brunialti MKC, Salomão R, Kayser C (2019) Effect of probiotics on gastrointestinal symptoms and immune parameters in systemic sclerosis: a randomized placebo-controlled trial. Rheumatology (Oxford) 58(11):1985–1990

Masuy I, Carbone F, Holvoet L, Vandenberghe A, Vanuytsel T, Tack J (2020) The effect of rikkunshito on gastrointestinal symptoms and gastric motor function: the first study in a Belgian functional dyspepsia population. Neurogastroenterol Motil 32(2):e13739

Mazurek D, Wyka J (2015) Down syndrome—genetic and nutritional aspects of accompanying disorders. Rocz Panstw Zakl Hig 66(3):189–194

McKeown-Eyssen G, Holloway C, Jazmaji V, Bright-See E, Dion P, Bruce WR (1988) A randomized trial of vitamins C and E in the prevention of recurrence of colorectal polyps. Cancer Res 48(16):4701–4705

Mirhafez SR, Farimani AR, Dehhabe M, Bidkhori M, Hariri M, Ghouchani BF et al (2019) Effect of phytosomal curcumin on circulating levels of adiponectin and leptin in patients with non-alcoholic fatty liver disease: a randomized, double-blind, placebo-controlled clinical trial. J Gastrointest Liver Dis 28:183–189

Mirhafez SR, Dehabeh M, Hariri M, Farimani AR, Movahedi A, Naderan RD et al (2021) Curcumin and piperine combination for the treatment of patients with non-alcoholic fatty liver disease: a double-blind randomized placebo-controlled trial. Adv Exp Med Biol 1328:11–19

Mochamat CH, Marinova M, Kaasa S, Stieber C, Conrad R et al (2017) Systematic review on the role of vitamins, minerals, proteins, and other supplements for the treatment of cachexia in cancer: a European Palliative Care Research Centre cachexia project. J Cachexia Sarcopenia Muscle 8(1):25–39

Mohamad Nor MH, Ayob N, Mokhtar NM, Raja Ali RA, Tan GC, Wong Z et al (2021) The effect of probiotics (MCP® BCMC® strains) on hepatic steatosis, small intestinal mucosal immune function, and intestinal barrier in patients with non-alcoholic fatty liver disease. Nutrients 13(9): 3192

Mohamed AA, Al-Karmalawy AA, El-Kholy AA, El-Damas DA, Abostate HM, Mostafa SM et al (2021) Effect of vitamin D supplementation in patients with liver cirrhosis having spontaneous bacterial peritonitis: a randomized controlled study. Eur Rev Med Pharmacol Sci 22:6908–6919

Moreillon JJ, Bowden RG, Deike E, Griggs J, Wilson R, Shelmadine B et al (2013) The use of an anti-inflammatory supplement in patients with chronic kidney disease. J Complement Integr Med 1:10

Ogobuiro I, Gonzales J, Tuma F (2022) Physiology, gastrointestinal. StatPearls Publishing, Treasure Island (FL)

Panahi Y, Kianpour P, Mohtashami R, Jafari R, Simental-Mendía LE, Sahebkar A (2016) Curcumin lowers serum lipids and uric acid in subjects with nonalcoholic fatty liver disease: a randomized controlled trial. J Cardiovasc Pharmacol 68(3):223–229

Panahi Y, Valizadegan G, Ahamdi N, Ganjali S, Majeed M, Sahebkar A (2019) Curcuminoids plus piperine improve nonalcoholic fatty liver disease: a clinical trial. J Cell Biochem 120(9): 15989–15996

Patel KR, Brown VA, Jones DJ, Britton RG, Hemingway D, Miller AS et al (2010) Clinical pharmacology of resveratrol and its metabolites in colorectal cancer patients. Cancer Res 70(19):7392–7399

Patursson P, Møller G, Muhic A, Andersen JR (2021) N-3 fatty acid EPA supplementation in cancer patients receiving abdominal radiotherapy—a randomised controlled trial. Clin Nutr ESPEN 43: 130–136

Qu J, Yu Z, Li Q, Chen Y, Xiang D, Tan L et al (2014) Blocking and reversing hepatic fibrosis in patients with chronic hepatitis B treated by traditional Chinese medicine (tablets of Biejia ruangan or RGT): study protocol for a randomized controlled trial. Trials 10(15):438

Redmond EJ, Murphy CF, Leonard J, Faulds K, Abdelfadil S, Crowley VE et al (2019) The influence of dietary supplementation with cranberry tablets on the urinary risk factors for nephrolithiasis. World J Urol 37(3):561–566

Remington R, Bechtel C, Larsen D, Samar A, Doshanjh L, Fishman P (2015) A phase II randomized clinical trial of a nutritional formulation for cognition and mood in Alzheimer's disease. J Alzheimers Dis 45(2):395–405

Rock CL, Doyle C, Demark-Wahnefried W, Meyerhardt J, Courneya KS, Schwartz AL et al (2012) Bandera nutrition and physical activity guidelines for cancer survivors. CA Cancer J Clin 62(4): 243–274

Saberi-Karimian M, Keshvari M, Ghayour-Mobarhan M, Salehizadeh L, Rahmani S, Behnam B (2020) Effects of curcuminoids on inflammatory status in patients with non-alcoholic fatty liver disease: a randomized controlled trial. Complement Ther Med 49:102322

Sanctuary MR, Kain JN, Chen SY, Kalanetra K, Lemay DG, Rose DR et al (2019) Pilot study of probiotic/colostrum supplementation on gut function in children with autism and gastrointestinal symptoms. PLoS One 14(1):e0210064

Saneian H, Khalilian L, Heidari-Beni M, Khademian M, Famouri F, Nasri P et al (2021) Effect of *l*-carnitine supplementation on children and adolescents with nonalcoholic fatty liver disease (NAFLD): a randomized, triple-blind, placebo-controlled clinical trial. J Pediatr Endocrinol Metab 34(7):897–904

Sanmukhani J, Satodia V, Trivedi J, Patel T, Tiwari D, Panchal B (2014) Efficacy and safety of curcumin in major depressive disorder: a randomized controlled trial. Phytother Res 28(4): 579–585

Savić Ž, Vračarić V, Milić N, Nićiforović D, Damjanov D, Pellicano R et al (2018) Vitamin D supplementation in patients with alcoholic liver cirrhosis: a prospective study. Minerva Med 109(5):352–357

Scheltens P, Twisk JW, Blesa R, Scarpini E, von Arnim CA, Bongers A et al (2012) Efficacy of Souvenaid in mild Alzheimer's disease: results from a randomized, controlled trial. J Alzheimers Dis 31(1):225–236

Sharifi-Razavi A, Karimi N, Jafarpour H (2022) Evaluation of selenium supplementation in acute ischemic stroke outcome: an outcome assessor blind, randomized, placebo-controlled, feasibility study. Neurol India 70(1):87–93

Shorey-Kendrick LE, McEvoy CT, O'Sullivan SM, Milner K, Vuylsteke B, Tepper RS, Haas DM, Park B, Gao L, Vu A, Morris CD, Spindel ER et al (2020) Impact of vitamin C supplementation on placental DNA methylation changes related to maternal smoking: association with gene expression and respiratory outcomes. Clin Epigenetics 13(1):177

Simrén M, Ohman L, Olsson J, Svensson U, Ohlson K, Posserud I, Strid H (2010) Clinical trial: the effects of a fermented milk containing three probiotic bacteria in patients with irritable bowel syndrome—a randomized, double-blind, controlled study. Aliment Pharmacol Ther 31(2): 218–227

Singer RF (2011) Vitamin C supplementation in kidney failure: effect on uraemic symptoms. Nephrol Dial Transplant 26(2):614–620

Singh AP, Singh R, Verma SS, Rai V, Kaschula CH, Maiti P et al (2019) Health benefits of resveratrol: evidence from clinical studies. Med Res Rev 39(5):1851–1891

Sivamaruthi BS, Suganthy N, Kesika P, Chaiyasut C (2020) The role of microbiome, dietary supplements, and probiotics in autism spectrum disorder. Int J Environ Res Public Health 17(8):2647

Sriphoosanaphan S, Thanapirom K, Kerr SJ, Suksawatamnuay S, Thaimai P, Sittisomwong S et al (2021) Effect of vitamin D supplementation in patients with chronic hepatitis C after direct-acting antiviral treatment: a randomized, double-blind, placebo-controlled trial. PeerJ 9:e10709

Suhail N, Bilal N, Khan HY, Hasan S, Sharma S, Khan F, Mansoor T et al (2012) Effect of vitamins C and E on antioxidant status of breast-cancer patients undergoing chemotherapy. J Clin Pharm Ther 37(1):22–26

Tamtaji OR, Heidari-Soureshjani R, Mirhosseini N, Kouchaki E, Bahmani F, Aghadavod E et al (2019) Probiotic and selenium co-supplementation, and the effects on clinical, metabolic and genetic status in Alzheimer's disease: a randomized, double-blind, controlled trial. Clin Nutr 38(6):2569–2575

Thompson KG, Kim N (2021) Dietary supplements in dermatology: a review of the evidence for zinc, biotin, vitamin D, nicotinamide, and polypodium. J Am Acad Dermatol 84(4):1042–1050

Tian HQ, Li HL, Wang B, Liang GW, Huang XQ, Huang ZQ et al (2010) Treatment of middle/late stage primary hepatic carcinoma by Chinese medicine comprehensive therapy: a prospective randomized controlled study. Chin J Integr Med 16(2):102–108

Tiwari S, Atluri V, Kaushik A, Yndart A, Nair M (2019) Alzheimer's disease: pathogenesis, diagnostics, and therapeutics. Int J Nanomed 14:5541–5554

Toprak O, Kurt H, Sarı Y, Şarkış C, Us H, Kırık A (2017) Magnesium replacement improves the metabolic profile in obese and pre-diabetic patients with mild-to-moderate chronic kidney disease: a 3-month, randomised, double-blind, placebo-controlled study. Kidney Blood Press Res 42(1):33–42

Urgesi R, Casale C, Pistelli R, Rapaccini GL, de Vitis I (2014) A randomized double-blind placebo-controlled clinical trial on efficacy and safety of association of simethicone and *Bacillus coagulans* (Colinox®) in patients with irritable bowel syndrome. Eur Rev Med Pharmacol Sci 18(9):1344–1353

Vernieri C, Nichetti F, Raimondi A, Pusceddu S, Platania M, Berrino F et al (2018) Diet and supplements in cancer prevention and treatment: clinical evidences and future perspectives. Crit Rev Oncol Hematol 123:57–73

Wells RE, Baute V, Wahbeh H (2017) Complementary and integrative medicine for neurologic conditions. Med Clin N Am 101(5):881–893

Wirth M, Benson G, Schwarz C, Köbe T, Grittner U, Schmitz D et al (2018) The effect of spermidine on memory performance in older adults at risk for dementia: a randomized controlled trial. Cortex 109:181–188

Witham MD, Lees JS, White M, Band M, Bell S, Chantler DJ et al (2020) Vitamin K supplementation to improve vascular stiffness in CKD: the K4 kidneys randomized controlled trial. J Am Soc Nephrol 31(10):2434–2445

Zhai XF, Liu XL, Shen F, Fan J, Ling CQ (2018) Traditional herbal medicine prevents postoperative recurrence of small hepatocellular carcinoma: a randomized controlled study. Cancer 124(10):2161–2168

Zhang CP, Tian ZB, Liu XS, Zhao QX, Wu J, Liang YX (2004) Effects of Zhaoyangwan on chronic hepatitis B and posthepatic cirrhosis. World J Gastroenterol 10(2):295–298

Zhang F, Shen A, Jin Y, Qiang W (2018) The management strategies of cancer-associated anorexia: a critical appraisal of systematic reviews. BMC Complement Altern Med 18(1):236

Zhao M, Tuo H, Wang S, Zhao L (2020) The effects of dietary nutrition on sleep and sleep disorders. Mediat Inflamm 2020:3142874

Zuniga KB, Chan JM, Ryan CJ, Kenfield SA (2020) Diet and lifestyle considerations for patients with prostate cancer. Urol Oncol 38(3):105–117

Chapter 17
Regulation of Noncoding RNA by Nutraceuticals: Implication in Neurological Disorders and Cancer

Anandan Balakrishnan, Arumugam R. Jayakumar, Bastian T. Sebastian, Arokiasamy Justin Thenmozhi, Ramamoorthy Rajalakshmi, Hussain Hussain, and Kanadan Anjali

Abstract Noncoding RNAs (ncRNAs) are RNAs transcribed from genes but are not translated. The ncRNAs such as lncRNAs and miRNAs play an important role in the development of nervous system functions and differentiation of neurons and glia. Dysfunctional activities of ncRNAs are also strongly implicated in the development and progression of various metabolic disorders, diabetic neuro-nephropathy, and cancer. These ncRNAs are also emerged as markers of various human disorders. Nutraceuticals (natural products or crude drugs) were used as drugs for the treatment

A. Balakrishnan (✉)
Departments of Dental Anatomy and Oral Histology, Mahe Institute of Dental Sciences, Mahe, Puducherry, India

A. R. Jayakumar
Division of Neuropathology, Miami VA Medical Centre, Miami, FL, USA

Department of Obstetrics, Gynaecology and Reproductive Sciences, University of Miami Miller School of Medicine, Miami, FL, USA

B. T. Sebastian
Department of Oral Pathology, Mahe Institute of Dental Sciences, Mahe, Puducherry, India

A. Justin Thenmozhi
Department of Biochemistry, School of Biological Sciences, Madurai Kamaraj University, Madurai, Tamil Nadu, India

Department of Biochemistry and Biotechnology, Faculty of Science, Annamalai University, Chidambaram, Tamil Nadu, India

R. Rajalakshmi
Department of Obstetrics, Gynaecology and Reproductive Sciences, University of Miami Miller School of Medicine, Miami, FL, USA

H. Hussain
Department of Internal Medicine and Infectious Disease, Larkin Community Hospital, South Miami, FL, USA

K. Anjali
Departments of Microbiology, Mahe Institute of Dental Sciences, Mahe, Puducherry, India

A. Justin Thenmozhi, T. Manivasagam (eds.), *Nutraceuticals for Alzheimer's Disease: A Promising Therapeutic Approach*, Nutritional Neurosciences,
https://doi.org/10.1007/978-981-99-0677-2_17

of various diseases, and nutraceuticals were also shown to have potent modulatory effect on ncRNAs. Natural products derived from plants can upregulate or downregulate the miRNA and lncRNA targets and inhibit the disease progression. This review summarizes an important role of long noncoding RNAs in the development and progression of neurological disorders such as Alzheimer's disease, Parkinson disease, depressive and neuropsychiatric disorders, as well as in and rare neurological conditions, glioblastoma and in cancer, and the protective effect of nutraceuticals.

Keywords Micro-RNA (miRNA) · Long noncoding RNA (lncRNA) · Cancer stem cells (CSCs) · PIWI-interacting RNAs (piRNAs) · Natural compounds

17.1 Introduction

Noncoding RNAs (ncRNAs) are RNAs which are transcribed from gene sequences but are not translated like coding RNAs (mRNA) (Santosh et al. 2014). In eukaryotes, 98% of transcripts are ncRNAs which are involved in the coordination and modulation of gene expression by binding with chromatin structure, DNA or RNA (Mattick 2001). These noncoding RNAs are further divided into short noncoding RNAs (sncRNAs) and long noncoding RNAs (lncRNAs) based on size. Both short and long ncRNAs are further subdivided into 3 types. In addition, there are certain newly identified ncRNAs (Fig. 17.1) and all are playing an important role in the regulation of cellular homeostasis (Taft et al. 2010; Santosh et al. 2014). The lncRNAs are made of >200 nucleotides which are actively involved in various biological processes such as epigenetic chromatin control, promoter specific gene regulation, mRNA stability, and X-chromosome inactivation (Hombach and Kretz 2016). In contrast, the length of sncRNAs varies between 18 and 200 nucleotides

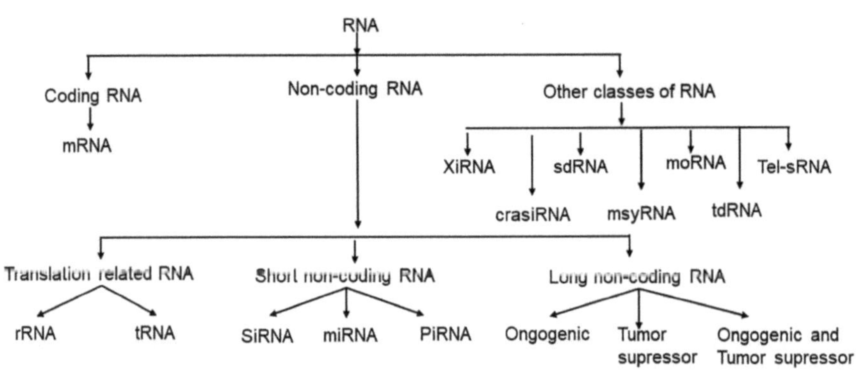

Fig. 17.1 Different classes of RNA. *miRNAs* micro-RNAs, *piRNAs* PIWI-interacting RNAs, *siRNAs* small interfering RNAs, *xiRNAs* X-inactivation RNAs, *sdRNAs* Sno-derived RNAs, *moRNAs* microRNA-offset RNAs, *tdRNA* tRNA-derived RNAs, *MSY-RNAs* MSY2-associated RNAs, *tel-sRNAs* telomere small RNAs, *crasiRNAs* centrosome-associated RNAs

and is important for translation, RNA processing, and RNA decay. Further, these sncRNAs are also involved in physiological (neuronal development) and pathological processes (metabolic disorders and tumorigenesis) (Li et al. 2021a).

The abundantly expressed concerved lncRNAs sequences adjacent to the protein coding gene loci of brain tissues was found to be involved transcriptional regulation or in nervous system development (Ponjavic et al. 2009). These lncRNAs and circular RNAs play an important role in CNS development, plasticity, and aging in various neurological conditions (Salvatori et al. 2020). Roberts et al. reported the involvement of lncRNA in the regulation of pluripotency of stem cells and its differentiation to form neurons and glia (Roberts et al. 2014).

Among the various ncRNAs, micro-RNAs (miRNAs) play an important role in the regulation of gene expression by binding with complementary mRNA transcripts and reduce protein synthesis (Ambros 2004; Bartel 2004; Alles et al. 2019). These miRNAs belong to snc-RNAs which are 22 nucleotides long, synthesized from intergenic and intragenic genomic regions as long primary transcripts and converted into mature miRNAs (Negrini et al. 2009). Some of the miRNAs can act as oncogenes (induce tumor formation) or tumor suppressor genes (block tumor formation) depending on the tissue target in which it is dysregulated (Garzon et al. 2009).

Natural compounds derived from plants such as curcumin, resveratrol, and quercetin were found to modulate the expression of many lncRNAs involved in the cancer and chronic diseases such as Alzheimer's diseases, diabetes, rheumatoid arthritis, ocular diseases, and cardiovascular diseases (Saghafi et al. 2019). Thus, this article focuses on the role of different ncRNAs including miRNAs in the development and progression of cancer and neurological disorders. Further, we discussed the therapeutic potential of natural compounds in the prevention and treatment of cancer and neurological disorders by reversing the activity of ncRNAs.

17.2 Advancements of Natural Compound in Neurological Disorders and Cancer by Controlling Noncoding RNAs

Noncoding RNAs (ncRNAs) transcribed from a significant portion of genome in different brain cells play an important role in the neural development, as well as in neuropathological changes (Bian and Sun 2011). In particular, the upregulation or downregulation of micro-RNAs (miRNA) is known to be involved in various neurological conditions including Alzheimer's disease (AD), Parkinson's disease (PD), amyotrophic lateral sclerosis, brain tumors, as well as in various rare neurological and neuropsychiatric disorders (Vieira et al. 2018). These ncRNAs are also used as biomarkers (Simionescu et al. 2022; Cente et al. 2022).

ncRNAs plays an important role in the physiological development of nervous system through various stages of neuronal differentiation by involvement in gene expression control at the transcriptional, epigenetic, and posttranslational levels (Salvatori et al. 2020). Further, if the genes (under the control of noncoding RNA) of the following secondary proteins (RMST, EVTs, PnKy, MALAT1, NEAT1, etc.) were mutated, this can activate different secondary messengers in various neurological diseases (Salvatori et al. 2020; Ramos et al. 2015). Therefore, alteration of noncoding RNA expression may broadly lead to formation of aberrant protein molecules (Salvatori et al. 2020). For instance, PnKy is a noncoding RNA expressed high in dividing neural stem cells and declines through neuronal differentiation. PnKy is also interacts with the splicing regulator polypyrimidine tract binding protein (PTBP-1) and regulates the appearance of multiple proteins and alternative splicing of a core set of targets that found in neurogenesis (Ramos et al. 2015). Numerous examples of mutated noncoding RNA have been identified in various neurological disorders including AD, peripheral neuropathy, central neuropathy, PD, brain malignancies, traumatic brain injury, spinal muscular atrophy, etc. (Salvatori et al. 2020; Abdolmaleki et al. 2017). In spinal muscular atrophy of SMN-1 gene mutations, noncoding RNA produces dysfunctional survivor motor neuronal (SMN) protein due to mutations of the splicing process, and consequently different phenotypes of spinal muscular atrophy can occur among the patients (Salvatori et al. 2020).

17.2.1 Alzheimer's Disease (AD)

AD is a neurodegenerative disorder described by the accumulation of amyloid beta (Aβ), the presence of transactivating DNA-binding protein-43 (TDP-43) inclusions, and the accumulation of hyperphosphorylated and ubiquitinated tau proteins (p-tau) and neurofibrillary tangles (NFTs) (Lei et al. 2021), while specific markers are not identified so far. Studies have shown that dysfunctional regulation of miRNA is responsible for the deposition of Aβ, tau protein, neuroinflammation, as well as synaptic failure, which may contribute to the development and progression of Alzheimer's disease (Kou et al. 2020). Further, decreased expression of miRNAs such as miR-9, miR-29, miR-101, miR-107, miR-124, miR-298, and miR-328 is shown to enhance the levels of Aβ in animal models of AD. Natural compounds such as Magnolol, a compound from *Magnolia officinalis*, cannabidiolic acid, *N-trans*-caffeoyltyramine, and cannabisin B from hemp seeds have been shown to reverse AD in mice model by inducing the expression of various miRNAs including miR 200c (Chen et al. 2022, Di Palo et al. 2022). In amyloid beta accumulation, a gene called Sortilin Related Receptor (SORL-1) produces SORL-1 antisense RNA (SORL1-AS), which drives a splicing shift of SORL-1 from synthesis of the canonical long protein variant A to differently spliced protein isoform (Salvatori et al. 2020). This step leads to accumulation of beta amyloid within the neurons and results in development of Alzheimer's disease (Salvatori et al. 2020). While these findings strongly suggest potential role of miRNAs in the pathogenesis of AD, their prospective impact on the development and progression as well as the therapeutic options for AD are warranted.

17.3 Natural Compounds in the Regulation of Noncoding RNAs in Other Neurological Conditions

Upregulation of miR-23a and miR-27a has an important role in neuronal protection through inhibiting apoptosis pathway in traumatic brain injury (Sabirzhanov et al. 2014). Upregulation or downregulation of many miRNAs was shown to play an important role in the development of PD by altering autophagy and α-synuclein mediated inflammation (Zhang et al. 2022). For example, miR-155-5p was heavily expressed in PD and triggers α-synuclein mediated inflammation (Thome et al. 2016). Natural product berberine was found to be effective in preventing PD by regulating the expression of miR-142-5p and inhibiting the NF-kB signaling pathway (Li et al. 2021b). Increased expression of miR-124, miR-335, miR-17-5p, miR-221, and miR-228 was found to be responsible for the development of various depressive disorders (Yang et al. 2020; Li et al. 2015; Shi et al. 2021). Luteolin is likely a choice of drug for the treatment of Breast Cancer Related Depression (BCRD) since it is known to inhibit hippocampal inflammation and neuronal cell pyroptosis by regulating miR-124-3p and related signaling pathways (Zhu et al. 2022). In addition, curcumin also shows antidepressant effect by regulating the miR-124 which in turn increases the transcription of brain derived neurotropic factor (BDNF) in stress induced depression (Yang et al. 2015b). Boswellic acid is plant derivative from genus Boswellia that showed improvement in inflammation associated with cognitive dysfunction in modulating the expression of miR-155 in mice model (Sayed et al. 2018).

Another important example of noncoding RNA that plays a role in one or more of the neurological conditions is the nuclear paraspeckle assembly transcript (NEAT-1 & 2) that is ubiquitously appears to have a scaffold role in creation of subnuclear bodies termed paraspeckles (Salvatori et al. 2020). The NEAT-2 is overexpressed in an early stage of pathogenesis of amyotrophic lateral sclerosis (ALS) (Salvatori et al. 2020). While in brain malignancies, there are multiple proteins dysregulation due to noncoding RNA gene mutation. For example, in glioblastoma multiform (GBM), H-19 is an oncogenic protein upregulated by those malignant neurons, which function to resist temozolomide medication (Mahinfar et al. 2022) where overexpression of miR-21 and miR-196 is known to contribute to the development of GBM (Guan et al. 2010). However, miR-378 inhibits the development of GBM along with curcumin and enhances the apoptosis (Li et al. 2017). Further, curcumin decreases the size of glioblastoma by inducing miR-146a (negative regulator of NF-kB signaling) and miRNA-378 and affects the tumor growth (Wu et al. 2015; Li et al. 2017). Another compound, apigenin, a flavonoid from fruits and vegetables, was found to regulate the expression of miR-16 and inhibits growth of glioma (Chen et al. 2016). Sulforaphane found in broccoli, cabbage, cauliflower, and kale was found to reduce tumor growth by enhancing miR-15b-5p level, apoptotic pathway (Gasparello et al. 2022). A nutraceutical derived from sweet potato, Delphinidin-3-rutin exhibits a suppressive effect on glioma cells by induction of miR-20b-5p/Atg7 mediated autophagy (Wang et al. 2022).

Additionally, small nuclear RNA host gene 12 (SNHG-12), ubiquitin-protein ligase (MDM2), micro-RNAs such as miR-7, miR-32a, etc. were shown to play various roles in the development and progression of GBM (Mahinfar et al. 2022).

17.4 Natural Compounds and Noncoding RNAs Role in Diabetic Mediated Neurological Disorders and Cancer

The neurodegenerative disorders in diabetes mellitus are vary. One of the most important ailments is the cognitive impairment. A study investigated the effect of resveratrol, a polyphenol from grapes exerts antioxidant effect (El-Sayed et al. 2022). The resveratrol enhances the expression of miRNA-21, which in turn increases neurogenesis and angiogenesis. This can increase the formation of new blood vessels to supply more blood to the neurons (El-Sayed et al. 2022). Diabetes mellitus also increases the levels of miR-146 and miR-9, which have shown an impairment of cognitive function in mice model through activation of different pathways such as NF-kB, TNF, apoptotic pathways. Quercetin a flavonoid derived from green tea, berries, and onion can downregulate miR-146a and miR-9 and enhance learning and memory improvement in diabetic mice (Ebrahimpour et al. 2020). Oleanolic acid is a plant product that upregulates the level of miRNA-142-5p and downregulates PTEN level in diabetic animal model to eventually attenuate the inflammation associated with the mesangial cell injury (Chen et al. 2019) (Table 17.1).

17.5 Phytochemicals and Natural Compounds in Cancer Treatment

In the clinical scenario, cancer is mainly treated by chemotherapeutic compounds along with other oncotherapeutic methods. Some of the compounds are naturally isolated and some are synthetic compounds. Terrestrial plants have broadly discovered sources of anticancer compounds, and they are called phytochemicals. The chemical structure-based classification of phytochemicals is (flavonoids) phenolics by 45%, terpenoids and steroids by 27%, alkaloids by 18%, and other chemicals by 10% (Koche et al. 2016).

Besides phytochemicals, other floras and faunas like bacteria, fungi, algae, lichens, and marine invertebrates and their extracts have anticancer properties.

Table 17.1 Natural compounds and noncoding RNAs in diabetic mediated neurological disorders and cancer

Natural compound	Disease	Micro-RNA regulation by the compounds	References
Cannabidiolic acid, Cannabisin B and *N-trans-*caffeoyltyramine (combination)	Alzheimer's disease	miR-708-5p↑, miR-181a-5p↑, miR-190a-5p↓, miR-199a-5p↓, and miR-143-3p↓	Angelucci et al. (2019), Chen et al. (2015b), Goh et al. (2019), Hanif et al. (2017), Lin et al. (2012), Slota and Booth (2019), Wen et al. (2022), and Xu et al. (2017)
Magnolol	Alzheimer's disease	miR-20c↑	
Berberine	Parkinson's disease	miR-142-5p↑	
Resveratrol	Diabetic neuropathy	miR-21↑, miR-18a-5p↑	
Quercetin	Impaired learning and memory by diabetes	miR-146a↓, miR-9↓	
Delphinidin-3-rutin	Glioma	miR-20b-5p↑	
Curcumin	Antidepressant Glioblastoma	miR-124↓ miR-378↑↑, miR-146a↑	
Luteolin	Neurotropic	miR-124-3p↑, miR-132↑, miR-32a↑	
Apigenin	U87 glioma cells	miR-16↑	
Boswellic	Neuroinflammation	miR-155↓	
Oleanolic acid	Mesangial cell injury	142-5p↑,	

↑ Upregulated by the respective compound, ↓ downregulated by the respective compound, ↑↑ overexpression of micro-RNA in the model to sensitize for particular compound

Doxorubicin is one of the bacterial extracts used for solid tumors of the ovary, uterus, breast cancer, osteosarcoma, esophagus, and hematological cancer. Pheophytin, phycocyanin, and fucoidans are algal products that have anticancer activity in lung cancer, skin cancer, and other cancer forms (Sharif et al. 2014). The fungal extracts polysaccharide-rich extracts from *Trametes Versicolor* and *Grifola Frondosa* are having an antiproliferative and anti-invasive effect in colon cancer cells (LoVo and HT-29 human colon cancer cells) (Daniel Roca-Lema et al. 2019). The lichen extract Physciosporin is a secondary metabolite, isolated from Pseudocyphellaria coriacea abridged metastasis of lung cancer cells by reducing the expression of the N-cadherin and KITENIN (KAI1 C-terminal interacting tetraspanin)-mediated AP-1 activity, the lichen extract benzoic acid, 2,4 dihydroxy, 6 methyl-methyl esters from *Rocella montagnei*, and other compounds from *Parmotrema reticulatum*, *Parmotrema hababianum* was identified as anticancer lichens on cervical cancer cells (Poornima et al. 2016). The marine sponge extract Cytarabine (Cyto star) has been used for non-Hodgkin lymphoma and Ecteinascidin, and Trabectedin (Yondelis) is the marine squirt extract being used in many cancer forms (Demain and Vaishnav 2011). The sea cucumber extract TBL-12 inhibits the proliferation, migration, and invasion of human prostate cancer cells by p38 mitogen-activated protein kinase and intrinsic caspase-mediated apoptosis pathway (Yuan et al. 2019). Some phytochemicals like curcumin and resveratrol have anticancer properties by targeting miR-34a, a cancer suppressor (Masika et al. 2016). Curcumin has the property of upregulating anticancer long noncoding RNA MEG-3 in hepatocellular carcinoma (Zamani et al. 2015). We know that particular phytochemical can effectively be anticancer to one or more cancer types; vinblastine has an anticancer property in breast cancer cells by suppressing miR-21a (Biersack 2016), miR-27b, miR-324-3p, miR-328, miR-148a, and miR-451 but above change is not observed in colon cancer cells (Zhu et al. 2008; Rodrigues et al. 2011). In some cases, cancer cells are resistant to chemotherapy and some are resistant to radiotherapy, this unfavorable resistance is favored by drug inactivation, drug target alteration, epithelial–mesenchymal transition, cell death inhibition, DNA damage repair, drug efflux, and by epigenetic regulation (Housman et al. 2014). Drug-resistant and radio-resistant cancers are obstacles to cancer treatment and made challenging incite to refractory tumor research.

17.6 An Overview of Micro-RNAs and Long-RNAs in Cancer Biology

As briefed in the introduction, the known micro-RNAs are 2300 and lnc-RNAs 172,216 transcripts with 96,308 genes according to "NONCODE," but their functions are obscure particularly the lnc-RNAs have delicate gene modulation and have not been explored to date. Moreover, the long noncoding RNAs are involved in many biological processes like the development of embryos, embryonic stem cell differentiation, disease, and induced pluripotent stem cells (iPSCs) (Rao 2017). The long noncoding RNAs were dysregulated in various cancer, for example, aHIF, Air, anril, Car intergenic 10, GAS5, NGAS1-as-RNA, H19, MALAT, MEG3, NEAT1, PINC, Tsix and Zfast, etc. are encoded from first chromosome to almost all 22 pair of autosomal chromosome including X-chromosome (Spizzo et al. 2012) and ultimately long non-RNAs control all six cancer hallmarks proliferation, growth suppression, motility, immortality, angiogenesis, and viability of cancer phenotypes (Schmitt and Chang 2016) (Fig. 17.2).

Some major classes of micro-RNAs like Piwi-interfering RNA have the ability in regulating cancer development and metastasis, despite the known 200,821 transcripts only 273 piRNAs were discovered out of which piR-36,743, piR-36,026, and piR-31,106 were dramatically upregulated and piR-34,736, piR-36,249, piR-35,407, piR-36,318, and piR-34,377 were significantly downregulated while among 100 identified piRNAs (Han et al. 2017). Another major group sno-RNA has an anticancer property in glioblastoma (Chen et al. 2015a) beside the evidence that sno-RNAs are involved in carcinogenesis, snoRNAs exhibit differential expression patterns in different human cancers and have the capability of regulating tumorigenesis, and metastasis, for example, U5O, h5sn2, RNU43, RNU44 are downregulated

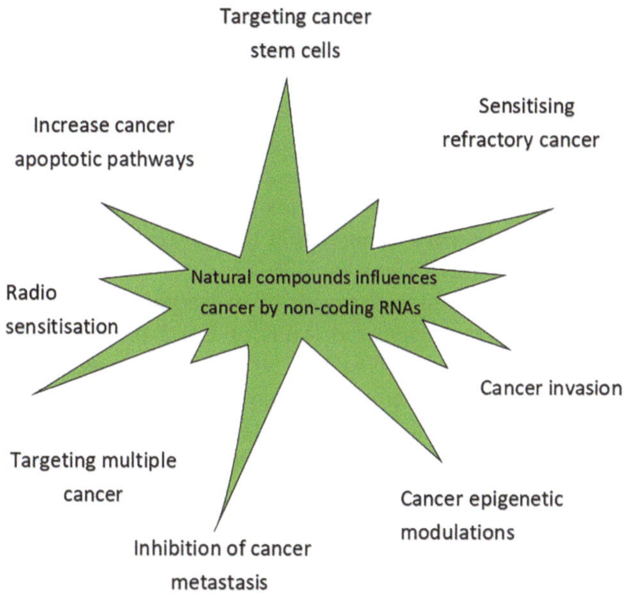

Fig. 17.2 Major cancer-related modulations by natural compounds

tumor suppressor in breast cancer and gliomas, besides upregulated sno-RNAs such as snoRD33, snoRD44, snoRD76 in non-small-cell lung cancer (Mannoor et al. 2012).

Initially micro RNA function is obscure but now its role in cancer is well understood and used for specific cancer diagnosis and prognosis. The circulating microRNAs and lnc-RNAs that are produced by exosomal release, exocytosis, microvesicles budding, necrosis, and apoptosis are used in cancer diagnosis. For example, miR-21 and miR-210 in B-cell lymphoma; miRNA-141 in prostate cancer; miR-25 and miRN-223 in lung cancer; miR-21, miR-92, miR-93, miR-126, and miR-29a ovarian in cancer; miR-17-3p and miR-92 in colorectal cancer; miR-92a in acute leukemia; miR-210, miR-155, and miR-196a in pancreatic cancer; miR-184 squamous cell carcinoma; miR-500 in hepatocellular carcinoma has been shown to be upregulated in cancer cells and can be detectable in circulating blood (Hamam et al. 2017). RNAs H19, MALAT1, ANRIL, HIFA-AS2, and HOTAIR are upregulated in breast cancer patients, can be detectable in uirne or plasma and serum (Yu et al. 2018).

17.7 Natural Compounds Have Anticancer Properties by Targeting Micro-RNA Etoposide

Etoposide is a natural compound derived from *Podophyllum peltatum,* and its semisynthetic form is a glycosylated podophyllotoxin with D-glucose. Etoposide is a topoisomerase inhibitor, and it is a known chemotherapeutic medication for testicular cancer, lung cancer Hodgkin's lymphoma, and many more (Anonymous 2011; Stuart et al. 2008). Human osteosarcoma cell lines, wild-type (wt) p53 U2-OS, mutant-p53 MG63, show better sensitivity to etoposide than p53-deficient MG63 and Saos-2 cells. The wild-type (wt) p53 U2-OS, mutant-p53 MG63 showed increased levels of unmethylated miR-34a, reduced expression of CDK4, and cell cycle arrest in the G1 phase (Novello et al. 2014). Another study by Kollinerová et al. shows miR-29b potentiates etoposide toxicity to Hela cells by downregulating the Mcl-1 protein (Kollinerová et al. 2017), Mcl-1 is an anti-apoptotic protein belonging to the Bcl-2 family (Xiang et al. 2018).

MCF-7 cells were sensitized to etoposide by overexpressed miR-195, miR-24-2, and miR-365-2 and showed decreased levels of Bcl-2 protein (Singh and Saini 2012), drugs regulating these miRNAs may be helpful in breaking resistance. The miR-195 arrests cancer cell proliferation, invasiveness, and migration of MCF-7 and MDA-MB-31 cells (Singh et al. 2015). miR-24-2 exhibit anti-apoptotic effect by targeting Bcl-2 in cancer cells (Srivastava et al. 2011), and miR-365 downregulated in hepatocellular carcinoma and stomach adenocarcinoma act as a tumor suppressor (Zhou and Liu 2013). Thus, these miRNAs can be used as therapeutic drug to treat various forms of cancers.

Hypoxia down-regulated the expression of miR-196b, which was induced by etoposide. The miR-196b overexpression increased the etoposide-induced apoptosis and reversed the protection of cell death observed under hypoxia. IGF2BP1 is the potential target of miR-196b. Indeed, miR-196b overexpression decreased IGF2BP1 RNA expression and protein level. The IGF2BP1 down-regulation by either miR-196b or IGF2BP1 siRNA led to an increase in apoptosis and a decrease in cell viability and proliferation in normal culture conditions (Rebucci et al. 2015). IGF2BP1 is essential in embryogenesis and tumor development (Huang et al. 2018a), while the miR-196b has a versatile antitumor role in chronic myeloid leukemia and pancreatic cancer cells through BCR-ABL1-HOXA9 and CADM1, respectively (Liu et al. 2013a; Wang and Zhou 2017). CADM1 is a cell adhesion molecule which is involved in metastasis of cancer cells (Wikman and Westphal 2014) (Table 17.2).

Etoposide is a refractory cancer drug for testicular cancer and is being used for various cancer, small cell carcinomas (multiple sites), adrenal cancer, gynecological cancers (GTD, others), sarcomas (Wilms tumor, soft tissue, Ewing's), CNS cancers, thymoma, Merkel cell, leukemias (CCO Formulary 2017). Furthermore, etoposide in being under clinical trial in a randomized phase II study of cisplatin and etoposide in combination with either Hedgehog inhibitor GDC-0449 or IGF-1R MOAB IMC A12 for patients with extensive stage cancer, and these novel miRNAs could extend etoposide research in other cancer with evidence (Anonymous 2019a). Hedgehog signal involved in cancer stem cells and self-renewal (Carballo 2018).

Table 17.2 Micro-RNA regulation in cancer by natural compounds

Natural compounds	Normal	In cancer	In treatment; sensitive and resistant
Etoposide *Podophyllum peltatum*	miR-34a ↔	miR-34a ↓	miR-34a ↑ in osteosarcoma cell lines and induces apoptosis through the p53 pathway
	miR-29b ↔	miR-29b ↓	miR-29b (Exo) ↑ in the HeLa cell line and induce cytotoxicity via downregulation of Mcl-1
	–	–	mi-R195, miR-24-2, miR-365-2 (Exo) (↑), augments the apoptosis of MCF-7 cells by Bcl2 downregulation
	miR-196b ↔	miR-196b ↓	miR-196b (Exo) and drug treatment ↑ mitigates cell proliferation and favors apoptosis in HepG2 cells by targeting IGF2BP1
Anacardic acid *Semecarpus anacardium*	–	–	miR-378g, miR-509, miR-513b-5p, miR-548j (↑) and Let-79-2-3p, miR-378j, miR-520d-5b, miR-1976, miR-551b-5p (↓) and many more in MCF-7 cells
	–	–	miR-378f, miR-1257, miR-1298-5p, miR-1304-5p (↑) and miR-23b-5p, miR-141-3p, miR-499a-5p, miR-1247-5p, miR-4284 (↓) in MDA-MB-231 cells
Paclitaxel	–	–	(Exo) miR-29c regulates resistance to paclitaxel in nasopharyngeal cancer, by targeting ITGB1
	LINC01118 ↔	LINC01118 ↑	lnc-RNA LINC01118 modulates paclitaxel resistance of epithelial ovarian cancer by regulating miR-134/ABCC1
	miR-22 ↔	miR-22 ↓	miR-22 expression, increase chemosensitivity to paclitaxel in MCF7 cells
	miR-155-5p ↔	miR-155-5p ↑	(Exo) miR-155-5p creates resistance in the MGC-803 gastric cancer cell to paclitaxel
	miR-34a ↔	miR-34a ↓	miR-34a reduce paclitaxel resistance in prostate cancer cells through suppression of the JAG1/Notch1 axis
Chrysin *Oroxylum indicum*	miR-34a, miR-22 ↔	miR-34a, miR-22 ↓	miR-34a, miR-22, miR-34a, miR-126, miR-18a, miR-21, miR-221, miR-9 and Let-7a were upregulated by nano-encapsulated chrysin in the gastric cancer cell
18β-Glycyrrhetinic acid *Glycyrrhiza glabra Linn*	miR-149-3p ↔	miR-149-3p ↓	miR-149-3p upregulation and Wnt-1 signaling by 18β-glycyrrhetinic acid suppress gastric cancer in vivo
Marine metabolite 1386A (marine fungi)	–	Many dysregulations	Let-7, miR-15, and miR-16 are downregulated and miR-27a, miR-21, miR-7, and miR-663 are upregulated by marine metabolite 1386A in MCF-7 cells. (An unknown species of marine fungi found in the South China Sea)

↔ Normal-physiologic, ↓ downregulation by cancer and drug, ↑ upregulation by cancer and drug, (Exo) indicates exogenous expression

17.7.1 Anacardic Acid

Anacardic acids are phenolic lipids (6-pentadecyl salicylic acid, 6-(8(Z), 11(Z), 14-pentadecatrienyl) salicylic acid, 6-(8,11,14-pentadecatrienyl)salicylic acid, 6-nonadecyl salicylic acid) derived from *Anacardium occidentale* (Cashew nuts), and *Semecarpus anacardium* (Marking nut) belongs to Anacardiaceae (Anonymous 2019b) (Rahman 2018). The marking nut plant is one of the medicinal plants prevalent in India, the plant has been described in the Siddha and Ayurveda system of alternative medicine literature (Murugesa Mudaliar 2003). Anacardic acid from cashew nut shells stimulates neutrophil extracellular trap (NET) that is responsible for bactericidal and complement select direct antimicrobial activities of the compound (Hollands et al. 2016). The miR-378g, miR-509, miR-513b-5p, miR-548j upregulated and Let-7a-2-3p, miR-378j, miR-520d-5b, miR-1976, miR-551b-5p and many more were downregulated in MCF-7 cells, in the same they have found miR-378f, miR-1257, miR-1298-5p, miR-1304-5p were upregulated and miR-23b-5p, miR-141-3p miR-499a-5p, miR-1247-5p, miR-4284 were downregulated in MDA-MB-231 cells (Deng et al. 2013). High miR-378 promotes cancer stem cell (CSC) properties, increased cell survival, and colony formation correlate with increased SOX2 (Deng et al. 2013) miR-548J functions as a metastasis promoter in breast cancer cells (Zhan et al. 2016), Let-7a-2-3p decreased expression with breast tumor grade and upregulated KEGG pathway targets have roles in cancer-related pathways, including cycle (MCM2), Jak-STAT (SOCS1), MAPK (STMN1), PPAR signaling (ME1) (Oztemur et al. 2015) and metastatic breast cancer cells in patient's bone marrow had increased expression of miR-23b-5p (Radde et al. 2016). Anacardic acid is proven to be antioxidant, anticancer, anti-inflammatory, antimicrobial, anti-obesity, and insecticidal (Hemshekhar et al. 2012), but human trial are not yet been studied, except preclinical study in which anacardic acids from *Anacardium occidentale* (Cashew nuts) gives better response in lung damage induced by exposure to diesel exhaust particles in mice (Carvalho et al. 2013).

17.7.2 Chrysin

Chrysin is a flavone sub-class of flavonoid and it is present in honey, propolis, passionflower (*Passiflora caerulea, Passiflora incarnata*),' and Indian trumpet flower (*Oroxylum indicum*) (family: Bignoniaceae). *Oroxylum indicum* (trumpet flower) is distributed throughout India, and its medicinal properties were enshrined in the Ayurveda literature (Deka et al. 2013). In multiple studies, miR-34a, miR-22, miR-126, miR-18a, miR-21, miR-221, miR-9, and Let-7a upregulated by nano-encapsulated chrysin in the gastric cancer cell (Mohammadian et al. 2015, 2016a, b, 2017a, b). Interestingly chrysin upregulated three tumorigenic miRNAs (miR-21, miR-221, miR-18), miR-18a and miR-25-106b in HCC was associated with poor survival in clinical samples and promoted proliferation in HCC cell lines

(Masika et al. 2016; Sanchez-Mejias et al. 2019). Moreover its tumor suppressive property (miR-34a, miR-22, miR-9, miR-26, and Let-7a) against other colon and breast cancer cells were investigated (Yang et al. 2015a). The miR-34a is a known tumor suppressor, and the other miR-9 act as a tumor suppressor shown in a study in which animal xenograft assays evidence that miR-9 acts as a tumor suppressor by targeting CXCR4 in vivo (Xiong et al. 2018) and the miR-26 acting as a tumor suppressor in bladder tumor, breast cancer, oral squamous cell carcinoma, anaplastic carcinomas by EZH2 as a target with oncogenic in glioma by PTEN as a target (Gao and Liu 2011). Chrysin may target gastric cancer stem cells by miRNA-34a and Let-7a (Table 17.3).

The protective effects of chrysin evident in toxic-agent (methotrexate, cisplatin, ethanol, etc.) induced toxicity amelioration, through various mechanisms in different tissues including the brain, heart, liver, kidney, lung, etc. (Samarghandian et al. 2017). Chrysin possesses potent neuroprotective effects and suppresses neuroinflammation. In addition, it improves cognitive decline by possessing anti-amyloidogenic and neurotrophic effects, and offered neuroprotective effect in experimental models of depression and epilepsy (Nabavi et al. 2015). Human clinical trials have not been reported yet.

17.7.3 Paclitaxel

Paclitaxel (Taxol) is derived from the Pacific yew tree, *Taxus brevifolia*. The samples were collected by USDA botanist Arthur Barclay on his expedition in 1962, and its anticancer property was explored by NCI (National Cancer Institute of U.S) Taxol (generic name paclitaxel) which is a popularly known drug that is approved by the Food and Drug Administration for the treatment of ovarian, breast, and lung cancer, as well as Kaposi's sarcoma. It is also used to treat gastroesophageal, endometrial, cervical, prostate, and head and neck cancers, in addition to sarcoma, lymphoma, and more (Weaver 2014).

The miR-29c regulates resistance to paclitaxel in nasopharyngeal cancer, by targeting ITGB1 (Huang et al. 2019), and miR-29c suppresses invasion and metastasis by targeting TIAM1 in nasopharyngeal carcinoma (Liu et al. 2013b). Brown seaweed Fucoidan inhibited human breast cancer progression by upregulating microRNA miR-29c and miRNA-29a is being upregulated by curcumin hepatocellular carcinoma (Wu et al. 2016), thus fucoidan or curcumin in future research with nasopharyngeal carcinoma could break paclitaxel resistance.

The lnc-RNA LINC01118 modulates paclitaxel resistance of epithelial ovarian cancer by regulating miR-134/ABCC1 (Shi and Wang 2018), (ABCC1/multidrug resistance-associated protein 1 (MRP1) transports (Munoz et al. 2007). The miR-134 clusters were downregulated in paclitaxel-resistant cancer cells than the sensitive cells, and the miR-17-92 cluster was inversely expressed in ovarian cancer (Zhu et al. 2016), and in a study the miR-134 can suppressor tumor by regulating suppressing EGFR and PI3K signaling in colorectal cancer (El-Daly et al. 2016).

Table 17.3 Long noncoding RNA regulation in cancer by natural compounds

Natural compounds	In normal	In cancer	In treatment; sensitive and resistant
Curcumin	MEG-3 ↔	MEG-3 ↓	MEG-3 ↑ by ↑ miR-29a and ↑ miR-186 in hepatocellular carcinoma cells
Curcuma longa	All lnc-RNAs are normal	All lnc-RNAs are dysregulated	Lnc-RNAs AF086415, AK095147, RP1-179N16.3, MUDENG, AK056098, and AK294004 reversed by curcumin in naso-pharyngeal carcinoma which was differentially expressed in the radioresistance cell of NPC
Rutin Buckwheat and citrus fruits	–	–	Differentially regulates 53 long noncoding RNAs in human SW480 colorectal cancer cells based and the details were not given in the article
Doxorubicin *Streptomyces peucetius* var. *caesius*	lnc-RNA CTA ↔	lnc-RNA CTA ↓	Dox resistance OS cells can be sensitized to Dox by (exo) upregulating lnc-RNA CTA and competitively downregulates miR-210
Paclitaxel and docetaxel *Taxus brevifolia*	LINC01118 ↔	LINC01118 ↑	lnc-RNA LINC01118 modulates paclitaxel resistance of epithelial ovarian cancer by
			Regulating miR-134/ABCC1
	–	–	Lnc-RNAs NONHSAG096479.1, NONHSAG048134.2
			NONHSAG048135.2, NONHSAG048143.2 and
			NONHSAG048143.2 creates docetaxel resistance by favoring drug-resistant transporter ABCB1
Magnolol *Magnolia officinalis*	GAS5 ↔	GAS 5 ↓	2-*O*-Methylmagnolol upregulates GAS5 long noncoding RNA in skin cancer A375 probably by caspase 3
Sulforaphane *Brassica oleracea* var. *italic*	LINC01116 ↔	LINC01116 ↑	LINC01351, LINC00883, LINC01059, and LINC01116 downregulated by sulforaphane in prostate cancer cells
	–	–	Sulforaphane provides chemoprotection to the fetus transplacentally exposed to Dibenzo [def,p]chrysene (DBC) and the lnc-RNA MSUR-1 upregulation by sulforaphane is mediated by Nrf2

(↔) Normal-physiologic, (↓) downregulation by cancer and drug, (↑) upregulation by cancer and drug, (Exo) indicates exogenous expression, (–) data not available

In osteosarcoma, miR-134 shows a tumor suppressive role by attenuating the expression of VEGFA and VEGR 1 (Zhang et al. 2018), miRNA-134 inhibits EMT of small cell lung cancer (Li et al. 2012) miRNA-134 could be a target of paclitaxel in cancer cells. In another case, the miR-22 expression increases chemosensitivity to paclitaxel in MCF-7 cells and it has been downregulated in breast cancer (Song et al. 2018).

The exosomal delivery of miR-155-5p converts the paclitaxel-resistant gastric cancer cells into paclitaxel sensitive cells and it could be a future target of resistance (Wang et al. 2019). MicroRNA-34a reduces paclitaxel resistance in prostate cancer cells by suppression of the JAG1/Notch1 axis (Liu et al. 2018a) we know that miR-34a is being upregulated by curcumin, chrysin, resveratrol in other cancer, further work in combination with curcumin, chrysin, resveratrol, and paclitaxel may break the resistance of prostate cancer cells to paclitaxel. Paclitaxel is being under clinical trial in refractory or relapsed solid tumors (Anonymous 2017a).

17.7.4 Glycyrrhetenic Acid

Glycyrrhetinic acid (Enoxolone) is mainly isolated from *Glycyrrhiza glabra linn* a tropical shrub that belongs to Leguminosae and used in Siddha and Ayurveda systems of medicine for thousands of years (Murugesa Mudaliar 2003). Glycyrrhetinic acid has several medical properties like antihyperglycemic, antiallergic, anti-inflammatory, antiviral, anticancer, and expectorant (Roshan et al. 2012). The miR-149-3p upregulation and Wnt-1 signaling by 18β-Glycyrrhetinic acid suppress gastric cancer in vivo study (Cao et al. 2016). The miR-149-3p shows to be tumor suppression (He et al. 2018) in dioscin-induced upregulation of miR-149-3p trough targeting Bax, Apaf-1, cleaved caspase-3/9, cleaved PARP, suppressing Bcl-2 levels in ASPC-1 and PANC-1 cell xenografts (Si et al. 2017), and tumor suppressive property of miR-149-3p is being reflected in bladder cancer cells by S100A4 (Yang et al. 2017a). In other cancer forms like bladder (BLCA), breast (BRCA), lung (LUSC), endometrial (UCEC), and prostate (PRAD), the miR-149-3p is differentially expressed and favors tumor progression, the miR-149 family miR-149-5p or miR-149-3p plays dual roles in the proliferation and apoptosis of various tumors (Bellazzo et al. 2018). Glycyrrhetinic acid is a generic drug in the name of enoxolone and is categorized as cicatrizant, an anti-inflammatory agent, drug acting on the gastrointestinal system (Anonymous 2017b).

17.7.5 Marine Metabolite 1386A

1386A is a metabolite of mangrove marine fungi indigenous to the South China Sea. Tang et al. study on MCF-7 cells shows cytotoxicity and influencing ambivalently both tumor-suppressing and oncogenic micro-RNAs. Let-7, miR-15, miR-16 were downregulated and miR-27a, miR-21, miR-7, miR-663 were upregulated by marine metabolite 1386A in MCF-7 cells (Tang et al. 2012a). It has been reported that miR-7 is a potential tumor suppressor in breast cancer, and it influences the CSCs of breast cancer cells. The miR-663 targets TGFβ1 transcripts, which are associated with the invasion and metastasis of gastric cancer through the activation of the TGFβR1-ALK5/SMAD3 pathway. In the same study, some are dysregulated, miR-320 family, miR-125b, miR-638. The miR-320a inhibits breast cancer cell proliferation and migration (Wang et al. 2015), and miR-320d suppresses the progression of breast cancer via lncRNA HNF1A-AS1 regulation and SOX4 inhibition (Shi et al. 2022). The overexpression of miR-125b is sensitizing paclitaxel-resistant breast cancer cells to paclitaxel by targeting Sema4C (Yang et al. 2014), in another case miR-125b promoted metastasis of MCF-7 and MDA-MB-231 cell (Tang et al. 2012b). The miR-125a-3p and miR-125a-5p show reduced expression in non-small-cell lung cancer and have inverse effects on the invasion and migration of lung cancer cells A549 and SPC-A-1 (Jiang et al. 2010). The miRNA-638 is the enhancer of autophagy in malignant phenotypes cells via directly suppressing DACT3 (Ren et al. 2017) but miR-638 was upregulated by its metabolite 1386A to enhance the autophagy. Even though many oncogenic micro-RNAs were upregulated and downregulated tumor suppressors, the marine metabolite 1386A reduced cell proliferation and arrested the growth of MCF-7 cells in vitro.

17.8 Natural Compounds Have Anticancer Properties by Targeting Long Noncoding RNA

Only a few natural compounds have studied for the anticancer action with long noncoding RNA regulation, because of its ambiguous functionality through 172,216 lns-RNAs transcripts were reported in humans. Almost 12 natural compounds influence cancer by oncogenic lnc-RNAs HOTAIR, ROR, MALAT-1, and H19, and enhancing cancer suppressors GAS5 and MEG3 (Mishra et al. 2019).

17.8.1 Curcumin

Curcumin is one of the multifunctional 134 acted as a tumor suppressor by regulating suppressing EGFR and PI3K signaling in colorectal cancer (El-Daly et al. 2016). In osteosarcoma, miR-134 shows a tumor suppressive role by attenuating the expression of VEGFA and VEGR 1 (Zhang et al. 2018), miRNA-134 inhibits EMT of small cell lung cancer (Li et al. 2012), miRNA-134 could be a target of paclitaxel in cancer cells. In another case, the miR-22 expression increases chemosensitivity to paclitaxel in MCF-7 cells and it has been downregulated in breast cancer (Song et al. 2018).

The exosomal delivery of miR-155-5p converts the paclitaxel-resistant gastric cancer cells into paclitaxel sensitive cells and it could be a future target of resistance (Wang et al. 2019). MicroRNA-34a reduces paclitaxel resistance in prostate cancer cells by suppression of the JAG1/Notch1 axis (Liu et al. 2018b) we know that miR-34a is being upregulated by curcumin, chrysin, resveratrol in other cancer, further work in combination with curcumin, chrysin, resveratrol, and paclitaxel may break the resistance of prostate cancer cells to paclitaxel. Paclitaxel is being under clinical trial in refractory or relapsed solid tumors (Anonymous 2017b).

17.8.2 Glycyrrhetenic Acid

Glycyrrhetinic acid (Enoxolone) is mainly isolated from *Glycyrrhiza glabra linn* a tropical shrub that belongs to Leguminosae and used in Siddha and Ayurveda systems of medicine for thousands of years (Murugesa Mudaliar 2003). Glycyrrhetinic acid has several medical properties like antihyperglycemic, antiallergic, anti-inflammatory, antiviral, anticancer, and expectorant (Roshan et al. 2012). The miR-149-3p upregulation and Wnt-1 signaling by 18β-Glycyrrhetinic acid suppress gastric cancer in vivo study (Cao et al. 2016). The miR-149-3p shows to be tumor suppression (He et al. 2018) in dioscin-induced upregulation of miR-149-3p trough targeting Bax, Apaf-1, cleaved caspase-3/9, cleaved PARP, suppressing Bcl-2 levels in ASPC-1 and PANC-1 cell xenografts (Si et al. 2017), and tumor suppressive property of miR-149-3p is being reflected in bladder cancer cells by S100A4 (Yang et al. 2017a). In other cancer forms like bladder (BLCA), breast (BRCA), lung (LUSC), endometrial (UCEC), and prostate (PRAD), the miR-149-3p is differentially expressed and favors tumor progression. The miR-149 family miR-149-5p or miR-149-3p plays dual roles in the proliferation and apoptosis of various tumors (Bellazzo et al. 2018). Glycyrrhetinic acid is a generic drug in the name of enoxolone and is categorized as cicatrizant, an anti-inflammatory agent, drug acting on the gastrointestinal system (Anonymous 2017b).

17.8.3 Marine Metabolite 1386A

1386A is a metabolite of mangrove marine fungi indigenous to the South China Sea. Tang et al. study on MCF-7 cells shows cytotoxicity and influencing ambivalently both tumor-suppressing and oncogenic micro-RNAs. Let-7, miR-15, miR-16 were reduced in expression, and miR-27a, miR-21, miR-7, miR-663 were upregulated by marine metabolite 1386A in MCF-7 cells (Tang et al. 2012a). It has been reported that miR-7 is a tumor suppressor in breast cancer, and it influences the CSCs of breast cancer cells. The miR-663 targets TGFβ1 transcripts, which are associated with the invasion and metastasis of gastric cancer through the activation of the TGFβR1-ALK5/SMAD3 pathway. In the same study, some are dysregulated, miR-320 family, miR-125b, miR-638. The miR-320a inhibits breast cancer cell proliferation and migration (Wang et al. 2015) and miR-320d suppresses the progression of breast cancer via lncRNA HNF1A-AS1 regulation and SOX4 inhibition (Shi et al. 2022). The overexpression of miR-125b is sensitizing paclitaxel-resistant breast cancer cells to paclitaxel by targeting Sema4C (Yang et al. 2014), in another case miR-125b promoted metastasis of MCF-7 and MDA-MB-231 cell (Tang et al. 2012b). The miR-125a-3p and miR-125a-5p expression is reduced in non-small-cell lung cancer and has inverse effects on the invasion and migration of lung cancer cells A549 and SPC-A-1 (Jiang et al. 2010). The miRNA-638 is the enhancer of autophagy in malignant phenotypes cells via directly suppressing DACT3 (Ren et al. 2017) but miR-638 was upregulated by its metabolite 1386A to enhance the autophagy. Even though many oncogenic micro-RNAs were upregulated and downregulated tumor suppressors, the marine metabolite 1386A reduced cell proliferation and arrested the growth of MCF-7 cells in vitro.

17.9 Natural Compounds Have Anticancer Properties by Targeting Long Noncoding RNA

Only a few natural compounds have studied for the anticancer action with long noncoding RNA regulation, because of its ambiguous functionality through 172,216 lns-RNAs transcripts were reported in humans. Almost 12 natural compounds influence cancer by oncogenic lnc-RNAs HOTAIR, ROR, MALAT-1, and H19 and enhance cancer suppressors GAS5 and MEG3 (Mishra et al. 2019).

Alkaloids isolated from the rhizome of *Curcuma longa* and belongs to Zingiberaceae. The rhizome is being used as a medication for versatile medical purposes like antibacterial, antiseptic, anthelminthic, sprain, chicken pox, and many more conditions by traditional practitioners in India (Deb et al. 2013; Murugesa Mudaliar 2003). Curcumin is one of the explored alkaloids with studies on both micro-RNAs and long-RNAs. MEG-3 upregulated miR-29a and miRAQ20-186 in hepatocellular carcinoma cells (Zamani et al. 2015). MEG-3 is a known tumor suppressor, i.e. suppression of pancreatic cancer by PI3K protein (Gu et al. 2017), controlling proliferation and metastasis of gastric cancer via p53 signaling pathway (Wei and Wang 2017) and tumor suppressive role in clinically non-functioning adenomas (neoplasia of gonadotrophic pituitary gland) by p53, BRCA1 and PTEN (Zhou et al. 2012). In the same case, curcumin concurrently upregulates the expression of tumor suppressors, miR-29a and mir-186. The miR-29a has tumor suppressor activity in pancreatic cancer (Trehoux et al. 2015), metastasis of pancreatic cancer, and gastric cancer (miR-29a-3p) by targeting MUC1, caveolin-2, and CDK (2, 4, 6), respectively (Liang et al. 2018a; Zhao et al. 2015), along with miR-186 which have tumor suppressive property in prostate cancer (Hua et al. 2016), prostate cancer metastasis (miR-186-5p), (Jones et al. 2018), and bladder cancer (Yao et al. 2015) by GOLPH3, AKAP12, and NSBP1 target, respectively. Curcumin could modulate miRNAs and in-RNAs simultaneously by MEG-3, miR-29a, and miR-186 on the hepatocellular carcinoma.

Lnc-RNAs AF086415, AK095147, RP1-179N16.3, MUDENG, AK056098, and AK294004 reversed by curcumin in nasopharyngeal carcinoma (NPC) which were differentially expressed in the radioresistance cell of NPC (Wang et al. 2014). Radio resistance is one of the drawbacks of cancer treatment by radiation therapy, curcumin can be the best drug in sensitizing cancer cells to radiation by the above lnc-RNA by reversing expression levels. Curcumin presumptuously used in India and Asia has potential anticancer properties in several cancer, and its first phase-1 human trial was reported by Cheng and colleagues 2001 (Hsu and Cheng 2007). Now curcumin, under clinical trial in memory effect (in persons with genetic risk of Alzheimer's disease) (phase-2) (Anonymous 2007), pancreatic cancer (Adenocarcinoma) (phase-2) (Anonymous 2013) and primary sclerosing cholangitis (phase-2) assessed by reducing ALP (alkaline phosphatase) by 40% (Anonymous 2014a) (Table 17.3).

17.9.1 Rutin

Rutin is flavonol glycoside present in many plants tobacco, buckwheat, viola, and many more (Anonymous 2015). Rutin upregulates 144 long noncoding RNAs (HSP90AA2P, AP000866.6, AC125257.1, GAS5, SLC25A24P2, ANKRD20A11P) and downregulates 54 long-noncoding RNAs (AL390038.1, LINC01126, LINC01106, SIAH2-AS1, AP001972.5, AL132800.1) in human SW480 colorectal cancer cells, in which reduced cell metabolism with a cell growth arrest at sub-G1 phase (Nasri Nasrabadi et al. 2019). Interestingly GAS5 is one of the tumor suppressor lnc-RNA which was upregulated by Rutin in SW480 colorectal cancer cells and inhibits cell growth, in other studies GAS5 showed tumor suppressive properties in breast cancer, colorectal cancer, ovarian cancer, and prostate cancer by acting through various targets. The clinical trial status of Rutin has interestingly been studied in autism spectrum disorders by reducing inflammatory mediators from mast cells and IL-6-induced autism-like behavioral deficits (Anonymous 2023).

17.9.2 Doxorubicin

Doxorubicin is derived from *Streptomyces peucetius* var. *caesius,* a bacterial product (Nakano et al. 2015) and used for solid tumors of the ovary, uterus, breast, osteosarcoma, esophagus, stomach, liver, childhood solid tumors, and hematological cancer forms (Carvalho et al. 2009). Doxorubicin resistant OS (osteosarcoma) cells can be sensitized to Dox by upregulating lnc-RNA CTA and competitively downregulating miR-210 in nude mice (Wang et al. 2017a). Exogenous overexpression of miR-210-3p inhibited the proliferation, migration and invasion of bladder cancer cells in vitro. In addition, the nude mouse xenograft model showed that miR-210-3p over-expressing inhibited bladder cancer growth and liver metastasis whereas silencing miR-210-3p caused an opposite outcome, which is mainly regulated by targeting fibroblast growth factor receptor-like 1 (FGFRL1) (Yang et al. 2017b). The miR-210 favors apoptosis which was associated with an upregulation of pro-apoptotic Bim expression and enhanced Caspase 2 in colorectal cancer (Tagscherer et al. 2016), and miR-210 knockdown promotes cell proliferation by upregulating E2F3 expression, thereby promoting the progression of pancreatic cancer (Sun et al. 2018) hence miR-210 is a tumor suppressor in all other cancer forms, but in OS miR-210 promotes cancer progression and development and sensitizing OS cells to doxorubicin by lnc-RNA CTA. Doxorubicin is being clinically trailed for 141 different conditions in European Union mostly in ovarian cancer, hepatocellular carcinoma, breast cancer, and refractory cancer (Anonymous 2018). This randomized phase III trial studies doxorubicin hydrochloride, cyclophosphamide, and paclitaxel to see how well they work with or without bevacizumab in treating patients with cancer that has spread to the lymph nodes (lymph node-positive) or cancer that has not spread to the lymph nodes but is at high risk for returning (high-risk, lymph node-negative breast cancer) (https://clinicaltrials.gov/ct2/show/NCT00433511) (Anonymous 2017c).

17.9.3 Paclitaxel and Docetaxel

lnc-RNA LINC01118 modulates paclitaxel resistance of epithelial ovarian cancer by regulating miR-134/ABCC1. The lnc-RNAs LINC01118 could facilitate cancer progression and metastasis of epithelial ovarian cancer and can be chemoresistant to paclitaxel by targeting miR-134 (downregulated) and ABCC1 protein (upregulated) (multidrug resistant transporters, C branch of ABC transporters). The miR-134 clusters were downregulated in paclitaxel-resistant cancer cells than the sensitive cells, miR-17-92 clusters were inversely expressed. miR-134 acts as a tumor suppressor by regulating suppressing EGFR and PI3K signaling in colorectal cancer. In other studies, miR-134 shows a tumor suppressive role in osteosarcoma by attenuating the expression of VEGFA, and VEGR 1, along with inhibition of epithelial to mesenchymal transition by targeting FOXM1 in non-small-cell lung cancer cells, miRNA-134 acts as anti-CSCs in glioblastoma. The tumor suppressor miR-134 being downregulated by lnc-RNA LINC01118 and sensitizing to paclitaxel in epithelial ovarian cancer is inversely related to another case indeed.

Docetaxel (Taxotere®), a semisynthetic compound analogous to paclitaxel (Taxol®) varies by two positions in chemical structure, docetaxel has significant activity in breast, non-small-cell lung, ovarian, and head, and neck cancers cell cycle arrest at G2/M, apoptosis, and cytotoxicity by promoting microtubule polymerization (Clarke and Rivory 1999). Docetaxel resistance is created by lnc-RNAs NONHSAG096479.1, NONHSAG048134.2, NONHSAG048135.2, NONHSAG048143.2, and NONHSAG048143.2 by which the ABCB1 transporters were the enhanced expression. ABCB1 is reported in drug-resistant cancer forms (Huang et al. 2018b).

17.9.4 Magnolol

Magnolol is tree bark and stem extract of *Magnolia officinalis*, a Chinese plant belonging to the Magnoliaceae family, and it is used for various medical conditions by traditional healers in China. Magnolol and its methoxylated 2-*O*-methylmagnolol (MM1) compound have the property of upregulating GAS5 (growth arrest-specific 5) in skin cancer cells in vitro and in vivo as well and they showed MM1 is more effective than Magnolol. MM1 has upregulated GAS5 twice the magnolol in A375 cells and the caspase 3 activity results that MM1 is more potent than magnolol in in vivo mouse experiment (Wang et al. 2017b). While GAS5 is identified as a tumor suppressor, in breast cancer cells GAS5 is downregulated, and it regulates autophagy by acting as ceRNA to miR23a via ATG3 both in vivo and in vitro (Gu et al. 2018), GAS5 is downregulated in colorectal cancer and pancreatic cancer and GAS5 overexpression could inhibit cancer cell proliferation and reduce EMT, metastasis by targeting miR-182-5p/FOXO3a axis (Cheng et al. 2018) and miR-221/SOCS3 (Liu et al. 2018a), respectively. The GAS5 alternatively suppresses ovarian cancer cells by inflammasome formation, and it has been hindered by downregulation of IL-1, IL-10, ASC, caspase 1, IL-1β, and IL-18 which are involved in pyroptosis pathways (Li et al. 2018). But GAS5 family lnc-RNA GAS5-007 is an oncogenic function in prostate cancer, and its function is controlled by androgen treatment in vitro (Zhang et al. 2017a).

Magnolol proven in the prevention and treatment of more than 15 cancers (Ranaware et al. 2018). Magnolol can prevent atherosclerosis and vessel restenosis, attenuate post-angioplasty restenosis, promote vessel dilation, and prevent platelet aggregation and thrombus formation (Ho and Hong 2012). Magnolol cloud is an abusable compound of GABA-ergic/cannabimimetic activities, by its metabolites tetrahydromagnolol and honokiol (Schifano et al. 2017).

17.9.5 Sulforaphane

Sulforaphane is isolated from broccoli (*Brassica oleracea* var. *italica*), a common vegetable in daily use. Sulforaphane has the property of tumor suppression in prostate cancer cell PC-3 cells, by a novel mechanism through lnc-RNAs LINC01351, LINC00883, LINC01059, and LINC01116 with inversely regulated genes GAPDH, MAP 1LC3B2 (autophagy), and H2AFY to the above lnc-RNAs by sulforaphane treatment. LINC01351, LINC00883, LINC01059, and LINC01116 were upregulated in prostate cancer cells and they are significantly reduced by sulforaphane treatment (Beaver et al. 2017). The LINC01116 enhanced tumor progressive and overexpressed in another tumor form, epithelial ovarian cancer (Fang et al. 2018), and non-small-cell lung cancer via the caspase-mediated pathway (Liang et al. 2018b), but in breast cancer, it acts like an endogenous sponge for miR-145 and upregulates ESR1 (estrogen receptor 1) (Hu et al. 2018).

Transplacental carcinogenesis with Dibenzo[def,p]chrysene (DBC): Timing of maternal exposures determines target tissue response in offspring (Shorey et al. 2012). They are ubiquitous and formed as a by-product of natural and anthropogenic combustion processes.

Sulforaphane has the potency to provide chemoprotection to the fetus transplacentally. The lnc-RNA MSUR-1 upregulation by sulforaphane is mediated by Nrf2 (Nuclear Factor Erythroid-2-Related Factor) (Patel et al. 2018). Sulforaphane has clinically trailed in the past decade, in recurrent prostate cancer (2018) (Anonymous 2018), COPD (Anonymous 2017a), breast cancer (Anonymous 2014b), and prostate cancer (Anonymous 2017b) by organizations from Australia and USA and have reached phase-2 clinical trials. In 2015, sulforaphane-rich broccoli sprout extracts in men with recurrent prostate cancer were clinically trailed in phase 2 (Alumkal et al. 2015).

17.10 Natural Compounds Attenuate Cancer Stem Cells by Targeting Noncoding RNAs

In cancer biology treatment, neither chemotherapy nor surgical or radiotherapy may cause a relapse of cancer a few years later cancer treatment probably by cancer stem cells (Mitra et al. 2015). The self-renewal of the cancer stem cells in various cancer forms is regulated by micro-RNAs as well as by long-noncoding RNAs. Firstly the micro-RNA contribution for cancer stem cells and self-renewal are, in breast cancer Let-7, miR-200 family, miR-34c, miR-16 are downregulated with upregulated miR-181 and miR-495; in prostate cancer, miRNA-34a and miR-Let-7b are downregulated with miR-143 upregulation; in glioblastoma miR-451 and miR-128 are downregulated; in hepatocellular carcinoma miR-181 family and miR-17 and miR-92 are upregulated (Ghosh and Mallick 2014); in colorectal cancer miR-349, miR-200b/c, miR-203, miR-137 are downregulated along with upregulated miR-221 (Mukohyama et al. 2017); in lung cancer, miR-27a is upregulated; in leukemia miR-22 and miR-126 is upregulated (Garofalo and Croce 2015), the downregulated miRNAs are cancer stem cell suppressive nature and upregulated are promoted self-renewal of CSCs.

Secondly, the lnc-RNAs influencing CSCs phenotype is, in gastric cancer ROR is upregulated; in colorectal cancer, HOTAIR is upregulated; in breast cancer, H19 is upregulated; in liver cancer, UCA1 is upregulated, the upregulated lnc-RNAs are having the potency of maintaining stem cell like property in respective cancer cells (Chen et al. 2017). Piperine and salinomycin reduce CSC potency by the Wnt pathway, celastrol and resveratrol arrest CSC property by notch, nanog (Burnett et al. 2012), and OCT4 pathway, sulforaphane acts through HSP90 and AKT pathways, and curcumin controls CSCs property by STAT3 phosphorylation in ALDH+/CD133+ colon CSCs (Taylor and Jabbarzadeh 2017).

Table 17.4 Natural compounds targeting noncoding RNAs may have controlling cancer stem cells

	Nat-compounds may influence CSCs by noncoding RNAs		
Natural compound	CSCs suppressive	CSCs enhancer	CSCs in cancer
Curcumin	a] miR-16 ↑, miR-181↑ b] miR-34a ↑ c] lnc-RNA ROR	– b] miR-27a ↓ –	a] In breast cancer Kronski et al. (2014) b] In colorectal cancer Mukohyama et al. (2017) c] Prostate cancer Liua et al. (2017)
Chrysin	miR-34a ↑ and Let-7a ↑	–	In gastric cancer cells Jafari and Abediankenari (2017) and Golestaneh et al. (2012)
Etoposide	34a ↑	–	Osteosarcoma cell Zou et al. (2017)
Anacardic acid	–	miR-378j ↓, miR-378g ↑ and miR-378f ↑	MCF-7 cells, MDA-MB-231 cells Deng et al. (2013)
Resveratrol	a] miR-34a↑ b] miR-141 ↑	– –	a] Colon cancer Bu et al. (2016) b] Prostate cancer Liu et al. (2017)
Boswellic acid	Let-7 family ↑ and miR-34a ↑	27a ↓	Colorectal cancer Winton (2013), Mizuno et al. (2018) and Zhang et al. (2017b)

↑ Upregulated by the respective compound, ↓ downregulated by the respective compound

Curcumin, chrysin, etoposide, resveratrol, and Boswellic acid may suppress cancer stem cells in respective cancer because miR-34a is reported as CSCs suppressive properties in more than three cancer types (breast cancer, prostate cancer, glioblastoma, and colorectal cancer and osteosarcoma) (Table 17.4). Let-7 family is reported as CSCs suppressive in prostate cancer and breast cancer, Let-7 is being upregulated by curcumin, chrysin, and boswellic acid, in gastric cancer and colorectal cancer (Table 17.4). The anacardic acid shows ambivalent regulation of the miRNA-378 family in MCF-7 cells and MDA-MB-231 cells (in Table 17.4). Boswellic acid and curcumin may control colorectal in a better way by upregulating CSCs suppressive miR-34a and Let-7 family and downregulating miR-27a a CSCs enhancer. In future research, the specific role of natural compounds respective to micro-RNA in CSCs suppression will be revealed. Natural products or extracts from the natural source (Fig. 17.3) could be a cure by targeting cancer stem cells.

17.11 Future Prospective

Natural compounds from natural sources (nutraceuticals) and its chemical extracts could be a promising cure for neural diseases such as Alzheimer's disease, Parkinson's disease, glioblastoma, neuroinflammation, etc.

The anticancer natural compounds are sourced from several plants, fungi, algae, and animals that are being anticancer by a plethora of cellular mechanisms. The recent studies involving the in vitro, in vivo, preclinical, and clinical trials of natural

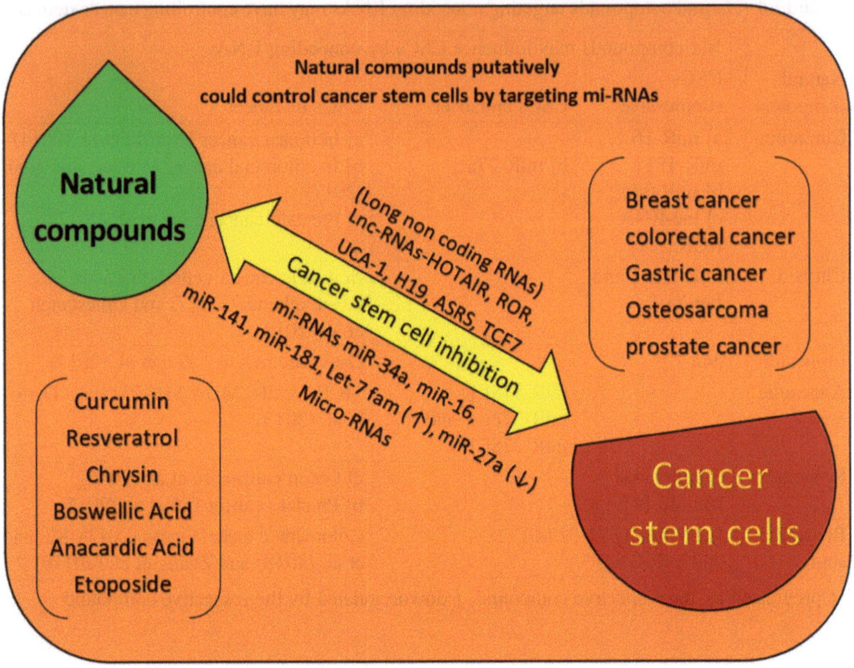

Fig. 17.3 The schematic diagram of natural compound in controlling cancer stem cells

compounds with the anticancer property are providing a less toxic effect and could be antitoxic to various toxic agents. Natural compounds are manifest to be anticancer through modulations of noncoding RNAs by enhancing cancer apoptotic pathways by caspase, PTEN, Bcl-2, Mcl-1, etc. Natural compounds can target multiple cancer, curcumin is anticancer in breast cancer, prostate cancer, colon cancer, and more. Natural compounds can inhibit cancer metastasis through modulating miRNAs and lnc-RNAs by attenuating the epithelial–mesenchymal transition of the cancer cells. They provide epigenetic modulation in cancer, curcumin alters methylation (DNMT3A, B, and DNMT1 inhibition) of MEG3 in hepatocellular carcinoma and more than 20 natural compounds are proven to epigenetic modulators in MCF-7 cells by DNA methylation, histone H3K9 and K27 studied (Vidakovic et al. 2018).

Drug-resistant cancer can be sensitized to natural compounds by miRNAs, such examples are miR34a in paclitaxel-prostate cancer resistance, LINC01118 in paclitaxel epithelial ovarian cancer resistance, miR-29b in doxorubicin-HeLa cell resistance, and many more in different cancer. Extended research in refractory cancers with specific miRNA and in-RNAs targeting compounds could be helpful in the future.

Providing cancer cure, the most appropriate words can be entitled to natural compounds by targeting cancer stem cells. The stemness of the cancer cells is insidious behind recurrent cancer or cancer relapse, indeed after surgical resection of cancer mass. Natural compounds can act as a double-edged sword, in regeneration (neurite formation, bone regeneration (Tohda et al. 2005) and ameliorates myocardial infarction (Kim et al. 2016) and degeneration of cancer mass by targeting miR-34a, Let-7 (putatively) in various cancer and ROR (prostate cancer) and attenuates cancer stem cells. Natural compounds could be clinically approved, a highly dynamic anticancer drugs.

References

Abdolmaleki A, Ferdowsi S, Asadi A, Panahi Y (2017) Long non-coding RNAs associated with brain disorders: a literature review. Gene Cell Tissue 8:e111802

Alles J, Fehlmann T, Fischer U, Backes C, Galata V, Minet M et al (2019) An estimate of the total number of true human miRNAs. Nucleic Acids Res 47(7):3353–3364

Alumkal JJ, Slottke R, Schwartzman J, Cherala G, Munar M, Graff JN, Beer TM et al (2015) A phase II study of sulforaphane-rich broccoli sprout extracts in men with recurrent prostate cancer. Investig New Drugs 33(2):480–489

Ambros V (2004) The functions of animal microRNAs. Nature 431(7006):350–355. https://doi.org/10.1038/nature02871

Angelucci F, Cechova K, Valis M, Kuca K, Zhang B, Hort J (2019) MicroRNAs in Alzheimer's disease: diagnostic markers or therapeutic agents? Front Pharmacol 10:665

Anonymous (2007) "Doxorubicin hydrochloride, cyclophosphamide, and paclitaxel with or without bevacizumab in treating patients with lymph node-positive or high-risk, lymph node-negative breast cancer", Sponsor: National Cancer Institute (NCI), Collaborators: Cancer and Leukemia Group B Eastern Cooperative Oncology Group North Central Cancer Treatment Group; Information provided by (Responsible Party)

Anonymous (2011) pubchem, Etoposide https://pubchem.ncbi.nlm.nih.gov/compound/etoposide. Web reference: "18-Month study of memory effects of curcumin (curcumin)", Sponsor: University of California, Los Angeles, U.S National Library of Medicine

Anonymous (2013) "Effects of the anti-inflammatory flavonoid luteolin on behavior in children with autism spectrum disorders", 2013, Sponsor: Attikon Hospital; Information provided by (Responsible Party): Dr. Konstantinos Francis, Attikon Hospital https://clinicaltrials.gov/ct2/show/record/NCT01847521. U.S National Library of Medicine, Clinical Trials.gov

Anonymous (2014a) "NONCODE", an integrated knowledge database dedicated to ncRNAs, especially lncRNAs. http://www.noncode.org/browse.php

Anonymous (2014b) "Trial of curcumin in advanced pancreatic cancer", 2014, Sponsor: M.D. Anderson Cancer Center, Collaborator: Sabinsa Corporation, Web reference: drugs.com. https://clinicaltrials.gov/ct2/show/record/NCT00094445

Anonymous (2015) "Paclitaxel in refractory or relapse solid tumours", U.S National Library of Medicine, Clinical Trial.gov, https://clinicaltrials.gov/ct2/show/record/NCT02379416

Anonymous (2017a) "Broccoli sprout extracts trial to see if NRF2 is enhanced by sulforaphane treatment in patients with COPD (BEST)", 2017, Sponsor: Johns Hopkins University; Collaborators: Temple University State University of New York at Buffalo; Information provided by (Responsible Party): Johns Hopkins University; U.S National Library of Medicine, Clinical Trial.gov. https://clinicaltrials.gov/ct2/show/record/NCT01335971

Anonymous (2017b.) "Clinical trials for doxorubicin hydrochloride (totally studied in 141 different clinical trials)" EU Clinical Trials Register https://www.clinicaltrialsregister.eu/ctrsearch/search?query=doxorubicin+hydrochloride&page=3

Anonymous (2017c) "Sulforaphane in treating patients with recurrent prostate cancer", 2017, Sponsor: OHSU Knight Cancer Institute; Collaborator: The Wayne D. Kuni and Joan E. Kuni Foundation; Information provided by (Responsible Party): Joshi Alumkal, OHSU Knight Cancer Institute, U.S National Library of Medicine, Clinical Trial.gov

Anonymous (2018) "Study to evaluate the effect of sulforaphane in broccoli sprout extract on breast tissue", 2018, Sponsor: Sidney Kimmel Comprehensive Cancer Center at Johns Hopkins; Collaborator: Breast Cancer Research Foundation; Information provided by (Responsible Party): Sidney Kimmel Comprehensive Cancer Center at Johns Hopkins. U.S National Library of Medicine, Clinical Trial.gov. https://clinicaltrials.gov/ct2/show/record/NCT00982319

Anonymous (2019a) "A study evaluating the safety and efficacy of curcumin in patients with primary sclerosing cholangitis", 2019, Sponsor: John E. Eaton, Collaborator: EuroPharma, Inc; Information provided by (Responsible Party): John E. Eaton, Mayo Clinic, https://clinicaltrials.gov/ct2/show/record/NCT02978339

Anonymous (2019b) "Biomarkers of sulforaphane/broccoli sprout extract in prostate cancer", 2019, Sponsor: University of Pittsburgh; Information provided by (Responsible Party): Bruce Jacobs, University of Pittsburgh. U.S National Library of Medicine, Clinical Trial.gov. https://clinicaltrials.gov/ct2/show/record/NCT03665922

Anonymous (2023) "Rutin", pubchem, U.S. National Library of Medicine and National Center for Biotechnology Information

Bartel DP (2004) MicroRNAs: genomics, biogenesis, mechanism, and function. Cell 116(2): 281–297. https://doi.org/10.1016/s0092-8674(04)00045-5

Beaver LM, Kuintzle R, Buchanan A, Wiley MW, Glasser ST, Wong CP, Johnson GS et al (2017) Long non-coding RNAs and sulforaphane: a target for chemoprevention and suppression of prostate cancer. J Nutr Biochem 42:72–83

Bellazzo A, Di Minin G, Valentino E, Sicari D, Torre D et al (2018) Cell-autonomous and cell non-autonomous downregulation of tumor suppressor DAB2IP by microRNA-149-3p promotes aggressiveness of cancer cells. Cell Death Differ 25(7):1224–1238

Bian S, Sun T (2011) Functions of noncoding RNAs in neural development and neurological diseases. Mol Neurobiol 44(3):359–373

Biersack B (2016) Non-coding RNA/microRNA-modulatory dietary factors and natural, products for improved cancer therapy and prevention: alkaloids, organosulfur compounds, aliphatic carboxylic acids and water-soluble vitamins. Non-coding RNA Res 1(1):51–63

Bu P, Wang L, Chen KY, Srinivasan T, Murthy PK, Tung KL et al (2016) A miR-34a-numb feed-forward loop triggered by inflammation regulates asymmetric stem cell division in intestine and colon cancer. Cell Stem Cell 18(2):189–202

Burnett J, Newman B, Sun D (2012) Targeting cancer stem cells with natural products. Curr Drug Targets 13(8):1054–1064

Cao D, Jia Z, You L, Wu Y, Hou Z, Suo Y, Zhang H, Wen S, Tsukamoto T (2016) 18-β-Glycyrrhetinic acid suppresses gastric cancer by activation of miR-149-3p-Wnt-1 signaling. Oncotarget 7(44):71960–71973

Carballo GB (2018) A highlight on sonic hedgehog pathway. Cell Commun Signal 16(1):11

Carvalho C, Santos RX, Cardoso S, Correia S, Oliveira PJ, Santos MS, Moreira PI (2009) Doxorubicin: the good, the bad and the ugly effect. Curr Med Chem 16(25):3267–3285

Carvalho ALN et al (2013) Anacardic acids from cashew nuts ameliorate lung damage induced by exposure to diesel exhaust particles in mice. Evid Based Complement Alternat Med 2013: 549879

CCO Formulary (2017) Cancer Care Ontario

Cente M, Matyasova K, Csicsatkova N, Tomikova A, Porubska S, Niu Y, Majdan M, Filipcik P, Jurisica I (2022) Traumatic microRNAs: deconvolving the signal after severe traumatic brain injury. Cell Mol Neurobiol 43(3):1061–1075

Chen L, Han L, Wei J, Zhang K, Shi Z, Duan R, Li S et al (2015a) SNORD76, a box C/D snoRNA, acts as a tumor suppressor in glioblastoma. Sci Rep 26(5):8588

Chen PY, Wu MJ, Chang HY, Tai MH, Ho CT, Yen JH (2015b) Up-regulation of miR-34a expression in response to the luteolin-induced neurite outgrowth of PC12 cells. J Agric Food Chem 63(16):4148–4145

Chen XJ, Mian-Yun W, Li D-H, You J (2016) Apigenin inhibits glioma cell growth through promoting microRNA-16 and suppression of BCL-2 and nuclear factor-κB/MMP-9. Mol Med Rep 14(3):2352–2358

Chen S, Zhu J, Wang F, Guan Z, Ge Y, Yang X, Cai J (2017) LncRNAs and their role in cancer stem cells. Oncotarget 8(66):110685–110692

Chen J, Cui Y, Zhang N, Yao X, Wang Z, Yang L (2019) Oleanolic acid attenuated diabetic mesangial cell injury by activation of autophagy via miRNA-142-5p/PTEN signalling. Cytotechnology 71:925–933

Chen T, Shou L, Guo X, Wei M, Zheng H, Tao T (2022) Magnolol attenuates the locomotor impairment, cognitive deficit, and neuroinflammation in Alzheimer's disease mice with brain insulin resistance via up-regulating miR-200c. Bioengineered 13(1):531–543

Cheng K, Zhao Z, Wang G, Wang J, Zhu W (2018) lncRNA GAS5 inhibits colorectal cancer cell proliferation via the miR-182-5p/FOXO3a axis. Oncol Rep 40(4):2371–2380

Clarke SJ, Rivory LP (1999) Clinical pharmacokinetics of docetaxel. Clin Pharmacokinet 36(2): 99–114

Deb N, Majumdar P, Ghosh AK (2013) Pharmacognostic and phytochemical evaluation of the rhizomes of *Curcuma longa* Linn. J PharmSciTech 2(2):81–86

Deka DC, Kumar V, Prasad C, Kumar K, Gogoi BJ, Singh L, Srivastava RB (2013) *Oroxylum indicum*—a medicinal plant of Northeast India: an overview of its nutritional, remedial, and prophylactic properties. J Appl Pharm Sci 3(Suppl 1):S104–SS11

Demain AL, Vaishnav P (2011) Natural products for cancer chemotherapy. Microb Biotechnol 4(6):687–699

Deng Z, Du WW et al (2013) The intermediate filament vimentin mediates microRNA miR-378 function in cellular self-renewal by regulating the expression of the Sox2 transcription factor. J Biol Chem 288(1):319–331

Di Palo A, Siniscalchi C, Crescente G, De Leo I, Fiorentino A, Pacifico S et al (2022) Effect of cannabidiolic acid, *N-trans*-caffeoyltyramine and cannabisin B from hemp seeds on microRNA expression in human neural cells. Curr Issues Mol Biol 44(10):5106–5116

Ebrahimpour S, Esmaeili A, Dehghanian F, Beheshti S (2020) Effects of quercetin-conjugated with superparamagnetic iron oxide nanoparticles on learning and memory improvement through targeting microRNAs/NF-κB pathway. Sci Rep 10:15070

El-Daly SM, Abba ML, Patil N, Allgayer H (2016) miRs-134 and -370 function as tumor suppressors in colorectal cancer by independently suppressing EGFR and PI3K signalling. Sci Rep 6:24720

El-Sayed NS, Elatrebi S, Said R, Ibrahim HF, Omar EM (2022) Potential mechanisms underlying the association between type II diabetes mellitus and cognitive dysfunction in rats: a link between miRNA-21 and resveratrol's neuroprotective action. Metab Brain Dis 37(7): 2375–2388

Fang N, Huang Z-L, Li H, Tan W-B, Zhang Q-G, Wang L, Wu J-L (2018) LINC01116 promotes the progression of epithelial ovarian cancer via regulating cell apoptosis. Eur Rev Med Pharmacol Sci 22(16):5127–5133

Gao J, Liu QG (2011) The role of miR-26 in tumors and normal tissues. Oncol Lett 2(6):1019–1023

Garofalo M, Croce CM (2015) Role of microRNAs in maintaining cancer stem cells. Adv Drug Deliv Rev 81:53–61

Garzon R, Calin GA, Croce CM (2009) MicroRNAs in cancer. Annu Rev Med 60:167–179. https://doi.org/10.1146/annurev.med.59.053006.104707

Gasparello J, Papi C, Zurlo M, Gambari L, Rozzi A, Manicardi A, Corradini R, Gambari R, Finotti A (2022) Treatment of human glioblastoma U251 cells with sulforaphane and a peptide nucleic acid (PNA) targeting miR-15b-5p: synergistic effects on induction of apoptosis. Molecules 27(4):1299

Ghosh Z, Mallick B (2014) Cancer stem cells and regulatory RNAs crosstalk: fostering possibilities for cancer therapies. Bioscience 64(12):1138–1149

Goh SY, Chao YX, Dheen ST, Tan EK, Tay SS (2019) Role of microRNAs in Parkinson's disease. Int J Mol Sci 20:5649

Golestaneh AF, Atashi A, Langroudi L, Shafiee A, Ghaemi N, Soleimani M (2012) miRNAs expressed differently in cancer stem cells and cancer cells of human gastric cancer cell line MKN-45. Cell Biochem Funct 30(5):411–418

Gu L, Zhang J, Shi M, Zhan Q, Shen B, Peng C (2017) lncRNA MEG3 had anti-cancer effects to suppress pancreatic cancer activity. Biomed Pharmacother 89:1269–1276

Gu J, Wang Y, Wang X, Zhou D, Wang X, Zhou M, He Z (2018) Effect of the LncRNA GAS5-MiR-23a-ATG3 axis in regulating autophagy in patients with breast cancer. Cell Physiol Biochem 48(1):194–207

Guan Y, Mizoguchi M, Yoshimoto K, Hata N, Shono T, Suzuki SO, Araki Y et al (2010) MiRNA-196 is upregulated in glioblastoma but not in anaplastic astrocytoma and has prognostic significance MiRNA-196 in glioblastoma. Clin Cancer Res 16:4289–4297

Hamam R, Hamam D, Alsaleh KA, Kassem M, Zaher W, Alfayez M, Aldahmash A, Alajez NM (2017) Circulating microRNAs in breast cancer: novel diagnostic and prognostic biomarkers. Cell Death Dis 8(9):e3045

Han YN, Li Y, Xia SQ, Zhang YY, Zheng JH, Li W (2017) PIWI proteins and PIWI-interacting-RNA: emerging roles in cancer. Cell Physiol Biochem 44(1):1–20

Hanif F, Muzaffar K, Perveen K, Malhi SM, Simjee SU (2017) Glioblastoma multiforme: a review of its epidemiology and pathogenesis through clinical presentation and treatment. Asian Pac J Cancer Prev 18(1):3–9

He Y, Yu D, Zhu L, Zhong S, Zhao J, Tang J (2018) miR-149 in human cancer: a systemic review. J Cancer 9(2):375–388

Hemshekhar M, Sebastin Santhosh M, Kemparaju K, Girish KS (2012) Emerging roles of anacardic acid and its derivatives: a pharmacological overview. Basic Clin Pharmacol Toxicol 110(2): 122–132

Ho JHC, Hong CY (2012) Cardiovascular protection of magnolol: cell-type specificity and dose-related effects. J Biomed Sci 19(1):70

Hollands A, Corriden R, Gysler G, Dahesh S et al (2016) Natural product anacardic acid from cashew nut shells stimulates neutrophil extracellular trap production and bactericidal activity. J Biol Chem 291(27):13964–13973

Hombach S, Kretz M (2016) Non-coding RNAs: classification, biology and functioning. Adv Exp Med Biol 937:3–17. https://doi.org/10.1007/978-3-319-42059-2_1

Housman G, Byler S, Heerboth S, Lapinska K, Longacre M, Snyder N, Sarkar S (2014) Drug resistance in cancer: an overview. Cancers (Basel) 6(3):1769–1792

Hsu C-H, Cheng A-L (2007) Clinical studies with curcumin. Adv Exp Med Biol 595:471–478

Hu HB, Chen Q, Ding SQ (2018) LncRNA LINC01116 competes with miR-145 for the regulation of ESR1 expression in breast cancer. Eur Rev Med Pharmacol Sci 22:1987–1993

Hua X, Xiao Y, Pan W, Li M, Huang X, Liao Z (2016) miR-186 inhibits cell proliferation of prostate cancer by targeting GOLPH3. Am J Cancer Res 6(8):1650–1660

Huang P, Li F, Li L, You Y, Luo S, Dong Z, Gao Q, Wu S, Brünner N, Stenvang J (2018a) lncRNA profile study reveals the mRNAs and lncRNAs associated with docetaxel resistance in breast cancer cells. Sci Rep 8(1):17970

Huang X, Zhang H et al (2018b) Insulin-like growth factor 2 mRNA-binding protein 1 (IGF2BP1) in cancer. J Hematol Oncol 11(1):88

Huang L, Hu C, Chao H, Wang R, Lu H, Li H, Chen H (2019) miR-29c regulates resistance to paclitaxel in nasopharyngeal cancer by targeting ITGB1. Exp Cell Res 378(1):1–10

Jafari N, Abediankenari S (2017) MicroRNA-34 dysregulation in gastric cancer and gastric cancer stem cell. Tumour Biol 39(5):1010428317701652

Jiang L, Huang Q, Zhang S, Zhang Q, Chang J, Qiu X, Wang E (2010) Hsa-miR-125a-3p and hsa-miR-125a-5p are downregulated in non-small cell lung cancer and have inverse effects on invasion and migration of lung cancer cells. BMC Cancer 10:318

Jones DZ, Schmidt ML, Suman S, Hobbing KR, Barve SS, Gobejishvili L, Brock G, Klinge CM et al (2018) Micro-RNA-186-5p inhibition attenuates proliferation, anchorage independent growth and invasion in metastatic prostate cancer cells. BMC Cancer 18(1):421

Kim YS, Jeong HY, Kim AR, Kim WH, Cho H, Um J, Seo Y et al (2016) Natural product derivative BIO promotes recovery after myocardial infarction via unique modulation of the cardiac microenvironment. Sci Rep 6:30726

Koche D, Shirsat R, Kawale M (2016) An overview of major classes of phytochemicals: their types and role in disease prevention. Hislopia J 9(1/2):976–2124

Kollinerová S, Dostál Z, Modrianský M (2017) MicroRNA hsa-miR-29b potentiates etoposide toxicity in HeLa cells via down-regulation of Mcl-1. Toxicol In Vitro 40:289–296

Kou X, Chen D, Chen N (2020) The regulation of microRNAs in Alzheimer's disease. Front Neurol 17(11):288

Kronski E, Fiori ME, Barbieri O, Astigiano S, Mirisola V, Killian PH, Bruno A et al (2014) miR181b is induced by the chemopreventive polyphenol curcumin and inhibits breast cancer metastasis via downregulation of the inflammatory cytokines CXCL1 and -2. Mol Oncol 8(3): 581–595

Lei P, Ayton S, Bush AI (2021) The essential elements of Alzheimer's disease. J Biol Chem 296: 100105

Li J, Wang Y, Luo J, Fu Z, Ying J, Yu Y, Yu W (2012) miR-134 inhibits epithelial to mesenchymal transition by targeting FOXM1 in non-small cell lung cancer cells. FEBS Lett 586(20): 3761–3765

Li J, Meng H, Cao W, Qiu T (2015) MiR-335 is involved in major depression disorder and antidepressant treatment through targeting GRM4. Neurosci Lett 8(606):167–172

Li W, Yang W, Liu Y, Chen S, Chin S, Qi X, Zhao Y (2017) MicroRNA-378 enhances inhibitory effect of curcumin on glioblastoma. Oncotarget 8(43):73938–73946

Li J, Yang C, Li Y, Chen A, Li L, You Z (2018) LncRNA GAS5 suppresses ovarian cancer by inducing inflammasome formation. Biosci Rep 38(2):BSR20171150

Li X, Peng J, Yi C (2021a) The epitranscriptome of small non-coding RNAs. Noncoding RNA Res 6(4):167–173. https://doi.org/10.1016/j.ncrna.2021.10.002

Li X, Su Y, Li N, Zhang FR, Zhang N (2021b) Berberine attenuates MPP^{+}-induced neuronal injury by regulating LINC00943/miR-142-5p/KPNA4/NF-κB pathway in SK-N-SH cells. Neurochem Res 46(12):3286–3330

Liang C, Shi S, Meng Q, Liang D, Hua J, Qin Y, Zhang B, Xu J, Ni Q, Yu X (2018a) MiR-29a, targeting caveolin 2 expression, is responsible for limitation of pancreatic cancer metastasis in patients with normal level of serum CA125. Int J Cancer 143(11):2919–2931

Liang Y, Ma Y, Li L, Shen X, Xin T, Zhao Y, Ma R (2018b) Effect of long non-coding RNA LINC01116 on biological behaviors of non-small cell lung cancer cells via the hippo signaling pathway. J Cell Biochem 119(7):6310

Lin LF, Chiu SP, Wu MJ, Chen PY, Yen JH (2012) Luteolin induces microRNA-132 expression and modulates neurite outgrowth in PC12 cells. PLoS One 7(8):e43304

Liu N, Tang LL, Sun Y, Cui RX, Wang HY, Huang BJ, He QM, Jiang W, Ma J (2013a) MiR-29c suppresses invasion and metastasis by targeting TIAM1 in nasopharyngeal carcinoma. Cancer Lett 329(2):181–188

Liu Y, Zheng W, Song Y (2013b) Low expression of miR-196b enhances the expression of BCR-ABL1 and HOXA9 oncogenes in chronic myeloid leukemogenesis. PLoS One 8(7): e68442

Liu C, Liu R, Zhang D, Deng Q, Liu B, Chao HP, Rycaj K, Takata Y, Lin K et al (2017) MicroRNA-141 suppresses prostate cancer stem cells and metastasis by targeting a cohort of pro-metastasis genes. Nat Commun 8:14270

Liu B, Wu S, Ma J, Yan S, Xiao Z, Wan L, Zhang F, Shang M, Mao A (2018a) lncRNA GAS5 reverses EMT and tumor stem cell-mediated gemcitabine resistance and metastasis by targeting miR-221/SOCS3 in pancreatic cancer. Mol Ther Nucleic Acids 13:472–482

Liu X, Luo X, Wu Y, Xia D, Chen W, Fang Z, Deng J, Hao Y, Yang X et al (2018b) MicroRNA-34a attenuates paclitaxel resistance in prostate cancer cells via direct suppression of JAG1/Notch1 axis. Cell Physiol Biochem 50(1):261–276

Liua TE, Chi H, Chen J, Chen C (2017) Curcumin suppresses proliferation and in vitro invasion of human prostate cancer stem cells by ceRNA effect of miR-145 and lncRNA-ROR. Gene 631:29–38

Mahinfar P, Mansoori B, Rostamzadeh D, Baradaran B, Cho WC, Mansoori B (2022) The role of microRNAs in multidrug resistance of glioblastoma. Cancers (Basel) 14(13):3217

Mannoor K, Liao J, Jiang F (2012) Small nucleolar RNAs in cancer. Biochim Biophys Acta (BBA) 1826:121–128

Masika J, Zhao Y, Hescheler J, Liang H (2016) Modulation of miRNAs by natural agents: nature's way of dealing with cancer. RNA Dis 3:e86

Mattick JS (2001) Non-coding RNAs: the architects of eukaryotic complexity. EMBO Rep 2:986–991. https://doi.org/10.1093/embo-reports/kve230

Mishra S, Verma SS, Rai V, Awasthee N, Chava S, Hui KM, Kumar AP et al (2019) Long non-coding RNAs are emerging targets of phytochemicals for cancer and other chronic diseases. Cell Mol Life Sci 76(10):1947–1966

Mitra A, Mishra L, Li S (2015) EMT, CTCs and CSCs in tumor relapse and drug-resistance. Oncotarget 6(13):10697–10711

Mizuno R, Kawada K, Sakai Y (2018) The molecular basis and therapeutic potential of Let 7 microRNAs against colorectal cancer. Can J Gastroenterol Hepatol 2018:5769591

Mohammadian F, Abhari A, Dariushnejad H, Zarghami F, Nikanfar A et al (2015) Upregulation of Mir-34a in AGS gastric cancer cells by a PLGA-PEG-PLGA chrysin nano formulation. Asian Pac J Cancer Prev 16(18):8259–8256

Mohammadian F, Abhari A, Dariushnejad H, Nikanfar A, Pilehvar-Soltanahmadi Y, Zarghami N (2016a) Effects of chrysin-PLGA-PEG nanoparticles on proliferation and gene expression of miRNAs in gastric cancer cell line. Iran J Cancer Prev 9(4):e4190

Mohammadian F, Pilehvar-Soltanahmadi Y, Zarghami F, Akbarzadeh A, Zarghami N (2016b) Upregulation of miR-9 and Let-7a by nanoencapsulated chrysin in gastric cancer cells. Artif Cells Nanomed Biotechnol 45(6):1–6

Mohammadian F, Pilehvar-Soltanahmadi Y, Mofarrah M, Dastani-Habashi M, Zarghami N (2017a) Down regulation of miR-18a, miR-21 and miR-221 genes in gastric cancer cell line by chrysin-loaded PLGA-PEG nanoparticles. Drug Res (Stuttg) 67(9):509–514

Mohammadian F, Pilehvar-Soltanahmadi Y, Alipour S, Dadashpour M, Zarghami N (2017b) Chrysin alters microRNAs expression levels in gastric cancer cells. Drug Res (Stuttg) 67(9):509–514

Mukohyama J, Shimono Y, Minami H, Kakeji Y, Suzuki A (2017) Roles of microRNAs and RNA-binding proteins in the regulation of colorectal cancer stem cells. Cancers 9:14

Munoz M, Henderson M, Haber M, Norris M (2007) Role of the MRP1/ABCC1 multidrug transporter protein in cancer. IUBMB Life 59(12):752–757

Murugesa Mudaliar KS (2003) Siddha Materia Medica (Medicinal plants division). Indian Medicine-Homeopathy Department, Chennai, pp 13–16

Nabavi SF, Braidy N, Habtemariam S, Orhan IE, Daglia M, Manayi A et al (2015) Neuroprotective effects of chrysin: from chemistry to medicine. Neurochem Int 90:224–231

Nakano I, Soe CZ, Codd R (2015) Isolation of doxorubicin from a bacterial culture using immobilised metal ion affinity chromatography. RSC Adv 5:46437–46442

Nasri Nasrabadi P, Zareian S, Nayeri Z, Salmanipour R, Parsafar S, Gharib E et al (2019) A detailed image of rutin underlying intracellular signalling pathways in human SW480 colorectal cancer cells based on miRNAs-lncRNAs–mRNAs-TFs interactions. J Cell Physiol 234(9):15570–15580

Negrini M, Nicoloso MS, Calin GA (2009) MicroRNAs and cancer—new paradigms in molecular oncology. Curr Opin Cell Biol 21(3):470–479. https://doi.org/10.1016/j.ceb.2009.03.002

Novello C, Pazzaglia L, Conti A, Quattrini I, Pollino S, Perego P, Picci P, Benassi MS (2014) p53-dependent activation of microRNA-34a in response to etoposide-induced DNA damage in osteosarcoma cell lines not impaired by dominant negative p53 expression. PLoS One 9(12): e114757

Oztemur Y, Bekmez T, Aydos A, Yulug IG, Bozkurt B, Dedeoglu BG (2015) A ranking-based meta-analysis reveals Let-7 family as a meta-signature for grade classification in breast cancer. PLoS One 10(5):e0126837

Patel KA et al (2018) Sulforaphane regulation of LncRNA in a transplacental Nrf2 knockout mouse model exposed to dibenzo[def,p]chrysene. Oregon State University, Corvallis

Ponjavic J, Oliver PL, Lunter G, Ponting CP (2009) Genomic and transcriptional co-localization of protein-coding and long non-coding RNA pairs in the developing brain. PLoS Genet 5(8): e1000617

Poornima S, Ponmurugan P, Ahmed KS, Ayyappadasan G, Aldhafiri FK, Santhanaraj B (2016) Evaluation of anti-cancer properties of lichens using albino wistar rats as an animal model. Cancer Res J 4(6):84–89

Radde BN, Alizadeh-Rad N, Price SM, Schultz DJ (2016) Anacardic acid, salicylic acid, and oleic acid differentially alter cellular bioenergetic function in breast cancer cells. J Cell Biochem 117(11):2521–2532

Rahman MS (2018) Preliminary cytotoxicity screening of some medicinal plants of Bangladesh. Dhaka Univ J Pharm Sci 7(1):47–52

Ramos AD, Andersen RE, Liu SJ, Nowakowski TJ, Hong SJ, Gertz CC, Lim DA (2015) The long noncoding RNA Pnky regulates neuronal differentiation of embryonic and postnatal neural stem cells. Cell Stem Cell 16(4):439–447

Ranaware AM, Banik K, Deshpande V, Padmavathi G, Roy NK, Sethi G et al (2018) Magnolol: a Neolignan from the Magnolia family for the prevention and treatment of cancer. Int J Mol Sci 19(8):2362

Rao MRS (2017) Long non-coding RNA biology. In: Advances in experimental medicine and biology (AEMB). Springer, Cham, p 1008

Rebucci M, Sermeus A, Leonard E, Delaive E, Dieu M, Fransolet M et al (2015) miRNA-196b inhibits cell proliferation and induces apoptosis in HepG2 cells by targeting IGF2BP1. Mol Cancer 8(14):79

Ren Y, Chen Y, Liang X, Lu Y, Pan W, Yang M (2017) MiRNA-638 promotes autophagy and malignant phenotypes of cancer cells via directly suppressing DACT3. Cancer Lett 390:126–136

Roberts TC, Morris KV, Wood MJ (2014) The role of long non-coding RNAs in neurodevelopment, brain function and neurological disease. Philos Trans R Soc Lond Ser B Biol Sci 369(1652):20130507

Roca-Lema D, Martinez-Iglesias O, de Ana Portela CF, Rodríguez-Blanco A, Valladares-Ayerbes M et al (2019) In vitro anti-proliferative and anti-invasive effect of polysaccharide-rich extracts from trametes versicolor and *Grifola frondosa* in colon cancer cells. Int J Med Sci 16(2): 231–240

Rodrigues AC, Li X, Radecki L, Pan YZ, Winter JC, Huang M, Yu AM (2011) MicroRNA expression is differentially altered by xenobiotic drugs in different human cell lines. Biopharm Drug Dispos 32(6):355–367

Roshan A, Verma NK, Kumar CS, Chandra V, Singh DP, Panday MK (2012) Phytochemical constituent, pharmacological activities and medicinal uses through the millennia of *Glycyrrhiza glabra* Linn: a review. Int Res J Pharm 3(8):45–55

Sabirzhanov B, Zhao Z, Stoica BA, Loane DJ, Wu J, Borroto C (2014) Downregulation of miR-23a and miR-27a following experimental traumatic brain injury induces neuronal cell death through activation of proapoptotic Bcl-2 proteins. J Neurosci 34(30):10055–10071

Saghafi T, Taheri RA, Parkkila S, Emameh RZ (2019) Phytochemicals as modulators of long non-coding RNAs and inhibitors of cancer-related carbonic anhydrases. Int J Mol Sci 20(12):2939

Salvatori B, Biscarini S, Morlando M (2020) Non-coding RNAs in nervous system development and disease. Front Cell Dev Biol 8:273. https://doi.org/10.3389/fcell.2020.00273

Samarghandian S, Farkhondeh T, Azimi-Nezhad M (2017) Protective effects of chrysin against drugs and toxic agents. Dose Response 15(2):1559325817711782

Sanchez-Mejias A, Kwon J, Chew XH, Siemens A, Sohn HS, Jing G, Sanchez-Mejias A, Kwon J, Chew XH, Siemens A, Sohn HS, Jing G et al (2019) A novel SOCS5/miR-18/miR-25 axis promotes tumorigenesis in liver cancer. Int J Cancer 144(2):311–321

Santosh B, Varshney A, Yadava PK (2014) Non-coding RNAs: biological functions and applications. Cell Biochem Funct 33(1):14–22

Sayed AS, Gomaa IE, Bader M, El Sayed NS et al (2018) Role of 3-acetyl-11-keto-beta-boswellic acid in counteracting LPS-induced neuroinflammation via modulation of miRNA-155. Mol Neurobiol 55:5798–5808

Schifano F, Guarino V, Papanti DG, Baccarin J, Orsolini L, Corkery JM (2017) Is there a potential of misuse for *Magnolia officinalis* compounds/metabolites? Hum Psychopharmacol 32(3): e2595

Schmitt AM, Chang HY (2016) Long noncoding RNAs in cancer pathways. Cancer Cell 29(4): 452–463

Sharif N, Munir N, Saleem F, Aslam F, Naz S (2014) Prolific anticancer bioactivity of algal extracts. Cell 3(4):8–21

Shi C, Wang M (2018) LINC01118 modulates paclitaxel resistance of epithelial ovarian cancer by regulating miR-134/ABCC1. Med Sci Monit 6(24):8831–8839

Shi Y, Wang Q, Song R, Kong Y, Zhang Z (2021) Non-coding RNAs in depression: promising diagnostic and therapeutic biomarkers. EBioMedicine 71:103569

Shi S, Hu X, Xu J, Liu H, Zou L (2022) MiR-320d suppresses the progression of breast cancer via lncRNA HNF1A-AS1 regulation and SOX4 inhibition. RSC Adv 12(41):26639

Shorey LE, Castro DJ, Baird WM et al (2012) Transplacental carcinogenesis with dibenzo[def,p] chrysene (DBC): timing of maternal exposures determines target tissue response in offspring. Cancer Lett 317(1):49–55. https://doi.org/10.1016/j.canlet.2011.11.010

Si L, Xu L, Yin L, Qi Y, Han X, Xu Y, Zhao Y, Liu K, Peng J (2017) Potent effects of dioscin against pancreatic cancer via miR-149-3P-mediated inhibition of the Akt1 signalling pathway. Br J Pharmacol 174(7):553–568

Simionescu N, Nemecz M, Petrovici AR, Nechifor IS, Buga RC, Dabija MG, Eva L, Georgescu A (2022) Microvesicles and microvesicle-associated microRNAs reflect glioblastoma regression: microvesicle-associated miR-625-5p Has biomarker potential. Int J Mol Sci 23(15):8398

Singh R, Saini N (2012) Downregulation of BCL2 by miRNAs augments drug induced apoptosis—a combined computational and experimental approach. J Cell Sci 125(Pt 6):1568–1578

Singh R, Yadav V, Saini N (2015) MicroRNA-195 inhibits proliferation, invasion and metastasis in breast cancer cells by targeting FASN, HMGCR, ACACA and CYP27B1. Sci Rep 3(5):17454

Slota JA, Booth SA (2019) MicroRNAs in neuroinflammation: implications in disease pathogenesis, biomarker discovery and therapeutic applications. Non-coding RNA 5:35

Song YK, Wang Y, Wen YY, Zhao P, Bian ZJ (2018) MicroRNA-22 suppresses breast cancer cell growth and increases paclitaxel sensitivity by targeting NRAS. Technol Cancer Res Treat 17: 1533033818809997

Spizzo R, Almeida MI, Colombatti A, Calin GA (2012) Long non-coding RNAs and cancer: a new frontier of translational research. Oncogene 31(43):4577–4587

Srivastava N, Manvati S, Srivastava A, Pal R, Kalaiarasan P, Chattopadhyay S et al (2011) miR-24-2 controls H2AFX expression regardless of gene copy number alteration and induces apoptosis by targeting antiapoptotic gene BCL-2: a potential for therapeutic intervention. Breast Cancer Res 13(2):R39

Stuart MC, Kouimtzi M, Hill SR (2008) MCL-1 inhibition in cancer treatment. In: WHO model formulary. World Health Organization, Geneva

Sun F-B, Lin Y, Li S-J, Gao J, Han B, Zhang C-S (2018) MiR-210 knockdown promotes the development of pancreatic cancer via upregulating E2F3 expression. Eur Rev Med Pharmacol Sci 22(24):8640–8648

Taft RJ, Pang KC, Mercer TR, Dinger M, Mattick JS (2010) Non-coding RNAs: regulators of disease. J Pathol 220(2):126–139. https://doi.org/10.1002/path.2638

Tagscherer KE, Fassl A, Sinkovic T, Richter J, Schecher S, Macher-Goeppinger S, Roth W (2016) MicroRNA-210 induces apoptosis in colorectal cancer via induction of reactive oxygen. Cancer Cell Int 16:42

Tang B, He WL, Zheng C, Cheang TY, Zhang XF, Wu H, Yang HL (2012a) Marine fungal metabolite 1386A alters the microRNA profile in MCF-7 breast cancer cells. Mol Med Rep 5(3): 610–618

Tang F, Zhang R, He Y, Zou M, Guo L, Xi T (2012b) MicroRNA-125b induces metastasis by targeting STARD13 in MCF-7 and MDA-MB-231 breast cancer cells. PLoS One 7(5):e35435

Taylor WF, Jabbarzadeh E (2017) The use of natural products to target cancer stem cells. Am J Cancer Res 7(7):1588–1605

Thome AD, Harms AS, Volpicelli-Daley LA, Standaert DG (2016) microRNA-155 regulates alpha-synuclein-induced inflammatory responses in models of Parkinson disease. J Neurosci 36(8): 2383–2390

Tohda C, Kuboyama T, Komatsu K (2005) Search for natural products related to regeneration of the neuronal network. Neurosignals 14(1–2):34–45

Trehoux S, Lahdaoui F, Delpu Y, Renaud F, Leteurtre E, Torrisani J (2015) Micro-RNAs miR-29a and miR-330-5p function as tumor suppressors by targeting the MUC1 mucin in pancreatic cancer cells. Biochim Biophys Acta 1853:2392–2403

Vidakovic M, Marinello J, Lahtela-Kakkonen M, Matulis D et al (2018) New insights into the epigenetic activities of natural compounds. OBM Genet 131(8):3033–3041

Vieira AS, Dogini DB, Lopes-Cendes I (2018) Role of non-coding RNAs in non-aging-related neurological disorders. Braz J Med Biol Res 51(8):e7566

Wang H-L, Zhou R (2017) MicroRNA-196b inhibits late apoptosis of pancreatic cancer cells by targeting CADM1. Sci Rep 7:11467

Wang Q, Fan H, Liu Y, Yin Z, Cai H, Liu J, Wang Z, Shao M et al (2014) Curcumin enhances the radiosensitivity in nasopharyngeal carcinoma cells involving the reversal of differentially expressed long non-coding RNAs. Int J Oncol 44(3):858–864

Wang B, Yang Z, Wang H, Cao Z, Zhao Y, Gong C, Ma L (2015) MicroRNA-320a inhibits proliferation and invasion of breast cancer cells by targeting RAB11A. Am J Cancer Res 5(9): 2719–2729

Wang TH, Chan CW, Fang JY, Shih YM, Liu YW, Wang TC, Chen CY (2017a) 2-*O*-Methylmagnolol upregulates the long non-coding RNA, GAS5, and enhances apoptosis in skin cancer cells. Cell Death Dis 8:e2638

Wang Z, Liu Z, Wu S (2017b) Long non-coding RNA CTA sensitizes osteosarcoma cells to doxorubicin through inhibition of autophagy. Oncotarget 8(19):31465–31477

Wang M, Qiu R, Yu S, Xu X, Li G, Gu R, Tan C, Zhu W, Shen B (2019) Paclitaxel-resistant gastric cancer MGC-803 cells promote epithelial-to-mesenchymal transition and chemoresistance in paclitaxel-sensitive cells via exosomal delivery of miR-155-5p. Int J Oncol 54(1):326–338

Wang M, Liu K, Bu H, Cong H, Dong G, Xu N, Li C, Zhao Y, Jiang F et al (2022) Purple sweet potato delphinidin-3-rutin represses glioma proliferation by inducing miR-20b-5p/Atg7-dependent cytostatic autophagy. Mol Ther Oncolyt 31(26):314–329

Weaver BA (2014) How taxol/paclitaxel kills cancer cells. Mol Biol Cell 25(18):2677–2681

Wei G-H, Wang X (2017) lncRNA MEG3 inhibit proliferation and metastasis of gastric cancer via p53 signaling pathway. Eur Rev Med Pharmacol Sci 21(17):3850–3856

Wen Q, Verheijen M, Wittens MMJ, Czuryło J, Engelborghs S, Hauser D, van Herwijnen MHM et al (2022) Lead-exposure associated miRNAs in humans and Alzheimer's disease as potential biomarkers of the disease and disease processes. Sci Rep 12(1):15966

Wikman H, Westphal L (2014) Loss of CADM1 expression is associated with poor prognosis and brain metastasis in breast cancer patients. Oncotarget 5(10):3076–3087
Winton DJ (2013) miR-34a sets the "sweet spot" for notch in colorectal cancer stem cells. Cell Stem Cell 12(5):499–501
Wu H, Liu Q, Cai T, Chen YD, Wang ZF (2015) Induction of microRNA-146a is involved in curcumin-mediated enhancement of temozolomide cytotoxicity against human glioblastoma. Mol Med Rep 12(4):5461–5466
Wu SY, Yan MD, Wu AT, Yuan KS, Liu SH (2016) Brown seaweed fucoidan inhibits cancer progression by dual regulation of mir-29c/ADAM12 and miR-17-5p/PTEN axes in human breast cancer cells. J Cancer 7(15):2408–2419
Xiang W, Yang CY, Bai L (2018) MCL-1 inhibition in cancer treatment. OncoTargets Ther 23(11): 7301–7314
Xiong WC, Han N, Ping GF, Zheng PF, Feng HL, Qin L, He P (2018) microRNA-9 functions as a tumor suppressor in colorectal cancer by targeting CXCR4. Int J Clin Exp Pathol 11(2):526–536
Xu X-H, Ding D-F, Yong H-J, Dong C-L, You N, Ye X-L, Pan M-L (2017) Resveratrol transcriptionally regulates miRNA-18a-5p expression ameliorating diabetic nephropathy via increasing autophagy. Eur Rev Med Pharmacol Sci 21(21):4952–4965
Yang Q, Wang Y, Lu X, Zhao Z, Zhu L, Chen S, Wu Q, Chen C, Wang Z (2014) MiR-125b regulates epithelial–mesenchymal transition via targeting Sema4C in paclitaxel-resistant breast cancer cells. Oncotarget 6(5):3268
Yang F, Hu Y, Liu HX, Wan YJ (2015a) MiR-22-silenced cyclin a expression in colon and liver cancer cells is regulated by bile acid receptor. J Biol Chem 290(10):6507–6515
Yang Y, Park SY, Nguyen TT, Yu YH, Nguyen TV, Sun EG, Udeni J, Jeong MH et al (2015b) Lichen secondary metabolite, physciosporin, inhibits lung cancer cell motility. PLoS One 10(9): e0137889
Yang D, Du G, Xu A, Xi X, Li D (2017a) Expression of miR-149-3p inhibits proliferation, migration, and invasion of bladder cancer by targeting S100A4. Am J Cancer Res 7(11): 2209–2219
Yang X, Shi L, Yi C, Yang Y, Chang L, Song D (2017b) MiR-210-3p inhibits the tumor growth and metastasis of bladder cancer via targeting fibroblast growth factor receptor-like 1. Am J Cancer Res 7(8):1738–1753
Yang W, Liu M, Zhang Q, Zhang J, Chen J, Chen Q, Suo L (2020) Knockdown of miR-124 reduces depression-like behavior by targeting CREB1 and BDNF. Curr Neurovasc Res 17(2):196–203
Yao K, He L, Gan Y, Zeng Q, Dai Y, Tan J (2015) MiR-186 suppresses the growth and metastasis of bladder cancer by targeting NSBP1. Diagn Pathol 10:146
Yu G, Zhang W, Zhu L, Xia L (2018) Upregulated long non-coding RNAs demonstrate promising efficacy for breast cancer detection: a meta-analysis. OncoTargets Ther 16(11):1491–1499
Yuan L, Huang X, Zhou K, Zhu X, Huang B, Qiu S, Cao K, Xu L (2019) Sea cucumber extract TBL-12 inhibits the proliferation, migration, and invasion of human prostate cancer cells through the p38 mitogen-activated protein kinase and intrinsic caspase apoptosis pathway. Prostate 79(8):826–839
Zamani M, Sadeghizadeh M, Behmanesh M, Najafi F (2015) Dendrosomal curcumin increases expression of the long non-coding RNA gene MEG3 via up-regulation of epi-miRs in hepatocellular cancer. Phytomedicine 22(10):961–967
Zhan Y, Liang X et al (2016) MicroRNA-548j functions as a metastasis promoter in human breast cancer by targeting Tensin1. Mol Oncol 10(6):838–849
Zhang R, Xu J, Zhao J, Bai J (2017a) Knockdown of miR-27a sensitizes colorectal cancer stem cells to TRAIL by promoting the formation of Apaf-1-caspase-9 complex. Oncotarget 8(28): 45213–45223
Zhang Y, Su X, Kong Z, Fu F, Zhang P, Wang D, Wu H, Wan X, Li Y (2017b) An androgen reduced transcript of LncRNA GAS5 promoted prostate cancer proliferation. PLoS One 12(8): e0182305

Zhang L, Lv Z, Xu J, Chen C, Ge Q, Li P, Wei D, Wu Z, Sun X (2018) MicroRNA-134 inhibits osteosarcoma angiogenesis and proliferation by targeting the VEGFA/VEGFR1 pathway. FEBS J 285(7):1359–1371

Zhang H, Yao L, Zheng Z, Koc S, Lu G (2022) The role of non-coding RNAs in the pathogenesis of Parkinson's disease: recent advancement. Pharmaceuticals (Basel) 15(7):811

Zhao Z, Wang L, Song W, Cui H, Chen G, Qiao F et al (2015) Reduced miR-29a-3p expression is linked to the cell proliferation and cell migration in gastric cancer. World J Surg Oncol 13:101

Zhou M, Liu W (2013) A novel onco-miR-365 induces cutaneous squamous cell carcinoma. Carcinogenesis 34(7):1653–1659

Zhou Y, Zhang X, Klibanski A (2012) MEG3 noncoding RNA: a tumor suppressor. J Mol Endocrinol 48(3):R45–R53

Zhu H, Wu H, Liu X, Evans BR, Medina DJ, Liu CG, Yang JM (2008) Role of microRNA miR-27a and miR-451 in the regulation of MDR1/P-glycoprotein expression in human cancer cells. Biochem Pharmacol 76(5):582–558

Zhu H, Yang S-Y, Wang J, Wang L, Han S-Y (2016) Evidence for miR-17-92 and miR-134 gene cluster regulation of ovarian cancer drug resistance. Eur Rev Med Pharmacol Sci 20(12): 2526–2531

Zhu Q, Meng P, Han Y, Yang H, Yang Q, Liu Z, Wang Y, Long M (2022) Luteolin induced hippocampal neuronal pyroptosis inhibition by regulation of miR-124-3p/TNF-α/TRAF6 axis in mice affected by breast-cancer-related depression. Evid Based Complement Alternat Med 6: 2715325

Zou Y, Huang Y, Yang J, Wu J, Luo C (2017) miR-34a is downregulated in human osteosarcoma stem-like cells and promotes invasion, tumorigenic ability and self-renewal capacity. Mol Med Rep 15(4):1631–1637

MIX
Papier aus verantwortungsvollen Quellen
Paper from responsible sources
FSC® C105338

If you have any concerns about our products,
you can contact us on
ProductSafety@springernature.com

In case Publisher is established outside the EU,
the EU authorized representative is:
Springer Nature Customer Service Center GmbH
Europaplatz 3, 69115 Heidelberg, Germany

Printed by Libri Plureos GmbH
in Hamburg, Germany